SYNCHROTRON RADIATION SOURCES

A PRIMER

Series on Synchrotron Radiation Techniques and Applications
(ISSN: 2010-2844)

Published

Vol. 1 Synchrotron Radiation Sources — A Primer
H. Winick

Vol. 2 X-ray Absorption Fine Structure (XAFS) for Catalysts & Surfaces
ed. Y. Iwasawa

Vol. 3 Compact Synchrotron Light Sources
E. Weihreter

Vol. 4 Novel Radiation Sources Using Relativistic Electrons: From Infrared to X-Rays
P. Rullhusen, X. Artru & P. Dhez

Vol. 5 Synchrotron Radiation Theory and Its Development: In Memory of I. M. Ternov
ed. V. A. Bordovitsyn

Vol. 6 Insertion Devices for Synchrotron Radiation and Free Electron Laser
F. Ciocci, G. Dattoli, A. Torre & A. Renieri

Vol. 7 The X-Ray Standing Wave Technique: Principles and Applications
eds. J. Zegenhagen & A. Kazimirov

Series on Synchrotron Radiation Techniques and Applications – Vol. 1

SYNCHROTRON RADIATION SOURCES

A PRIMER

Editor

Herman Winick

Stanford Synchrotron Radiation Laboratory
Stanford Linear Accelerator Center

World Scientific
Singapore • New Jersey • London • Hong Kong

Published by

World Scientific Publishing Co. Pte. Ltd.

5 Toh Tuck Link, Singapore 596224

USA office: 27 Warren Street, Suite 401-402, Hackensack, NJ 07601

UK office: 57 Shelton Street, Covent Garden, London WC2H 9HE

Library of Congress Cataloging-in-Publication Data

Synchrotron radiation sources : a primer / editor, Herman Winick
p. cm. -- (Series on synchrotron radiation techniques and applications ; vol. 1)
Includes bibliographical references and index.
ISBN-13 978-981-02-1856-0 -- ISBN-10 981-02-1856-7
ISBN-13 978-981-02-2424-0 (pbk) -- ISBN-10 981-02-2424-9 (pbk)
1. Synchrotron radiation sources. I. Winick, Herman.
II. Series.
QC793.5.E622S978 1994
539.7'35--dc20 94-21246
CIP

British Library Cataloguing-in-Publication Data

A catalogue record for this book is available from the British Library.

Cover design created by Irwin Poliakoff, Center for Advanced Microstructures and Devices at Louisiana State University.

PREFACE

With the rapid and continuing growth of synchrotron radiation research, and with new facilities in design, construction, or coming on-line around the world, there is a need for a reference book that describes the various technical components of a synchrotron radiation source in a manner that would be useful to both technical and non-technical people. We concentrate on the sources of synchrotron radiation. We do not cover the beam lines, which convey the radiation to experimental stations, or the experimental use of synchrotron radiation. For discussion of the beam lines the interested reader is referred to proceedings of national and international synchrotron radiation instrumentation conferences (published in *Nuclear Instruments and Methods* and the *Review of Scientific Instruments*), annual activity reports published by many synchrotron radiation laboratories, and the scientific literature.

Specialists will find this book a useful compendium of the technical aspects of the various components of a synchrotron light source. The book also addresses the needs of those without specialized technical background, who may have responsibility for the design, construction, operation, or development of such a facility, including technicians, engineers, and physicists who have technical backgrounds in related fields, but no direct experience with a synchrotron radiation source. Project managers, laboratory directors, and government officials involved with synchrotron radiation facilities will also find this book helpful.

This book is a single-volume reference on synchrotron radiation sources. Individual chapters are devoted to the various technical components. Most chapters start with an elementary description of the function of that component and how it relates to the overall facility. The description is written in a manner that is useful to non-technical people. A more technical description is given of the advantages and drawbacks of the different approaches that have been successfully used at different facilities. Other parts of the book (the Introduction and Overview, Glossary of Terms and the Index) will be useful to all.

In addition to review of each chapter by the editor, comments from specialists on each chapter of this book were solicited by the authors and the editor. It is a pleasure to acknowledge the contributions of the following reviewers:

Allen, M.
Baltay, M.
Bell, B.
Blumberg, L.
Bordas, J.
Boyce, R.
Byrd, J.
Carr, R.
Cerino, J.
Chao, A.
Corbett, J.
Cornacchia, M.
Cox, A.
Craft, B.
Crook, K.
Cross, J.
Damm, R.
Donaldson, A.

Dylla, F.
Eriksson, M.
Evans, I.
Eyberger, C.
Fox, J.
Genin, R.
Golde, A.
Gough, D.
Green, M.
Haid, D.
Halbach, K.
Hettel, R.
Hofmann, A.
Holden, T.
Hoyer, E.
Hoyt, E.
Hseuh, H. C.
Humphrey, R.
Ipe, N.
Kennedy, K.
Kersevan, R.
Klotz, W.-D.
Kuske, B.
Lee, M.
Marks, S.
Mayer, T.
Murphy, J.
Nuhn, H.-D.
Perkins, C.
Poole, M.
Ruggiero, A.
Sabersky, A.
Safranek, J.
Savage, W.
Schwarz, W.
Scott, B.
Sebek, J.
Singh, O. V.
Talmadge, J.
Tartar, R.
Thomlinson, W.
Walker, R.
Wang, C.-X.
Wiedemann, H.
Wolf, Z.
Yotam, R.

Herman Winick

CONTENTS

Preface v

1. INTRODUCTION AND OVERVIEW **1**
Herman Winick

1. Introduction to Synchrotron Radiation Sources 1
1.1. What is Synchrotron Radiation? 1
1.2. What is a Storage Ring Synchrotron Radiation Source? 2
1.3. A Brief Description of the Main Technical Components 3
1.4. Cost of a Synchrotron Radiation Facility 4
1.5. Survey of Synchrotron Radiation Facilities 5
1.6. Generations of Synchrotron Radiation Sources 6
1.6.1. First-generation storage rings 6
1.6.2. Second-generation storage rings 7
1.6.3. Third-generation storage rings 7
1.7. Sources for Specialized Applications 8
1.7.1. X-ray lithography 8
1.7.2. Micromechanics 8
1.7.3. Medical applications 8
1.8. Properties of Synchrotron Radiation 8
1.8.1. Emission patterns of synchrotron radiation 9
1.8.2. Radiated energy and power 10
1.8.3. Spectral distribution 11
1.8.4. Polarization 12
1.8.5. Time structure 12
1.8.6. Electron beam emittance, photon beam brightness, and coherence 12
1.9. Future Sources: Storage Rings and Free Electron Lasers (FELs) 13
1.9.1. Future storage rings 13
1.9.2. FELs based on storage rings 14
1.9.3. FELs based on linacs 14
References 15

2. LATTICES **30**
Max Cornacchia

2.1. Introduction 30
2.2. Overview of Electron Dynamics in a Storage Ring 31
2.2.1. The storage ring 31

2.2.2. Collective and individual motion, frame of reference 31
2.2.3. Lattice definition 33
2.3. Equations of Motion and Solution 33
2.3.1. Basic magnetic elements in a lattice 33
2.3.2. The synchronous orbit 34
2.3.3. Equations of the synchronous orbit and their solutions 34
2.3.4. The β-function 35
2.3.5. The emittance 35
2.3.6. Tunes and resonances 36
2.3.7. Effects of insertion devices on the particle motion 40
2.3.8. Off-energy particle motion, dispersion, beam size, and momentum compaction 42
2.3.9. Chromaticity correction and dynamic aperture 45
2.4. Emission of Radiation and the Equilibrium Emittance 47
2.4.1. Emission of radiation, damping times and equilibrium emittance 47
2.4.2. Antidamping of betatron oscillations 48
2.4.3. Damping of betatron oscillations 49
2.4.4. The horizontal emittance 49
2.4.5. The vertical emittance 49
2.4.6. Design criteria for synchrotron light sources 50
2.4.7. Strong focusing lattices 50
2.5. Characteristics of Lattices for Synchrotron Radiation Sources 51
2.5.1. General considerations 51
2.5.2. The cell 51
2.5.3. Types of cells 53
References 56

3. INJECTOR SYSTEMS **59**
Gottfried Mülhaupt

3.1. Introduction 59
3.1.1. Requirements 60
3.1.2. Examples 61
3.2. Electron Guns 63
3.3. Linacs 65
3.4. Microtrons 66
3.5. Booster Synchrotron 68
3.6. Principles of Injection and Extraction from a Circular Accelerator 69
3.6.1. Transverse phase space 69

3.6.2. Longitudinal phase space 72
3.7. Kickers 73
3.8. Septum Magnets 75
3.9. Positron Production 77
3.10. Transport Lines 79
3.11. Modes of Operation 81
3.12. Procurement and Installation Considerations 83
3.13. Trends in Synchrotron Light Source Injector Designs 84
References 85

4. R.F. SYSTEMS **87**
D. J. Thompson and D. M. Dykes

4.1. Introduction 87
4.1.1. The purpose of an r.f. system 87
4.1.2. Other effects of the r.f. system on the beam 88
4.1.3. The choices to be addressed 88
4.2. Longitudinal Motion and Phase Stabilty 90
4.3. Voltage Requirement 93
4.3.1. Energy gain per turn 93
4.3.2. Peak cavity voltage 93
4.4. Choice of Frequency 95
4.4.1. Theoretical issues 95
4.4.2. Practical issues 96
4.4.3. Summary 97
4.4.4. Dual frequency systems 97
4.5. Power Requirement 97
4.5.1. The different components of the power 97
4.5.2. The beam power 98
4.5.3. The cavity dissipation 98
4.5.4. Reflected power 98
4.5.5. Feeder losses 100
4.6. Cavities 100
4.6.1. Introduction 100
4.6.2. Superconducting cavities 101
4.6.3. Low frequency cavities 102
4.6.4. High frequency cavities 103
4.6.5. Computer codes for cavity design 106
4.6.6. Input coupling 107
4.6.7. Cavity tuning 108
4.6.8. Higher order mode damping 109

	4.6.9.	Power input limitations	109
	4.6.10.	Cavity construction techniques	109
4.7.	Power Supplies		109
	4.7.1.	Introduction	109
	4.7.2.	Low frequency: Tetrodes	110
	4.7.3.	High frequency: Klystrons	111
	4.7.4.	Wide band: inductive output tubes	113
	4.7.5.	Choice of number of power sources	114
	4.7.6.	Transmission	114
	4.7.7.	Voltage control	115
	4.7.8.	Phase control	115
	4.7.9.	Frequency control	116
	4.7.10.	DC power supplies	117
	4.7.11.	RF protection	117
References			118
Bibliography			118

5. MAGNET DESIGN **119**

Neil Marks

5.1.	Introduction		119
	5.1.1.	Dipole magnets	119
	5.1.2.	Quadrupole magnets	121
	5.1.3.	Sextupole and other lattice magnets	122
	5.1.4.	A.C. and D.C. magnets	122
	5.1.5.	Design and fabrication considerations	123
	5.1.6.	Economic issues	124
5.2.	Fundamentals of Magnet Design		125
	5.2.1.	Field distributions in free space	125
	5.2.2.	Symmetry constraints	127
5.3.	Design of the Practical Poles and Yokes		128
	5.3.1.	Dipoles	128
	5.3.2.	Superconducting dipoles	132
	5.3.3.	Quadrupole magnets	133
	5.3.4.	Sextupole magnets	136
	5.3.5.	Correction magnets	138
	5.3.6.	Magnet ends	140
5.4.	Coils		142
	5.4.1.	Excitation	142
	5.4.2.	Choice of current density	143
5.5.	Materials and Fabrication		143

5.5.1. Magnet steel 143
5.5.2. Conductor material 146
5.5.3. Construction 146
5.6. References 148

6. MAGNET POWER SUPPLIES 149
Rudolf Richter

6.1. Introduction 149
6.2. DC-Current Supplies 150
6.2.1. Power part 150
6.3. Correction Schemes 158
6.3.1. Additional correction with a main magnet 158
6.3.2. Additional correction by the same supply 158
6.4. Regulation and Control 159
6.4.1. The principles of the regulation 159
6.4.2. Strategy for the current calibration 159
6.4.3. Regulation requirements 161
6.5. Booster Supplies 163
6.5.1. General 163
6.5.2. Ramping supplies 163
6.5.3. The white circuit 164
References 169

7. MAGNETIC MEASUREMENTS 171
Richard P. Walker

7.1. Introduction 171
7.2. Magnetic Measurement Techniques 172
7.2.1. Magnetic induction 172
7.2.2. Hall effect 173
7.2.3. Nuclear magnetic resonance 175
7.2.4. Force on a current carrying conductor 176
7.2.5. Magneto-optical effect 176
7.3. Storage Ring Dipole Magnets 176
7.3.1. Point-by-point field measurements 176
7.3.2. Integrated field measurements 180
7.4. Quadrupoles, Sextupoles and Higher Order Multipole Magnets 181
7.4.1. Harmonic coil method 181
7.4.2. Hall plate methods 190
7.4.3. Other techniques 190

7.5. Insertion Devices 191
7.5.1. Point-by-point field measurements 191
7.5.2. Integrated field measurements 193
7.5.3. Other techniques 194
7.6. A.C. and Pulsed Magnets 194
7.7. Magnetic Materials Testing 194
7.8. References 195

8. VACUUM SYSTEMS **197**
John Noonan and Dean Walters

8.1. Introduction 197
8.1.1. Background 197
8.1.2. Beam interactions with molecules and beam lifetiiue 198
8.2. Vacuum System Design 201
8.2.1. Vacuum state equation 202
8.2.2. Vacuum materials 202
8.2.3. Vacuum pumps 204
8.2.4. Vacuum monitoring 207
8.2.5. Vacuum system components 209
8.2.6. Vacuum sealing methods 210
8.2.7. Contamination control 210
8.3. Insertion Device Vacuum Chambers 212
8.4. Conclusions 212
8.5. References 213

9. ACCELERATOR CONTROLS AND MODELING **215**
Jeff Corbett and Clemens Wermelskirchen

9.0. Introduction 215
9.1. Control System Overview 215
9.2. Control System Basics 216
9.3. General Control Systems 217
9.4. Control System Architectures 218
9.4.1. Front-end systems 219
9.4.2. Centralized control systems 220
9.4.3. Distributed control system 220
9.5. Networks 221
9.6. Application Programs 221
9.6.1. Machine Parameter Maintenance 221

	9.6.2. Operator interface	222
	9.6.3. Monitoring, alarms, logging	222
	9.6.4. Automatic Procedures	223
9.7.	Accelerator Modeling Overview	224
9.8.	Model Components	224
9.9.	Accelerator Model Interface	226
9.10.	Model-Based Calculations	227
	9.10.1. Transport and response matrices	227
	9.10.2. Beta functions	228
	9.10.3. Closed orbit perturbations	230
	9.10.4. Dispersion	231
	9.10.5. Synchrotron integrals	232
9.11.	Model Calibration	233
	9.11.1. Optics calibration	233
	9.11.2. Beam based alignment	236
9.12.	Model-Based Control	237
	9.12.1. Orbit control	238
	9.12.2. On-line optics control	240
References		241

10. BEAM DIAGNOSTICS **244**

Peter Kuske

10.1.	Introduction	244
	10.1.1. Function and role of beam diagnostics	244
	10.1.2. Intensity	245
	10.1.3. Beam position	245
	10.1.4. Distribution of particles	246
	10.1.5. Additional accelerator parameters	247
10.2.	Measurement of the Beam Intensity	248
	10.2.1. In transfer lines	248
	10.2.2. In synchrotron and storage ring	249
	10.2.3. Determination of current related parameters	249
10.3.	Measurement of Beam Positions	250
	10.3.1. Sensors for beam position monitoring	250
	10.3.2. Signal processing	251
	10.3.3. Interfacing of BPMs	253
	10.3.4. Comparison of BPM systems	254
	10.3.5. Beam positions measured with fluorescent screens	256
10.4.	Measurement of the Distribution Functions	257
	10.4.1. Imaging with synchrotron radiation	257

10.4.2. Bunch length measurements 259
10.4.3. Examples for diagnostic beamlines 259
10.4.4. Alternatives for beam size measurements 260
10.5. High Precision Determination of the Energy 261
10.5.1. Polarisation and depolarisation of the spins 261
10.5.2. Monitoring the depolarisation 262
10.5.3. An example 262
10.6. Beam Loss Monitor 262
10.7. Measurement of Other Accelerator Parameters 263
10.7.1. Tune measurements 264
10.7.2. Tune related parameters 265
10.7.3. Beam position related parameters 266
10.7.4. Lifetime and beam size related parameters 268
10.7.5. Observation of instabilities 270
References 271

11. MAGNET SUPPORT AND ALIGNMENT **274**
Robert E. Ruland

11.0. Introduction 274
11.1. Magnet Supports 274
11.1.1. Spacers 275
11.1.2. Manual adjustment systems 277
11.1.3. Motorized adjustment systems 286
11.2. Alignment 288
11.2.1. Survey reference frames 288
11.2.2. Layout description reference frame 292
11.2.3. Fiducialization 293
11.2.4. Prealignment of girders 294
11.2.5. Absolute positioning 295
11.2.6. Relative positioning (smoothing) 297
11.2.7. Survey and Alignment Toolbox 299
11.3. Ground Motion 299
11.3.1. Natural sources 300
11.3.2. Cultural noise 301
11.3.3. Countermeasures 301
Acknowledgments 302
References 302

12. BEAM INSTABILITIES **306**
M. Furman, J. Byrd and S. Chattopadhyay

12.1. Introduction 306
12.1.1. Stability 307
12.1.2. Overview of instabilities and their effects 307
12.1.3. Damping mechanisms 309
12.2. Wake Fields and Impedances 310
12.2.1. Definitions 310
12.2.2. Properties and basic uses of impedances 311
12.2.3. Resonator impedance model 314
12.2.4. Impedance beyond cutoff 316
12.2.5. Impedance calculations and measurement techniques 318
12.2.6. Broad-band impedance model 318
12.3. Landau Damping 320
12.4. Single-Bunch Issues 323
12.4.1. Calculation of instabilities 323
12.4.2. Parasitic power loss 324
12.4.3. Longitudinal effects 325
12.4.4. Transverse effects 328
12.5. Coupled-Bunch Instabilities 331
12.5.1. Basics 331
12.5.2. Longitudinal coupled-bunch instability 334
12.5.3. Transverse coupled-bunch instability 338
12.5.4. Coupled-bunch instability cures 338
12.6. Trapped Ions and Beam Lifetime Issues 339
12.6.1. Trapped ions 339
12.6.2. Intrabeam and Touschek scattering 340
12.7. Acknowledgments 341
12.8. References 342

13. ORBIT STABILIZING AND MULTIBUNCH FEEDBACK SYSTEMS **344**
John N. Galayda, Youngjoo Chung and Robert O. Hettel

13.1. Introduction 344
13.2. Nature of Beam Stability Problems 344
13.2.1. Beamline apertures 344
13.2.2. Beam stability criteria 345
13.2.3. Categories of beam motion 346
13.2.4. Sources of beam motion 348
13.2.5. Stabilizing systems 349

13.3. Beam Position Monitors 349
13.3.1. Photon beam position monitors 350
13.3.2. Radiofrequency beam position monitors (RFBPMs) 352
13.4. Corrector Magnets 353
13.4.1 Corrector location 353
13.4.2. Dynamic range and response 353
13.4.3. Vacuum chamber eddy currents 354
13.5. Feedback Orbit Control 354
13.5.1. General feedback systems 354
13.5.2. Frequency domain response of feedback systems 355
13.5.3. Eddy current compensation—frequency domain 356
13.5.4. Digital signal processing and Z-transform 357
13.5.5. Proportional, integral and derivative (PID) control algorithm 358
13.5.6. Eddy current compensation with digital signal processing 359
13.6. Local Beam Steering Systems 360
13.6.1. 3-magnet, 1-monitor feedback 360
13.6.2. 4-magnet, 2-monitor feedback 360
13.6.3. Feedback system design 361
13.7. Global Orbit Feedback Systems 362
13.7.1. Global feedback basics 363
13.7.2 Global harmonic feedback 363
13.7.3. Global feedback using digital signal processing 364
13.8. Feedback Control of Multibunch Instabilities 365
13.8.1. Basics of feedback for instabilities 365
13.8.2. Kicker bandwidth 368
13.8.3. Transverse kick strength and kicker power 368
13.8.4. Longitudinal kick strength 369
13.8.5. Measuring y'_β or ε 370
13.8.6. Suppression of "stable beam" signals 371
13.8.7. Multibunch feedback 372
13.8.8. Broadband longitudinal feedback 374
References 375

14. WIGGLER AND UNDULATOR INSERTION DEVICES 377
Ross D. Schlueter

14.1. Introduction 377
14.1.1. Motivation for the development of insertion devices 377
14.1.2. Applications of insertion devices 378

	14.1.3.	Insertion device radiation overview	379
14.2.		Synchrotron Radiation From Wigglers and Undulators	381
	14.2.1.	Wiggler radiation	381
	14.2.2.	Undulator radiation	382
	14.2.3.	Free electron lasers	383
14.3.		Pure Permanent Magnet Insertion Devices	384
	14.3.1.	Advantages	384
	14.3.2.	Design methodology	384
	14.3.3.	Design variations	386
14.4.		Hybrid (Permanent Magnet Plus Iron) Insertion Devices	386
	14.4.1.	Advantages	386
	14.4.2.	Design methodology	388
	14.4.3.	Design variations	388
14.5.		Warm Electromagnet Insertion Devices	389
	14.5.1.	Advantages	389
	14.5.2.	Design methodology	390
	14.5.3.	Design variations	391
14.6.		Superconducting Electromagnet Insertion Devices	392
	14.6.1.	Advantages	392
	14.6.2.	Design	393
14.7.		Circularly Polarized Radiation From Insertion Devices	393
	14.7.1.	Helical insertion devices	394
	14.7.2	Asymmetric insertion devices	397
	14.7.3.	Interference devices producing circularly polarized radiation	397
14.8.		Other Novel Insertion Device Design Ideas	398
14.9.		Insertion Device Fields and Errors	401
	14.9.1.	Field requirements	401
	14.9.2.	Insertion device design implications	403
14.10.		Power and Power Density Considerations	406
14.11.		References	406

15. CONVENTIONAL FACILITIES **409**

V. Saile and J. D. Scott

15.1.	Introduction	409
15.2.	Planning the Conventional Facility	410
15.3.	Site Selection	412
15.4.	Choosing the Architectural and Engineering (AE) Firm	413
15.5.	Design Parameters	415
15.6.	Structural and Vibrational Considerations	418

15.6.1. Hutches, clean rooms and other enclosures built on the experimental hall floor 424
15.6.2. Movement of heavy equipment within the facility 424
15.7. Utilities; Quality, Distribution and Access 425
15.8. Safety and Security 427
15.9. Construction Oversight by Owner 429
15.10. Occupation and Operation 430
References 431

16. SAFETY **432**
Thomas Dickinson

16.1. Introduction 432
16.1.1. Safety management 432
16.1.2. Built in safety facilities 433
16.1.3. Research operations 433
16.1.4. Laws, codes, and government regulation 433
16.1.5. Ordinary safety issues 434
16.1.6. Radiation protection 434
16.1.7. Interlocks 435
16.2. Safety Management 435
16.2.1. Safety analysis 435
16.2.2. Organizational structure 436
16.2.3. Safety and design reviews 436
16.3. Built in Safety Facilities 437
16.3.1. Experimental support labs 437
16.3.2. Setup and storage space 437
16.4. Research Operations 439
16.4.1. Parameters of the NSLS affecting research operations 439
16.4.2. Research operations staff: safety professionals 439
16.4.3. Research operations staff: research operators 440
16.4.4. Experiment safety review 440
16.4.5. Control of experimental hazards 441
16.4.6. The experiment safety approval form 441
16.4.7. Oversight of experimental activities 444
16.4.8. Configuration control 445
16.4.9. User safety orientation 445
16.5. Radiation Protection 447
16.5.1. Bremsstrahlung radiation 447
16.5.2. Neutron radiation 448

16.5.3. Practical considerations for fixed shielding 449
16.5.4. Synchrotron radiation 449
16.5.5. Radiation units 450
16.5.6. Radiation levels significant to personnel 450
16.5.7. Radiation monitoring 451
16.5.8. Radiation measuring instruments 451
16.5.9. Configuration control of shielding 452
16.6. Personnel Protection Interlocks 453
16.6.1. Interlock reliability requirement 453
16.6.2. Technical design features 453
16.6.3. Fail safe design 453
16.6.4. Redundant design 454
16.6.5. Contribution of redundancy to reliability 454
16.6.6. Common cause failures and diversity 455
16.6.7. An example interlock system 455
16.6.8. Interlock testing 456
16.6.9. Virtues of simplicity 456
16.6.10. Programmable logic controllers vs relays 457
16.7. The Integrated Safety Program 457
References 458

GLOSSARY 459

INDEX 493

CHAPTER 1: INTRODUCTION AND OVERVIEW*

HERMAN WINICK◊

Stanford Synchrotron Radiation Laboratory, Stanford Linear Accelerator Center
Stanford University, Stanford, CA 94309

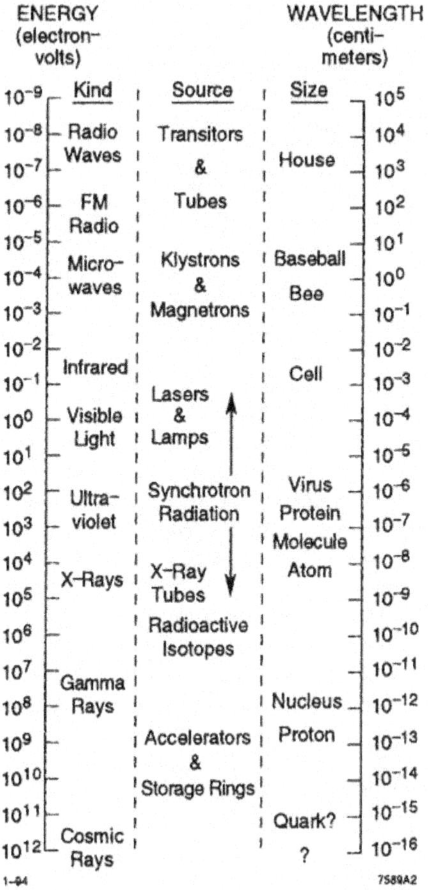

Figure 1.1. The Electromagnetic Spectrum and the region occupied by synchrotron radiation.

1. Introduction to Synchrotron Radiation Sources

1.1 What is Synchrotron Radiation?

Synchrotron radiation is the electromagnetic radiation emitted when charged particles travel in curved paths. For high-energy electrons curving in the magnetic fields of storage rings, this radiation is extremely intense over a broad range of wavelengths extending from the infrared through the visible and ultraviolet range, and into the soft and hard x-ray parts of the electromagnetic spectrum. (see Figure 1.1). The high intensity, broad spectral range and other properties (collimation, polarization, pulsed-time structure, partial coherence, high-vacuum environment) make synchrotron radiation a powerful tool for basic and applied studies in biology, chemistry, medicine, and physics, and their many subfields, as well as

---

* Work supported by the Office of Basic Energy Sciences, Department of Energy contract DE-AC03-76SF00515.

◊ Author e-mail: winick@slac.stanford.edu and FAX: (415) 926-4100.

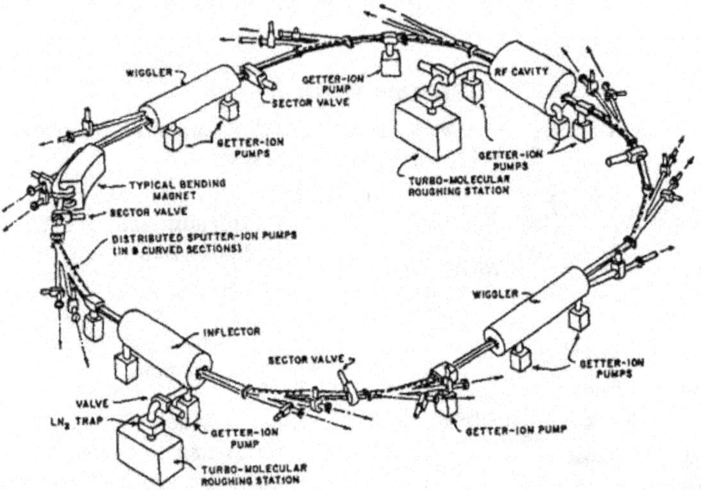

Figure 1.2. Schematic of a small synchrotron radiation source (Courtesy of J. Godel, BNL).

in applications to technology such as x-ray lithography, micromechanics, materials characterization, and trace element analysis.[1]

1.2 *What is a Storage Ring Synchrotron Radiation Source?**

A storage-ring synchrotron radiation source[2] is an arrangement of components (see Figure 1.2) that enables a current of electrons (or positrons) to circulate at essentially the speed of light on a closed orbit for periods of several hours while synchrotron radiation is emitted from all curved parts of this orbit. [Note some synchrotron radiation facilities store positrons, rather than electrons, to avoid certain problems that can occur with electrons. For convenience we will refer to electrons as the circulating particles.] Some of the radiation from the bending magnets, and most or all of the radiation from specially designed wiggler and undulator insertion devices, leaves the ring through tangential ports called beam lines that allow the radiation to pass to experimental stations located outside the ring. Photographs and drawings of several synchrotron radiation facilities are given at the end of this chapter and in other chapters of this book (see chapters 15 and 16). A comprehensive presentation of the physics of electron storage rings is given in a recent book.[3] Directions for future sources were the topic of a workshop on Fourth-Generation Light Sources.[4]

---

* After finalizing this manuscript, the author became aware of the book *Synchrotron Radiation Facilities in Asia*, edited by T. Ohta, S. Suga, and S. Kikuta, published by Ionics Publishing Co. Ltd., Kawada Building 2-3-4 Koishikawa, Bunkyo-ku, Tokyo 112, Japan. These proceedings of the Asian Forum on Synchrotron Radiation held in Kobe, Japan on May 13, 1994 contain many excellent and detailed descriptions of facilities in Asia and is recommended for those planning new facilities.

1.3 A Brief Description of the Main Technical Components

The main elements controlling the circulating electron beam are the magnets (see Chapter 5), of which there are usually several types. The *bending magnets*, or dipoles, bend the beam through short arcs that together make up a total of 360 degrees of bending so that the orbit is closed. If there were only uniform magnetic field bending magnets the electron beam would grow steadily in transverse dimensions due to the mutual repulsion of charges of the same sign. To maintain small transverse dimensions, focusing magnets—called *quadrupole magnets*— are used. Other magnets (*sextupoles, steering magnets*) provide corrections of various kinds to permit large currents to be stored, and to maintain a constant orbit so that the radiation source points do not move. In some designs two or more of the magnet functions are combined; e.g., some rings have bending magnets that have field gradients in the radial direction to provide focusing as well as bending.

Several different arrangements of magnets are commonly used in synchrotron radiation sources. The magnet arrangement, called the lattice (see Chapter 2), determines the basic features of the circulating beam, such as its emittance and the transverse dimensions of the electron beam at the synchrotron radiation source points. Of course the lattice also determines the number and length of straight sections available for wiggler and undulator insertion magnets (see Chapter 14), devices which provide radiation with enhanced flux, brightness, and spectral range compared to the ring bending magnets.

The energy lost by the circulating beam to synchrotron radiation is replenished by longitudinal electrical kicks as the beam traverses one or more special chambers, called rf cavities, in which electromagnetic fields oscillating at radio frequencies (rf) are maintained (see Chapter 4). These rf fields are generated by transmitters, or radio frequency power supplies, located adjacent to the ring. The rf energy is conducted from the transmitter to the cavity by waveguide or coaxial cable. The oscillatory nature of the rf results in a bunching of the circulating electron beam. The magnet lattice and the parameters of the rf system determine the length of these bunches, and hence the duration of the photon pulses. The rf frequency determines the minimum spacing of these bunches. From the point of view of the beam dynamics, the rf system and the magnet lattice are the main components of a synchrotron radiation source.

The stored beam circulates in a vacuum chamber (see Chapter 8) in which the pressure is typically maintained in the range of 10^{-9} to 10^{-10} torr (1 torr = 1 mm of Hg = 0.0075 Pa), or about one thousandth of one billionth of normal atmospheric pressure, which is 760 torr. At this low pressure, comparable to the pressure in outer space, collisions between the circulating electrons and residual gas molecules are sufficiently rare that the stored-beam intensity decreases slowly, typically reaching half its initial intensity after 5–50 hours; that is, the stored-beam intensity decays approximately exponentially with a lifetime of many hours. To achieve this desired low pressure in the presence of the high flux of synchrotron radiation, which liberates gas from the surfaces of vacuum chamber components, high-speed vacuum pumps and careful attention to component selection, cleaning, bakeout, and assembly are required. Smoothness of the vacuum chamber is also important to reduce interactions with the circulating beam that can cause instabilities. A major concern in many rings, particularly high-energy rings, is the high power and power density in the photon beams. This could result in heat loading, distortion, and potential failure of the storage-ring vacuum vessel and beam line components.

The ring is equipped with a variety of measurement devices or beam diagnostics (see Chapter 10) that provide information about the intensity of the circulating beam, its position at several locations around the ring, the dimensions of the circulating packets of electrons, and the frequency of transverse and longitudinal oscillations that take place as the beam circulates (called betatron and synchrotron oscillations, respectively). With the aid of hardware controls, application programs, and models of the ring stored in the control computer (see Chapter 9) the operator uses information from the various diagnostic devices to optimize the performance of the ring.

The storage ring is filled with electrons (or positrons) from a source called an injector (see Chapter 3). Several different types of accelerators can be used as an injector. An injector can be a linear accelerator (also called a linac), a microtron, or a cyclic accelerator called a booster synchrotron, which is fed by small linac or microtron. Injection can be at the full operating energy of the storage ring (the most desirable, but most expensive alternative) or at a fraction of this energy. In the latter case, the storage ring is used briefly as an accelerator to increase the energy of the stored-beam energy from its value at injection to the final operating energy.

Other chapters of this book cover:

- magnet power supplies (see Chapter 6) that provide precisely regulated electrical currents to the different types of magnets;
- measurements carried out on the magnets (see Chapter 7) before installation to assure that they meet specifications set by the lattice design;
- magnet supports and alignment techniques (see Chapter 11) to assure that the magnets are positioned to the accuracy set by the lattice design;
- instabilities that could limit the stored-beam intensity and stability (see Chapter 12);
- beam stabilizing and steering systems (see Chapter 13);
- the various building, utilities, services that come under the heading of conventional facilities (see Chapter 15);
- and the basic elements of a comprehensive safety system (see Chapter 16) to protect personnel and safeguard the environment.

1.4 Cost of a Synchrotron Radiation Facility

The cost of a synchrotron radiation facility varies depending on its basic parameters: energy, emittance, number of straight sections, circumference, injector type and energy, as well as the local conditions under which it is built, such as cost of the land, and labor costs. For example, the basic components for a 50-m circumference, 1 GeV, second-generation storage ring can be purchased from industry for about $25 million. Buildings, a few beam lines, and utilities, would approximately double this price. The cost

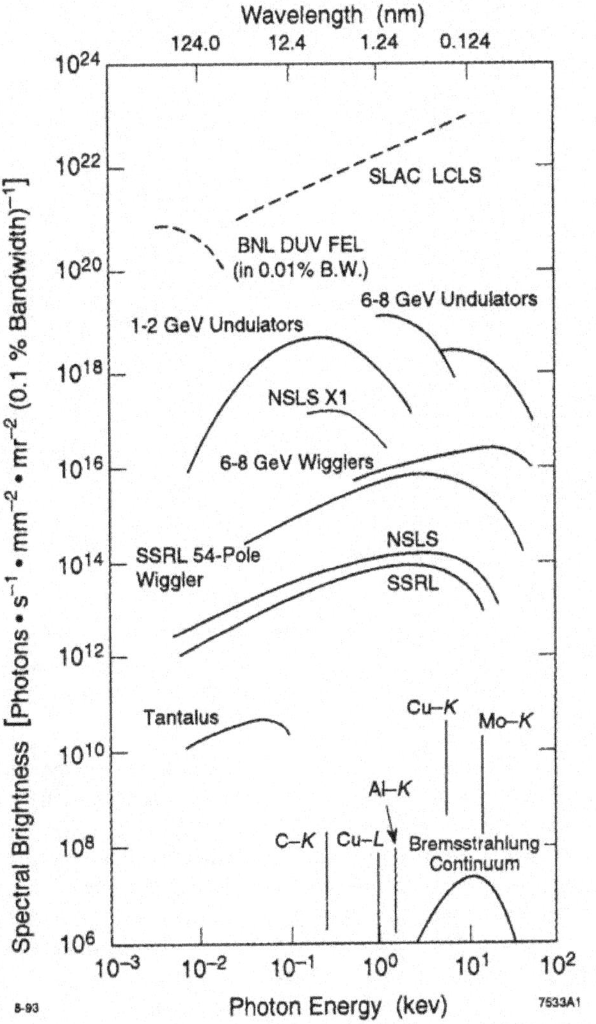

Figure 1.3. Spectral brightness from various radiation sources. The dotted lines at the top are free electron laser proposals

could be reduced with a smaller, lower-energy machine, by using a simple design consisting of a single magnet such as SURF II at NIST, or by making the project partly a student training exercise as is done at the Duke and Dortmund facilities. At the other extreme, a high-energy, 6–8 GeV third-generation x-ray facility with a circumference of about 1–1.5 km, an initial complement of beam lines, and a full-energy positron injector system can cost about $1 billion, or more.

1.5 Survey of Synchrotron Radiation Facilities[2]

The explosive growth of synchrotron radiation research and facilities for such research over the past two decades is due to the immense capability of synchrotron radiation as a tool for basic and applied research and technology. Synchrotron radiation provides more than five orders of magnitude higher flux (photons/s/mrad/unit bandwidth) and ten orders of magnitude higher brightness (photons/s/unit solid angle/unit source area/unit bandwidth) compared with more conventional sources such as VUV lamps and x-ray tubes (see Figures 1.3 and 1.4).

About thirty-two storage rings are in operation around the world (in early 1994) as synchrotron radiation sources for basic and applied research. Another fifteen rings are in construction and thirteen are in the design or proposal stage. About forty-three laboratories in sixteen countries are involved. These rings are located in government laboratories and universities. It is now customary to classify these synchrotron-radiation sources by generation as described below. Tables 1.1, 1.2, and 1.3 list worldwide synchrotron radiation sources by generation. In addition, about eight rings are in operation in industrial settings and used for technological purposes, primarily for x-ray lithography to fabricate semiconductor devices with line widths of less than 0.5 microns. Designs are under development by several groups for storage rings optimized for micro-machining (the fabrication of precision mechanical and electromechanical components on a 0.1–10 mm scale) and diagnostic radiography (less-invasive coronary angiography).

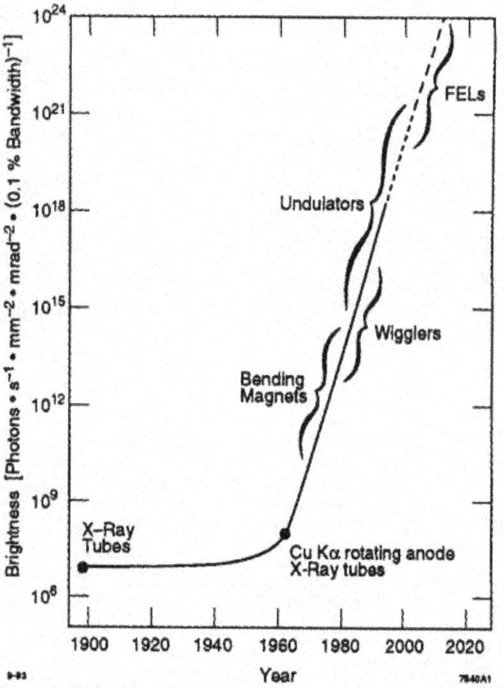

Figure 1.4. Spectral brightness of x-ray sources as a function of time.

Synchrotron radiation sources range from compact devices a few meters in circumference operating at a few hundred million electron volts (MeV), to behemoths up to 1500 meters in circumference operating at 8 billion electron volts (GeV, for giga electron volts). Synchrotron radiation facilities are not restricted to the most technologically advanced countries. An increasing number of facilities are located in less-developed countries, where synchrotron radiation sources are seen as an effective way to provide advanced capability for basic research and student training as well as for industrial applications such as x-ray lithography and materials characterization.

1.6 Generations of Synchrotron Radiation Sources

1.6.1 First-Generation Storage Rings

First-generation storage rings (see Table 1.1) were originally built for high-energy physics research, in most cases involving the study of reactions that occur when counter-rotating electron and positron beams collide. In many cases the synchrotron radiation programs, which started in a parasitic mode, have grown to the point where the facility is now partly or fully dedicated to synchrotron radiation research. Long straight sections in rings such as DORIS, SPEAR and VEPP-3 can be used for long undulators, optical klystrons (as implemented on VEPP-3) or bypasses accommodating many insertion devices (as implemented on DORIS).

By our definition, the very large, high-energy (15–33 GeV) colliders such as PEP, PETRA and TRISTAN are first-generation rings. However, with their ability to operate at reduced energy with very low emittance and with their long straight sections (100–200 m) for damping wigglers and long undulators, they offer opportunities to achieve brightness exceeding that of third-generation rings.

Experience with two high-brightness undulator beam lines on PEP and studies of PEP's potential as a light source have been reported.[5] An undulator beam line is now in construction on PETRA,[6] aimed primarily at producing very high brightness at photon energies above 50 keV. Plans for modifications to TRISTAN[7] include the use of a 70-m undulator delivering a brightness 2 to 3 orders of magnitude higher than third-generation rings. Studies of PEP[8] and TRISTAN[9] as drivers for 3- to 4-nm FELs with peak power approaching 1 GW have been reported. However, PEP and TRISTAN are being rebuilt as dual ring *B* factories. *B* factories involve collisions between currents in excess of one ampere in each of two storage rings, one with energy around 3 GeV and the other around 9 GeV. These rings could serve as sources of extremely high synchrotron radiation flux from bending magnets or insertion devices.

1.6.2 Second-Generation Storage Rings

Second-generation storage rings (see Table 1.2) are designed specifically as dedicated light sources. The first of these is the 380-MeV SOR ring at the University of Tokyo, that started operation in the mid-1970s. In general these rings have a large number of beam lines and experimental stations, primarily from bend magnets, and they serve many users. In some cases, the user community exceeds 2000. The immense research productivity of second- (and first-) generation rings, and the successful experience with wigglers and undulators has led to a new generation of more advanced rings.

1.6.3 Third-Generation Storage Rings

Third generation storage rings (see Table 1.3) are distinguished by lower electron-beam emittance and many straight sections for insertions. They offer much higher brightness, particularly from undulators. Most of these rings employ full-energy injection, which enhances orbit reproducibility and minimizes time spent injecting the electron beam. Some store positrons to eliminate the deleterious effects caused by the trapping of positively charged dust particles or ions in electron beams.

Most third-generation rings fall into two broad categories; 1–2 GeV rings with 100 to 200-m circumference designed primarily for the spectral region below about 2 keV (VUV and soft x rays) and 6–8 GeV rings with a circumference of 800–1500 m, designed primarily for harder x rays, above about 2 keV. Rings now in operation or construction as FEL drivers may also be considered third-generation rings.

1.7 Sources for Specialized Applications

1.7.1 X-ray Lithography

Storage rings are generating interest as sources of soft x rays to produce electronic circuits with sub-micron features by the lithographic process.[10] At present circuits are produced with visible or UV lithography with features as small as about 1/3 micron, limited by diffraction effects. Storage rings provide intense photon beams at shorter wavelengths, pushing diffraction limits to below 0.1 micron. Seven industrial storage rings primarily used for x-ray lithography are in operation in Japan and one in the US.

1.7.2 Micromechanics

A related application, micromechanics,[11] involves the lithographic production of high-aspect ratio, miniature devices (gears, motors, transducers) on the 10-micron to 10-mm scale using x-rays with energy around several kilovolts. Future developments in this area could lead to the need for commercial storage-ring sources.

1.7.3 Medical Applications

Dedicated rings optimized for less invasive coronary angiography[12] may be needed once clinical-quality images are obtained with synchrotron radiation. Rings operating at about 2 GeV with multipole high-field superconducting wigglers can provide the required flux at 33 keV, the K-absorption edge of iodine. Contrast is enhanced by rapidly recording images at two x-ray wavelengths—one just above and another just below the k-absorption edge of iodine. Reducing the concentration of iodine required to obtain clinical quality images opens the possibility of obtaining images without the inter-arterial catheter now routinely used, and hence reducing the risks associated with such a catheter.

1.8 Properties of Synchrotron Radiation

Whenever charged particles undergo an acceleration they emit electromagnetic radiation. For example, electrons oscillating at radio frequencies in an antenna emit radio waves. In this case the velocity and acceleration are collinear. When electrons are subjected to an acceleration perpendicular to their velocity (that is, a centripetal acceleration such as is the case when electrons pass through a magnetic field), their direction is changed—they begin to travel in a circular path. When the electron kinetic energy is low compared with the rest mass energy given by mc^2 (i.e., when the electron velocity is non-relativistic or low compared to the velocity of light) the radiation is emitted in a rather non-directional pattern (called a dipole pattern) as shown in Figure 1.5, Case I.

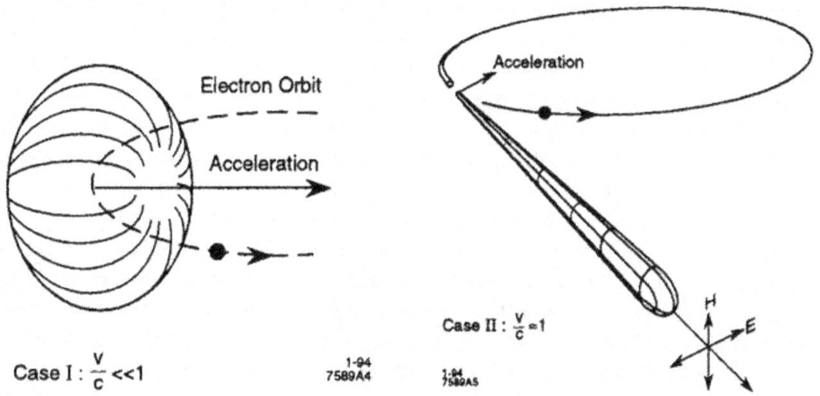

Figure 1.5. Emission patterns of radiation from electrons in circular motion; Case I: at a low velocity compared to the velocity of light; and Case II: approaching the velocity of light.

1.8.1 Emission Patterns of Synchrotron Radiation

Qualitative and quantitative changes occur at electron energies well above the rest mass energy, and the emitted radiation is called synchrotron radiation. The amount of energy radiated increases dramatically and relativistic effects cause the pattern to be folded into a sharp forward cone with an opening angle given by $mc^2/E = \gamma^1$ (Figure 1.5, Case II). Since the electron's rest mass energy, mc^2, is only 0.511 MeV, at an energy of 511 MeV the opening angle of the cone is only .001 radian (1 milliradian), or .057 degrees, or 3.45 minutes. At 5100 MeV or 5.1 GeV the opening angle is ten times smaller.

In a *bending magnet* the sharp cone sweeps around like a well-focused searchlight, producing a continuous swath of radiation in the bending plane. The vertical opening angle of this swath remains small, given by mc^2/E. A *wiggler magnet* is a succession of alternating polarity magnetic poles, each of which bends the electron beam through an angle large compared with mc^2/E. The magnet is designed such that these alternating deflections cancel out and no net bending is produced. Thus the magnet can be placed in a straight section of a storage ring with little disturbance to the orbit. The flux and brightness of the radiation are enhanced by a factor approximately equal to the number of poles. If the magnetic field in the wiggler is higher than the ring bending magnet field, the wiggler spectrum extends to higher photon energy.

An *undulator magnet* is similar to a wiggler in that it is also a succession of alternating magnetic poles. However, in this case the angle of bend in each pole is of the order of mc^2/E, so that the small angular divergence due to the emission pattern of synchrotron radiation is not significantly increased. Thus, the intrinsic brightness of synchrotron radiation is preserved in both the vertical and horizontal planes. Furthermore, interference effects in the emission of radiation in the succession of essentially collinear source points result in a spectrum that is enhanced at certain wavelengths, and is thus different from the smooth, continuous spectrum produced in a bend magnet or wiggler.

The emission patterns from bending magnets, wigglers, and undulator magnets are compared in Figure 1.6.

1.8.2 Radiated Energy and Power

The energy lost due to synchrotron radiation in the bending magnets in a single turn by one electron is given by

$$\Delta E/\text{turn[keV]} = 88.5\ E^4[\text{GeV}]/R[\text{m}], \tag{1.1}$$

where E is the electron energy in GeV, and R is the radius of curvature in meters of the orbit in the bending magnets.

Since the particle energy and bending radius are related to the magnetic field by

$$E[\text{GeV}] = 0.3\ B[\text{T}]\ R[\text{m}], \tag{1.2}$$

Eq. (1.1) can also be expressed as

$$\Delta E/\text{turn[keV]} = 26.6\ E^3[\text{GeV}]B[\text{T}], \tag{1.3}$$

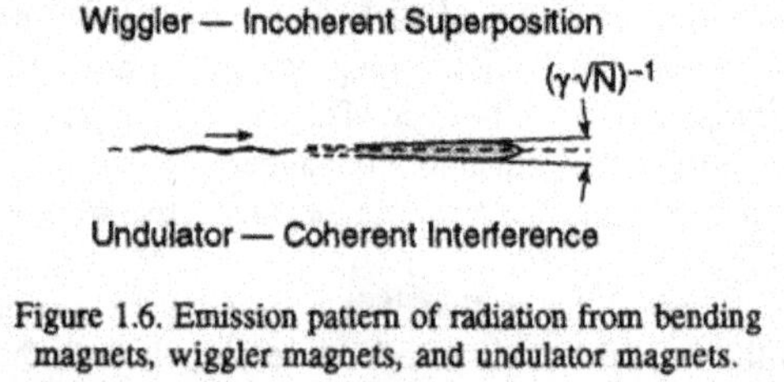

Figure 1.6. Emission pattern of radiation from bending magnets, wiggler magnets, and undulator magnets.

where B is the magnetic field in tesla.

Thus, an electron with 1-GeV energy in a storage ring in which the orbit in the bending magnets has a radius of 3.33 meters (corresponding to a bending magnet field of 1 tesla) loses 26.6 keV per turn. Multiplying the energy loss per turn by the stored current in amperes gives the total power radiated in the bending magnets in kilowatts. This is the power that must be replenished by the rf system. Thus, in this example, for a stored current of 0.5 A the total radiated power is 13.3 KW or about 2 W into each milliradian of horizontal bending angle. Since the vertical opening angle is only about 0.5 milliradian, the power density is about 4 W per square milliradian. By comparison, powerful electron impact x-ray tubes, in which about 60 KW of electron-beam power strike a rapidly rotating anode, radiate a total of about 10 W of x-rays into a solid angle of about 2π steradians or less than 2 microwatts per square milliradian.

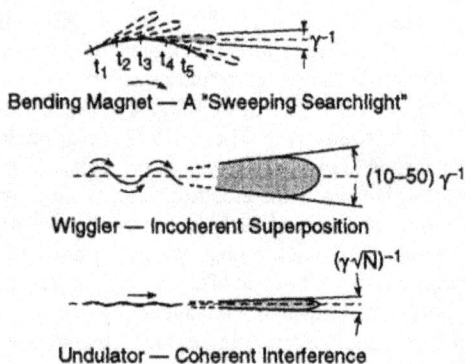

For a 7-GeV ring, the comparison is more dramatic due to the 3rd power energy dependence in Eq. (1.3), and the reduction in vertical opening angle due to the higher electron energy. It can easily be seen that for the same magnetic field and stored current the total radiated power would be about 4.5 MW. For this reason, such high-energy rings use a lower field in the bending magnets and store a lower current, around 0.1–0.2 A.

Additional radiation is produced in insertion devices. The energy loss in a wiggler or undulator insertion device is given by

$$\Delta E[\text{keV}] = 0.633E^2[\text{GeV}]<B^2[\text{T}]>L[\text{m}]\ , \qquad (1.4)$$

where $<B[T]^2>$ is the average value of the square of the magnetic field in tesla over the length L of the insertion device in meters. Thus, an electron traversing a one meter long insertion device with a magnetic field of 1 tesla in a 1 GeV ring loses 0.633 keV per pass. For a current of 0.5 A the radiated power is 317 W. All of this radiation is directed to the experimental stations, whereas only a small fraction of the radiation emitted in the bending magnets can be used at the experimental stations. [Note: Eq. (1.1) would apply to protons if it is multiplied by the fourth power of the ratio of the mass of the electron to the mass of the proton. This ratio, 1/1836, when raised to the fourth power becomes 0.88×10^{-13}. For the parameters of the SSC (E = 20,000 GeV, R = 10.1×10^3m) Eq. (1.1) gives an energy loss per turn of 123 keV. With a circulating current of 73 mA, the total radiated synchrotron radiation power would be 9.8 kW.]

1.8.3 Spectral Distribution

For bending magnet and wiggler sources, the spectrum of the radiation is a smooth continuum (see Figure 1.7) defined by a single parameter, called the critical energy, given by

$$e_c[\text{keV}] = 0.665\ B[\text{T}]E^2[\text{GeV}] \qquad (1.5)$$

Half the power is radiated above the critical energy and half is radiated below. Thus, for a 1-GeV ring with 1-tesla bending magnets the critical energy is 0.665 keV. For an undulator, the spectrum consists of quasi-monochromatic peaks as shown in Figure 1.7. See Chapter 14 for a more detailed discussion of the spectral properties of the radiation from wiggler and undulator insertion devices.

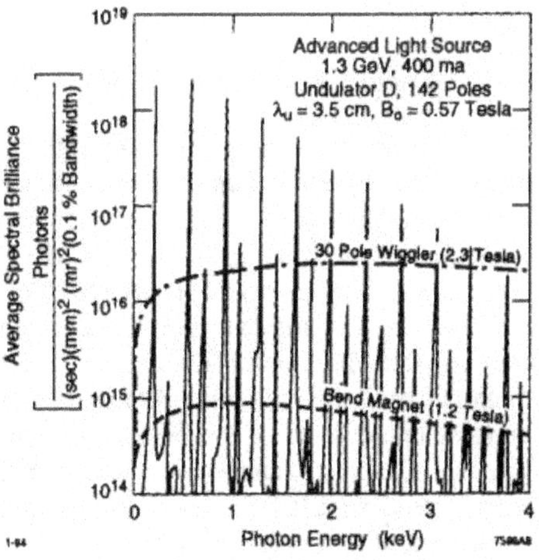

Figure 1.7. Spectrum of radiation emitted by electrons in bending magnets, wigglers and undulators. (Courtesy K.-J. Kim, LBL)

1.8.4 Polarization

Synchrotron radiation is highly polarized. In the orbital plane the electric field vector of the emitted radiation is in the direction of the instantaneous acceleration. Thus radiation from bending magnets is linearly polarized in the plane of the orbit. Out of the orbit plane the polarization becomes elliptical and eventually circular, with opposite helicity above and below the plane. If the magnetic fields in a wiggler and undulator are confined to one transverse plane, the alternating poles cancel this elliptical polarization out of the plane and the radiation is linearly polarized everywhere. For insertion devices with helical magnetic fields the radiation is elliptically polarized. These, and other techniques for producing elliptical and circular polarization using insertion devices are described in Chapter 14.

1.8.5 Time Structure

Since the electrons in a storage ring are bunched by the rf system (see Chapter 4) an observer looking at the circulating beam tangentially sees pulses of light as bunches pass the observation point. Typically each bunch is 0.5 to 5 cm long. Since these bunches pass the observation point at essentially the speed of light (3×10^{10} cm/s), a 3-cm-long bunch would take 100 picoseconds to pass, resulting in a light pulse of the same duration. Shorter bunches would pass correspondingly more quickly. The magnet lattice (see Chapter 2) and the frequency of the rf system (see Chapter 4) determines the bunch length. The rf frequency determines the maximum number of bunches (usually on the order of 100 or more) that can be stored, and the minimum spacing (usually a few nanoseconds). However, it is not necessary to fill each bunch. If needed by experimenters, only one or a few bunches can be filled resulting in a dark period of several hundred nanoseconds to a microsecond or more between pulses.

1.8.6 Electron Beam Emittance, Photon Beam Brightness, and Coherence

In many experiments, it is the high flux (photons/s/mrad/unit bandwidth) on the sample that is important. Such experiments are well served by all storage rings. However, an increasing number of experiments benefit from very high brightness (photons/s/unit solid angle/unit source area/unit bandwidth) or high coherent power, which is proportional to brightness. Photon beam brightness and coherence are largely determined by the transverse size and angular divergence of the electron beam. The product of the beam transverse size and angular divergence in each direction transverse to the direction of motion is called the emittance. Emittance may also be thought of as a measure of the transverse temperature of the beam. Horizontal emittance is determined by the magnet lattice (see Chapter 2) and is measured in meter-radians or nanometer-radians (nm-rad). Vertical emittance is largely determined by coupling to the horizontal and can be as low as a few percent of the horizontal emittance.

Although the beam size and angular divergence vary around the ring, their product, the emittance, is a constant. First- and second-generation rings typically have horizontal emittances of about 100–200 nm-rad. Undulators on these rings can produce photon beam brightness of about 10^{16} to 10^{17} photons/s/mm^2/mrad^2/0.1% bandwidth. Third-generation rings typically have horizontal emittances in the 5–25 nm-rad range. Undulators on these rings provide photon-beam brightness up to about 10^{19}.

The increase in photon beam brightness as emittance decreases is ultimately limited by diffraction effects. The diffraction-limited emittance is given by $\lambda/4\pi$ where λ is the photon wavelength. Thus for a wavelength of 1.2 nm (1-keV photon energy) the diffraction-limited electron-beam emittance (that is, the lowest that is useful in producing 1.2 nm light) is about 0.1 nm-rad. As mentioned above, present synchrotron radiation sources have much higher emittances. For example a third-generation ring with 5-nm-rad emittance produces diffraction-limited light at photon wavelengths longer than about 60 nm, corresponding to photon energies below about 20 eV.

Clearly, even lower emittance would further increase brightness and coherence and approach diffraction limits at higher photon energies. The quest for lower emittance beams has therefore been central to the evolution of light sources and is key to improved scientific opportunities in brightness-limited applications. Low emittance rings pose more severe challenges to stability and reproducability of the stored beam, since the source points for synchrotron radiation beams must be kept constant to a fraction of the beam size.

1.9 Future Sources: Storage Rings and Free Electron Lasers (FELs)

Many future sources will be similar to existing storage ring sources since second- and third-generation sources provide excellent radiation to large scientific and technological communities. However, some experiments require higher brightness, coherence, peak power, or an extension of some other property of the radiation. Since present sources are far from fundamental limits on the brightness and power of the radiation that can be produced, it is likely that future sources will be developed that will exceed present sources in performance. Concepts for future sources have been pursued at several laboratories. The Workshop on Fourth-Generation Light Sources[4] surveyed a wide range of possible directions. A workshop, Towards Short-Wavelength Free-Electron Lasers,[13] reviewed the use of storage rings and linacs to drive FELs at wavelengths below 200 nm. Such sources would produce radiation with brightness several orders of magnitude higher than undulators on third-generation rings.

1.9.1 Future Storage Rings

Concepts for future rings reaching emittances in the 10^{-11} m-rad range while preserving dynamic aperture were presented at the Workshop on 4th Generation Light Sources.[4] In one design[14] a large ring with many short focussing, defocussing (or FODO) cells and long, dispersion-free straight sections filled with damping wigglers was considered. Another design[15] used short cells with combined-function bend magnets, including a sextupole component to provide local chromatic correction. Low emittance, with large momentum and betatron acceptance were achieved in a ring of moderate size. Quasi-isochronous, or low-momentum compaction, rings[16] would reduce bunch duration to about 1 ps.

Recently designs of possible future storage-ring sources have been produced at Daresbury (Diamond, 3 GeV and Sinbad, 0.6 GeV), Orsay (Soleil, 2.15 GeV), and at the Paul Scherrer Institute (Swiss Light Source, 2.1 GeV). These designs include features such as lower emittance, superconducting bending magnets incorporated in the lattice, and long straight sections for FELs.

Also, large circumference, high-energy, colliding beam rings such as PEP, PETRA and TRISTAN offer possibilities to reach extremely low emittance, as well as accommodating very long undulators in 100 to 200-m-long straight sections. Since emittance decreases quadratically with energy in a storage ring, by operating these 15–33 GeV rings at one-fourth to one-eighth of their full energy, the emittance can be reduced by a factor of 16 to 64. Further reduction of emittance by a factor of about 4 can be achieved by operating in stronger-focusing, low-emittance modes. In this way, emittances of less than 1-nm-rad may be possible.

1.9.2 FELs Based on Storage Rings

Storage rings have been used to drive FELs in the visible and UV at Orsay,[17] Tsukuba (Electrotechnical Laboratory),[18] and Novosibirsk.[19] The shortest wavelength reached to date is 240 nm at Novosibirsk. A 170-nm FEL is being developed at the Photon Factory.[10] New rings optimized as drivers of FELs at wavelengths down to about 30 nm are in construction at Dortmund[20] and Duke Universities.[21] Shorter wavelength (3–4 nm) FELs have been proposed on PEP[8] and TRISTAN.[9]

1.9.3 FELs Based on Linacs

The development of high-brightness electron guns greatly improves the prospects for linac-based FELs to reach VUV and x-ray wavelengths.[13,22] High-gradient rf photocathode guns[23] delivering 1–2 nC in pulses a few picoseconds long with small energy spread and a normalized emittance, ε_n as low as 2 mm-mrad (rms) have been operated. In a linac, the emittance decreases as the electron energy increases. Thus, at an energy E, the emittance is given by $\varepsilon = \varepsilon_n (mc^2 / E) = \varepsilon_n / \gamma$. This facilitates reaching the diffraction-limited emittance. For example, a linac equipped with a 3-mm-mrad gun can produce an emittance of 0.3 nm-rad if the beam is accelerated to 5 GeV ($\gamma=10^4$). This is the diffraction limited emittance for a wavelength of 4 nm. FEL performance improves as the electron beam emittance approaches the diffraction limit.

Proposed short-wavelength FELs include the following: A BNL design[24] for 75- to 300-nm sub-harmonically seeded FELs is based on an 80-MeV superconducting linac recirculated twice to achieve 250 MeV. A Los Alamos design[25] uses a 1-GeV linac to drive a series of FEL oscillators from 125 nm down to 1 nm using multifaceted grazing incidence mirrors forming ring resonators.

Also, a study[26] has been carried out on the use of the SLAC linac at energies below 7 GeV to drive a water window (2.3–4.4 nm) FEL and at 15–25 GeV to drive shorter wavelength devices, down to a few Angstroms, using self-amplified spontaneous emission. These devices would deliver several gigawatts of peak power with an rms pulse duration of 0.1–0.2 ps at 120 Hz. The first device would deliver about 10^{14} coherent photons in a single sub-picosecond pulse, enough to make a hologram of a live biological sample before it is changed or damaged.

Technical problems that must be faced to develop short-wavelength, linac-based FEL sources include preservation of electron-beam emittance during beam transport, acceleration and bunch compression; control of beam energy spread; and steering. The successful experience with the linear collider project at SLAC[27] shows that these needs can be met. Also, long (>30 m) and precise undulators are needed to reach saturation. Again, the successful experience with undulators up to 5-m long at third-generation light sources and even longer (up to 25 m) undulators for previous FEL projects[28] shows that these needs can be met.

References

1. Handbook on Synchrotron Radiation, Vols. 1–4 (North Holland).
2. V. Suller, Proc. 3rd European Part. Accel. Conf., Berlin, March 1992. S. Krinsky, Proc. 1991 IEEE Part. Acc. Conf. (91CH3038–7),11–15. A. Jackson, Proc. 1989 Int. Conf. on High-Energy Acc., 1701–10. H. Winick, Proc. 1989 IEEE Part. Acc. Conf. (89CH2669–0), 7–11. M. Cornacchia and H. Winick, Proc. 15th Int. Conf. on High-Energy Accel.; J. Mod Phys. 2A, (1993) 468–472.
3. H. Wiedemann, Particle Accelerator Physics vol. 1 (Springer-Verlag, 1993) 462 pages.
4. M. Cornacchia and H. Winick (editors), SSRL Report 92/02. Available from T. Slater, SSRL.
5. A. Bienenstock, G. Brown, H. Wiedemann, H. Winick, Rev. Sci. Instr. **60(7)** (1989) 1393–8.
6. W. Brefeld, ref, 4, p. 115.
7. H. Kitamura, S. Yamamoto, S. Kamada, Rev. Sci. Instr. **60** (1989) 1407. S. Kamada, K. Ohmi, ref 4, p. 106.
8. H.-D. Nuhn, R. Tatchyn, H. Winick, A. S. Fisher, J. C. Gallardo, and C. Pellegrini, Nucl. Instr. & Meth., **A319**, (1992) 89–96.
9. H. Kitamura, ref 4 p. 179. H. Kitamura, Photon Factory Act. Rep. 1992 pages R9–11, S2–15.
10. T. Tomimasu, *Rev. Sci. Instr.* **60** (1989) 1622. T. Tomimasu, *Nucl. Instr. & Meth.* **A308** (1991) 6–12.

11. H. Guckel, *Proc. SPIE Conf. on Precision Eng. and Optomechanics* Aug 10–11, 1989, p.151–8.

12. W.-R. Dix et al, *Nucl. Instr. & Meth.* **A314** (1992) 307.

13. I. Ben-Zvi, H. Winick (editors); *"Towards Short-Wavelength Free-Electron Lasers,"* Brookhaven Nat. Lab. Workshop, May 21–22, 1993, BNL report 49651.

14. L. Emery, in ref 4, p 149.

15. W. Klotz, G. Mülhaupt, in ref 4, p 138.

16. C. Pellegrini, D. Robin, *Nucl. Instr. & Meth.* **A301** (1991) 27. A. Amiry, C. Pellegrini, ref 4, p 195.

17. R. Prazers, C. Bazin, M. Bergher, M. Billardon, M. Couprie, J. Ortega, M. Velghe, Y, Petroff, *Rev. Sci. Inst.* 60 (1989) 1429.

18. T. Yamazaki, K. Yamada, S. Sugiyama, H. Ogaki, N. Sei, T. Mikado, T. Noguchi, M. Chiwaki, R. Suzuki, M. Kawai, M. Yokoyama, K. Owaki, S. Hamada, K. Aizawa, Y. Oku, A. Iwata, M. Yoshiwa, *Nucl. Instr. & Meth.* **A331** (1993) 27–33.

19. N.Vinokurov, I. Drobyazko, G. Kulipanov, V. Litvinenko. I. Pinayev, V. Popik, I. Silvestrov, A. Skrinsky, S. Sokolov, *Rev. Sci. Inst.* **60** (1989) 1435–8.

20. K. Wille, *Nucl. Instr. & Meth.* **A272** (1988) 59.

21. Y. Wu, V. Litvinenko, E. Forest, J. Madey, *Nucl. Instr. & Meth.* **A331** (1993) 287–292.

22. M. Poole, *Rev. Sci. Instr.* **63** (1992) 1528. R. Sheffield, *Proc. 1991 IEEE Part. Accel. Conf.* (91CH3038–7) 1110–14.

23. B. Carlsten, *Nucl. Instr. & Meth.* **A285** (1989) 313–319.

24. I. Ben-Zvi, L. Mauro, S. Krinsky, M. White, L. Yu, K. Batchelor, A. Friedman, A. Fisher, H. Halama, G. Ingold, E. Johnson, S. Kramer, J. Rogers, L. Solomon, J. Wachtel, X. Zhang, *Nucl. Instr. & Meth.* **A318** (1992) 201.

25. B. Newnam et al, *SPIE vol. 1552* (1991) 154.

26. C. Pellegrini, J. Rosenzweig, H.-D. Nuhn, P. Pianetta, R. Tatchyn, H. Winick, K. Bane, P. Morton, T. Raubenheimer, J. Seeman, K. Halbach, K.-J. Kim, J. Kirz, *Nucl. Instr. & Meth.* **A331** (1993) 223–227.

27. J. Seeman, *Advances of Accel. Physics & Technologies,* H. Schopper (editor), World Scientific Pub. Co., 1993, p.219.

28. G. Deis, A. Harvey, C. Parkinson, D. Prosnitz, J. Rego, E. T. Scharlemann, K. Halbach, *IEEE Trans. Mag.* **24(2)** (1988) 1090.

Table 1.1. First-Generation Synchrotron Radiation Sources

Location	Ring	Energy (GeV)	Notes
China			
Beijing	BEPC	1.5–2.8	Partly Ded.
Denmark			
Aarhus	ASTRID	0.6	Partly Ded.
France			
Orsay	DCI	1.8	Dedicated
Germany			
Bonn	ELSA	1.5–3.5	Partly Ded.
Hamburg	DORIS III	4.5–5.3	Dedicated
	PETRA II	7–13	Beam line in cons.*
Italy			
Frascati	DAΦNE	0.51	Parasitic*
Japan			
Sendai	TSSR	1.5	Proposed
Tsukuba	Accum. Ring	6.5	Partly Ded.
	TRISTAN MR	6–30	Planned Use
Netherlands			
Amsterdam	AmPS	0.9	Planned Use
Eindhoven	EUTERPE	0.4	Planned Use
Russia			
Novosibirsk	VEPP-2M	0.7	Partly Ded.
	VEPP-3	2.2	Partly Ded./FEL use
	VEPP-4	5–7	Partly Ded.
Ukraine			
Kharkov	N-100	0.1	Dedicated
	HP-2000	2.0	Partly Ded.*
USA			
Gaithersburg,MD	SURF II	0.28	Dedicated
Ithaca,NY	CESR	5.5	Partly Ded.
Stanford,CA	SPEAR	3–3.5	Dedicated

* In Construction as of 4/94

Table 1.2. Second-Generation Synchrotron Radiation Sources

Location	Ring	Energy (GeV)	Notes
Brazil			
Campinas	LNLS-1	1.15	Dedicated*
China (PRC)			
Hefei	HESYRL	0.8	Dedicated
England			
Daresbury	SRS	2	Dedicated
Germany			
Berlin	BESSY I	0.8	Dedicated
India			
Indore	INDUS-I	0.45	Dedicated*
Japan			
Okasaki	UVSOR	0.75	Dedicated
Tokyo	SOR-Ring	0.38	Dedicated
Tsukuba	TERAS	0.8	Dedicated
Tsukuba	Photon Factory	2.5–3	Dedicated
Russia			
Moscow	Siberia I	0.45	Dedicated
	Siberia II	2.5	Dedicated*
Zelenograd	TNK	1.2–1.6	Dedicated*
Sweden			
Lund	MAX I	0.55	Dedicated
USA			
Baton Rouge, LA	CAMD	1.2	Dedicated
Stoughton, WI	Aladdin	0.8–1	Dedicated
Upton, NY	NSLS I	0.75	Dedicated
	NSLS II	2.5–2.8	Dedicated

* In construction as of 4/94

Table 1.3. Third Generation Synchrotron Radiation Sources

Location	Ring	Energy (GeV)	Notes
Brazil			
Campinas	LNLS-2	2.0	Design/Proposed
China (ROC-Taiwan)			
Hsinchu	SRRC	1.3	Dedicated
England			
Daresbury	Sinbad	0.6	Design/Proposed
	Diamond	3.0	Design/Proposed
France			
Grenoble	ESRF	6	Dedicated
Orsay	SuperACO	0.8	Dedicated
	SOLEIL	2.15	Design/Proposed
Germany			
Dortmund	DELTA	1.5	FEL Use*
Berlin	BESSY II	1.5–2	Dedicated*
India			
Indore	INDUS-II	2	Design/Proposed
Italy			
Trieste	ELETTRA	1.5–2	Dedicated
Japan			
Hiroshima	HISOR	1.5	Design/Proposed
Ichihara	Nanohana	2.5	Design/Proposed
Kashiwa	ISSP	2.0	Design/Proposed
Nishi Harima	SPring-8	8	Dedicated*
Tsukuba	NIJI IV	0.5	FEL Use
Korea			
Pohang	PLS	2	Dedicated*
Spain			
Barcelona	Catalonia SR	2.5	Approved for const.
Sweden			
Lund	MAX II	1.5	Dedicated*
Switzerland			
Villigen	SLS	1.5–2.1	Design/Proposed
Ukraine			
Kiev	KLS	0.8	Design/Proposed
USA			
Argonne,IL	APS	7	Dedicated*
Berkeley,CA	ALS	1.5	Dedicated
Durham,NC	FELL	1–1.3	FEL Use*
Raleigh,NC	NCSTAR	2.5	Design/Proposed

* In construction as of 4/94

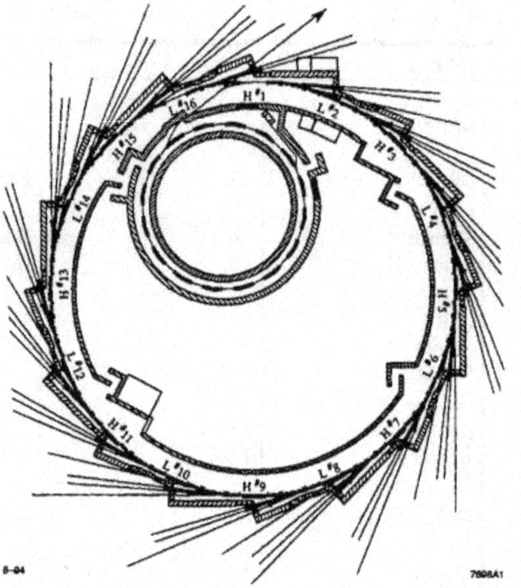

BESSY II facility in Berlin, Germany. (Courtesy of E. Jaeschke.)

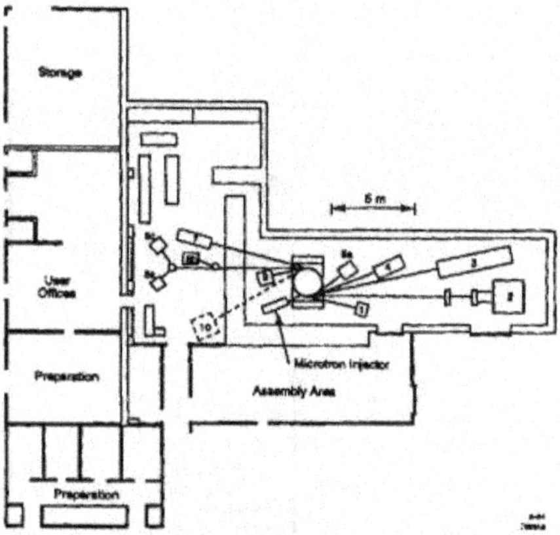

Synchrotron Ultraviolet Radiation Facility (SURF II) at the National Institute of Standards and Technology (NIST) in Gaithersburg, MD. (Courtesy of R. Madden.)

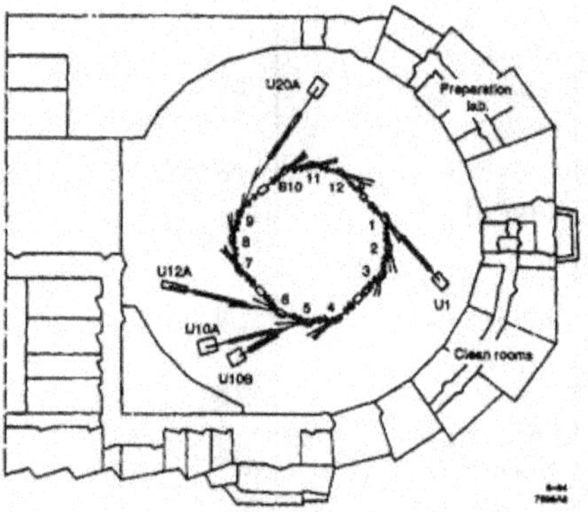

Heifei Synchrotron Radiation Lab in Hefei, PRC. (Courtesy of Z. Bao.)

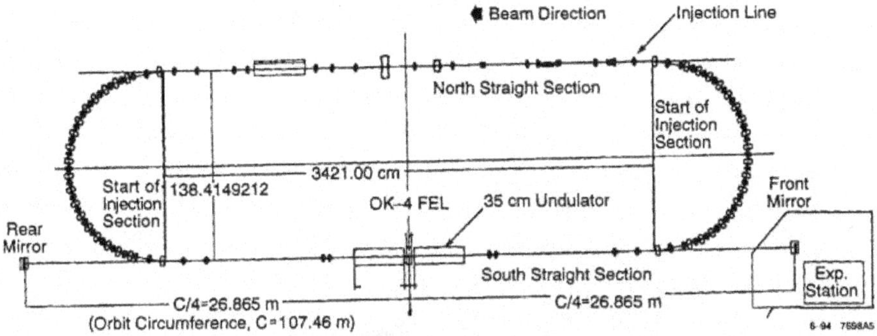

Free Electron Laser Laboratory (FELL) facility at Duke University, Durham, NC. (Courtesy of V. Litvinenko.)

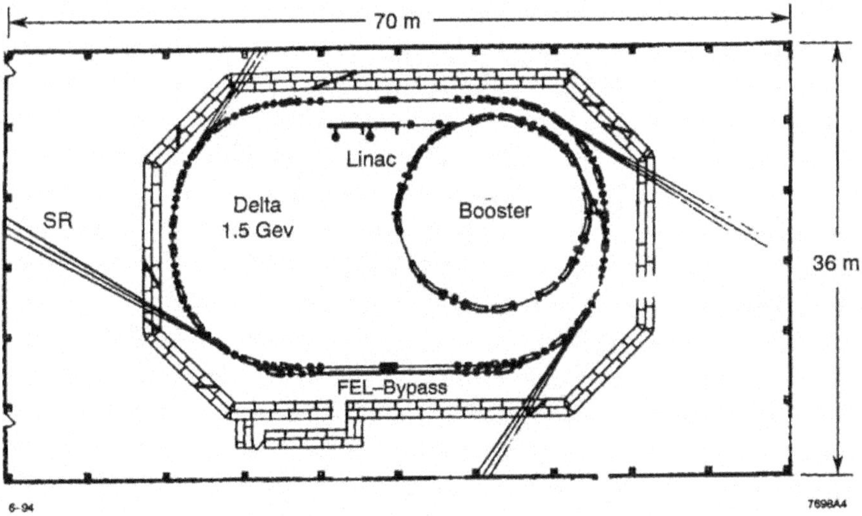

DELTA facility at Dortmund University, Dortmund, Germany. (Courtesy of K. Wille.)

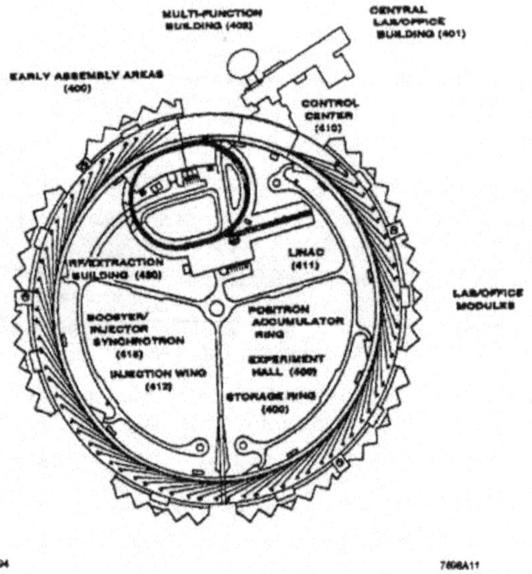

Advanced Photon Source (APS) at Argonne National Laboratory, Argonne, IL. (Courtesy of J. Galayda.)

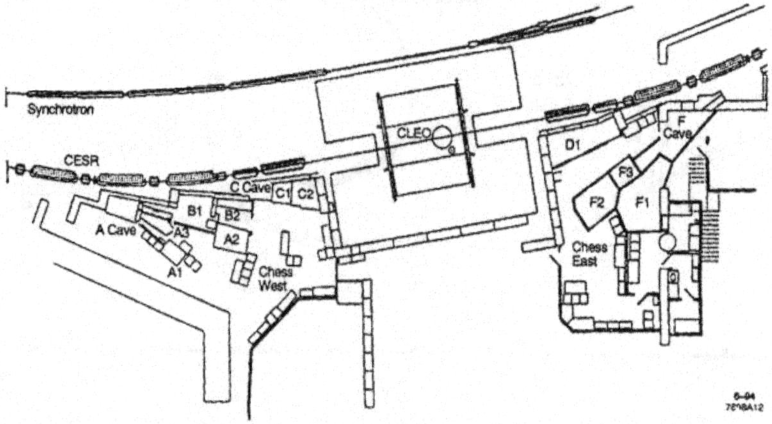

Cornell High-Energy Synchrotron Source (CHESS) at Cornell University, Ithaca, NY. (Courtesy of B. Batterman.)

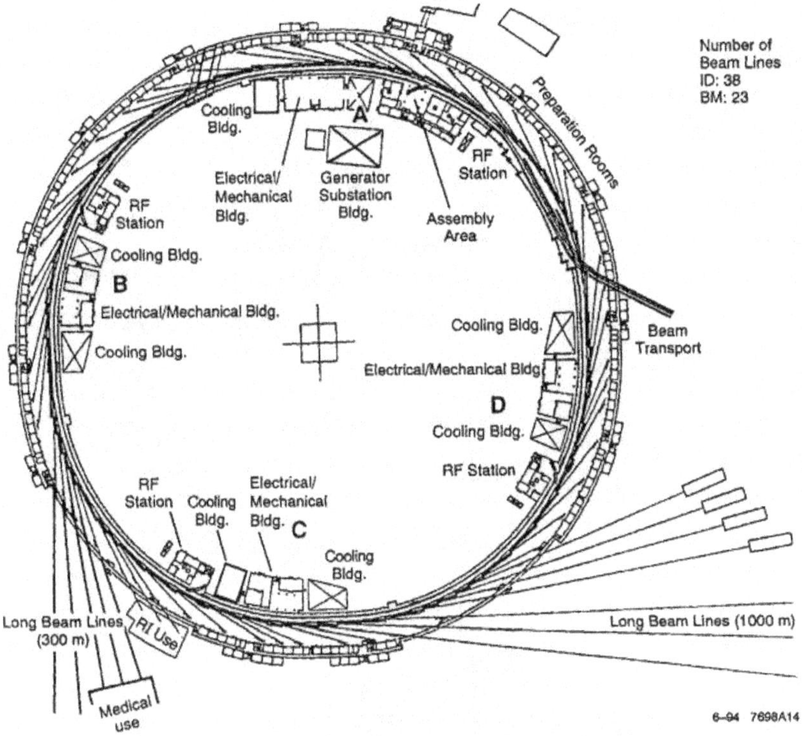

SPring-8 facility in Nishi-Harima, Japan. (courtesy of H. Kamitsubo.)

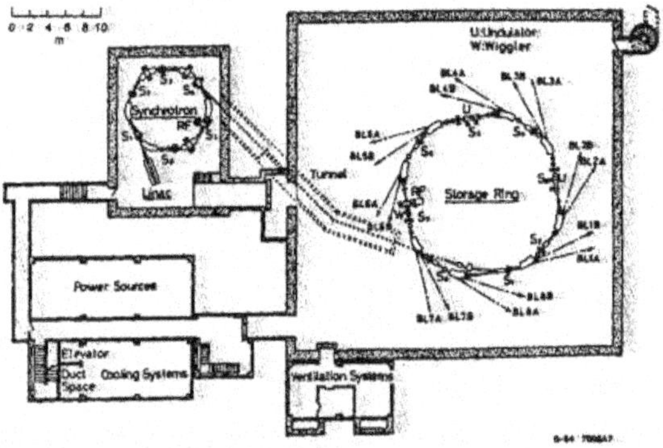

UVSOR facility in Okasaki, Japan. (Courtesy of G. Isoyama.)

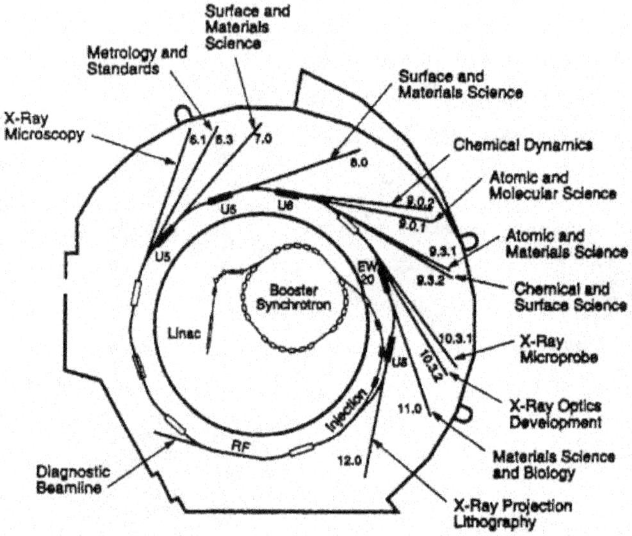

Advanced Light Source facility (ALS) at Lawrence Berkeley Laboratory, Berkeley, CA. (Courtesy of A. Robinson.)

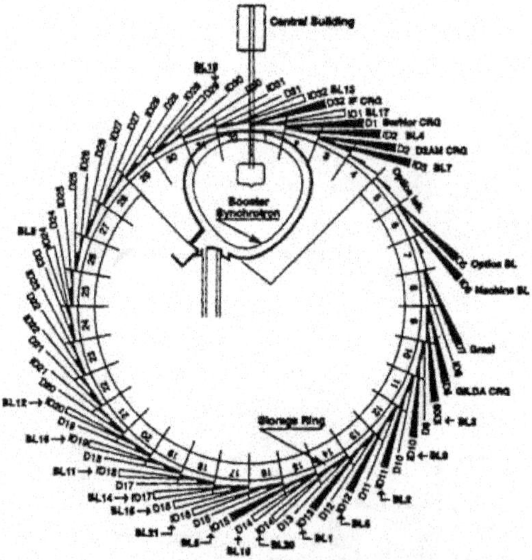

European Synchrotron Radiation Facility (ESRF) in Grenoble, France. (Courtesy of Y. Petroff.)

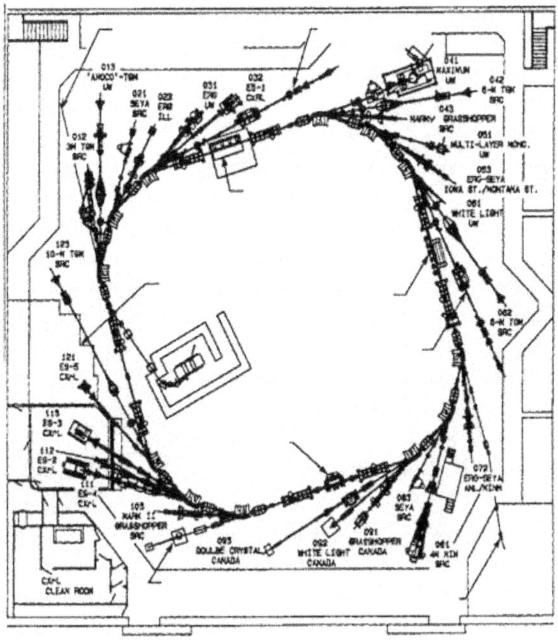

Aladdin facility at the University of Wisconsin, Stoughton, WI. (Courtesy of E. Rowe.)

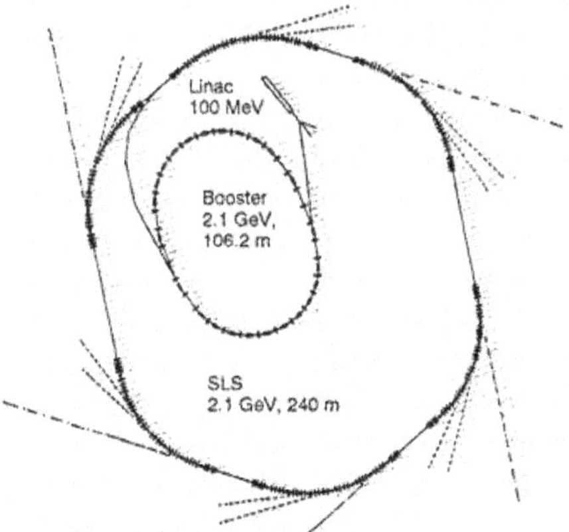

Swiss Light Source (SLS, proposed), Villigen, Switzerland. The design includes several superconducting bending magnets incorporated into the lattice. (Courtesy of L. Rivkin.)

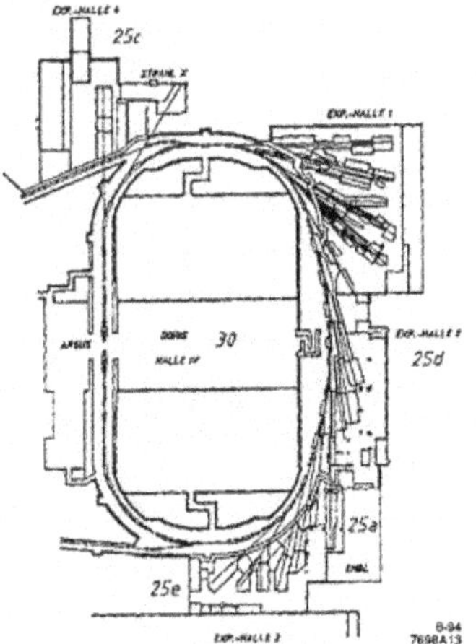

HASYLAB at DESY in Hamburg, Germany. One of the long straight sections of the ring has been replaced by an arc accommodating seven insertion device sources. (Courtesy of W. Brefeld.)

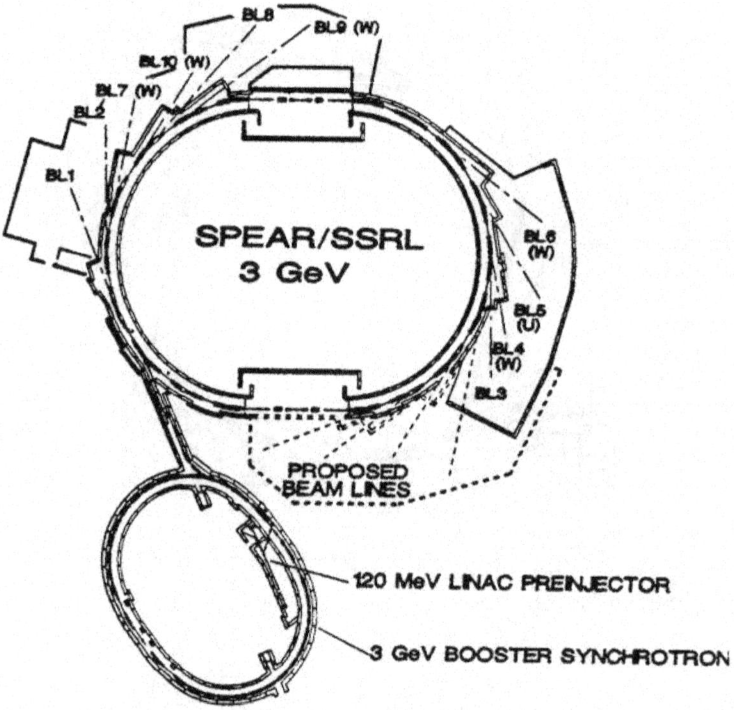

SPEAR/SSRL facility at SLAC, Stanford, CA

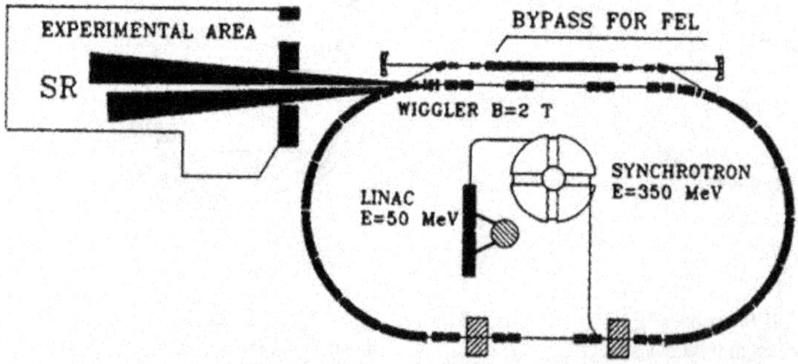

VEPP-3 facility at the Budker Institute; Novosibirsk, Russia
(Courtesy of G. Kulipanov)

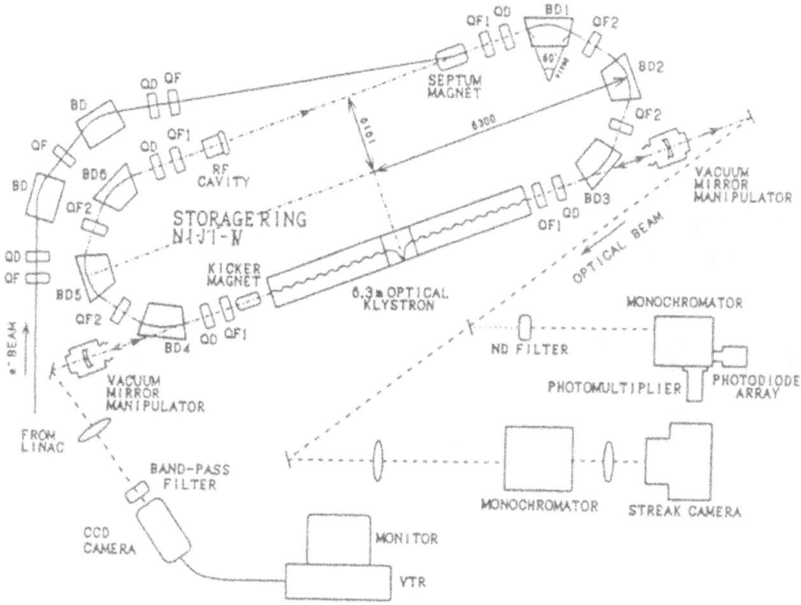

NIJI-IV FEL facility at the Electrotechnical Laboratory; Tsukuba, Japan
(Courtesy of T. Tomimasu)

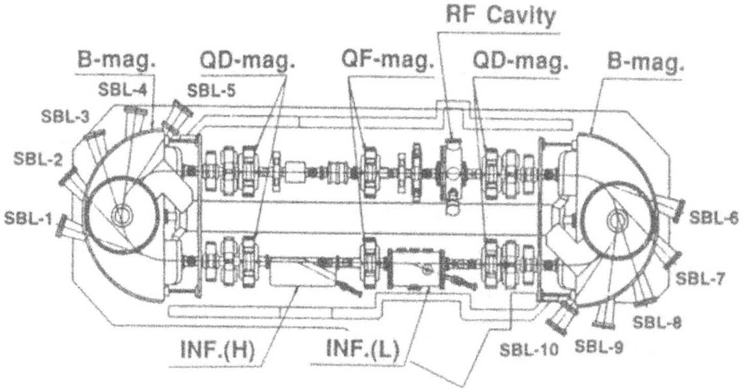

Super-ALIS facility for x-ray lithography
NTT Corporation; Atsugi, Japan (Courtesy of M. Nakajima)

The 1.5–2 GeV Elettra facility in Trieste, Italy (courtesy of A. Wrulich).

The 1.3 GeV facility at the Synchrotron Radiation Research Center in Hsinchu, Taiwan (courtesy of E. Yen).

Third Synchroton Radiation Facilities

	ESRF	APS	SPring-8
Facility	European Synchrotron Radiation Facility	Advanced Photon Source	Super Photon ring-8 GeV
Energy	6.0 GeV	7.0 GeV	8.0 GeV
Photon Energy	19.2 keV	19.0 keV	28.9 keV
Beamlines (ID)	29	34	38
(BM)	24	34	23
Circumference	844 m	1,104 m	1,436 m
Begin Operation	1994	1996	1998

CHAPTER 2: LATTICES*

MAX CORNACCHIA+
Stanford Linear Accelerator Center
Stanford Synchrotron Radiation Laboratory
Stanford, CA 94309, USA

2.1. Introduction

This chapter introduces the reader to the motion of electrons[1] in a storage ring, and to the connection between electron beam dynamics and the properties of synchrotron radiation.

The system of magnetic lenses that guides and focuses an electron beam is called the *lattice*. The choice of a lattice for a synchrotron radiation source is, arguably, the single most important decision in the history of a project. The lattice determines the emittance of the electron beam, the brightness of the photon beam, the beam lifetime, the quality of the experimental conditions, the number of insertion devices that can be accommodated in the straight sections, and the size and cost of the accelerator. The best choice of lattice is not a straightforward affair, involving complex performance and cost trade-offs, and a certain amount of intuition and subjectivity. The various types of lattices, and future directions in lattice design, will be covered, however briefly, in this review.

Synchrotron radiation is emitted from the bending sections of the electron trajectory and in the straight sections, where insertion devices might be installed (see Chapter 12). From the *source points* the radiation is channeled into a *beam line* for experimental use. The lattice and the electron beam energy define the trajectory and, together with the natural divergence of the radiation, the size and divergence of the source.

After a broad overview in Section 2.2, the magnetic forces acting on the electrons and the associated differential equations of motion are discussed in Section 2.3. The solutions of the equations are given without derivation; the method of solution is outlined, and references for deeper studies are given.

Section 2.4 shows how the dynamics of electron motion in magnetic lattices and the emission of radiation define the beam emittance. Examples of lattices for machines in operation or under construction are given in Section 2.5. Throughout this chapter, the electrons are assumed to be ultra-relativistic.

*Work supported by the Department of Energy contract DE–AC03–76SF00515. SSRL is funded by the Department of Energy, Office of Basic Energy.
+Author e-mail: CORNACCHIA@SSRL01.SLAC.STANFORD.EDU.

2.2. Overview of Electron Dynamics in a Storage Ring

2.2.1. The Storage Ring

A storage ring accumulates and stores electrons that have been pre-accelerated and transported from an Injection System (see Chapter 3). The electrons are injected and stored in packets called bunches, which are held together in the direction of motion by the bunching effect of the radio-frequency system (see Chapter 4). The latter also provides the energy lost by radiation and, if acceleration is needed, the energy gain required by the particles to keep in step with the magnetic field

The electrons circulate inside a doughnut-shaped chamber, in which a high vacuum is maintained, delimited by metallic walls (see Chapter 8). The chamber is surrounded by magnets alternating with empty, or drift, spaces. The magnets curve the electron trajectories (dipole field) and keep them close together in the plane perpendicular to the direction of motion (quadrupole field).

2.2.2. Collective and Individual Motion, Frame of Reference

Figure 1 gives simplified top and cross sectional views of electron bunches, frozen in time, circulating in a vacuum chamber surrounded by magnets. The picture is much out of proportion. The circumference of a ring is on the order of 50–250 m for UV and soft x-radiation sources (beam energy: 0.5–3.0 GeV) and 800–1500 m for hard x-ray sources (6–8 GeV). The bunch length is on the order of centimeters or smaller. Tens to hundreds of bunches may circulate in a storage ring.

The motion[2] of the electrons is described in a reference system with an azimuthal axis tangent to the orbit, and the transverse horizontal (x) and vertical (y) coordinates, lying in the plane perpendicular to the orbit (indicated in Fig. 2). The azimuthal coordinate s is the independent variable, and is the distance along the orbit from a reference point s_0. Lattice designers and orbit scientists spend much of their time studying the functions $x(s)$, $x'(s) = dx/ds$, $y(s)$, and $y'(s)$.

The trajectory of an individual electron in a storage ring is qualitatively shown in Fig. 2. It consists of oscillations around an orbit that closes on itself after one revolution, appropriately called the *closed orbit*. The oscillations are called *betatron oscillations* and take place in both the horizontal and vertical planes. The orbit has horizontal and vertical components.[3] In an accelerator with neither vertical bends nor magnetic errors or misalignments, the orbit lies in the horizontal plane (x-s in Fig. 2), and the vertical closed orbit is zero everywhere. This ideal situation does not occur in practice, and there always is an orbit component in the vertical plane.

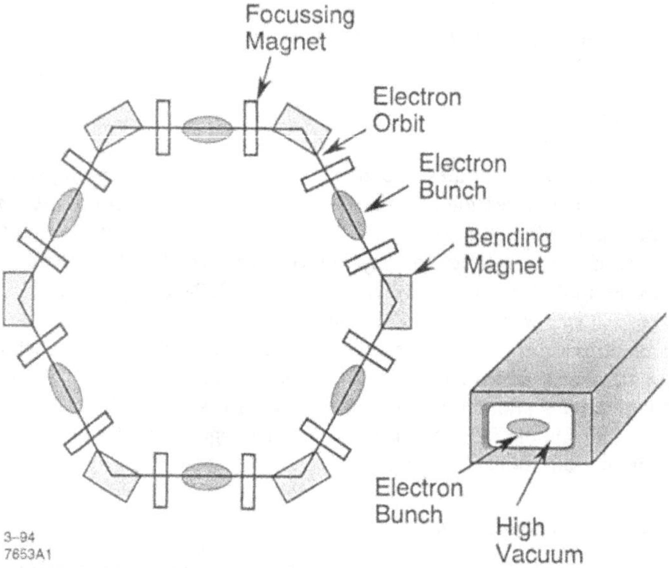

Fig.1. Simplified top and cross sectional views of electron bunches circulating in a storage ring.

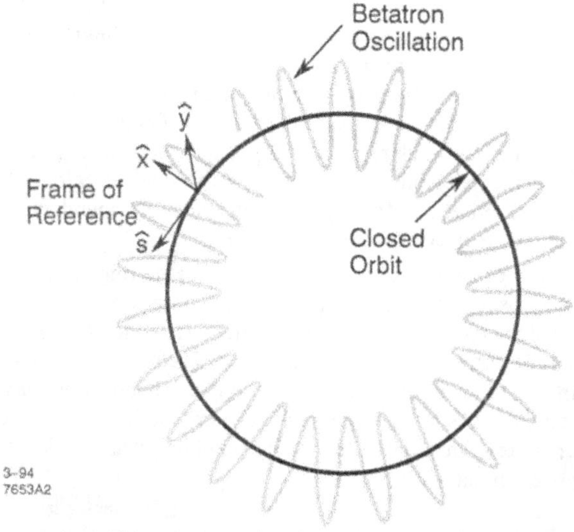

Fig.2. Descriptive view of the closed orbit and a betatron oscillation. The open section is meant to emphasize that the betatron oscillation is not closed.

A large number of electrons (10^{10} or more per bunch) oscillate around a closed orbit with all possible phases and amplitudes. The amplitudes are within a given range defined by the transverse size of the vacuum chamber or, as explained in Section 2.3, by the maximum stable amplitude.

Not all the particles in a bunch have the same energy. Due to the quantum emission of radiation, there is a distribution of energies. To each energy, there corresponds a closed orbit, around which off-energy particles execute betatron oscillations.

2.2.3. Lattice Definition

The lattice of a storage ring is defined to be the sequence of magnetic lenses designed to insure that electrons circulate for a period of several hours (corresponding to billions of revolutions) while maintaining the appropriately small dimensions of the beam. The former is requisite to guarantee the users long periods of uninterrupted emission of radiation, the latter to provide a small and hence bright source of light.

The magnetic properties of the lattice, together with the electron energy, determine the transverse size and divergence of the beam which, after convolution with the divergence of the radiation (see Chapter 14), define the photon beam size.

2.3. Equations of Motion and Solution

2.3.1. Basic Magnetic Elements in a Lattice

The basic lattice of a storage ring consists of a sequence of dipole (bending) and quadrupole (focusing or defocusing) magnets joined by field-free regions, or *drift spaces*. The sequence closes on itself to allow the electrons repeated revolutions around a determined reference orbit and within a confined region around this orbit.

The *bending magnets* are characterized by a magnetic field that is perpendicular to the direction of motion and is uniform in the region occupied by the beam (see Chapter 5). A bending magnet causes a charged particle to follow a circular trajectory along its length. Straight trajectories are joined by sections of circles. The bending magnets are positioned in such a way that there exists a trajectory that is a closed curve that satisfies given geometrical constraints. An electron on this trajectory repeats its motion every revolution. This trajectory is the *closed orbit*, already introduced in Section 2.2.2. In the absence of magnetic, alignment, and other imperfections, it is also the ideal (or design, or reference) orbit. The beam lines are positioned to receive the light emitted from, and are centered tangential to, the closed orbit in the bending magnets and insertion devices.

If the bending magnets were the only elements of the lattice, particles with spatial coordinates different from those of the ideal orbit would move progressively away from this orbit. Since a beam of electrons contains a distribution of particles having different positions and angles, as well as energies, eventually the whole beam would spread out and be lost. For this reason, focusing elements are required to keep together this collection of particles having different coordinates. These elements are *quadrupole magnets*, and they are characterized by a magnetic field whose components are linear functions of the x and y coordinates.[4]

The components of the magnetic field of the basic lattice components, dipoles, and quadrupoles, are:

$$\begin{aligned} &B_y = B_0, \\ &B_x = 0, && \text{for dipoles, and} \\ &B_y = Gx, \\ &B_x = Gy, && \text{for quadrupoles,} \end{aligned} \tag{1}$$

where B_0 and G are constant. In a quadrupole, the field is zero at $x = y = 0$. This point defines the magnetic axis in the azimuthal direction. A quadrupole for which a particle away from the magnetic axis is deflected back onto(away from) it is called focusing(defocusing). It is a consequence of Maxwell's equations that a quadrupole field that is horizontally focusing is vertically defocusing, and vice versa. A sequence of focusing and defocusing quadrupoles, appropriately designed, can focus the beam in both planes. This is demonstrated in the theory of strong focusing,[5] on which all modern synchrotrons are based.

2.3.2. *The Synchronous Orbit*

For a given bending field, there is one value of the electron energy for which the particle follows the ideal orbit. We call this the *synchronous* energy and the particle the synchronous particle. The energy is given by the expression equating the centrifugal force to the Lorentz force:

$$E_0 = ecB_0\rho_0, \tag{2}$$

where E_0 is the electron energy, e the electron charge, c the speed of light in a vacuum, B_0 the bending field, and ρ_0 the radius of curvature in the field of the dipole magnets.[6]

In a more commonly used form, Eq. (2) can be written as

$$E_0[GeV] = 0.3B[T]\rho_0[m], \tag{2a}$$

where GeV, T, and m denote giga-electron-volt, Tesla, and meter, respectively.

2.3.3. *Equations of the Synchronous Orbit and Their Solutions*

An electron that, at a given initial azimuthal position s_0, has the same energy as the synchronous particle, but is displaced (in position or angle in the transverse coordinates x, x', y, y') with respect to the ideal orbit, executes *betatron oscillations* around this orbit. These oscillations occur in the horizontal and vertical planes and are defined by the following differential equations of motion:

$$\begin{aligned} &x'' + K_x(s)x = 0, \\ &y'' + K_y(s)y = 0. \end{aligned} \tag{3}$$

The focusing strengths $K_{x,y}(s)$ are proportional to the quadrupole fields (focusing or defocusing) and also include relatively small effects of the dipoles not discussed here.[7]

Because the magnets have constant fields along the direction of motion, these functions are dichotic ($K_{x,y}(s)$ = constant = 0 in magnet-free regions, or $\neq 0$ in a magnetic field). Equation (3) was solved in the original, classical paper, in which the principles of strong focusing were described.[5]

$$\begin{aligned} x(s) &= \sqrt{\varepsilon_x \beta_x(s)} \cos[\phi_x(s) + \phi_{0x}], \\ y(s) &= \sqrt{\varepsilon_y \beta_y(s)} \cos[\phi_y(s) + \phi_{0y}] \end{aligned} \tag{4}$$

Here, x and y are the transverse displacements from the closed orbit defined earlier. The meaning of the constants ε_x and ε_y is discussed in Section 2.3.5, together with the functions β_x and β_y. The betatron phases ϕ_x and ϕ_y are functions of the distance s along the closed orbit, and ϕ_{0x} and ϕ_{0y} are the initial phases. The $\phi_{x,y}$ are given by

$$\phi_{x,y} = \int_0^s \frac{ds}{\beta_{x,y}(s)}. \tag{5}$$

The motion is a pseudo-harmonic oscillator, with instantaneous amplitudes proportional to the square root of the β-functions and instantaneous wavelength $\lambda_{x,y}(s) = 2\pi\, \beta_{x,y}(s)$.

2.3.4. *The β–Function*

The reader who has been exposed to accelerator terminology will have heard the term β–function used often. It was seen in Eq. (4) that these functions (horizontal and vertical) are related to the maximum amplitude of the oscillations at a given location s:

$$x, y_{\max}(s) = \sqrt{\varepsilon_{x,y} \beta_{x,y}(s)}.. \tag{6}$$

Similarly, the maximum angle of the oscillation at a location s is given by

$$x', y'_{\max}(s) = \sqrt{\varepsilon_{x,y} / \beta_{x,y}(s)}.. \tag{7}$$

The β-functions are periodic in s and follow the periodicity of the lattice. Together with the constants $\varepsilon_{x,y}$, they determine the maximum amplitude of the betatron oscillations. The units in use are meter-radian for $\varepsilon_{x,y}$ and meter/radian for the β-functions. Examples of β-functions are given in Fig. 10.

2.3.5. *The Emittance*

Figure 3 shows the locus of all possible positions and angles (x,x' or y,y') of a particle that is going around the accelerator, as it would be monitored by an observer placed at an azimuth s. All the points fall in an ellipse whose area, it can be shown, is equal to the constant $\varepsilon_{x,y}$ multiplied by π. The shape and orientation of the ellipse changes as a function of s. In an optical system without acceleration, emission of radiation, collective effects, or horizontal-vertical coupling, $\varepsilon_{x,y}$ remains constant as the particle revolves around the accelerator (a consequence of Liouville's theorem). In reality, this "constant" is perturbed by radiation emission and acceleration, and this leads to the statistical concepts discussed in Section 2.4.

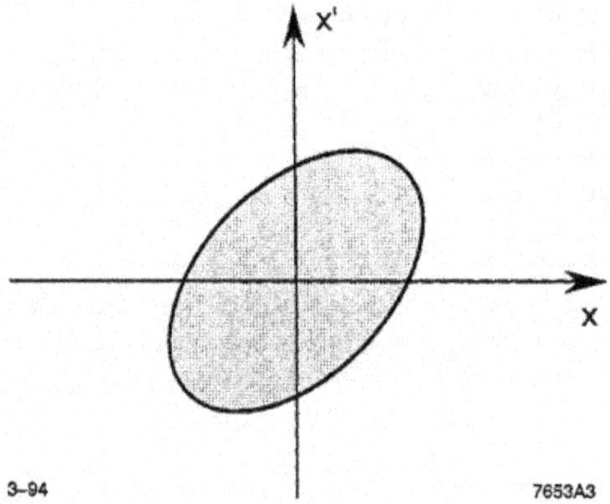

Fig.3. Phase space ellipse.

When the ellipse represents the motion of that particle in a bunch with the highest value of $\varepsilon_{x,y}$ is the *emittance* of the beam. Its importance is immediately recognized: multiplying by the value of the β-function at a given location and taking the square root (Eq. 6) gives the value of the maximum amplitude of the oscillations in the beam, its size.[8]

In an electron storage ring the distribution of betatron oscillation amplitudes is Gaussian, and it is normal practice to define the emittance of the beam as the values of the constants $\varepsilon_{x,y}$ that are related to the standard deviation of the distribution of amplitudes and angular divergences. The relationships are given by the expressions (derived from Eqs. (6) and (7)),

$$\sigma_x = \sqrt{\varepsilon_x \beta_x}, \qquad \sigma_y = \sqrt{\varepsilon_y \beta_y}, \tag{8a}$$

$$\sigma'_x = \sqrt{\varepsilon_x / \beta_x}, \qquad \sigma'_y = \sqrt{\varepsilon_y / \beta_y}, \tag{8b}$$

where $\sigma_{x,y}$ and σ'_{xy} are the distribution standard deviations of position and angle.

Table 1 shows the horizontal emittance of a few representative synchrotron radiation sources. One should note how the electron beam emittance of the various generations of storage rings has evolved towards smaller and smaller values, producing photon beams with smaller and smaller beam sizes and divergences, i.e., brighter photon beams.

2.3.6. *Tunes and Resonances*

2.3.6.1. Definition of Tunes

The numbers of horizontal and vertical betatron oscillations per ring revolution are called the tunes and are denoted by the symbols ν_x and ν_y. From Eq. (5), the tunes are given by

Table 1.		
Facility	Energy (GeV)	Emittance ($\times 10^{-9}$ meter-radian)
SRS	2.0	108
BESSY I	0.8	38
NSLS VUV	0.7	138
NSLS x-ray	2.5	102
SPEAR	3.0	135
Photon Factory	2.5	130
ALS	1.5	4
APS	7.0	8
ESRF	6.0	7
SPRING-8	8.0	7

$$\nu_{x,y} = \frac{1}{2\pi}\int_0^C \frac{ds}{\beta_{x,y}(s)} \quad (9)$$

The integral is extended to the entire lattice length C. The tunes play an important role in the stability of the motion. Their values vary, depending on the optics, but their integer part is on the order of 5–15 units in UV and soft x-ray light sources. In larger, hard x-ray machines, like for instance ESRF, they have the values 36.2 (horizontal) and 11.2 (vertical). Sometimes the symbol $Q_{x,y}$ is used for the tunes.

2.3.6.2. Survey and Magnetic Imperfections, Linear and Nonlinear Resonances

The analysis of the motion shows[9] that there are certain values of the tunes that potentially threaten the stability of the motion. Considering for a moment the transverse plane only (i.e. neglecting energy oscillations) they are those that satisfy the relationship

$$m\nu_x \pm n\nu_y = p, \quad (10)$$

where m, n, and p are integer values. Equation (10) expresses the phenomenon that, if there are magnetic perturbations in the accelerator (unavoidable), the perturbing effect (colloquially called the kick in accelerator jargon) can add up at each revolution, causing the amplitude of the oscillations to grow. For this to happen, the numerical relationship of Eq. (10) must be satisfied, otherwise the perturbations tend to cancel each other over a sufficiently large number of turns. Since m and n can take any integer values, it appears very difficult to find a pair of tunes values that escape Eq. (10). Fortunately, it happens that the perturbing effect becomes weaker and weaker as the *order of the resonance*, defined as the sum of $|m|$ and $|n|$, becomes larger. In general, in electron accelerators, one does not worry about resonances for which $|m|+|n|>5$. This is because radiation damping tends to neutralize the resonant amplitude growth when it and the damping rates are of the same order (see Section 2.4.3).

It is not necessary for Eq. (10) to be perfectly satisfied for a magnetic perturbation to be felt. There is a region around a resonance line, defined by Eq. (10), where the trajectory can be perturbed. Fortunately, this band (called *stop-band width*) becomes narrower the higher the order of the resonance.

A. Linear Resonances, Orbit and Focusing Perturbations. The resonances for which $|n|+|m|\leq 2$ are driven by *linear imperfections* in the lattice. The resonances $\nu_{x,y}$ = integer are particularly disruptive. They are driven by magnetic imperfections of the dipole type, by survey imperfections in the transverse locations of the quadrupoles, and by rotational errors in the placement of the dipoles. These resonances cause closed orbit distortions and are responsible for the movement of the photon beam that is so disruptive to the experimentalists. The orbit distortions act like $1/(\nu_{x,y}^2-p^2)$ (where p is any integer) and, although the tunes are normally set at a respectable distance from an integer value, orbit distortions can be, and are, driven at any tune values. For this reason, *dipole correctors* are used to correct the orbit distortions and are a necessary part of any accelerator (Chapter 13 is devoted to the important aspects of orbit correction). To reduce the amplitude of the orbit distortions, tight tolerances are set for the random relative variation of the bending field (on the order of a few times 10^{-4} rms), for the transverse positioning of the quadrupoles (typically 0.10–0.15 mm rms) and for the rotation angle of the bending magnets (0.5–1.0 mrad rms),

Magnetic imperfections of the quadrupole type drive second order resonances. They perturb the β-functions, couple the horizontal and vertical motion (see Section 2.3.6.2.C), and, if strong enough, may lead to an unstable lattice. The tolerances on the variation of the field gradient (G in Eq. (1)) from quadrupole to quadrupole are specified to limit this effect, and are typically on the order of 10^{-3}.

B. Non-linear Resonances. Those resonances for which $|n|+|m|>2$ are driven by nonlinear fields, for example the two third integer resonances:

$$\begin{aligned} &3\nu_x = p, \\ &2\nu_y \pm \nu_x = p. \end{aligned} \tag{11}$$

They are driven by sextupoles fields that have the form

$$\begin{aligned} &B_y = S(x^2-y^2), \\ &B_x = S2xy, \end{aligned} \tag{12}$$

where S is the sextupole strength, normally expressed in T/m^2. Sextupole magnets are part of any storage ring lattice because, as we shall see in Section 2.3.9.1, they are needed to correct the chromatic aberrations of the quadrupoles.

Higher order resonances are driven by magnetic fields with higher order non-linearities. For instance, octupole fields (those that have cubic dependence on the displacement, $B_y(x,0)$ = constant x^3) drive fourth order resonances, for which $|n|+|m|=4$. Decapole fields (quartic dependence on displacement) drive fifth order resonances, and so on. Some resonances require a slight rotation of the magnetic axis in order to be driven, and this often sets the survey tolerances.

At the construction stage the magnet builder requires a set of tolerances from the accelerator physicists for the purity of the magnetic field.[10] This is typically on the order of a few times 10^{-4}. The subject of beam-stability in the presence of non-linear fields is discussed further in Section 2.3.9.2.

Two more points need to be mentioned concerning Eq. (10). It can be shown that only the + sign (*sum resonances*) on the left-hand side of the equation leads to indefinite growth in both the horizontal and vertical directions. The – sign resonances (*difference resonances*) lead to a transfer of oscillation amplitudes from the horizontal into the vertical, and vice versa, but the motion is bounded. The behavior is much like that of a coupled pendulum, with the maximum amplitudes beating between the two directions of transverse motion. Sum resonances are in general much more dangerous.

For a resonance condition to be established, the perturbation (dipoles, quadrupoles, non-linear fields) must have a p-th (integer of Eq. (10)) Fourier component, analyzed as a function of the azimuth, that is non-zero. This is the harmonic that drives the resonance. Linear and non-linear resonances are corrected by canceling out, with appropriate magnets, the more dangerous harmonics of the field errors.

Figure 4 shows the working diagram of the ALS storage ring.[11] This is a plot of resonance lines defined by Eq. (10), with axes given by ν_X and ν_Y. The *working point* is the point having the tune values as coordinates. The accelerator physicist chooses this working point to be at a suitable distance from resonance lines. It is worthwhile to mention the order of magnitude of the tolerable departure of the tunes from the design (or, in an existing machine, experimentally found, optimum) values. This tolerance varies greatly from storage ring to storage ring, but it could be as tight as 0.001 in tune. Remember that tune values are in the tens of units. Thus, this tolerance is rather tight and is reflected in the high stability required from the power supplies.

It is shown in Chapter 4 that, due to the restoring force of the radio-frequency field, particles oscillate in energy, describing *synchrotron oscillations*. The number of oscillations per revolution is denoted by the symbol ν_S (synchrotron wave number), and is on the order of 0.01 (100 turns per oscillation period). If the three-dimensional motion is considered (two transverse and one energy variable), then more resonances appear that involve the energy oscillations. The extended numerological condition for resonance is

$$m\nu_x \pm n\nu_y \pm k\nu_s = p. \tag{13}$$

When Eq. (13) applies, the resonance is called a synchrotron-betatron resonance. It may be driven, for instance, when the value of the dispersion at the location of the radio-frequency accelerating cavities is non-zero.

C. Horizontal-Vertical Coupling. It is important to note that in Eq. (3) horizontal and vertical motions were not coupled, i.e., the horizontal differential equation of motion did not depend on the vertical coordinates, and vice versa. This is only true in an ideal lattice in which the horizontal and vertical components of the magnetic field are perfectly aligned and in absence of field imperfections (see Eq. (1)). In practice, a small amount of coupling is always present.

Particularly important is the coupling due to a *rotated quadrupole*, i.e., a quadrupole that, because of survey tolerances, is slightly (on the order of one mrad or less) rotated around its magnetic axis. This imperfection excites the coupling resonances $\nu_x \pm \nu_y = p$. The sum resonance must be avoided. Although normally not "fatal," special attention is required also for the difference resonance $\nu_x - \nu_y = p$. This resonance couples horizontal and vertical motion. Since the vertical beam emittance is only a few percent of the horizontal one, this resonance may appreciably increase the vertical beam size and thus reduce the brightness of the photon beam.

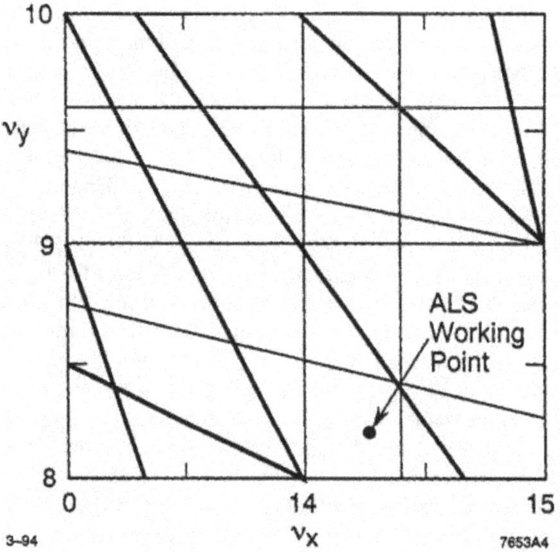

Fig. 4. Tune diagram of the ALS, showing resonances of the type $m\nu_x \pm n\nu_y = 12p$, up to order 6. (The ALS has a twelve-fold periodicity, and p is any integer.)

To combat the linear coupling effect, most storage rings are provided with rotated quadrupoles (i.e., quadrupoles that are rotated by 45° around the magnetic axis) placed at strategic positions to cancel the effect of the rotation errors of the lattice quadrupoles.

2.3.7. *Effects of Insertion Devices on the Particle Motion*

Synchrotron light facilities are making ever increasing use of wigglers and undulators (see Chapter 14), to the extent that these devices are becoming significant parts of the beam optical system. Theoretical studies,[12] confirmed by experimental observations, have shed light on the perturbations to the trajectory caused by the magnetic field of such devices. The analytical expressions for the field are given in Chapter 14, namely a dipole field (but rich in higher harmonic content) of alternating polarities that imposes an oscillatory trajectory to the electrons (Fig. 5a). In the general form[13] the expressions for the field in a planar undulator are[14]

$$\begin{aligned} B_y &= B_0 \cosh k_x x \cosh k_y y \cos ks, \\ B_x &= \frac{k_x}{k_y} B_0 \sinh k_x x \sinh k_y y \cos ks, \\ B_z &= -\frac{k_x}{k_y} B_0 \cosh k_x x \sinh k_y y \sin ks, \end{aligned} \tag{14}$$

where $k_x^2 + k_y^2 = k^2 = (2\pi/l)^2$, and l is the length of the magnetic period.

If the poles are flat, and in the approximation that they are infinitely large, $k_x = 0$. Shaping the poles to provide horizontal focusing gives $k_x^2 > 0$.

The amplitudes of the orbit oscillations are, typically, on the order of a few microns in undulators and hundreds of microns in wigglers. Figures 5a and b give an impression of the trajectory.

The first requirement of the field is that the trajectory not be perturbed outside the length of the insertion device (Fig. 5b). This is so in a perfectly designed and built magnet, but inevitable field errors cause a perturbation to the orbit that, if uncorrected, propagates around the ring. Correcting dipole magnets and beam position monitors are normally added at the beginning and end of the insertion device to cancel any orbit distortion. More sophisticated corrections may also be included. These corrections are particularly important in modern light sources, in which the undulator field is often changed during the experiment, and it is important that one insertion device does not perturb the orbit for other users.

Even a perfectly built insertion device, however, has focusing terms that must be accounted for in the design of the optics and contains significant non-linear components, as implicit in Eqs. (14).

The focusing effect of insertion devices results in linear[15] tune shifts. In parallel pole devices (no horizontal focusing) the linear tune shift only occurs in the vertical plane. This tune variation can be significant, particularly in wigglers, and must be corrected with quadrupole magnets, preferably locally (i.e., in the same straight section that houses the insertion device). For a given insertion device the tune shift scales inversely with the square of the energy. For this reason it is higher in UV and soft x-ray sources (1.5–2.0 GeV) than in hard x-ray facilities (6–8 GeV). On the other hand, larger, higher energy facilities tend to accommodate more undulators. The tune shifts can be on the order of 0.01–0.03 in the lower energy storage rings and a factor of 20 or so smaller in hard x-ray sources.

The non-linear terms of Eq. (14) drive mainly third and fourth order resonances (see Section 2.3.6.2.B) and cause tune dependence on the betatron amplitude; the optics designer must make sure that they have a negligible effect on the stability of the beam.

Wigglers have stronger fields than undulators, but often also longer periods (see Chapter 14). In wigglers the tune shifts are higher than in undulators, but the non-linear effects are weaker, since the deflecting field of undulators has fast azimuthal variations (short poles) that tend to enhance the non-linear components.

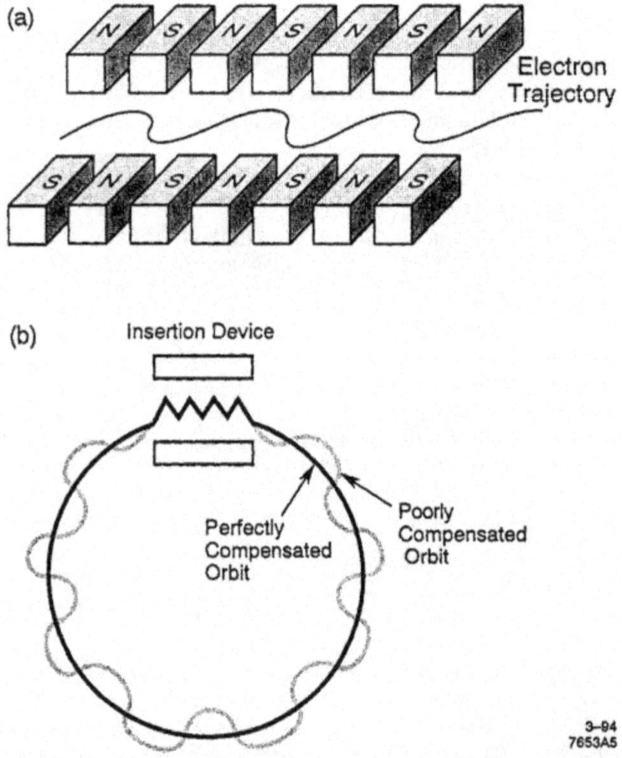

Fig. 5. Beam trajectory in an insertion device. The amplitudes of the oscillations are highly exaggerated.

2.3.8. Off-Energy Particle Motion, Dispersion, Beam Size, and Momentum Compaction

In this section, non-synchronous orbits, namely those of electrons having energy different from the one defined by Eq. (2), are discussed. The off-energy dynamics is not just of theoretical interest. It is important because, due to the quantum emission of radiation and the action of the radio-frequency system, the particle energy fluctuates around an average value. It is this energy fluctuation that, as we shall see, largely determines the electron beam emittance.[16]

Four important functions describe the motion of off-energy particles. Two are the *dispersion*, normally denoted by the symbol η, and its derivative η' with respect to the independent variable s. The others are the horizontal and vertical *chromaticities*.

2.3.8.1. The Dispersion

If the momentum of a particle changes, the bending radius in the dipoles changes according to Eq. (2), and the closed orbit also changes. A particle whose energy differs from the reference value follows a different orbit. The differential equations of motion (Eq. (3)) now becomes:

$$x'' + K_x(s)x = \frac{1}{\rho_0(s)}\frac{\Delta E}{E_0}, \qquad y'' + K_y(s)y = 0. \tag{15}$$

They differ from Eq. (3) by the presence of a driving term in the x-axis and by a small (but important, see Section 2.3.9) change in the focusing terms K_x and K_y. The latter reflects the fact that a change in energy (denoted as the relative change $\Delta E/E_0$ with respect to the synchronous energy E_0) changes the focusing strength of the quadrupoles. The term $(1/\rho_0(s))(\Delta E/E_0)$ represents the perturbation introduced by the fact that the energy of the particle does not match the strength of the bending field, $\rho_0(s)$ being the bending radius in the dipoles of the synchronous particle with energy E_0. The vertical plane does not have such perturbation, unless vertical bends are present in the lattice. The function $1/\rho_0(s)$ follows the periodicity of the bending magnets. One of the solutions of Eq. (15) is periodic with the lattice periodicity, i.e., satisfies the conditions $x(0) = x(C)$, $x'(0) = x'(C)$, where C is the length of the orbit after one revolution and can be expressed in terms of the dispersion $\eta(s)$ and its derivative $\eta'(s)$ defined as:

$$x_c(s) = \eta(s)\frac{\Delta E}{E_0}, \qquad x_c'(s) = \eta'(s)\frac{\Delta E}{E_0}. \tag{16}$$

The dispersion is expressed in units of meters. Its derivative is dimensionless.

The solutions of Eq. (15) are those of the non-homogeneous and the homogenous forms, the latter given by Eq. (4). In a general form that includes energy deviation and betatron oscillation, the horizontal motion of an electron can be described as the sum of a term which is a periodic function of s and of an oscillatory term:

$$x(s) = \eta(s)\frac{\Delta E}{E_0} + \sqrt{\varepsilon_x\beta_x(s)}\cos[\phi_x(s) + \phi_{0x}]. \tag{17}$$

The slope, dx/ds, is given by

$$x'(s) = \eta'(s)\frac{\Delta E}{E_0} - \alpha(s)\sqrt{\frac{\varepsilon_x}{\beta_x(s)}}\cos[\phi_x(s) + \phi_{0x}] - \sqrt{\frac{\varepsilon_x}{\beta_x(s)}}\sin[\phi_x(s) + \phi_{0x}], \tag{18}$$

with

$$\alpha(s) = -\frac{1}{2}\frac{d\beta}{ds}.$$

2.3.8.2. The Beam Size and Divergence

Having introduced a function for the motion of off-energy particles, we are in a position to generalize the beam size and divergence expressed by Eqs. (6) and (7). Those equations ignored the contribution of the spread in energy that is always present in a beam. Like the distribution of betatron amplitudes, the distribution of the energy spread is Gaussian. If $<\Delta E>$ is the root-mean-square of the energy deviation, and, as is normal practice, the emittance ε_x also defines the rms of betatron amplitudes, then, since these quantities are uncorrelated, they contribute quadratically to the overall beam size:

$$\sigma_x = \sqrt{\varepsilon_x \beta_x + \eta^2 \left(\frac{\langle \Delta E \rangle}{E_0} \right)^2}\,,$$
$$\sigma'_x = \sqrt{\frac{\varepsilon_x}{\beta_x} + \eta'^2 \left(\frac{\langle \Delta E \rangle}{E_0} \right)^2}\,. \qquad (19)$$

Typically, the relative energy spread is on the order of 10^{-3}, and the dispersion is measured in meters. One meter dispersion gives a contribution of 1 mm to the beam size. For comparison, an emittance of 5×10^{-9} meter-radians at a location at which β_x is, say, 10 m, gives a beam size of 0.22 mm. This is one of the reasons why insertion devices are normally located in "dispersion-free regions" where $\eta = \eta' = 0$, or is at least very small.

2.3.8.3. The Momentum Compaction Factor

Let us now introduce a quantity that is of fundamental importance for the longitudinal motion. This parameter is the *momentum compaction*. It is a measure of how the time taken by the particle to complete one turn in the accelerator varies with energy. In high-energy electron accelerators the velocity of the particle is nearly constant with energy, and the revolution time is determined by the longer (or shorter) path a higher (lower) energy particle has to travel. Only the curved sections contribute to a lengthening of the orbit with energy, and higher energy particles have a larger bending radius. The momentum compaction factor is defined as

$$\alpha_c = \frac{\Delta T / T_0}{\Delta E / E_0}\,, \qquad (20)$$

where ΔE is the energy difference from the synchronous energy E_0, and T_0 is the revolution period of the synchronous particle. The momentum compaction is determined by the properties of the lattice. In fact, it is the average of the dispersion in the bending section divided by the average machine radius.[17] The stronger the focusing, the lower this value α_c is. An approximation often used is $\alpha_c \approx 1/\nu_x^2$. The small value of the momentum compaction function in synchrotron radiation sources has important implications for the longitudinal motion (see Chapters 4 and 12).

2.3.9. *Chromaticity Correction and Dynamic Aperture*

The focusing (or defocusing) action of the quadrupoles is inversely proportional to the particle energy. In analogy with optical lenses, this effect is called a *chromatic effect*. It leads to a dependence of the tunes on energy. This dependence is measured by the horizontal and vertical chromaticities, ξ_x and ξ_y:

$$\Delta\nu_x = \xi_x \frac{\Delta E}{E_0}, \qquad \Delta\nu_y = \xi_y \frac{\Delta E}{E_0}, \tag{21}$$

where the $\Delta\nu_{x,y}$ are the shifts in tunes from those of the synchronous particle and are caused by a change in energy $\Delta E/E_0$.

Because the focusing action decreases with energy, the uncorrected chromaticities are negative numbers. Corresponding to a spread in energy within a beam of particles, Equation (21) implies that a spread in tunes follows, and this may have adverse effects if it results in crossing resonance lines (Section 2.3.6.2). Sextupole magnets, non-linear elements already introduced in Eq. (12), are used to correct the chromaticities. Most accelerators operate with zero or slightly positive chromaticities. The next section gives a simple treatment of how sextupoles are used to control the chromaticities.

2.3.9 1. How Sextupoles Correct the Chromaticities

Consider the field of a sextupole magnet (Eq. (12)):

$$B_y = S(x^2 - y^2),$$
$$B_x = 2Sxy.$$

If a particle is off-energy, its horizontal displacement x consists of two terms: a betatron oscillation x_β and an orbit shift x_E (Eq. (17)). The vertical displacement is a pure betatron oscillation y_β. The field seen by the particle can be decomposed into the components of the displacements

$$\begin{aligned} B_y &= Sx_\beta^2 + \mathbf{2Sx_E x_\beta} + Sx_E^2 - Sy_\beta^2, \\ B_x &= 2Sx_\beta y_\beta + \mathbf{2Sx_E y_\beta}. \end{aligned} \tag{22}$$

The terms in bold in Eq. (22) have the form of a quadrupole field, a field that is linear in the betatron displacements x_β and y_β. The "strength" of the quadrupole is $2Sx_E$, and is proportional to the particle energy via its closed orbit displacement x_E. This fact is utilized to offset the (linear) energy dependence of the focusing strength of the quadrupoles. Figure 6 shows the quadratic dependence of the horizontally deflecting field.

Since the horizontal and vertical machine chromaticities are both negative, but the equivalent quadrupole of Eq. (22) has opposite focusing and defocusing effects in the horizontal and vertical axes, two families of sextupoles are required, both placed in dispersive regions. The horizontally correcting sextupoles are located in regions in which the horizontal β-function is high and the vertical β-function is low. The converse is true for the vertical chromaticity correcting sextupoles. Since, from Eq. (16), $x_E = \eta(s)\frac{\Delta E}{E_0}$, it

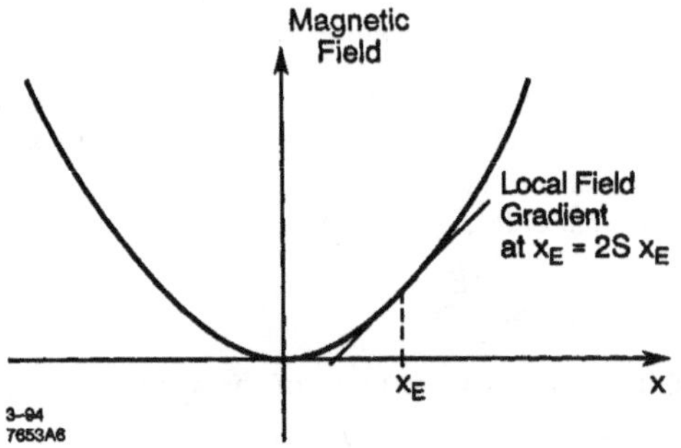

Fig.6 Sextupole field and local field gradient for an orbit displaced by X_E.

is convenient, in order to reduce the sextupoles strength, to place the sextupoles at locations where the dispersion is high.

Equation (22) indicates that, besides the "useful" terms (in bold characters) that correct the chromaticities, unwanted, non-linear terms crop-up that perturb the motion. Some storage rings include more than two families of sextupoles, the additional families being used to neutralize some of the resonances create by the unwanted terms of Eq. (22).

2.3.9.2. The Dynamic Aperture Problem

Sextupoles are non-linear elements, and, while they correct for the linear part of the chromatic aberrations, they can also disrupt the motion and cause particle loss. Low emittance lattices are characterized by strong sextupoles, and the problem of the *dynamic aperture* is one of the most important design issues.

The dynamic aperture is defined to be the maximum betatron oscillation that can be sustained in the accelerator for a sufficient number of turns. In electron storage rings, the time scale is on the order of the damping time (see Section 2.4.2).

The amplitude may be limited by the transverse size of the vacuum chamber (physical aperture), or by the perturbing effect of the non-linear fields (dynamic aperture). The problem of determining the maximum stable amplitude of the oscillations in the presence of non-linear perturbations is not amenable to an exact mathematical solution. The *dynamic aperture limit* is estimated by computer simulation of the motion of the particles in the presence of the non-linear field of the sextupoles and other perturbing non-linearities, like those caused by magnetic imperfections. It is desirable to design a lattice and chromaticity correction sextupoles such that the maximum amplitude of the betatron oscillations is determined by the physical aperture of the chamber, and not by the non-linear perturbations.

The dynamic aperture is often plotted in a graph that depicts the maximum amplitudes of the vertical betatron oscillations that are stable as a function of the maximum stable horizontal amplitudes. Figure 7 shows the dynamic aperture of the Advanced Light Source (ALS), as computed for the Conceptual Design Report.[18]

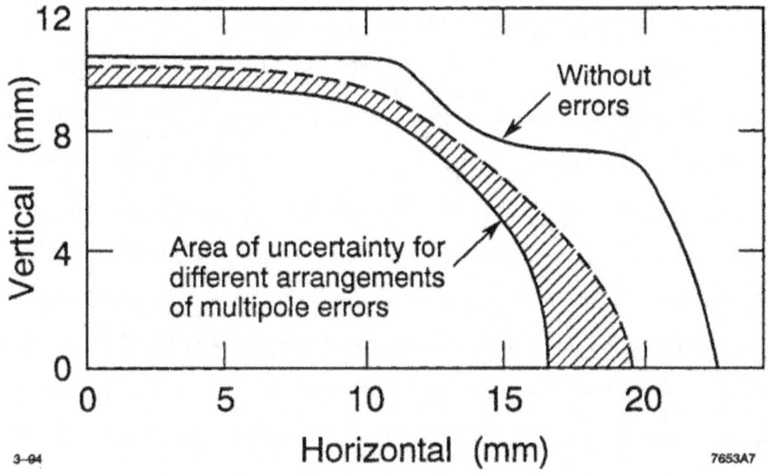

Fig.7. Dynamic aperture in presence of multipole errors in the ALS.

The dynamic aperture is sensitive to the degree of symmetry of an accelerator, high periodicity usually being associated with a larger dynamic aperture. Unfortunately, even in a machine designed with high periodicity, the regular lattice pattern is broken by magnetic imperfections, orbit errors, and the presence of different types of insertion devices in the straight sections. Figure 7 shows the dynamic aperture of the ALS lattice in which the only non-linear elements are the chromaticity sextupoles ("without errors" curve). The maximum stable amplitudes are reduced when magnet misalignments and field imperfections are included in the computation.

2.4. Emission of Radiation and the Equilibrium Emittance

2.4.1. Emission of Radiation, Damping Times and Equilibrium Emittance

A charged particle on a curved path emits radiation. In a storage ring, the orbit is curved in bending magnets and insertion devices. The radiation is emitted in the direction tangential to the direction of the motion and is concentrated in a narrow cone with an apex angle of about a milliradian (see Chapters 1 and 14).

The emittance (horizontal or vertical) is not defined by the characteristics of the beam upon injection into the storage ring. Instead, after characteristic times, called the *damping times* (horizontal and vertical), the beam size and angular spread are determined by the lattice and the emission of radiation. In other words, the beam loses all memory of its previous dynamical history. The equilibrium emittance is the result of a balance between an anti-damping mechanism that tends to increase the beam size and a damping process that tends to reduce it.

If there is a transient perturbation (power supply, residual gas effects, fast magnetic kicks, etc.), the value of the emittance settles back to equilibrium after a damping time.

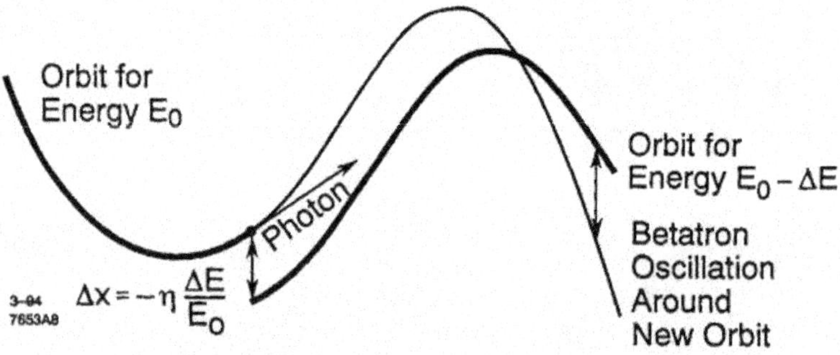

Fig.8. Sketch of a trajectory following the emission of a photon of energy ΔE.

2.4.2. *Antidamping of Betatron Oscillations*

Consider an electron that follows a horizontal betatron oscillation around the closed orbit defined by its energy (Eq. (17)). Suppose that, at a given azimuthal position s, the electron emits a photon. The photon can be emitted anywhere within a cone of aperture $1/\gamma$ centered around the tangent of motion (where γ is the ratio of the particle energy to its rest energy). Since the effect of this angular aperture is small compared to the average perturbation[19] we are justified in modeling the photon as if it were emitted tangential to the motion.

Figure 8 shows the trajectory of an electron emitting a quantum of energy ΔE at a point at which its betatron amplitude and angle are x and x' with respect to the closed orbit relevant to its energy E_0.

Since the emission occurs in the direction of the particle momentum, there is no change in displacement or slope following the emission of radiation. The particle energy, however, changes by an amount $-\Delta E$. The closed orbit associated with the new energy, $E_0 - \Delta E$, is now different, as shown in Fig. 8. After the emission of radiation, the particle trajectory is different from the one it would have followed if, hypothetically, no radiation emission had taken place. The motion is still described by the sum of periodic (closed orbit) and oscillatory terms (Eqs. (17) and (18)), but the coordinates of the closed orbit are now those relevant to the new energy. Since the position and angle of the particle have not changed following the emission of radiation, and since Eqs. (17) and (18) are still valid, the amplitude and angle of the betatron oscillation must change.

A rigorous description of the statistical nature of the phenomenon[17,19] reveals that the quantum nature of the radiation emission leads to an increase of the betatron oscillation amplitudes of the ensemble of particles.

It is intuitive, and it can be rigorously shown, that *the greater the amplitude of the oscillations are greater, the greater the value of the dispersion and its derivative at the location where the radiation is emitted.*

Fortunately, this is not the whole story, as another phenomenon takes place that tends, instead, to decrease the amplitude of the betatron oscillations. This is discussed in the next section.

2.4.3. Damping of Betatron Oscillation

It is shown in Chapter 4 that the radio-frequency accelerating system restores the energy that the particles lose due to the emission of radiation. We have also mentioned that the radiation is predominantly emitted in a direction which is tangent to the direction of motion. For a particle executing betatron oscillations (horizontal or vertical), the radiation is emitted at an angle with respect to the closed orbit (Fig. 8). A particle loses part of its transverse momentum (p_x or p_y) if, in the course of the betatron oscillations, there are components of the momentum in the x or y direction. The electric field of the radio-frequency accelerating system restores the momentum in the direction of the ideal orbit, and thus tends to "align" the electrons in the direction of this orbit.

The overall effect is a transfer of momentum from the horizontal and vertical oscillations into the azimuthal direction, causing a reduction of the horizontal and vertical betatron amplitudes.

2.4.4. The Horizontal Emittance

The horizontal emittance of the beam in a storage ring is determined by the equilibrium between the two actions described in Sections 2.4.2 and 2.4.3, namely the *antidamping effect* of the quantum emission and the *damping* effect of the restoring radio-frequency field. Apart from collective effects and other perturbing factors, the emittance is completely determined by the energy, bending field, and lattice functions.

Once the beam is injected into a storage ring, it takes some time for the emittance to settle to the equilibrium value. This time depends on the rate of emission of radiation and is usually on the order of a few milliseconds. It is referred to as the *betatron damping time*. This time plays an important role in the injection process (see Chapter 3).

2.4.5. The Vertical Emittance

In a machine for which the bending occurs only in the horizontal plane, the vertical motion is privileged, since it experiences the damping effect of the radio-frequency field, but not the antidamping effect of the quantum emission of radiation. Without vertical bending there is no vertical dispersion. Theoretically, the vertical emittance should be almost zero in a storage ring with no vertical bends. In practice, the vertical emittance is small but finite, due to several factors, for instance vertical bends are present due to quadrupole misalignment and associated correction magnets, horizontal-vertical betatron coupling caused by quadrupoles that are rotated, due to the limitations of survey accuracy, around their magnetic axis.

Most electron storage rings are characterized by a very small vertical/horizontal emittance ratio, on the order of 0.01–0.03. The synchrotron light spot looks like an ellipse with a much larger horizontal axis.

Sometimes it becomes necessary to intentionally increase the coupling, and the vertical beam size, in order to reduce the charge density when intra-beam scattering (see Chapter 12) is important.

2.4.6. Design Criteria For Synchrotron Light Sources

Let us now qualitatively discuss how the horizontal emittance depends on the lattice characteristics. For a given energy, the energy loss per turn to synchrotron radiation is inversely proportional to the bending radius (Chapter 1). This energy loss *increases* the amplitude of the horizontal betatron oscillations via the anti-damping mechanism discussed in Section 2.4.2. We have also seen that this increase is greater, the greater the dispersion and its derivative at the location where the radiation is emitted.

The rate of energy loss is proportional to the fourth power of the particle energy (Chapter 1). It can be shown that this leads to a dependence of the emittance on the *square of the energy*, for a given lattice.

Based on these factors, the main criteria that dictate the design of low emittance synchrotron radiation lattices are:

1) A *large bending radius* to reduce the rate of energy loss by radiation. A large bending radius leads to larger rings. High energy storage rings for hard x-ray machines like the APS and the ESRF must have larger circumferences to achieve emittances comparable to low energy machines for UV and soft x-ray production.
2) A *small value of the dispersion* function and its derivative where the radiation is emitted (bending magnets and insertion devices). This requires large bending radii and/or *strong focusing* lattices to keep the value of the dispersion small. The new, 3rd generation, sources, with their design emphasis on small emittance, are, in fact, characterized by strong focusing lattices.

There is a limit on how strongly focusing a lattice can be made. This is the subject of the next section.

2.4.7. Strong Focusing Lattices

We saw in the previous section that low emittance lattices require that the dispersion function and its derivative be kept small. The dispersion is created by the (bending magnets) excitation of a driving term on the right hand side of Eq. (15). This driving term is due to the fact that the energy does not match the design value defined by Eq. (2). According to Eq. (3), this perturbation propagates like a harmonic oscillator until it senses another perturbation by a bending magnet, and so on.

One way to keep the dispersion low is to have a large bending radius, with many short bending magnets. This approach leads to rather large accelerators. It has been proposed as a method to reduce the emittance limit of third generation light sources.[20]

The recent wave of third generation light sources (ALS, APS, ESRF, SRRC, ELETTRA, etc.) has chosen to reduce the amplitude of the dispersion function by increasing the focusing action of the lattice, given by the term $K_x(s)$ in Eq. (2). This is achieved by increasing the strength of the quadrupoles and by spacing them close together.

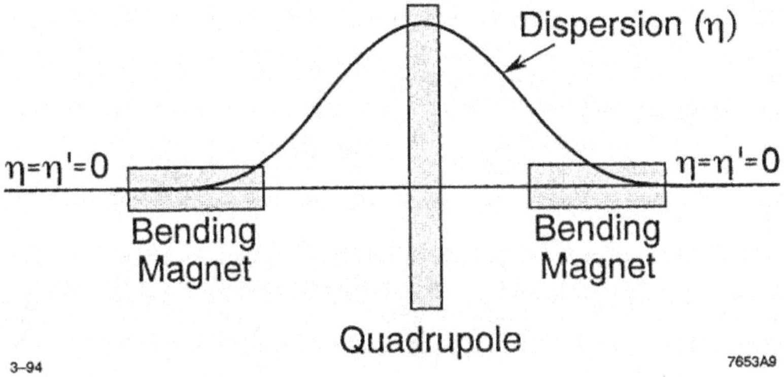

Fig.9. The dispersion in simplified, Chasman-Green type, cell.

2.5. Characteristics of Lattices For Synchrotron Radiation Sources

2.5.1. General Considerations

The basic building block of a lattice is the *cell*. This is a sequence of magnetic elements characterized by certain requirements and by the fact that the lattice functions (β-function, dispersion, their derivatives, etc.) have the same values at the beginnings and ends of the cells. If this is the case, then one can build a lattice by aligning the cells one after another always repeating the lattice functions with one exception: the more unit cells from which one builds a lattice, the smaller the bending radius of the dipoles must be, since the total bending angle in the ring (2π) is unchanged. This leads to a strong, *cubic dependence* of the emittance on the number of cells. Remembering that the emittance acts like the square of the energy yields the following important expression for the horizontal emittance:

$$\varepsilon_x = k\frac{E^2}{N_c^3}, \tag{23}$$

where E and N_c are the electron energy and the number of identical cells, respectively. The constant k depends on the type of cell.

The first task of a lattice designer is, then, to build a cell. When this task is accomplished, the designer has to decide on the *periodicity* of the accelerator, i.e., how many cells to build into the lattice.

2.5.2. The Cell

In designing a cell, the lattice designer must keep several considerations in mind. If the goal of the accelerator is to achieve as small an emittance as possible, the designer will need to keep the dispersion function as small as possible in the bending magnets. Fig. 9 depicts a simplified cell, two bending magnets with a quadrupole in between.

If the quadrupole strength is chosen judiciously, a dispersion function that has coordinates $\eta = \eta' = 0$ at the beginning of the cell ends up with the same coordinates at the end of the cell. The bending magnets create the dispersion (due to the fact that particles of different energy are subject to different deflections). To prevent it from growing too large, the dipoles must be kept short and the quadrupole placed as close as possible to the magnets. This in turn requires strong quadrupole strengths, increases the chromatic aberrations, requires stronger sextupoles, and creates potential dynamic aperture problems.

The other requirement to be kept in mind is the need to inject the beam into the storage ring. This is discussed in Chapter 3. We mention here that the injection process requires special elements (fast kicker magnets and a septum magnet). The cell must include sufficient free space to accommodate these elements along with special optics requirements to be considered.

The presence of straight sections (free of magnetic elements) to accommodate insertion devices is an essential feature of the new generation of light sources. The length of the sections depends on the length of the insertion devices and on possible limitations on the circumference of the ring. In most third generation storage rings the length varies between 3 and 7 meters. One of the reasons for the upper limit is the need to prevent the β-function from growing too large. Symmetry considerations show that the β-function has a minimum in the middle of a straight section (waist). Away from the waist, where $\beta = \beta^*$, the function grows as

$$\beta^* = \beta^* + \frac{s^2}{\beta^*}. \tag{24}$$

Why does one keep the β-function from growing too large? There are two reasons: 1) beam size (Eq. (6)), with the associated hardware requirements of larger aperture, more costly magnets, etc., and 2) the fact that chromatic aberrations and sensitivity to survey errors become more important as the β-functions in the quadrupoles increase. In future light sources, these limitations could be overcome if focusing (horizontal and vertical) could be introduced in the magnetic fields of insertion devices. This would open up the possibility of installing very long undulators.

It has been common practice in 3rd generation light sources to design the optics of the cells such that the dispersion and its derivative are zero in the straight sections. This approach minimizes the beam size and divergence, as shown in Eq. (19). A non-zero dispersion in the field of wiggler magnets also leads to an emittance increase, according to the mechanism described in Section 2.4.2. In both cases the result is a reduction of photon beam brightness.

The process of determining the magnet strengths and positions to satisfy design conditions for the lattice functions is called *matching*. Often, to match the dispersion and its derivative to achieve zero values in the straight sections requires ingenuity and cost trade-offs in the number of quadrupoles and their strengths. Recently, the need for a rigorous zero value of the dispersion in the region of the insertion devices has come under scrutiny, and has been challenged as a too high a price to pay for a relatively small improvement in beam quality.[21]

2.5.3. *Types of Cells*

In this section the types of cells most commonly used for synchrotron light sources are reviewed[22] and critiqued, together with indications for future directions.

2.5.3.1. The FODO Cell

The simplest of the cell layouts is the so-called FODO structure. Its name stands for Focusing-Drift-Defocusing-Drift. Figure 10a shows the version built at the Daresbury SRS2 facility. The drift spaces may contain bending magnets or, in some cases, be left empty for dispersion matching purposes. Straight sections for insertion devices can be accommodated by special insertion optics matched to a sequence of FODO cells.

This type of cell is inherited from high energy accelerators, where it is commonly used. The Photon Factory (Tsukuba) adopted FODO cells for the regular part of the lattice, as did the Daresbury Light Source. FODO optics are also the choice of the Duke University FEL Storage Ring and of damping rings for linear colliders, which share many design criteria with synchrotron light sources.

FODO lattices are easily tuned and well understood. More recent third generation low emittance sources, however, have shied away from this concept. One of the reasons, in the author's view, is the engineering compactness of the design required to achieve a small emittance and the difficulty in providing sufficient space for radiation ports at bending magnets.

2.5.3.2. The Double-Bend Achromat (DBA)

Rina Chasman and Ken Green of Brookhaven National Laboratory proposed for the National Synchrotron Light Source (NSLS) a cell specifically designed for synchrotron radiation sources.[23] In various forms, it has become very popular. The NSLS x-ray ring version is shown in Fig. 10b. It consists of two bending dipoles separated by a single quadrupole. In other designs, a doublet may replace the single quadrupole. The straight section contains quadrupoles to match the β-functions. The dispersion is zero in the straight sections, increases as it goes through the dipoles, and is focused by the middle quadrupole(s). The Chasman-Green lattice is compact and economizes the number of magnets. The dispersion in the bending sections can be kept small (as demanded by a low emittance optics) by using short dipoles and keeping the space between dipoles to a minimum. This ingenious layout has served the scientific community well. Examples of light sources employing the Chasman-Green lattice are the NSLS (Brookhaven) VUV and x-ray rings, MAX at Lundt, NIJI III at SUMITOMO Industries (Japan), HISOR (Hiroshima), LSU CAMD (Louisiana University), the proposed LNLS facility in Campinas, and others.

Its drawbacks, for very small emittance applications, are represented, in the author's view, by the limited length of the region where the dispersion is non-zero and by the small decoupling[24] between the horizontal and vertical β–functions. This may lead to strong sextupole requirements, and the dynamic aperture may be a problem for very low emittance rings. It is also rather limited in tunability.[25]

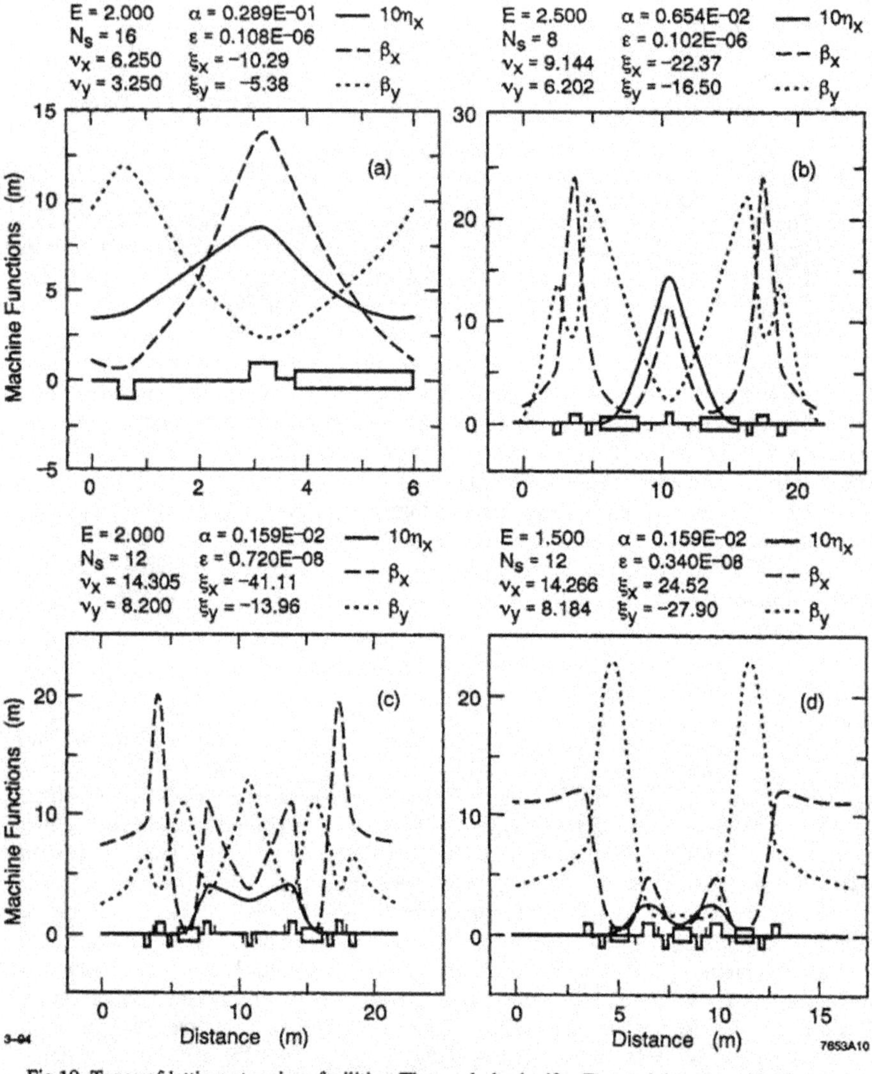

Fig.10. Types of lattices at various facilities. The symbols signify: E = particle energy (GeV), N_S = number of cells, ν_x = horizontal tune, ν_y = vertical tune, α = momentum compaction factor, ε = horizontal emittance (meter-radian), ξ_x = horizontal chromaticity, ξ_y = vertical chromaticity, η_x = dispersion function (m), β_x = horizontal β function (m), β_y = vertical β function (m).

The problem of the compactness of the Chasman-Green cell and the restricted area where sextupoles can be placed may be overcome by enlarging the region between the two dipoles (Fig. 10c, ELETTRA version). Typically, three to four quadrupoles between the dipoles (instead of one or two as in a simple Chasman-Green layout) give higher dispersion and more decoupling between horizontal and vertical β-functions. The benefits are weaker sextupoles (as implied by Eq. (22)), an improved dynamic aperture, more space for diagnostics, and an overall more flexible lattice. This cell is known as "Enlarged Chasman-Green," although the distinction between a simple Chasman-Green and an enlarged version is often a matter of subjective definition.

The Enlarged Chasman-Green lattice is particularly suitable for hard x-ray facilities, and is used in the ESRF, APS, and SPRING-8 facilities. It was also chosen for SuperAco, ELETTRA, BESSY II, MAX II, and the proposed SOLEIL and SIBERIA 2 designs.

2.5.3.3. The Triple Bend Achromat (TBA)

In the Triple Bend Achromat, a third bending magnet is symmetrically placed between two outer ones. The addition of the third dipole has the advantage that the bending radius may be increased, since more magnets contribute to the total bending. This reduces the dispersion in the dipoles and makes it easier to achieve a low emittance. This lattice, shown in Fig. 10d in the ALS version, has good tunability, particularly if a pair of quadrupoles (rather than one) is placed on each side of the center dipole. The TBA was the choice of the BESSY I and Hefei facilities and, later, was adopted by the ALS and SRRC. In the ALS and the SSRC the dipoles have a vertically focusing gradient superimposed on the bending field. This "combined function" magnet helps to reinforce the vertical focusing and also reduces the horizontal emittance by decreasing the horizontal damping time.[26] The latter can be shown by analyzing the emission of radiation in a superimposed quadrupole-dipole field, where the orbit of an electron with an excess of energy is in a lower bending field and therefore radiates less.[27] ELETTRA, a double-bend-achromat structure, also utilizes combined function dipoles.

Higher energy rings (ESRF, APS) have avoided using combined function magnets, mostly because the strong gradient that is required makes the construction difficult.

2.5.3.4. Other Types of Lattice

This brief discussion does not presume to cover the whole variety of lattices for synchrotron radiation sources.[28]

A few interesting new concepts have emerged in recent times. The proposed SLS facility at PSI[29] (Villigen, Switzerland) introduces two very long straight sections (18 m) for long undulators, in addition to four, 7-m-long ones. The SLS lattice layout departs from the more conventional types described above. An interesting new feature is the incorporation of six superconducting magnets that act as wavelength shifters, extending the versatility of the facility (1.5–2.1 GeV energy) to the hard x-ray region of the spectrum (up to 50 keV photons).

SINBAD, a 700 MeV proposed ring at Daresbury, also introduces two very long (14 m) straight sections for novel insertion devices. The other Darebury facility (3 GeV) under consideration for construction is DIAMOND, a Triple-Bend-Achromat in which the central dipole is superconducting.

The lattice for the proposed 3 GeV Light Source ROSY[30] in Rossendorf (Germany) also departs from the conventional FODO, DBA, and TBA. The basic cell consists of five bending magnets (three with 20° and two with 15° bending angles). The bending magnets are of the combined function type (vertically focusing), allowing a more compact machine and further reduction of emittance. The ring consists of four achromats.

One of the concepts proposed to overcome the problem of the small dispersion, strong sextupoles, and dynamic aperture of low emittance light sources utilizes combined function magnets that include bending, quadrupole, and sextupole components for chromaticity correction.[20] With distributed sextupole fields a very small dispersion is permissible, while still maintaining a large dynamic aperture. The small dispersion is achieved using long magnets with large bending radius. A remarkably small emittance of 0.7 nanometer-radians at 6 GeV was computed for a ring consisting of 262 FODO-like cells using only combined function magnets. The length of each cell is about 1.8 m.

The possibility of creating and sustaining very short (sub-picosecond) bunches in storage rings has received attention recently. A storage ring operating in this manner would be very attractive for high-energy physics colliders, and there are also experiments in synchrotron radiation that would greatly benefit from such a machine. The theoretical feasibility and design criteria for such lattice have been investigated,[31] and the results are encouraging. This lattice is designed such that particles of different momenta have the same (or nearly the same) revolution period, leading to very short bunches.

Finally, we conclude this brief review by remarking that not all synchrotron radiation sources need to be large or of high energy. In addition to the compact machine designed for lithography applications (not included in this chapter) the 240 MeV SURF II machine at the National Bureau of Standards is a small, 5-m-circumference storage ring. It consists of a single magnet that bends and focuses the beam. It was converted into a storage ring in 1973 from a 180 MeV synchrotron.[32]

References

1. In the narrative, the word electron is used to signify electron or positron. The dynamics of a single particle are not affected by the sign of its charge when the magnetic field direction is changed accordingly.
2. The reader may be confused by the use of the word orbit and the accompanying adjectives. In this chapter, orbit always refers to an electron path that closes on itself after one revolution around the accelerator, thus, it is always a "closed" orbit. When this orbit describes the motion in an ideal lattice without magnetic imperfections or misalignment, this closed orbit is also the "ideal" orbit. The word trajectory is used to describe an oscillation around the closed orbit, called a betatron oscillation, which does not close after one revolution.
3. Defined by the azimuthal and horizontal/vertical axis.
4. Dipole and quadrupole fields may co-exist in a single *combined function magnet.* These magnets have found applications in the most recent third generation light sources, like the ALS, ELETTRA, and the SRRC light source. For the sake of clarity, in this description we treat the dipole and quadrupole magnets separately. The function of combined function magnets will be discussed in Section 2.5.3.3.

5. E. D. Courant, H. S. Courant, and Snyder, "Theory of the Alternating-Gradient Synchrotron," Annals of Physics **3**, 1–48 (1958). The principle of strong focusing opened the way to a new generation of synchrotrons and storage rings in which the beam size, and thus magnet aperture and cost, could be kept much smaller than in the previous, weak focusing type, accelerators.
6. Recall that we are using the ultra-relativistic approximation $E \sim c|P|$, where P is the particle momentum.
7. See, for instance, K.G. Steffen, "High Energy Beam Optics," by Interscience Publishers, a division of John Wiley & Sons, New York.
8. Additional factors, as we shall see later on, also contribute to the beam size
9. See, for instance, G. Guignard, Physics of Particle Accelerators, Chapter on "Nonlinear Dynamics" pp. 820–890, AIP Conference Proceedings 184, Melvin Month & Margaret Dienes, Editors.
10. Defined as the relative variation of the field with respect to the ideal value.
11. Reproduced from the ALS Conceptual Design Report, courtesy of Alan Jackson.
12. L. Smith, "Effects of Wigglers and Undulators on Beam Dynamics," 1986, Proceedings of the International Particle Accelerator Conference, Novosibirsk, USSR.
13. K. Halbach, *Nucl. Inst. Meth.* **169**, p. 1, 1980 and *Nucl. Inst. Meth.* **107**, p. 109, 1981.
14. Equation (14) includes a single harmonic in the azimuth s. This is appropriate for closely spaced magnets. Otherwise, more harmonics must be included.
15. The term linear implies that the tune shift does not depend on the betatron amplitude and is driven by fields of the gradient, or quadrupole, type (see Eq. (1)).
16. Other effects that, under certain circumstances, contribute to increasing the beam emittance (collective instabilities, intra-beam scattering, ion trapping) are discussed in Chapter 12.
17. M. Sands, "The Physics of Electron Storage Rings—An Introduction," SLAC Report 121.
18. Reproduced from the ALS Conceptual Design Report, courtesy of Alan Jackson.
19. H. Wiedemann, "Particle Accelerator Physics", Springer-Verlag, p. 355.
20. W.D. Klotz and G. Mulhaupt, "A Novel Low Emittance Lattice for a High Brilliance Electron Beam," Proceedings of the Workshop on Fourth Generation Light Sources, Feb. 24–27, 1992, SSRL report 92–02.
21. For instance, the design MAX II (Lund) storage ring accepts a small (13 cm) dispersion in the center of the insertion device straight section.
22. The cell layouts of Fig. 10 are taken from J. Murphy, "Synchrotron Light Source Data Book," BNL Report 42333, Version 3.0, by courtesy of the author.
23. R. Chasman, K. Green, E. Rowe, *IEEE Trans. Nucl. Sci.* **22**, 1765 (1975). This type of achromat was originally proposed by P. Panofski for utilization as a spectrometer. Private communication in K. Steffen, "High Energy Beam Optics" Interscience Publishers, 1965, p. 113.

24. In decoupled β-functions $\beta_y << \beta_x$, where the sextupoles controlling the horizontal chromaticity are placed, and $\beta_y >> \beta_x$ in the vertical chromaticity sextupoles.
25. The tunability may be loosely defined as the capacity to change the tunes and/or lattice functions, while preserving the beam stability without appreciable changes of the source size and divergence.

26. G. Vignola, "Preliminary Design of a Dedicated 6 GeV Synchrotron Radiation Storage Ring," *Nucl. Inst. Meth. Phys. Res.* **A236**, 414 (1985).
27. H. Wiedemann, "Particle Accelerator Physics", Springer-Verlag, p. 347.
28. For a complete list, see J. Murphy, "Synchrotron Light Source Data Book", BNL report 42333.
29. Conceptual Design of the Swiss Synchrotron Light Source, Paul Scherre Institut, CH–232 Villigen PSI, Switzerland.
30. D. Einfeld and M. Plesko, "A Modified QBA Optics for Low Emittance Storage Rings", *Nucl. Inst.Meth.* **A335** (1993) 402–416.
31. C. Pellegrini and D, Robin, "Quasi-Isochronous Storage Rings," NIM A301 (1991), pp. 27–36.
32. E.M. Rowe, M.A. Green, W.S. Trzeciak and W.R. Winter, Jr. "The Conversion of the NBS 180 MeV Electron Synchrotron to a 240 MeV Electron Storage Ring for Synchrotron Radiation Research," Proceedings of the 9th International Conference of Particle Accelerators, SLAC, 1965, p. 689.

CHAPTER 3 : INJECTOR SYSTEMS

GOTTFRIED MÜLHAUPT
European Synchrotron Radiation Facility
B.P. 220
Avenue des Martyrs
38043 Grenoble, France

3.1 Introduction

Due to the rather stringent requirements on spatial and temporal stability of the produced light, all dedicated light sources are based on storage rings. In storage rings the particles are stored at a fixed average energy E_o and a finite average lifetime τ_o of the stored particle beam. Therefore, systems are needed which, from time to time, produce new free electrons, accelerate them to the operation energy of the storage ring E_o and inject them into the storage ring. The function and the technical components of these so-called injection systems are described in this section.

These machines are essentially particle accelerators as have been used for many decades for high energy nuclear and particle physics, normally modified to achieve lower costs by reduced duty cycle and to accommodate adequate beam emittances and time structures.

A large variety of types of injector systems exist in present synchrotron light sources. There are injectors which deliver the beam at the full operating energy of the light source, and injectors which deliver only a fraction of the operating energy and where the beam has to be post accelerated to full operating energy inside the light source storage ring. The injector system can consist of a single accelerator (e.g. a linac) or of several different types of accelerators in series (e.g. a microtron as a preinjector, a synchrotron as a main accelerator and an accumulator ring to achieve the final peak currents needed before transfer to the light source). The choice of system is often not only determined by the technical needs of the light source, but also by consideration of the expertise and/or accelerator equipment already existing on site prior to the construction of the light source, by the possibility of parallel use of the injector for purposes other than injection into the light source (e.g. for high energy physics needs) or simply by limited funds.

Nevertheless, in any case a typical injection system for a synchrotron light source consists of 5 basic stages :

Stage 1 : an electron gun to create a stream of free electrons (gun)

Stage 2 : a first acceleration part where the electrons created as a continuous stream of particles from the gun are bunched into short electron packages (bunches) and accelerated to relativistic velocities (i.e. to energies $E >> m_oc^2$)

e-mail : *mulhaupt@esrf.fr*
fax : *(33) 76 88 20 54*

Stage 3 : an acceleration at constant speed v = c to the final energy

Stage 4 : an extraction system from the injector composed of fast switching magnets (kickers) and special bending magnets (septa) and a transfer line to the storage ring

Stage 5 : a system for injection into the storage ring composed of kickers and septum magnets

The components and accelerators which are used to achieve these stages vary according to

- the maximum energy necessary or affordable
- the type of parasitic use (if any)
- the experience and equipment existing at the site before construction of the light source.

After a short description of the requirements, a description of two examples of injection systems is given in general terms, while the technical details of the different components including their performance limits are described in later paragraphs. A more general discussion is then resumed under 3.11, 3.12 and 3.13.

3.1.1 Requirements

The electrons stored in a storage ring show a finite lifetime τ_0 due to particle losses by scattering on the rest gas inside the storage ring vacuum system, by scattering between beam particles, by the statistical nature of synchrotron radiation emission and by collective effects. The lifetime τ_0 varies according to operating conditions – normally between 1 and 100 h – and describes the decay of the stored beam current I with time (see footnote (a)).

$$I = I_0 \, e^{-t/\tau_0}$$

In order to maintain a sufficiently high average current over the total operating cycle (normally weeks or months), injection of new particles is needed to replace the lost particles after a certain fraction of the lifetime. The time needed for these injections must be small compared to the beam lifetime. This defines the number of particles N to be delivered per unit time into the storage ring of circumference C to be

$$e\frac{dN}{dt} >> \frac{I_0 \, C}{\tau_0 \, c}$$

(e = elementary charge of an electron)
(c = velocity of light)

(a) The lifetime τ_0 is not normally constant during the decay of the beam. In most cases the product of actual current I and actual lifetime τ is constant : $I.\tau = \text{const}$. In this case the real decay of the stored current I is better described by $I(t) = \frac{I_0}{1+\frac{t}{\tau_0}}$ with I_0, τ_0 being the current and lifetime at t = 0.

The efficiency of the injection process which is especially crucial for the topping up mode (see sec. 3.11) into the SR depends, on the one hand, on the storage ring properties, but also on the emittance and energy spread of the beam delivered from the injection system : the smaller the emittance and energy spread of the beam delivered from the injection system, the easier the process of injection into the limited aperture of a storage ring and the higher the injection efficiency. Consequently, injection systems must provide beams with reasonably low emittance and energy spread. In addition, the emittance and energy spread must be constant from pulse to pulse. The ways of achieving these requirements vary with the different types of injector machines. They are described further below.

Synchrotron light sources are, in most cases, operated as a service to a large number of external users whose usual center of professional activity is often far away from the light source. In addition, typical synchrotron light experiments are short (days or weeks) due to the high data flux available. It is therefore mandatory that the light source is operational according to a predetermined time schedule with an up-time of more than 90%. In the case of the "topping up" mode of operation, the injection system must have the same reliability. In the case of the normal injection mode, the reliability requirement is slightly relaxed as short interventions on the hardware of the injection system can be done between two injections without affecting the synchrotron light production for users. These up-time requirements define the quality of hardware of the injector components and require, in cases with equipment with a short lifetime, either an operation redundancy or very fast exchange possibilities.

The capital costs of an injection system usually represent a significant fraction of the capital cost of the total light source accelerator complex (about 10 - 40%). The minimization of the costs of the injector system, without compromising the ability to meet the above mentioned requirements, is therefore a dominant concern in the design of an injection system.

Given the significant cost of the injection system and its normally low duty cycle, parasitic use of the injection system for other types of scientific work has been considered and, in a few cases, routinely implemented. This parasitic use tends to impose additional requirements on the design if the injection system is to be used for forefront research work. Presently there are two directions for parasitic use of injection systems : usage as stretcher rings to allow experiments in nuclear or particle physics [1] and usage as high brilliant beam source for free electron laser studies.[2] The impact of this parasitic use on injector design and their compatibility with cost minimization is discussed in 3.13.

3.1.2 Examples

a) Example of an injection system for low energy light sources $E \leq 500$ MeV

Low energy light sources have been built as regional or national facilities due their relatively low cost [3 4 5]. They have mainly been built for the use of bending magnet radiation and designed with rather large beam emittances in order to ensure sufficient Touschek lifetime (see footnote (b))

(b) The elementary process called Touschek effect is an elastic scattering of two electrons of an electron bunch exchanging their transverse momenta into longitudinal ones. These longitudinal momenta can be larger than the $\frac{\Delta p}{p}$ acceptance provided by the machine. The particles can therefore be lost. The contribution of this special loss mechanism to the overall lifetime is called Touschek lifetime.

Since large emittance sources are not too sensitive to source point drifts, their injector systems were often designed for energies E which were lower than the operating energy of the storage ring, thereby needing further acceleration in the storage ring itself. This further decreases the injector costs.

For low energy synchrotron light sources with large emittance electron beams, trapped ions can have a strong and detrimental influence on the lifetime and coupling between horizontal and vertical emittance. However, for these low energy rings the incremental costs of a positron option which would, in principle, avoid any problems of ion trapping are extremely large and are therefore seldom used.

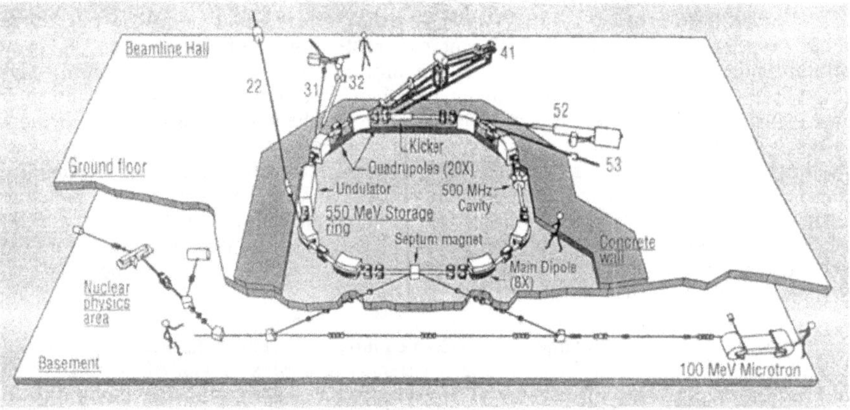

Fig. 1 Layout of the MAX accelerator system (Lund/Sweden).

Fig. 1 shows the layout of the MAX I accelerator system (Lund/Sweden) : Stage 1 is achieved by an external 100 KV gun. Stages 2 and 3 are achieved by a buncher cavity after the gun and the subsequent acceleration in a racetrack microtron up to the energy of 100 MeV. The extraction stage from a microtron is easy since the electrons of the different energies in a microtron follow different paths. The injection into the storage ring is (as for all other examples) achieved by a septum magnet and a kicker magnet allowing an off axis accumulation of successive pulses from the injection system (see 3.6).

After injection and accumulation of the necessary beam current at the energy level of 100 MeV, the beam is further accelerated in the MAX I storage ring up to the final operation energy of 550 MeV. Please also note the use of the same septum to achieve a slow extraction with large duty cycle into a nuclear physics research area.

b) Example of an injection system for a high energy synchrotron light source

Present high energy synchrotron light sources are either machines originally built for high energy physics, more or less adapted to the new type of use, 2nd generation dedicated synchrotron light sources, or dedicated 3rd generation machines designed for high brilliance beams from undulators/wigglers for a national or international user community.

In the following the injection system into the ESRF, which was the first 3rd generation light source in the hard X-ray domain, is described.

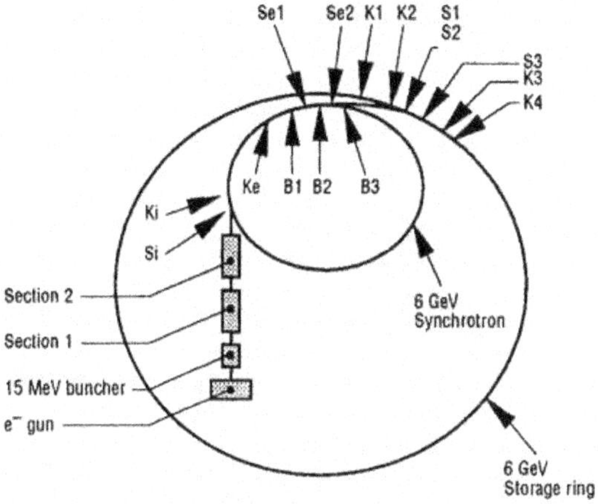

Si : Synchrotron injection septum magnet
Ki : Synchrotron injection kicker magnet
} on 200 MeV energy level

Ke : Synchrotron extraction kicker magnet
B1, B2, B3 : Slow bump magnets
Se1 : Synchrotron thin extraction septum (in vacuum)
Se2 : Synchrotron thick extraction septum (in air)
K1, K2, K3, K4 : Storage Ring injection kicker magnets
S1, S2 : Storage Ring injection thick septum magnet (in air)
S3 : Storage ring injection thin septum magnet (in vacuum)
} on 6 GeV energy level

Fig. 2 ESRF Injection system.

As shown in Fig. 2, stage 1 again consists of a thermionic electron gun with DC acceleration to 100 KeV. Stage 2 is achieved with a pre-buncher cavity and a buncher section with acceleration to 15 MeV. Stage 3 is done in two steps : first, acceleration in a linear accelerator to E = 200 MeV and then injection of this 200 MeV beam into a booster synchrotron which accelerates to the final energy of 6 GeV. Stage 4 consists of an extraction system from the synchrotron, a transport line to the storage ring and stage 5 of a septum/kicker arrangement for injection into the storage ring, all working at 6 GeV energy level. The buildings are arranged such that an e^-/e^+ converter and a further 400 MeV e^+ linac could be implemented optionally should problems concerning trapped ions arise.

3.2 *Electron guns*

Electron guns have to deliver the required pulsed current of electrons with the necessary emittance and energy for the envisaged application :

Up to now only thermionic guns are used in synchrotron radiation facilities . In these guns the electrons are liberated from the surface by heating a cathode material to high temperatures (see Fig. 3 for emissivity of different usual cathode materials).

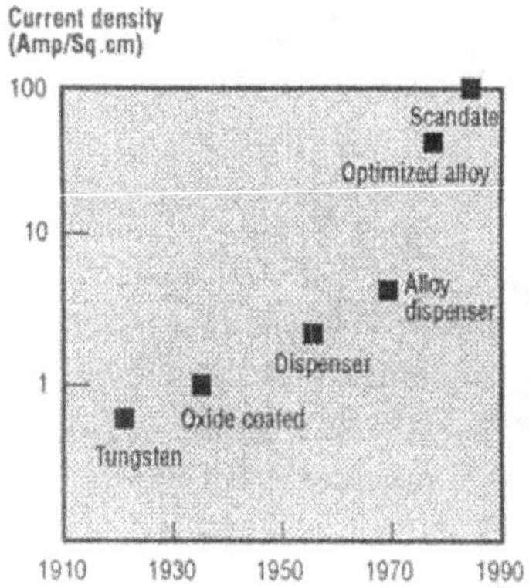

Fig. 3 Historical perspective of thermionic cathode emission capabilities (from L.R.Falce, Semicon Associates, 1801 Old Frankfurt Vike, Lexington, KY 40510).

In applications where, due to geometrical boundary conditions, only small surfaces can be used (as in classical microtrons or in RF guns) Lanthanhexaborid (L_aB_6) cathodes are normally used. These cathodes can deliver high current densities at moderate temperatures but the lifetimes are usually only ~1000 h and the emissivity over the surface varies with time. The restrictive geometrical boundary conditions in these applications do not normally allow the use of sophisticated optics in order to ensure a low emittance beam.

In applications where the gun construction is not limited by geometrical boundary conditions as in linac guns or guns for racetrack microtrons, cathodes of extended surface can be used and sophisticated electrostatic and magnetic focusing employed to guarantee a low emittance beam. Since the design of the gun optic has to take into account the space charge forces of the extracted beam, the optic is only optimal for the calculated beam current. Fig. 4 shows a gun with Pierce type optics designed for a FEL experiment.

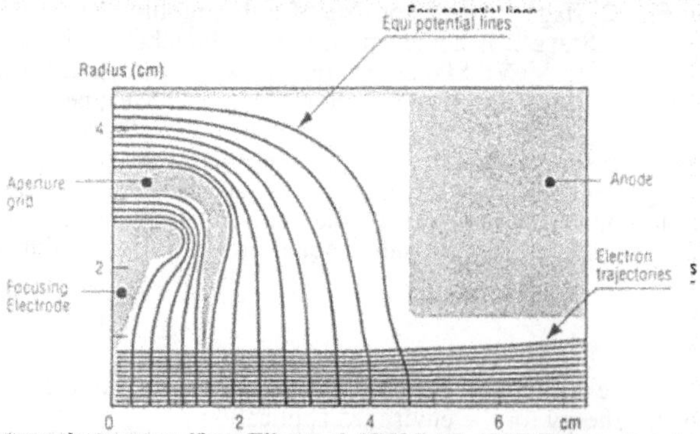

Fig. 4 Pierce type electron gun (from Elias et.al. 1985 PAC, *IEEE Transactions on Nuclear Science*, **Volume NS 32 No. 5** (1985) 1733).

An interesting recent development is the so-called RF gun[6] : these guns integrate a cathode into a set of RF cavities which work on the RF frequency of the immediately downstream linac or microtron. As the electric fields in an RF cavity can be significantly higher than in a classical DC gun, the acceleration is faster and therefore the effects of space charge forces are reduced. In addition, the gun already creates a beam which is nearly relativistic and is bunched on the frequency of the following accelerator and this renders the separate buncher/acceleration stage 2 unnecessary. Fig. 5 shows the set up of a complete thermionic RF gun for a linac pre-injector.

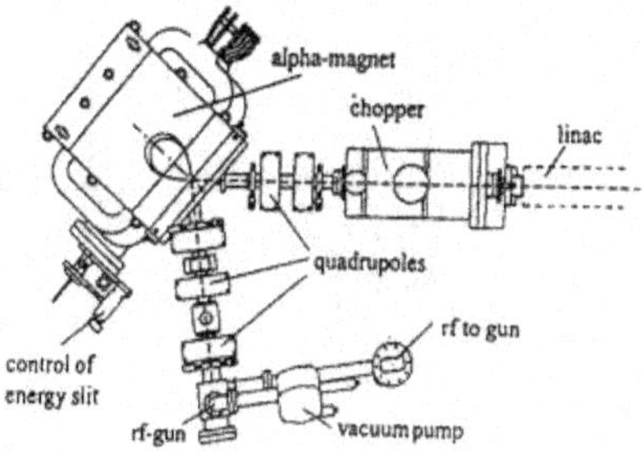

Fig. 5 RF-gun arrangement (courtesy of H.Wiedemann/SSRL).

The modulation of the emitted current can be achieved by pulsing the cathode anode voltage in diode type, applying RF pulses to RF guns or by pulsing a grid on the cathode surface (as in most linac or racetrack microtron guns). A gridded gun also easily allows the production of a single bunch with bunch lengths of around 1 nsec in the injector (see 3.11).

A disadvantage of thermionic RF guns is the difficulty to create single bunches. Up until now grids could not be implemented on the cathodes and consequently chopper systems had to be employed for single bunch or multisingle bunch formation. An elegant way to create very short high current pulses from an RF gun is to replace the thermionic cathode by a photo cathode illuminated by short laser pulses. These RF photo cathode guns are more expensive and still show a low cathode life time.

3.3 Linacs

Linear accelerators (linacs) are described in detail in ref.[7]. They accelerate the particles along a linear path at the rate of about 15 MeV/m. The accelerating field is provided by a large series of UHF cavities having a phase relation between them such that the average phase advance of the RF field equals the velocity of the accelerated particles. These cavities are assembled in sections and fed by (normally pulsed) high power klystrons. A sketch of the commercially produced [8] Linac of the ESRF is given in Fig. 6.

Linacs are modularly built accelerators : any number of Modulator / Klystron / Section Modules can be added to reach any given final energy. There is no physical or practical limitation for the length and consequently the final energy of a Linac. Costs are roughly proportional to energy. Linacs can be pulsed from a single shot up to several hundreds of Hertz.

For use with synchrotron light sources only e^- or possibly e^+ linacs are of interest. They can and, indeed sometimes, are used as injectors up to the final operating energy of the light source Storage Ring.[9, 10]

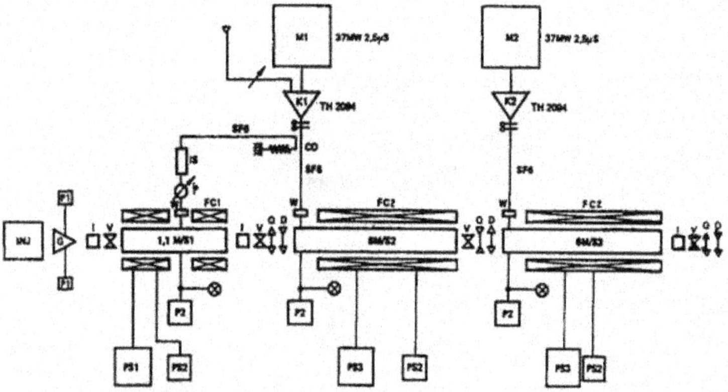

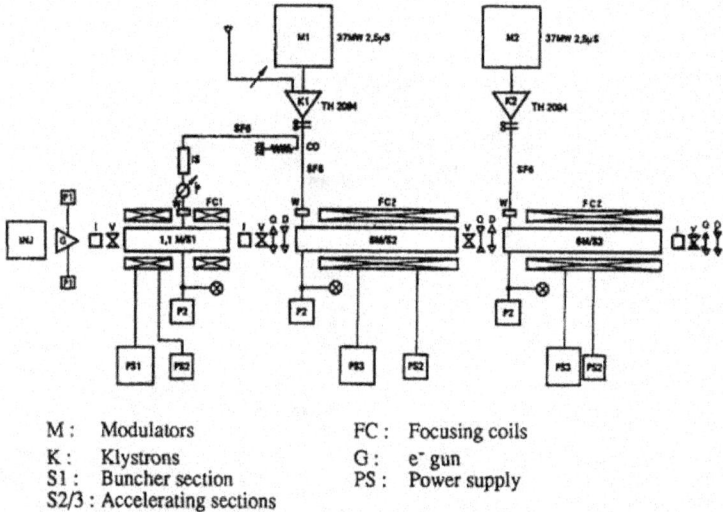

M :	Modulators	FC :	Focusing coils
K :	Klystrons	G :	e^- gun
S1 :	Buncher section	PS :	Power supply
S2/3 :	Accelerating sections		

Fig. 6 200 MeV Preinjector synoptic for the ESRF Grenoble.

Their main advantages are :

- full freedom to accommodate sophisticated gun systems (e.g. for low emittance guns [10], RF guns [11], and high current guns for single bunch operation or for e^+ production)

- the same acceleration principle is used for non-relativistic energies up to the maximum energy needed

- high current capability for e^-/e^+ conversion[12]

The main disadvantages are :

- large energy spread of the output beam compared to a microtron or synchrotron due to the absence of phase averaging

- higher investment costs per MeV output energy (compared to a microtron or synchrotron) and higher running costs (mainly klystron costs)

3.4 Microtrons

Microtrons are described in detail in ref. [13]. Classical microtrons use a kind of RF gun situated in a homogenous time constant magnetic field such that the beam from the gun is recycled several times through the same cavity (Fig. 7). Due to the increasing energy the radii of the successive paths increase. The necessary volume of the magnetic field (which increases with E^2) sets therefore a practical limit for the final energy of classical microtrons to a few tens of MeV.

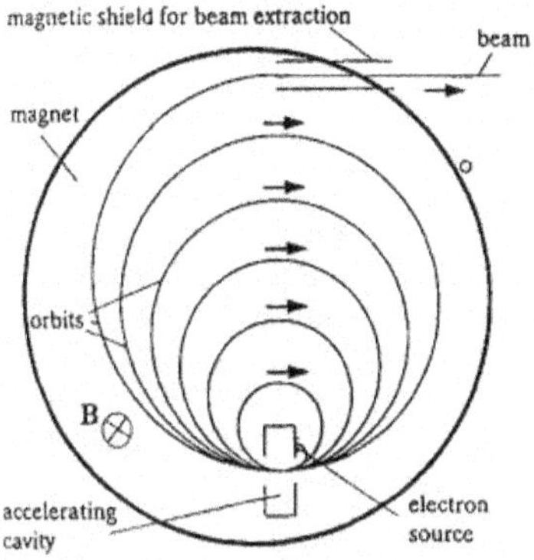

Fig. 7 Schematics of a classical microtron (courtesy of H.Wiedemann, *Particle Accelerator Physics*).

Higher energies of up to several hundreds of MeV can be reached with the racetrack microtron. Here the single cavity is replaced by a short linac and the magnet is split into two 180° magnets (Fig. 8). This type of microtron has the advantage that the energy increase per turn can be higher and that the electron gun can be situated outside the magnet fields almost as on a linac without geometrical constraints. However here also the necessary volume of the magnetic field increases proportionally to E^2 of the final energy.

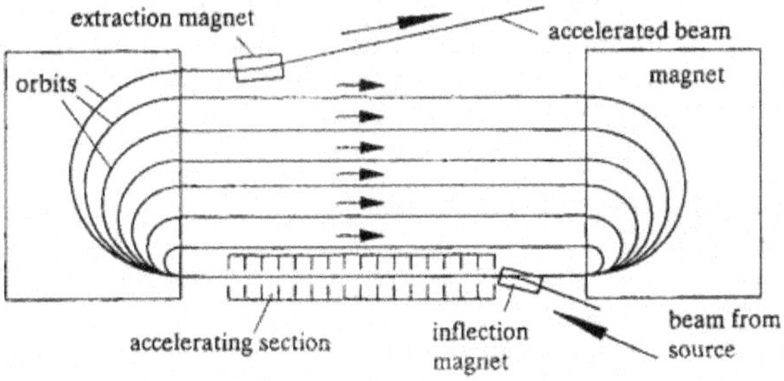

Fig. 8 Schematics of a racetrack microtron (courtesy of H.Wiedemann, *Particle Accelerator Physics*).

The microtron shares with the linac the advantage of being able to accelerate directly from non-relativistic energies but the need for magnet volume to accommodate the different turns limits the maximum energy to a few hundred MeV. The main advantage of the microtron is the fact that the recycling of the beam through the same accelerating section already allows a kind of synchrotron oscillation and therefore an averaging of phases which results in a reduced momentum spread of only about one per mille for the output beam. The main disadvantage of the microtron is the fact that the accelerating section is loaded simultaneously with the currents of all turns, thereby limiting the output current to a few tens of milliamperes. The microtron is therefore a "clean" and cheap injector accelerator for energies up to about 100 MeV but it is not suited for high current applications such as e^-/e^+ conversion.

3.5 Booster Synchrotron

Booster Synchrotrons are described in detail in ref. [14]

The synchrotron, like the microtron, uses the principle of recycling the beam several times through the same accelerating section, but instead of having a fixed magnetic field with the orbit radius changing with energy, a fixed path (constant orbit radius) is used and the magnet field increases with time as the energy of the electrons increases. This minimizes the necessary magnetic field volume of the recycle path and allows an increase in the number of recycling turns from a few tens of revolutions in a microtron to several 10 000 revolutions, thereby reducing the RF power needs for the accelerating section. This allows final energies of many GeV in a synchrotron to be reached.

It is also worthwhile noting that in a booster synchrotron all the circulating particles have, at any given moment, only one mean energy as dictated by the field integral of the magnet guide field, while in a microtron different energy levels on the different paths coexist and simultaneously load the accelerating section. The output pulse length of a microtron can therefore be much longer, even up to cw operation,[15] limited only by RF power needs and cooling possibilities, while in a synchrotron the maximum output pulse length is equal to the time needed for the electrons to travel once around the circumference of the machine (unless special pulse stretching techniques are applied at the expense of the pulse current).

In a synchrotron, particles travel for a long time around the circumference. In order to confine the particles to a finite beam size, not only magnetic dipole fields for bending but also quadrupole (and sextupole) magnets for focusing are required. The field of all of these magnets must track precisely proportionally to each other in such a way that the focusing properties stay constant whilst the energy of the particles changes during the acceleration cycle.

Due to remanence effects in the field of classical steel magnets and contributions from the earth's magnetic field, this tracking cannot be easily maintained at very low excitation (< 0.01T) and, due to saturation, cannot be maintained at high excitation either.

Depending on the sensitivity of the individual machine under consideration, this sets a limit for the ratio :

$$\frac{E^{ouput}}{E^{input}} \leq 50$$

unless quite sophisticated correction methods are applied and/or the magnet lattice is drastically simplified.[16]

Presently used injector synchrotrons have separated function lattices (similar to storage ring lattices) which allow horizontal and vertical betatron oscillations to be radiation damped. This acts in addition to the adiabatic damping of transverse beam dimensions due to the energy increase. This does not only allow the use of the synchrotron as a storage ring for diagnostic (or other) purposes but it has the additional advantage that the output emittance is exclusively determined by the synchrotron lattice properties and is no longer dependent on the starting conditions (emittance, injection position or angle and energy spread of the beam from the preinjector).

3.6 Principles of injection and extraction from a circular accelerator

While injection and extraction into and out of single pass accelerators (linacs for example) do not present any problem, the injection and extraction system into and out of circular accelerators, where the beam travels along the same trajectory for many turns, employs fast switching magnets which have to be triggered at the right moment.

3.6.1 Transverse phase space

Fig. 9a shows the principle of injection into a booster synchrotron : the beam of the pre-injector is brought to the booster synchrotron in such a way that it crosses the synchrotron closed orbit at an angle α. At the crossing point a fast switching kicker magnet deflects the beam right onto the axis of the synchrotron. After one turn in the synchrotron this injection kicker is switched off so that the beam continues to circulate on the synchrotron closed orbit in the middle of the aperture. After extraction of the beam at full energy the injection kicker can be switched on again for the next injection. The kicker pulse must have a flat top (in order not to create betatron oscillations) and a fall time τ^F which is short compared to the orbital period $T_0 = \frac{C}{c}$ (where C = circumference of the synchrotron, c = velocity of light) in order to effectively fill the circumference. In order to allow the switch off times of about 10 nsec with voltages which are easy to handle, the impedances of the kickers and the magnetic fields must be low.

In most cases a single kicker cannot produce a sufficient integrated field in order to bend the injected beam at a bending angle α large enough to let the injected beam enter the synchrotron without interfering with other magnetic elements. In these cases the bending is done with both a fast switching kicker magnet with a small bending angle and a dc or slowly pulsed bending magnet S (Fig. 9b). These so-called septum magnets should be positioned as close to the synchrotron beam aperture as possible in order to allow weak kicker fields. They are normally constructed as C-magnets with one excitation coil positioned in the gap (see Sec. 3.8 and Fig. 15).

The extraction, e.g. from a synchrotron, at high energies with a kicker/septum arrangement still normally requires kicker strengths which are difficult to achieve. In these cases one can use the fact that the synchrotron beam had been adiabatically and/or radiation damped during the acceleration, and thereby occupies only a small fraction of the aperture of the synchrotron at the extraction energy. One can therefore employ a slow beam bump ($\tau \sim$ few msec) to bring the beam closer to the extraction septum shortly before reaching the extraction energy. Fig. 10 shows the geometry and a cross section of such a slow bump magnet. Note that thin stainless steel chambers (d = 0.3 mm) can be used in these slow bump magnets since eddy current effects are small for slow bumps.

Injection schemes according to Fig. 9 a and b allow single pulse on axis injection as needed, e.g. in a synchrotron. As soon as the beam has to be accumulated by successive injections, a slightly different injection scheme has to be employed (Fig. 9c). Obviously, a kicker deflecting the incoming beam by an angle α to the axis would deflect a beam already circulating on the axis by the same amount and consequently throw the already circulating beam out of the acceptance limits. Any accumulation of fresh incoming beam joining a beam already circulating in the accelerator/storage ring uses the fact that, at the moment of injection, the circulating beam mainly occupies the center part of the aperture, leaving the outer part of the aperture unpopulated (or weakly populated). In this situation a local injection bump, created by at least 2 kicker magnets, could be excited with an amplitude which would not scrape off the center of the circulating beam but only the weakly populated outer phase space areas. However, the output of the septum magnet would now be inside the aperture so that the freshly injected beam could also circulate, provided the injection bump is switched off before the freshly injected beam (after one or a few turns) hits the septum magnet. This accumulation process works better the larger the ratio of the phase space density of the injected beam to the phase space density of the scraped off parts of the circulating beam.

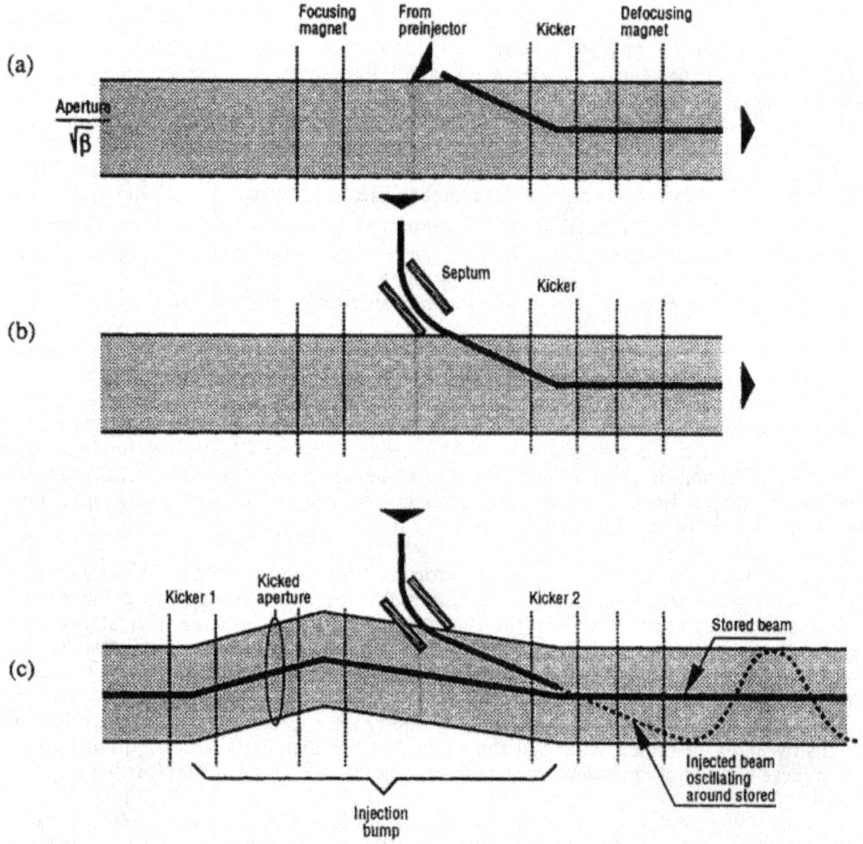

Figs. 9a, 9b and 9c Injection into transverse phase space.
(Please note that in this drawing the transverse scale is much larger than the horizontal)

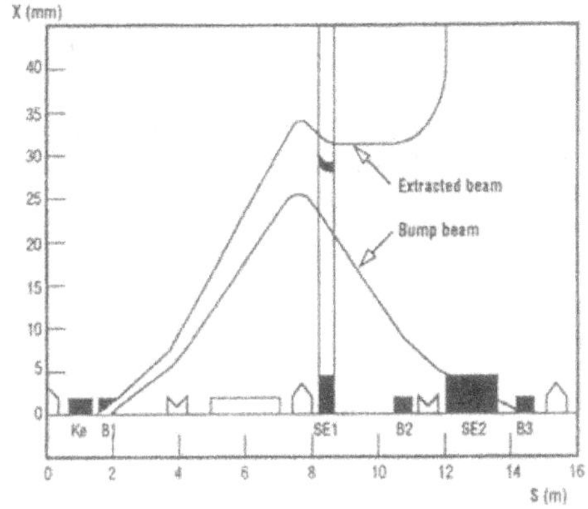

Fig. 10a Schematic view of extraction region with slow bumpers Bi, B2 and B3 to minimize the necessary strength of the fast kicker Ke (see also Fig. 2 for key).

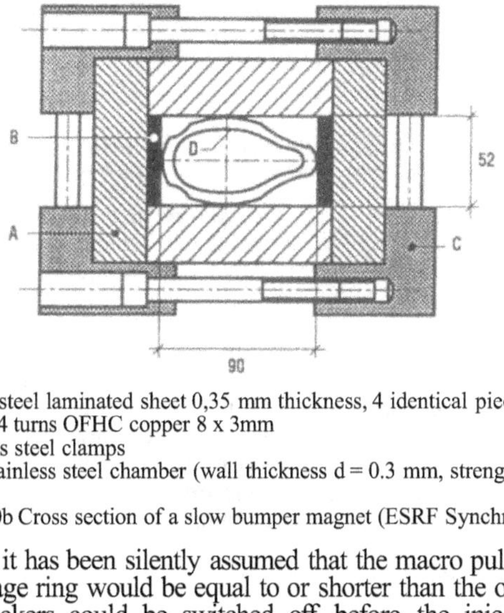

Silicon steel laminated sheet 0,35 mm thickness, 4 identical pieces
Coil: 14 turns OFHC copper 8 x 3mm
Stainless steel clamps
QDE stainless steel chamber (wall thickness d = 0.3 mm, strengthened with ribs).

Fig. 10b Cross section of a slow bumper magnet (ESRF Synchrotron).

Up until now it has been silently assumed that the macro pulse delivered from the injector into the storage ring would be equal to or shorter than the orbital period T_0 such that the injection kickers could be switched off before the injected beam has fully completed the first turn in the storage ring. This single turn injection is indeed used for nearly all light sources with a circumference larger than a few hundred meters (i.e. E > 1 GeV).

However, in low energy light sources and especially those with a very short circumference such as those using superconducting magnets,[17] the potential of the injector to deliver longer pulses could only be partly used in single turn injection. In those cases a multi turn injection can be employed : as shown in Fig. 11 for the case of a storage ring with a tune of Q = n + 0.25, *the off-axis freshly injected* beam would be right on axis after completion of one turn, after two turns it would be opposite the septum sheet, after the third turn it would again be on the axis and only after the fourth turn would it hit the septum. If the kicker is therefore kept on for 4 turns and then switched off before the fifth turn, a macro pulse of t = 4 T_0 could be injected and the injection speed could therefore be multiplied by 4. The optimization of this process depends on the tune and on the ratio of the phase space kicked beyond the septum sheet and the total acceptance. This principle can be extended to even more sophisticated resonance injection schemes employed, for example, in compact storage rings for lithography application.[17]

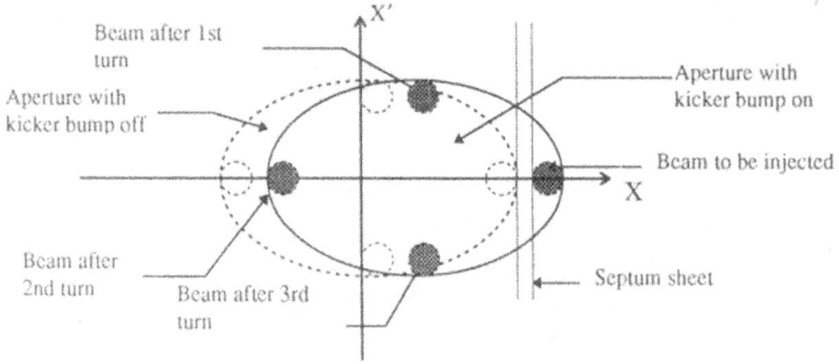

Fig 11. Evolution of the injected beam in the horizontal phase space with a fractional tune of 0.25. (Note that in this case the kicker bump has to be switched off at the latest before completing the fourth turn).

After the beam is captured inside the aperture it is finally diluted over the total phase space and in the presence of a damping mechanism (as for example radiation damping) the betatron amplitudes shrink down to the equilibrium beam size of the "stored" beam.

Phase space dilution allows injection repetition frequencies which are significantly higher than that principally allowed by the damping time (e.g. in low energy injection processes). It should also be noted that a closed injection bump which would not create any closed orbit distortion outside of the bump is often not the practical optimum for injection efficiency : by powering the downstream kicker higher than the upstream kicker one can decrease the betatron amplitude of the freshly injected beam at the expense of introduced betatron oscillations of the stored beam.

3.6.2 Longitudinal phase space

Any beam delivered from a pre-accelerator into a following host accelerator is in the longitudinal phase space defined by a time structure and an energy distribution. For high injection efficiency these parameters have to fit the energy acceptance and RF bucket size of the synchrotron or storage ring into which the beam is to be injected. As long as pre-acceleration and the host accelerator employ the same RF frequency, this match is normally fulfilled and with the correct relative phase between the two RF systems close to 100% efficiency is normally achieved.

Electron linacs and microtrons work with RF frequencies in the few GHz regime, while synchrotrons or storage rings, in all cases, use lower frequencies (usually in the several hundred MHz domain). In these cases injection from a linac or microtron into a synchrotron or storage ring can be done

- either by matching the time structure of the linac/microtron beam to the synchrotron/storage ring bucket size and frequency by a chopper or modulation system on the linac/microtron gun.[18] In this case the chopper/modulation frequency (and possibly even the linac/microtron RF frequency) is phase locked to the synchrotron/storage ring RF frequency and the energy distribution is shaped by a slit system in a dispersive part of the transport line. This system can give an injection efficiency of above 80% and minimizes losses in the booster/storage ring (with the beneficial aspect of reduced radiation levels) and transient beam loading of the cavities by non captured charges.

- or by the "brute force" method : in this mode the energy acceptance and the bucket size and structure of the synchrotron or storage ring select all particles which can be accepted out of the unmodulated linac/microtron preinjector. This method can give an injection efficiency of 20-30%, higher radiation levels and (especially in the case of injection into a synchrotron) uncontrollable beam loading of the cavities by non captured charges.[19]

3.7 Kickers

Ideally, kickers produce homogeneous magnetic fields with the time dependence of a rectangular pulse. The kicker power supply system which comes the closest to this ideal are delay line kickers, the schematics of which are indicated in Fig.12 :

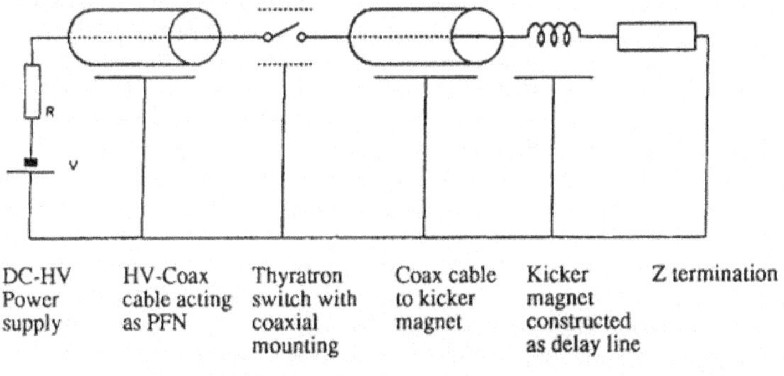

Fig. 12 Delay line type kicker magnet and power supply.

The principal idea is to load a coaxial cable of wave resistance Z with a DC-HV power supply and then discharge the stored energy via the thyratron switch over a transmission cable of wave resistance Z, the kicker magnet with wave resistance Z into a load of resistance Z. This should allow the transmission of a rectangular current pulse without reflections or pulse shape modification. The length of cable B determines the length of the current pulse and the voltage V determines the magnet field strength. These systems are used whenever high magnetic field strength, short rise times, low pulse ripple and a large spatial separation of the kicker and power supply are required. However, the delay line type kicker magnets in particular are expensive items and they do not apply only magnetic but also electric fields to the beam.

For many applications much simpler systems are used where the kicker magnet consists only of one or a few turns of conductor (very often inside the vacuum system). These purely inductive kicker magnets are also powered through a thyratron switch by discharging a charged delay line or a Pulse Forming Network (PFN) (Fig. 13). In these simplified systems the pulser must be positioned very close to the magnet in order to minimize stray inductance of the connection cable.

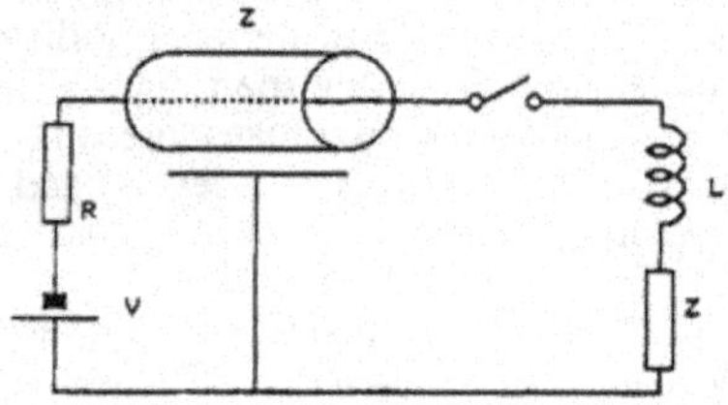

Fig. 13 "Air coil" type kicker magnet and power supply.

For storage ring applications, the design of a kicker magnet is complicated by the requirement to keep the impedance of the kicker/vacuum tank system (as seen by the high frequency parts of the beam wall currents) very small to avoid beam instabilities. There are two solutions :

– the magnet is kept outside the vacuum. In this case the vacuum chamber is to be made from non-conductive material (ceramics) in order to avoid a weakening and deformation of the field by eddy currents due to the ac components in the field (up to several MHz dependent on the rise time). In order to maintain the electric boundary conditions for the high frequency components of the beam wall currents (up to several 10 GHz) the ceramic chamber maintains the cross section of the other parts of the storage ring and is coated on the inside with a very thin layer of metal. The thickness of this coating is such that the GHz components of the beam wall currents can flow almost undamped while the MHz components of the kicker field will see an insufficient skin depth to develop eddy currents.

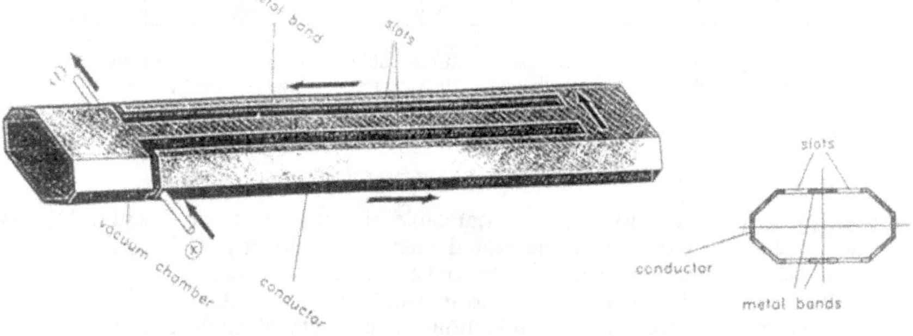

Fig. 14 Delta kicker.

– a very recent elegant solution[20] is shown in Fig. 14. Here the conductive environment of the beam continues but the metal wall is slotted in such a way that conductive filaments are formed to carry the kicker currents and that the continuation of the beam wall currents are secured. An outer vacuum envelope also containing the magnetic return flux makes the whole "magnet" vacuum tight.

As an example, the set of parameters of the above shown Delta injection kicker is given in Table 1.

Table 1 Parameter set of Delta kicker.

Parameter		Value
Length	l	0.45 m
Bending angle	σ	2.60 mrad
Field at 1.5 GeV	B	29 mT
Peak current	I_p	3630 A
Peak voltage	U_p	13420 V
Rise time	τ	170 ns
Calibration	dB/dI	8mT/kA
Inductivity	L	0.4 μH
Loss factor for s = 2 cm	k (σ)	1.6×10^{10} V/C
Good field region	Δx	± 18 mm

Kicker magnets, especially those used in storage rings, normally act in groups of 2 or 4 in order to create a localized kicker bump (cf. 3.6, Fig. 9c). This requires that all kicker pulses are triggered individually according to the traveling time of the beam between the kicker magnets and that the pulses are of identical shape. Any deviation results in increased betatron amplitudes of parts of the beam.

3.8 Septum magnets

The fields of kicker magnets deflect the ingoing/outgoing beam by the same deflection angle as the circulating beam.

On the other hand, the septum magnet is usually the first or last bending magnet of a transferline from or to a circular accelerator which has to bend the incoming/outgoing beam without affecting the circulating beam. As explained in Sec. 3.6 the geometrical boundary conditions and the limited kicker strength require the septum magnet to be positioned as close as possible to the aperture of a circular accelerator. The ideal septum magnet would therefore provide a homogeneous dipole field spatially confined to the channel in which the beam is to be injected or extracted and falling to zero in a step function outside the injection/extraction channel in order not to affect the neighboring aperture region of the circular accelerator.

The cross section of a septum magnet is shown schematically in Fig. 15. It shows the inner and outer conductor of a single turn coil in the gap of a C type ferromagnetic flux return yoke. With homogeneous current distribution in the conductor, infinite permeability of the return yoke material and without eddy currents, the magnetic field outside the injection/extraction channel would be zero. In reality, leakage fields remain in the aperture of the circulating beam in the order of 10^{-3} ...10^{-4} of the field inside the injection/extraction channel.

A septum magnet should allow a minimum separation of the homogeneous field in the injection/extraction channel and the unperturbed area of the machine aperture. This minimum separation can be achieved by placing the septum magnet inside the machine vacuum vessels and by minimizing the thickness of the septum sheet ($d < 1$ mm is possible). Since the septum bends the injected/extracted beam successively from the machine aperture, only one end of the septum normally requires this minimum thickness. Therefore, quite often septum magnets are separated into two or more sections where only the one closest to the aperture is of a vacuum type with a thin septum sheet, while the more separated parts could have their magnets in air with a separated vacuum chamber for the injection/extraction channel.

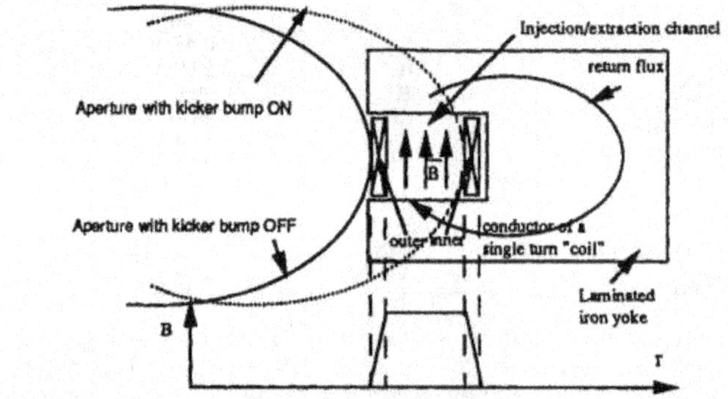

Fig. 15 Schematics of a septum magnet with its magnetic field B in the mid plane.

Septum magnets can either be dc powered (necessitating forced liquid cooling of the outer conductor due to the necessarily high current density) or pulsed (which normally allows the cooling of the septum sheet by conduction over the cooled return yoke).

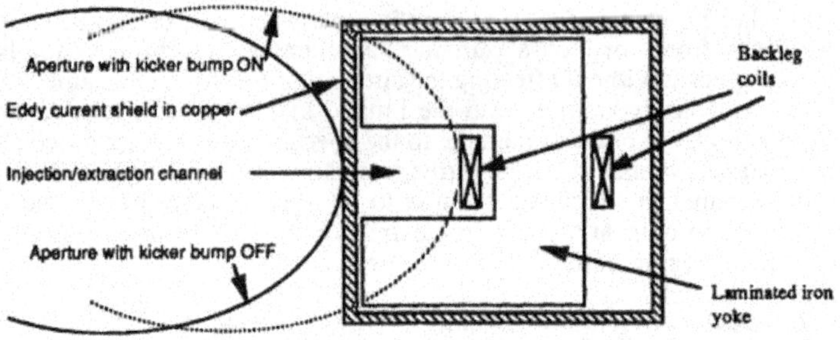

Fig. 16 Cross section of an eddy current septum magnet.

Powering the septum with short current pulses creates an interesting different type of septum, the eddy current septum (Fig. 16). Here the coil can be constructed without any geometrical constraint as a backleg coil while all magnetic field components which tend to traverse the highly conducting copper eddy current enclosure are blocked by the induced eddy currents. In this case the "septum sheet" only carries eddy currents and no primary excitation current. Therefore, the eddy current septum sheets can be thin and uncooled. Another septum which is often used is the "Lambertson" Septum magnet where the injected beam is bent in the vertical plane and injected into the horizontal or vice versa.

Fig. 2 schematically shows the injection into the ESRF storage ring, employing a thin vacuum septum (S3), two septa magnets in air with separated stainless steel vacuum chambers (thickness of steel d = 0.3 mm) matched to the pulse width of 2 msec and the 4 delay line type injection kickers to form a symmetrical localized injection bump (Fig. 17).

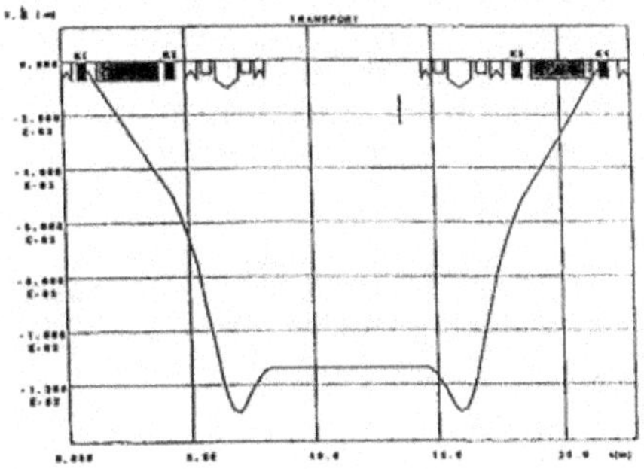

Fig. 17 Schematic view of storage ring injection bump.

3.9 *Positron production*

Depending on beam emittances, time structure of the stored beam and vacuum conditions, the electron beam in the storage ring can ionize residual gas particles by the knockout of electrons from the atomic shell and these positively charged ions can subsequently be trapped in the space charge potential of the negatively charged electron beam. This can lead to undesirable effects such as increased coupling, emittance growth, reduced lifetime or instabilities. Significant efforts have been put into understanding the trapping mechanism and in quantifying the necessary emittance, time structure and vacuum conditions which would allow the operation of synchrotron light sources with electrons without detrimental effects of ion trapping.[21] Very promising results are reported :[22] Until now instabilities or lifetime reductions due to trapped ions could not be found either at the ESRF or at the Trieste Sincrotrone. Furthermore, in both cases the natural coupling between horizontal and vertical emittances is measured to be in the few percent range and is not dependent on beam current. These first results confirm the theoretical expectation that in low emittance beams residual gas ions will be unstable in the potential of the beam.

The trapping of very heavy ionized particles like dust remains, nevertheless, a potential danger. These particles might be too heavy to be overfocussed but the reported absence (until now) of evidence of trapped dust particles might be due to the limited experience with the 3rd generation rings up to now or due to disintegration mechanisms of this material under the power density of the electron beam and the associated synchrotron radiation.

In order to principally avoid those ion trapping related problems, several synchrotron light sources are operated with positrons instead of an electron beam.

Unfortunately the incremental investment costs of producing e^+ instead of e^- with comparable phase space density is in the region of several millions of US dollars.

Presently the only technically exploited method to produce positrons is by pair production : here an e^- beam of energy E hits a high Z target. The produced bremsstrahlung photons create e^-/e^+pairs above a photon energy threshold of $2m_0c^2$ (~1 MeV). The e^+ which are emitted into nearly a 2π space angle with a large energy spread are then focused and accelerated. Since the number of created e^+ is proportional to the incident e^- beam current and the incident energy E, only high current accelerators such as linacs are used to produce the incident e^- beam. After the converter rapid acceleration is needed in order to prevent the e^+ beam from growing unacceptably in transverse size. Also here the highest gradients of dE/ds are achieved in linacs. A schematic example of a 200 MeV positron pre-accelerator is given in Fig. 18.

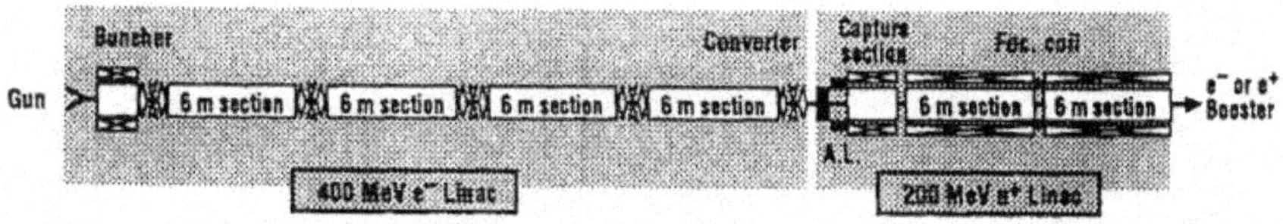

Fig. 18 200 MeV e^+ linac with e^-/e^+ conversion at 400 MeV. (A.L : Adiabatic Lens)

It shows a 400 MeV high current electron linac, the beam of which is focused onto a 7 mm thick tungsten target plate. The e^+ emerging are simultaneously focused and accelerated into the entrance of a further 200 MeV low current linac.

In practical terms, the present best achieved conversion rates from the e^- current incident on the converter target to the e^+ beam leaving the 200 MeV linac are

$$\frac{N^{e^+}}{N^{e^-}} = 0.04 \text{ per GeV of incident electron energy}$$

for the non energy analyzed e^+ beam and

$$\frac{N^{e^+}}{N^{e^-}} = 0.024 \text{ per GeV of incident electron energy}$$

for the e^+ beam within an energy spread of $\frac{\Delta P}{P} = \pm 1\%$.

From the above it is clear that e^+ linacs can only deliver currents which are an order of magnitude less than the currents which can be achieved with electrons. In order to avoid a corresponding decrease of injection speed into the storage ring, a correspondingly higher repetition rate of injection must be used. In cases where the repetition rate is limited by slow damping times or where an intermediate synchrotron is designed for too slow a repetition rate, one can employ low energy intermediate storage rings[23] which were pioneered by DESY for PETRA injection[24]. With double frequency RF systems these intermediate low energy storage rings also allow efficient bunching.

Fig. 19 [25] shows the basic parameters of the positron accumulator ring for the APS project.

Table 2 PAR lattice Parameters

Circumference	30.6667	m
Energy	450	MeV
Tunes, Q_x, Q_y	2.170, 1.217	
Largest β_x, β_y, η_x	4.70, 13.71, -3.21	m
β_x, β_y, η_x at injection point	2.00, 9.80, 0.0054	m
Momentum compaction	0.247	
τ_x, τ_δ	20.8, 14.7	ms
J_x, J_δ	1.242, 1.758	
Energy loss per turn	3.56	keV
ε_0	0.36	π-μm
σ_δ	0.041	%
RF 1st, 12th harmonic		
Voltage	40,30	kV
Synch. tune	1.86, 5.90	x 10^{-3}
RMS bunch length	0.884, 0.280	ns

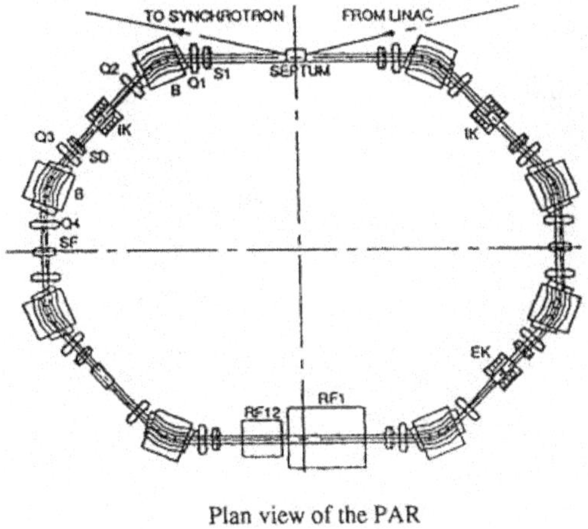

Plan view of the PAR

Fig. 19 Positron Accumulator Ring (from M.Borland/Argonne National Laboratory).

3.10 Transport lines

Transport lines between different accelerators have the task of transporting the beam from the exit of one accelerator A to the entry point of a second one B in such a way that the extraction and injection losses can be minimized. In addition, the transport line should permit the measurement of all beam parameters relevant to the downstream injection process.

For this to be possible, the entry and exit of the line have to be matched to the respective accelerator parameters in the position and angle of the transfer line axis, in the focusing properties $\beta_{x,z}$, $\beta'_{x,z}$ and in the dispersion $D_{x,z}$, $D'_{x,z}$. In order to facilitate emittance measurements, a dispersion free section which includes at least 2 quadrupoles and a beam profile monitor is desirable. In transport lines from accelerators with a large energy distribution like linacs, dispersive sections should allow the inclusion of slit systems in order to measure and if necessary reduce the energy spread of the beam at the entry of accelerator B. Fig. 20 shows a simple transport line from a 200 MeV linac to a synchrotron.

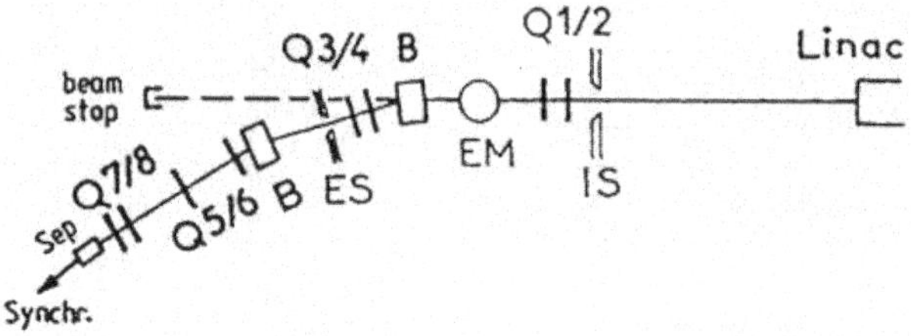

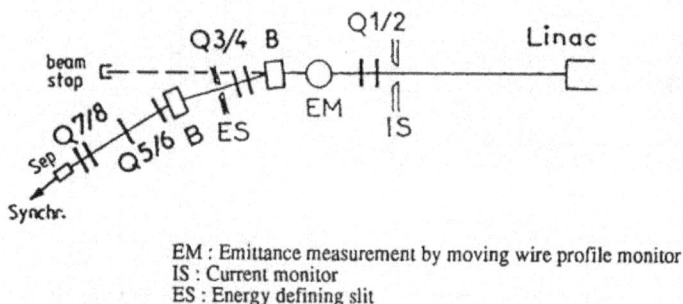

EM : Emittance measurement by moving wire profile monitor
IS : Current monitor
ES : Energy defining slit

Fig. 20 ESRF 200 MeV transport line : linac to synchrotron

Fig. 21 Last air coil septum magnets of the 6 GeV booster to the storage ring transfer line of the ESRF.

Transport lines finally leading the beam into a high brilliance light source storage ring can profit partly from the fact that the limited dynamic aperture of these machines requires an injector beam with a rather small emittance and energy spread. Since the beam only passes through the transport line once, the necessary vacuum pressure can be several orders of magnitude higher than in the storage ring. The aperture of these lines can be minimized and therefore cost reductions can be made on magnets, power supplies and vacuum components. The minimization of apertures allows easy implementation of differential pumping sections at the entry into the storage ring. Fig. 21 shows the last air coil septum magnets of the 6 GeV booster to the storage ring transport line of the ESRF. Note the size of the vacuum chambers inside these magnets of 14 x 10 mm.

3.11 Modes of Operation

The electrons must be delivered to the storage ring with a pre-defined time structure :

- as the energy lost by a particle in the storage ring due to synchrotron light emission is replaced by particle acceleration in an RF field, particles can be stably captured only in a given phase window $\Delta\varphi$ around the stable phase point φ_0 (see chapter 4).

  Therefore, the particles have to be delivered by the injection system preferably in bunches of a length σ_b.

$$\sigma_b < \frac{\Delta\varphi}{2\pi} \cdot \lambda^{RF}$$

- as, for practical reasons, the frequency of the storage ring-RF system is normally chosen to be much higher than the circumferential frequency of the electrons in the storage ring, there are many stable phase areas around the circumference of the storage ring (buckets) where electrons could be stored. The number of RF buckets on the circumference is called the harmonic number h.

There are different modes of having these buckets filled with electrons, depending on specific user requests or on beam stability requirements :

- single bunch mode i.e. a single bunch is stored in the storage ring (i.e. the distance between successive bunches is equal to the storage ring circumference)

- few bunch mode i.e. a few bunches are stored in the storage ring with any pre-programmed distance between successive bunches

- multibunch mode, i.e. all RF buckets on the circumference are filled with bunches (i.e. number of bunches $N = \frac{C}{\lambda^{RF}}$)

- gated multibunch mode i.e. a train of successive bunches shorter than the circumference is filled leaving an unfilled gap in the circumference : $N < \frac{C}{\lambda^{RF}}$

Accumulation in the single bunch mode requires the injector to deliver the single bunch at the end of the transfer line synchronously with the passage of the already stored single bunch in the storage ring. This requires the single bunch to be triggered in the injector with a time accuracy $\Delta t^{trigger}$, which should be short compared to a bucket length ($\Delta t^{trigger} << T_0^{RF} = \frac{\lambda^{RF}}{c}$).

The trigger system which allows such injection schemes normally consists of a combination of a "slow timing system" with a time accuracy of $\Delta t << \frac{C}{c}$ governing the trigger for

- linac/microtron RF pulse

- injection/extraction kickers and septa

- RF amplitude modulation of a possible booster synchrotron

and a "fast timing system" often called the "bunch clock" governing the triggers for the single bunch creation (normally acting on the grid of the linac/microtron gun or a chopper system after the gun, which is phase locked to the RF and has a time accuracy $\Delta t << T_0^{RF}$).

The bunch clock also provides the circumferential triggers for the storage ring and possibly a booster synchrotron for diagnostic purposes.

In the few bunch mode one can use successive single bunch injector pulses to populate the different buckets or, if the injector allows a macro pulse length longer than the distance of two successive bunches in the storage ring, one can apply multi-triggering of the linac/microtron gun during an injector macro pulse[26] in order to increase the injection speed.

The injection into the storage ring can take place in time intervals comparable to the lifetime τ_0 of the storage ring beam (e.g. each $\frac{\tau_0}{2}$ hours, when the stored beam has lost about 40% of its initial charge : normal injection mode) or in time intervals which are short compared to the lifetime τ_0 (e.g., each $\frac{\tau_0}{1000}$ hours, when the stored beam is still almost at its initial charge : Topping up mode).

The normal injection mode has the advantage that,

- the decay function over time of the stored current is smooth for many hours

- the injection system has to be switched on only for short periods during the lifetime τ_0 of the stored beam. This saves operation costs and avoids any electromagnetic interference of booster components with the stored beam

- the beam shutters, which protect the experimental area from high energy background radiation (produced by unavoidable electron losses during the injection process), can be closed during the short period of injection

and the disadvantage of

- producing variable heat loads in the storage ring and on the optical components of the experimental set ups which affect the stability of the stored beam and the properties and lifetime of the optical components

- reducing the averaged photon flux for a given experiment

The topping up mode of injection where one keeps the stored e^- current close to its maximum value with frequent short injections, avoids the above mentioned disadvantages but

- the experiments have to cope with a non smooth decay function of the stored current

- the beam shutters must stay open all the time with consequences of enforced radiation protection measures in the experimental area

- the injector has to run all the time during which the storage ring is operated. This results in higher operation costs

- additional effort is needed to avoid electro-magnetic interference between time varying currents and fields of the injector (magnets, septa, kicker) and the stored beam and the experimental set ups

3.12 Procurement and installation considerations

Procurement and implementation philosophy for a light source injector depends on the specific boundary conditions within which the injector is to be created.

Turn key systems have been produced by specialized industry up to energies of 1.5 GeV.[27] The customer specifies the requested beam parameters and modes of operation as well as the infrastructure boundary conditions. The manufacturer is responsible for meeting these specifications. In certain contracts the client takes over responsibility as a sub supplier for the provision of certain components which he could provide either at a lower cost than the main contractor or for which the client has already developed expertise, or intends to (e.g. SRRC provided the RF systems to the turn key injector for the SRRC project because SRRC had to produce a similar system for the storage ring in any case).

Nearly all accelerator components can at present be purchased from industry. This is certainly true for magnets (including kickers and septa), cavities, transmitters, klystrons, waveguides, vacuum chambers, vacuum pumps and diagnostics, power supplies [dc, ac and pulsed], modulators, linac sections, electron guns etc.). Today, the depth of industrial component production has significantly increased : In earlier times industry was only responsible for producing, for example, magnets according to mechanical specifications while magnet design, engineering design, acceptance testing and magnetic measurements were carried out by the client. Today, magnets can be produced by industry according to magnetic field specifications. Acceptance testing and magnetic measurements can be contracted to commercial testing companies.[28]

Items which are not easily available from industry are very specialized items, mainly one-offs like, for example, the bunch clock or special beam diagnostics usually only representing a minute fraction of the investment costs for the accelerator.

In cases where complete light sources are erected on green fields [29], significant savings in investment expenses, manpower efforts and maintenance costs can be achieved by applying the same or at least similar philosophies to the equipment and software requirements of the light source.

- The RF systems of storage rings and booster synchrotrons can be very similar (same frequency, same cavities, same transmitters, same control software, same diagnostics).
- Beam diagnostics such as BPMs, DCCTs and optical monitoring in the booster and the storage ring can use the same pick ups, the same electronics and the same software.
- All parts of the facility should be specified to meet the standard interfaces of the building infrastructure.
- The same alignment techniques can be used for all accelerators, front ends and beamlines.
- The same control hardware and software should be used for all parts of the accelerator complex (preferably including insertion devices, front ends and beamlines).

The main advantage of general approaches to similar technical fields is the reduced cost for maintenance due to the reduced need for permanent expertise in the staff.

Delivery times of long lead acceleration components can extend to 20 months. If there are no budget, site or authorization obstacles, a classical 1 GeV injector could be designed, built and commissioned in about 30 months, and multi GeV injectors could be completed in about 40 months (starting from a "green field").

3.13 Trends in synchrotron light source injector designs

There are two main objectives in synchrotron light source injector designs : the minimization of costs of a machine which is absolutely necessary but which is usually operated for a very small fraction of time, and the creation of a potential for interesting parasitic use of this otherwise nearly unused machine.

Due to their function, synchrotron light source injectors usually operate only for short periods of time with shut off times similar to the lifetime of the beam in the storage ring if no other "parasitic" use is foreseen. The cost optimization of any accelerator normally minimizes the sum of the capital costs and the running costs for its assumed lifetime. In the case of an injector which is not used parasitically, the weight of the respective operation costs is practically negligible. This mainly has consequences for the construction of the magnets and their coils : the size of the coils and magnets are minimized at the expense of higher power dissipation.

At present there do not seem to be many possibilities for further significant cost reductions for linacs and microtrons. Studies are being carried out in the context of linear collider projects to see if an increase of RF frequency from L or S band (1.5 - 3 GHz) to the X band (10 GHz) could reduce the cost of a linac.

For booster synchrotrons there seems to be a possibility of further reducing the apertures in the magnets which in turn would reduce magnet and power supply costs. Given the progress in closed orbit control, ideas have come up to reduce the margins for closed orbit deviation and to apply higher focusing so that the remaining aperture requirements would allow simple 1 turn coils in magnets[30] with apertures between 10 and 20 mm instead of the present 30 to 50 mm range. This would not only lower the magnet costs, but also the power supply (due to reduced impedance and stored energy) and vacuum chamber costs. These proposed low aperture synchrotrons [31] would require a beam of low emittance and small energy spread, already well pre-bunched on the synchrotron RF frequency, to be injected on axis which could favor the use of racetrack microtrons as pre-injectors.

Practically all dedicated synchrotron light sources are organized for the simultaneous use of the light produced by many user groups (multi user facilities). Consequently, experiments needing exclusive control of the light source due to frequent machine access or needing use of modes of operation which cannot be used simultaneously by the other users are considered "unpopular" and consequently find no or "non optimum" experimental conditions in the multi user synchrotron light facilities. The practically unused injectors could provide a kind of test bench for some of those experiments. It might be worthwhile noting that, as an example, the ESRF 6 GeV booster synchrotron, despite being fully optimized for its role of injection, can operate at 1.5 GeV as a storage ring similar to 3rd generation VUV/soft X-ray light sources with an emittance of 8nm. The uncooled vacuum system would allow lifetimes of several hours and storage of more than 100mA at this energy. Dispersion free straight sections of 2m lengths are available. Even the build up of a bypass of 40m length is possible.

The possible use of an injector as such a test bench for single user experiments strongly depends on the capacity of the injector to provide high brilliance beams. The present trend to reduce apertures to achieve lower magnet, power supply and vacuum system costs by designing low emittance booster synchrotrons also independently goes in this direction. It might therefore be worthwhile considering such possible use of injectors as test bench for single user experiments on high brilliance beams right at the injector design stage more seriously.

References

1. e.g. MAX I (Lund) is shared between synchrotron light use and use as a stretcher for nuclear physics.
2. e.g. Studies have been undertaken to use the 1.5 GeV injector linac of Trieste Sincrotrone as a driver for FEL type experiments.
3. TANTALUS (Wisconsin).
4. UVSOR (Tokyo).
5. MAX I (Lund).
6. e.g. M.Borland, *SLAC Report 402* (1991).
7. e.g. L. Smith, *Encyclopedia of Physics*, **Vol. 44**, 1959 or H.Wiedemann, *Particle Accelerator Physics*, Springer Verlag. 1993.
8. CGR MeV / General Electric

9. e.g. Photon factory at KEK.
10. Sincrotrone Trieste.
11. SSRL/Stanford.
12. ESRF/LURE/Photon Factory/SPring 8/APS.
13. S.P.Kapitza and V.N.Melekhin, *The Microtron*, Harwood Academic, London, 1987.
14. e.g. R.Wilson, *Encyclopedia of Physics*, **Vol. 44**, 1959 or Courant et al. *Phys.Rev.* *88*, 1190 (1952)
15. e.g. MAMI/MAINZ.
16. e.g. older combined function synchrotrons (which have the horizontal betatron oscillation antidamped used $\frac{E^{output}}{E^{input}}$ of up to 150 (DESY I, Hamburg), or even 300 for the 6 GeV CEA/Cambridge/Massachusetts which started life with a 20 MeV linac injector.
17. e.g. AURORA : H.Yamada, *J.Vac.Sci.Tech.* **B8** (6) (1990) p.1628.
18. e.g. SRS-linac/synchrotron injection and SLAC linac/synchrotron injection.
19. e.g. Bessy microtron/synchrotron injection and ESRF linac/synchrotron injection.
20. Blokesch et al., NIM to be published.
21. e.g. A.Wrulich, *ESRF Foundation Phase Report* CII-3.3 (1987).
22. from BESSY, ESRF, ALS : cf. e.g. *ESRF Machine Related Activities Report* (1993) Ch. III, p.5.
23. e.g APS.
24. e.g A. Febel et al, *PAC 1979, IEEE Trans. Nucl. Sci.* **Vol NS-26** (1979)
25. taken from M.Borland, *Proc. of 1993 PAC*, Washington, p. 2028-2030 (1993).
26. e.g; the multitriggering of the linac gun for 8, 16 and 32 bunch operation of the ESRF.
27. e.g. 0.8 GeV separated function booster synchrotron with microtron preinjector for BESSY I (manufacturer SCX) ; 1.5 GeV linac for Trieste Sincrotrone (manufacturer CGR MeV) ; 1.3 GeV separated function booster synchrotron with linac preinjector for SRRC (SCX) ; injector together with storage ring for Baton Rouge : Maxwell Brobeck.
28. e.g. the ESRF has subcontracted the bulk of mechanical control and magnetic measurements of several hundreds of magnets to "APAVE" (France) and "TÜV" (Germany).
29. e.g. BESSY, ESRF, SRRC, Spring 8, Sincrotrone Trieste.
30. i.e. like the combined function magnets of the SLAC linear collider arcs.
31. e.g. G.Mülhaupt, ESRF internal report, *ESRF/MACH-INJ/94-13*.

CHAPTER 4: R.F. SYSTEMS

D.J. THOMPSON
and
D.M. DYKES
Daresbury Laboratory, Daresbury,
Warrington, England. WA4 4AD

4.1 Introduction

4.1.1 The purpose of an r.f. system

In an electron (or positron*) storage ring, the energy lost by the electrons as synchrotron radiation (and to higher order modes) is replaced by means of an electric field established across a so-called "accelerating gap" or gaps. Because the field must always be in the same direction when the electrons pass, it must alternate in time; this means that the electrons become bunched, those passing across the gap when the field is insufficient or in the wrong direction being decelerated and lost. The electrons orbit typical storage rings in times ranging from tens of nanosecs to a few microsecs, depending on the circumference. The frequency of the alternating field must be integrally related to the orbital frequency and is usually some multiple of it, so the frequency is normally in the radio-frequency (r.f.) band (between a few MHz and a few hundred MHz). The r.f. system comprises the accelerating gap or gaps together with the means to establish the necessary field across them and also to deliver to the beam power equal to that being radiated as synchrotron radiation.

If the energy of injection is less than the operating energy of the storage ring, the r.f. field must also increase the electron energy during ramping. However because ramping is always carried out slowly (over minutes) the r.f. power needed to do this is negligible compared to the power emitted as synchrotron radiation at full energy. Moreover because the electrons travel with virtually the speed of light at all times in the storage ring, their revolution frequency and hence the r.f. system frequency remains constant, even if the injection energy is much less than the operating energy. The need for ramping therefore has no effect on the design of the r.f. system other than, in some cases, the need to be able to control some of the r.f. parameters over a wider range than would otherwise be the case.

In practice, each accelerating gap (there can be more than one around the storage ring circumference) is made part of a resonant cavity, and the cavity is energised from a power amplifier at an appropriate frequency so as to establish the necessary voltage across the gap. The dynamic interaction of the electron beam with the cavity is such that so long as the voltage is larger than the minimum necessary, a state of "phase stability" exists and those electrons captured at injection automatically receive, on average over many orbits, the energy increment per turn which they require to counterbalance their energy loss. The power required to be delivered to the cavity is the sum of two components - the power needed to sustain the required voltage across the accelerating gap, i.e. to make good the resistive losses in the walls of the cavity, and the power needed to be transmitted to the beam, to replace the power being radiated. All these topics will be discussed more fully in the following sections.

The r.f. system as a whole consists of one or more accelerating cavities installed in the ring, energised by one or more high frequency power amplifiers, which in turn are fed from a master oscillator which controls the frequency. Each cavity will have a tuner, with

* This chapter will, for convenience, refer only to electrons. Use of positrons does not affect the r.f. system.

**D.M.DYKES@dl.ac.uk (e-mail) 0925 603124 (fax number)

an automatic tuning system with controlled offset; and each cavity will be coupled to the power amplifier (or one of the power amplifiers) by a transmission line system which may be coaxial or waveguide, depending on the frequency and power involved and which may include an isolator or circulator (devices which prevent power reflected from the cavity reaching and possibly damaging the power amplifier tube).

Provision will be made for adjusting the phasing between cavities and between cavity and master oscillator, for controlling the cavity voltage and tuning offset and for monitoring all important parameters. Because the powers involved are large, ranging from a few kilowatts for a small VUV source to a megawatt or more for a source such as the ESRF, the APS or SPring-8, there will be cooling systems for the cavities and the power amplifier. The cavities will be fitted with vacuum pumps. The drive system will include a fast switch, because switching off the r.f. power is the fastest way of dumping the electron beam, should this be necessary to ensure personnel safety or to avoid damage to hardware (e.g. if a mis-steered beam is hitting or causing synchrotron radiation to hit an uncooled surface in the vacuum chamber).

An example of an r.f. system with two cavities fed from one power source is shown in simplified diagrammatic form in Fig. 4.1. The main components will be described in more detail in the following Sections.

4.1.2 Other effects of the r.f. system on the beam

The r.f. system affects the electron beam in various ways, apart from imparting the necessary energy for it to remain on its orbit. The main effects are on:

Beam lifetime - because electrons are lost through intra-beam coulomb scattering (the Touschek effect) and through the excitation of synchrotron oscillations by the quantum nature of the synchrotron radiation. The rates of loss from both these causes are functions of, amongst other things, the r.f. voltage and frequency.

Bunch length - which may be important for some users of the radiation, and which depends on the same parameters.

Beam stability - which can be affected by higher order mode electromagnetic fields excited by the beam in the cavities, and by the fields induced by the beam in other vacuum chamber components, whose amplitudes are a function of the bunch length. Thresholds for some instabilities depend on the values of the synchrotron oscillation frequency and the r.f. energy acceptance, both of which are functions of the r.f. voltage and frequency.

These effects will be discussed in more detail in sections 4.3, 4.4 and 4.6 below, as well as in chapter 12.

4.1.3 The choices to be addressed

When designing an r.f. system, a number of choices need to be made, all of which are closely inter-related. The principal issues are:

Frequency : This can range from a few MHz to a few hundred MHz.

Cavities : The number of gaps (N); whether these are provided in N single-cell (i.e. single-gap) cavities or N/n multicell cavities (n gaps per cavity); the shape of the cavities, their material, the method of feeding in the power, of tuning and of cooling them.

Power source: The size of each unit; the power amplifier tube to use; the means of transmitting the power to the cavities; the type of power supply for the amplifier tube, the means of controlling the output power, phase and frequency.

Though the correct design of the r.f. system is crucial to achieving the required optimum performance from a storage ring, there is no unique answer to these questions and the variety of solutions is illustrated in Table 4.1 which lists the r.f. parameters of a number of synchrotron radiation sources now operating.

Storage Ring	Ring Energy (GeV)	Parameters and features of r.f. systems								
		Frequency (MHz)	Energy loss per turn (keV)	No. of cavities	Type of cavity	Material of cavity	Coupling device	Power tube	No. of amplifiers	Amplifier power
NSLS VUV	0.75	53	14.7	1	λ/4	copper-clad steel	loop	tetrode	1	50kW
BESSY	0.8	62/500	20	1/1	λ/4/1 cell	copper	loop	tetrode/ klystron	1/1	20/60kW
SUPER-ACO	0.8	100	21.3	1	1-cell		loop	tetrode	1	
HESYRL	0.8	204	16.3 (BM)	1	double lambda/4 re-entrant	copper-clad steel	loop	tetrode	1	30kW
SRRC	1.3	500	91	4	1-cell	copper	loop	klystron	2	300kW
ALS	1.7	500	92 (BM) 129 (total)	2	1-cell	copper	loop	klystron	1	300kW
DCI	1.8	25	243 (BM)	1	1-cell	aluminium	loop	tetrode	3	100kW
ELETTRA	2	500	258 (BM) 320 (total)	4	1-cell not re-entrant	copper	loop	klystron	4	60kW
SRS	2	500	254 (BM)	4	1-cell	copper	window	klystron	1	300kW
Pohang Light Source	2	500	225 (BM) 356 (total)	4	1-cell	copper	loop	klystron	4	60kW
NSLS X-ray	2.5	53	504	3	λ/4	copper-clad steel	loop	tetrode	4	100kW
SIBERIA 2	2.5	180	704	2	1-cell	copper	loop	tetrode	1	300kW
Photon Factory	2.5	500	480	4	1-cell	copper	loop	klystron	4	180kW
SPEAR	3	358	553 (BM)	2	5-cell	aluminium	loop	klystron	2	500kW
ESRF	6	352	4600 (BM) 6100 (total)	4	5-cell	copper	loop	klystron	2	1->1.3MW
APS	7	353	5450 (BM)	15	1-cell	copper	loop	klystron	3	1MW
SPring-8	8	509	9200 (BM) 12900 (total)	32	1-cell	copper	loop	klystron	4	1MW

Table 4.1 RF Parameters of SR Sources

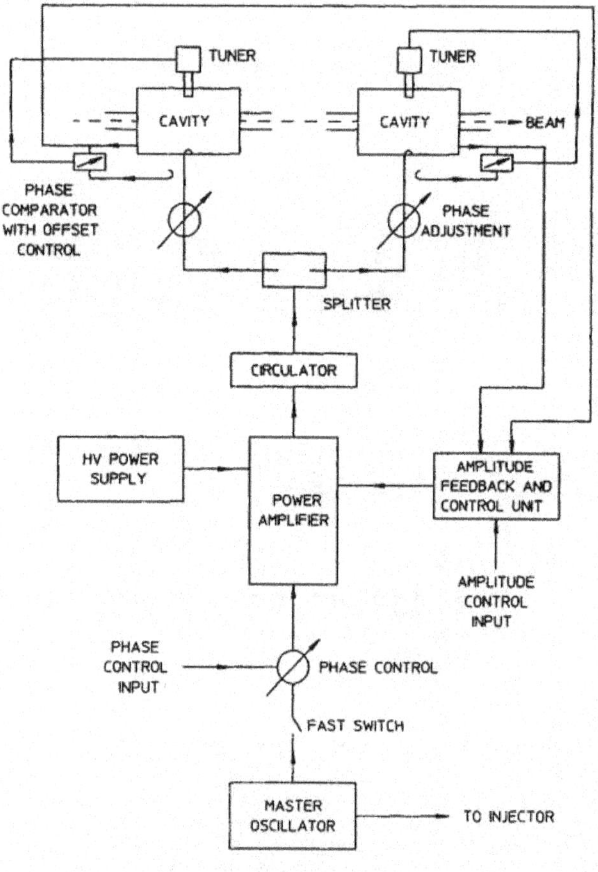

Fig. 4.1 Illustrative diagram of an r.f. system.

The remainder of this chapter first explains the concepts of longitudinal motion and phase stability, then how to calculate the required cavity voltage, and following this, it discusses in some detail the engineering choices available when designing the r.f. system.

4.2 Longitudinal motion and phase stability

The integrated electric accelerating field crossed by the particles on each orbit must vary periodically with time at the orbital frequency or, more usually, some multiple thereof. Except for dual frequency systems (see section 4.4.4 below) it will in practice vary sinusoidally. All normal synchrotrons and storage rings work because of the principle of phase stability, which means that it is not essential for each electron to cross the accelerating gap at precisely the moment when the field is such as to give the correct energy increment. The special case of isochronous rings (not so far used as synchrotron radiation sources) is beyond the scope of this book.

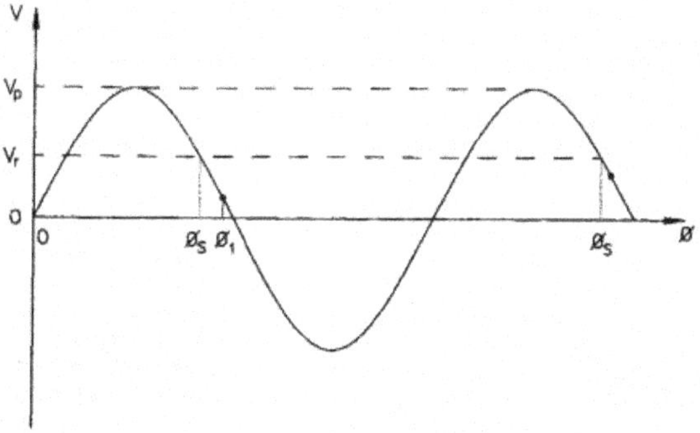

Fig. 4.2 Principle of phase stability.

In Fig.4.2 the cavity voltage V is shown varying sinusoidally with respect to a phase angle ϕ. (This is equivalent to the variation with time, which is ϕ divided by 2π times the frequency). If the electron passes through at a phase ϕ_S it will receive the correct energy gain eV_r. This is numerically equal to V_r if the energy is measured in electron-volts. If it passes through late, at $\phi_1 > \phi_S$, it will receive too little energy, and consequently move inwards on to an orbit of smaller radius. This means, as its velocity is effectively constant, that its revolution period will be less and it will take less time than the r.f. period to arrive back at the gap - it will then be nearer to, or on the other side of, the correct phase. An electron traversing the gap early will do the opposite. Note that this only applies to electrons crossing whilst the r.f. voltage is decreasing. Thus electrons crossing within a certain "phase acceptance" of the nominal stable phase angle ϕ_S perform stable so-called "phase", "longitudinal" or "synchrotron oscillations" about ϕ_S. These cause corresponding energy and radial position oscillations, as is evident from the above description, and the potential well within which they are stable, equivalent to the phase acceptance referred to above, is often called the "r.f. bucket".

A number of important beam dynamics properties can be defined in relation to this longitudinal motion. Analysis shows[1] that if ε is the difference between an electron's actual energy and that of a "reference particle" (one which has the correct energy and crosses the cavity at the correct time and phase), and τ is the corresponding time displacement from the nominal crossing time (i.e. from the centre of the bunch), then

$$\frac{d\varepsilon}{dt} = \frac{1}{T_0}\left[e\left(\frac{dV}{dt}\right)_0 \tau - D\varepsilon \right] \tag{4.1}$$

where e is the electronic charge, T_0 is the orbit period, $D = (dV_r/d\varepsilon)_0$, and the suffix o indicates evaluation at the nominal crossing time and energy.
Introducing the momentum compaction factor α (see chapter 2) defined by

$$\frac{\ell}{L} = \alpha\frac{\varepsilon}{E_0} \tag{4.2}$$

where L is the orbit length and ℓ the change due to an energy change ε, it can be shown that

$$\frac{d\tau}{dt} = \alpha \frac{\varepsilon}{E_0} \tag{4.3}$$

and these equations can be solved for τ to give

$$\frac{d^2\tau}{dt^2} + \frac{2}{T_\varepsilon}\frac{d\tau}{dt} + \Omega^2\tau = 0 \tag{4.4}$$

or, alternatively, for ε, which satisfies the same equation (except for a phase difference of $\pi/2$). These are the equations for damped harmonic oscillations with an oscillation (angular) frequency Ω and damping time constant T_ε. These important parameters of the motion are given by

$$T_\varepsilon = \frac{2T_0}{D}$$

or, after further manipulation

$$T_\varepsilon = \frac{2E_0 h}{fJ_\varepsilon V_r} \tag{4.5}$$

and

$$\Omega^2 = \frac{\alpha e\left(\frac{dV_0}{dt}\right)_0}{T_0 E_0} \tag{4.6}$$

or, putting the synchrotron oscillation frequency $f_s = \Omega/2\pi$

$$f_s = f\sqrt{\frac{-\alpha V_p \cos\phi_s}{2\pi E_0 h}} \quad \text{Hz} \tag{4.7}$$

where f is the cavity frequency (in Hz), V_p (in volts) is the peak cavity voltage per turn (transit time corrected, see section 4.3.2) i.e. numerically equal to the energy gain which the electron would receive if it crossed the cavity at the peak of the waveform, ϕ_s is the equilibrium phase angle of the beam, defined as in Fig. 4.2, E_0 is the electron energy (in electron volts), h is the harmonic number (the ratio of the cavity frequency f to the orbital frequency f_0) and J_ε is the damping partition coefficient for energy oscillations, approximately equal to 2 for normal rings (see chapter 2). Note that Eq. 4.4 applies only to small oscillations and the last equations (4.5, 4.7) to sinusoidal gap voltages.

Further more general analysis, of large amplitude oscillations,[2] shows that there is a maximum stable amplitude of oscillation which is given, in energy terms, by a maximum stable energy deviation ε_{max}, where

$$\varepsilon_{max} = E_0\sqrt{\frac{V_r F(q)}{\pi h \alpha E_0}} \quad \text{eV} \tag{4.8}$$

where q is the "overvoltage factor" V_p / V_r .

$$F(q) = 2[\sqrt{(q^2 - 1)} - \cos^{-1}(1/q)] \tag{4.9}$$

$$\approx 2q - \pi \qquad \text{for large q}$$

[As an example: for a 2 GeV ring, with f = 500 MHz, peak cavity voltage = 700 kV, energy loss per turn = 257 keV, harmonic number = 300, and α = 0.015, then cos ϕ_s = -0.93, q = 2.7, f_s = 25 kHz, and ε_{max}/E_0 = 0.5 %.]

4.3 Voltage requirement

4.3.1 Energy gain per turn

The function of the r.f. system is to make good the energy lost by the electrons as they orbit the storage ring. The energy lost in the bending magnets as synchrotron radiation is very simply calculated from

$$U_0 = 88.4 E_0^4 / R \quad \text{keV per turn} \tag{4.10}$$

where E_0 is the electron energy, in GeV this time, and R is the bending radius in metres. [In the same example of the 2 GeV ring, with a bending radius of 5.5 m, the energy loss per turn is 257 keV.]

To this must be added the energy lost by radiation from insertion devices and also that lost in the walls of the vacuum vessels (including the r.f. cavities) through which the beam passes, because of the fields which the beam induces due to its bunched structure.

The losses in the insertion devices can easily be calculated if their parameters have been specified (see chapter 14), otherwise an approximate allowance may have to be estimated. The losses in the vacuum chamber, known as the parasitic losses, can only be estimated as they depend on the effective impedances of all the cavity-like objects through which the beam passes, summed over the frequency components present in the bunched beam. These impedances must in any case be kept small to avoid overheating vacuum vessel components such as valves and bellows or, in the case of the r.f. cavities which in practice comprise most of the effect, to avoid instabilities induced by higher order modes (see chapter 12).

By maintaining a very smooth vacuum chamber wherever possible, and by avoiding the excitation of higher order modes in the cavities (see section 4.6 below) the parasitic loss can be kept sufficiently low (a few percent of the main radiation loss) that an accurate prediction of its magnitude is not necessary.

The total of these various energy losses is referred to as the "required energy gain per turn", referred to as V_r and expressed in units of volts (or kV) when referring to the cavity voltage, or electron-volts (or keV) when referring to the energy lost (or gained) by the electrons.

4.3.2 Peak cavity voltage

The complete r.f. system may comprise more than one cavity, and the cavities may have more than one cell, i.e. more than one gap. The peak instantaneous voltage summed over all the cavity gaps must exceed V_r by the "overvoltage factor" and by the "transit time factor". The latter arises because an electron takes time to traverse the cavity gap and during this time the gap voltage changes. Because this factor depends on the gap dimension, i.e. on the cavity design, it is usual to allow for it in the shunt impedance (see section 4.6 below). To do this, the cavity shunt impedance is defined as

$$Z = V_p^2 / 2P \tag{4.11}$$

where V_p is the effective ("transit time corrected") peak voltage, i.e. it is the energy which the electron would gain if it crossed the centre of the cavity at the moment of voltage maximum, and P is the power input needed to establish that voltage.

The peak cavity voltage V_p, defined in this way and understood to refer to the sum of all the cavity gaps, must exceed V_r in order to achieve a finite potential well and hence a satisfactory quantum lifetime, Touschek lifetime, synchrotron oscillation frequency and bunch length. The overvoltage factor q is the ratio of V_p to V_r and must be chosen after consideration of these four effects.

i) Quantum lifetime. The quantum lifetime is the time for the beam intensity to fall to 1/e of its initial value on account of the following effect alone. It arises from the energy distribution of the emitted photons. When an electron emits a photon, its energy is reduced and a phase oscillation is initiated. In some cases the energy loss is greater than the maximum amount for stability (Eq. 4.8) and then the electron is lost. The rate of loss is strongly dependent on the r.f. voltage. The quantum lifetime T_q depends on the depth of the potential well or "bucket", i.e. on the maximum energy acceptance ε_{max}, according to the equation:

$$T_q = T_\varepsilon e^\xi / 2\xi \tag{4.12}$$

where $\xi = \varepsilon^2_{max}/2\sigma^2_\varepsilon$,

T_ε is the radiation damping time for energy oscillations (see Eq. 4.5)
ε_{max} is the maximum stable energy deviation (see Eq. 4.8) and
σ_ε is the standard deviation of the energy spread which is given by

$$\sigma_\varepsilon = E_o \gamma_o \sqrt{\frac{3.83\text{x}10^{-13}}{J_\varepsilon R}} \tag{4.13}$$

where γ_o is the electron energy expressed in units of the electron rest mass. T_ε and σ_ε are both independent of the r.f. parameters. ε_{max} depends directly (though not linearly) on q, as well as on the frequency, as shown in Eq. 4.8.

The overvoltage factor should be large enough that T_q is considerably greater than the gas scattering lifetime and, for low energy machines, the Touschek lifetime, so that the quantum lifetime has a negligible effect on the overall beam lifetime. This is not difficult, because T_q has an approximately exponential dependence on the over-voltage factor q, increasing very rapidly once q is greater than the minimum value necessary to give a reasonable lifetime. The necessary value of q usually lies between 2 and 10 for typical machine parameters.

[For the same example of the 2 GeV ring as in sections 4.2 and 4.3.1, T_ε = 5 msec, σ_ε/E_o = 0.07% and T_q = 61 hours. If V_p is changed to 650 kV, T_q falls to 4 hours, but with V_p = 750 kV, it becomes almost 1000 hours.]

ii) Touschek lifetime (T_t). The Touschek effect is particle loss from the potential well due to excitation of energy oscillations by intra-beam coulomb scattering. It is strongly energy-dependent, being a particularly serious limitation at electron energies below about 1 GeV. It depends on the particle density in the bunch, so is most significant in low emittance machines and at high currents. Again it is a function of the maximum energy deviation, and hence of the r.f. voltage. The relationship is complex (see Section 12) but at low machine

energies or in very low emittance machines it may be necessary to increase the over-voltage factor beyond what is needed for an adequate quantum lifetime, in order to increase the Touschek lifetime.
iii) Synchrotron oscillations. The number of synchrotron oscillations per orbit (Q_s) is given by

$$Q_s = \sqrt{\frac{-\alpha f V_p \cos\phi_s}{2\pi f_o E_o}} \tag{4.14}$$

It is important that this number be small, because if there is betatron - synchrotron coupling through the dispersion of the machine, then the betatron resonances, which anyway occupy much of the tune space, will have sidebands separated by Q_s and the larger Q_s the greater will be the restriction on available tune space in which to locate the working point. This leads to a preference for a low r.f. voltage, but if the lattice design is such as to give a very low momentum compaction factor, and if the synchro-betatron coupling is small, the choice may not be of crucial importance. A value of Q_s of 0.1 is normally regarded as dangerously or undesirably high. (See chapter 12 for a fuller discussion of instabilities.)
iv) Bunch length. The natural bunch length (i.e. without bunch lengthening caused by instabilities) is given by

$$\sigma_\ell = c\sigma_t = c\alpha\sigma_\varepsilon / 2\pi f_s E_o \quad \text{cm} \tag{4.15}$$

where c = velocity of light in cm/sec and σ_t is the length expressed in seconds.

From the point of view of the machine builder and operator, a long bunch length is a good thing because it implies a lower peak current and hence smaller induced fields in the vacuum chamber components. This means less heating in components with high impedance, such as perhaps a bellows, and it means less tendency to instabilities, so that a higher current can be stored.

However, in most synchrotron radiation facilities some of the users will wish to carry out time-resolved measurements which require the shortest possible bunch lengths. If this is the case then a higher voltage is preferable.
iv) Summary. The r.f. voltage must be high enough to give a good quantum lifetime, as given by equation 4.12, and may need to be even higher to benefit the Touschek lifetime or to achieve a short bunch length. However it should be no higher than necessary, both for economic reasons, and because this will make the synchrotron oscillation frequency unnecessarily high. If the synchrotron oscillation sidebands are significantly reducing the available tune space, it may be necessary, in the case of a machine with a low injection energy, to adjust the r.f. voltage during ramping in order to control the position of the sidebands whilst maintaining an adequate lifetime.

4.4 Choice of frequency

4.4.1 Theoretical issues

Except that the orbit circumference must be an exact number of wavelengths, there are no theoretical criteria which lead to a clear choice of frequency. Rather there are a number of matters which must be taken into account, together with the practical issues described below, when making the choice. These matters are:
i) Overvoltage factor. As explained in section 4.3.2 above, the peak cavity voltage must be greater than the required energy gain per turn in order to achieve a satisfactory beam lifetime. This ratio is frequency-dependent, as shown by equations 4.12 and 4.8, and

choice of a higher frequency results in a need for a higher voltage, though not in proportion. In one example[3] a factor of ten change in frequency implied a factor of four in voltage. However this does not necessarily mean that a low frequency is best, because power may be easier or cheaper to provide at a higher frequency.
ii) Synchrotron oscillations. The need for the number of synchrotron oscillations per orbit (Q_S) to be small was discussed in section 4.3.2 above. Equation 4.14 showed that this favours a low frequency and a low voltage. However, this may not be too important, if the momentum compaction factor is very low, as is the case in many lattices now popular.
iii) Bunch length. Section 4.3.2 discussed the natural bunch length (i.e. without bunch lengthening caused by instabilities) and explained that from the point of view of the machine builder and operator, a long bunch length is a good thing because it implies a lower peak current and hence smaller induced fields in the vacuum chamber components; but that at many synchrotron radiation facilities some of the users wish to carry out time-resolved measurements which require the shortest possible bunch lengths. If this is important then a high frequency as well as a high voltage should be chosen.
iv) Instabilities. There are two fundamentally different types of instability - multibunch and single-bunch (see chapter 12). The threshold for single-bunch instabilities will in general be higher if the bunch length is long, which implies a low frequency. Multibunch instabilities are likely to be easier to counteract if there are fewer bunches, which again implies a lower frequency. However it is quite easy to arrange to fill only a fraction of the r.f. buckets in the multibunch case, and sufficiently high single-bunch currents may be achievable despite a high frequency. Once again these criteria do not define an optimum frequency, but must be taken into account when choosing the frequency.

4.4.2 *Practical issues*

The main components of an r.f. system are the cavity and the power amplifier chain. The design of both varies greatly with the frequency, and they are considered in turn.
i) The cavity. A cavity has two main parameters: its resonant frequency and its shunt impedance. The shunt impedance Z, which is discussed in detail in section 4.6, defines the power input needed to establish a given voltage across the accelerating gap. It, rather than the Q, is therefore the figure of merit for a cavity for a storage ring. An upper limit on frequency is created by the need for holes at opposite ends of the cavity for the electron beam to pass through. If the frequency is so high that power can be propagated down the beam pipe, the cavity clearly does not function as such, and to keep Z high, the wavelength used must be greater than twice the widest dimension of the pipe. In practice, the horizontal dimension of the beam pipe is usually of the order of 10 cm. and this limits the maximum frequency to less than 1000 MHz if a high shunt impedance is to be attained.

For a given geometry, the cavity dimensions increase linearly with wavelength, and the shunt impedance with the square root, because of the increasing skin depth. However, in practice space for the cavity is always limited, both longitudinally and transversely, and vacuum and cost considerations influence the designer towards modest dimensions, so the geometry must change with frequency. Under conditions of constant overall dimensions the shunt impedance decreases with frequency, tending to result in a roughly similar power requirement (because of the reduced voltage required with a lower frequency) regardless of frequency. The cavity design will tend to be more complex at lower frequencies, but this complexity, and the shunt impedance actually realisable will depend rather critically on the actual space available in relation to the frequency being considered.

The lower limit of frequency will occur when it is impossible to obtain a reasonable shunt impedance, (or, put another way, to adequately cool the cavity when energised at the necessary gap voltage) with a cavity which will fit into the space available.

The choice of cavity and power source will interact in that a lower shunt impedance implies that more power is needed, but for a given frequency and hence cavity shunt impedance, the total shunt impedance can be increased by increasing the number of cavities - if space is available.
ii) The power source. There are various types of r.f. power amplifier tubes for the frequencies and powers needed for synchrotron radiation sources (see section 4.7 below). However, in the frequency range of interest, power tubes tend to be developed for specific purposes and are only readily available if there are sufficient applications (or "sockets") world-wide. A synchrotron radiation source requires at most a few tubes, and so the choice of frequency is strongly influenced by the availability of tubes of the required specification. In general this means choosing either a frequency used by one or more of the very large particle physics accelerators with many sockets - such as 500 MHz or, for very high (MW) powers, 350 MHz; or a frequency used for high power radio or television transmitters.

4.4.3 Summary

Although the choice of frequency is important and influences both the design and the behaviour of the machine in many ways, there is no optimum. The availability of hardware and even local expertise can be a major factor. Bunch length may play an important role if it is a key user requirement, and so also may the number of cavities and the cavity dimensions if total available straight section space is restricted. The variety of solutions to the problem of choosing the frequency is illustrated in Table 4.1 above, which lists a number of synchrotron radiation sources world-wide with the parameters of their r.f. systems.

4.4.4 Dual frequency systems

In a few cases, machines have been built or proposed with two separate r.f. systems operating at different frequencies (though harmonically related). The objective, unless the two systems are merely to be used as alternatives giving different bunch lengths and different maximum currents, is, by operating them together, to achieve a non-sinusoidal accelerating voltage (integrated round the orbit) and hence modify the longitudinal beam dynamics. For instance[4] by adding a third harmonic cavity, which can have a considerably lower voltage than the main one, a waveform with a substantial shoulder or a double peak can be produced. The parameters of the longitudinal motion, including the bunch length, depend on the slope of the r.f. waveform at the time the bunch traverses the cavity (see Eq. 4.1), and so by adjusting the phasing, bunches which are either longer or shorter than without the harmonic cavity can be produced and the synchrotron oscillation frequency can be varied. The purpose is usually to lengthen the bunch or reduce Q_S so as to combat instabilities, but it has also been proposed as a way of reducing the bunch length for the benefit of users. In its simplest but somewhat limited form the harmonic cavity is unpowered, its field being generated by the beam and the phasing being controlled by detuning the cavity.

4.5 Power requirement

4.5.1 The different components of the power

The total r.f. power required to be delivered at the output of the amplifier is the sum of the power transmitted to the beam, the power dissipated in the cavity, the power reflected by the cavity and the power dissipated in the feeder system. The beam power includes the parasitic losses as well as the power radiated as synchrotron radiation.

4.5.2 The beam power

The power transmitted to the beam (P_b) is simply given by

$$P_b = I_b . V_r \quad \text{(watts)} \tag{4.16}$$

where I_b is the beam current (amps) and V_r the total energy loss per turn (volts) as defined in section 4.3.1 above.

4.5.3 The cavity dissipation

Section 4.3.2 above explained how to decide upon the peak accelerating voltage (V_p), by consideration of the quantum lifetime, the Touschek lifetime and the required natural bunch length. Then the cavity dissipation (P_c) is given by

$$P_c = V_p^2 / 2Z \quad \text{(watts)} \tag{4.17}$$

where Z (ohms) is the transit time corrected shunt impedance of the cavity. Both V_p and Z refer to the total voltage and total number of cavities around the orbit. The cavity shunt impedance depends on the number, shape and material of the cavities and is discussed in detail in section 4.6 below.

4.5.4 Reflected power

When r.f. power is fed to a load, a proportion will be reflected back if the impedance of the load, as seen through the input coupling device, does not match that of the transmission line. In the case of the r.f. system of a storage ring, the load impedance will change from time to time for two reasons and allowance must be made for the consequent reflected power. The first reason is that that part of the load due to the "beam loading" (the beam power in 4.5.1 above) varies with the beam current and so the equivalent impedance V_r/I_b varies (V_r must be kept constant). The second is that the cavity must be kept detuned with respect to the generator if the beam - cavity - generator system is to be stable.

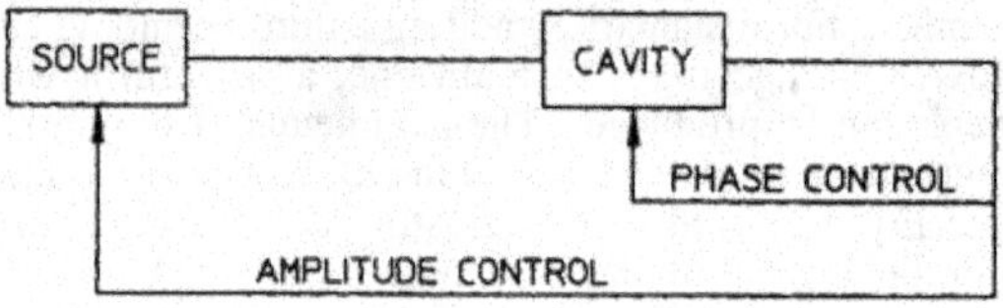

Fig. 4.3 Simple r.f. system - block diagram.

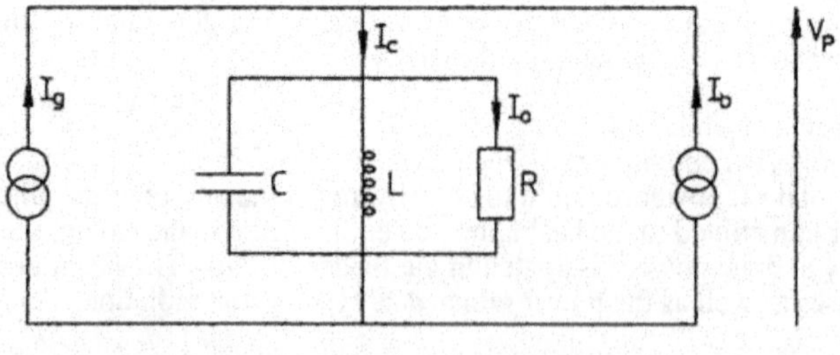

Fig. 4.4 Simple r.f. system - equivalent circuit.

To understand the reason for this last statement, consider a simple r.f. system as in Fig. 4.3 with equivalent circuit as in Fig.4.4. Here $\mathbf{I_g}$ is the (vector) generator current, **Ib** the beam current, and the R/L/C combination represents the cavity, across which is established the required peak voltage $\mathbf{V_p}$. $\mathbf{I_o}$ is the current supplying the cavity losses. The corresponding currents and voltages are shown in the vector diagram, fig.4.5. **Bold** type represents vector quantities, *italic* the real parts. To control V_p and $\mathbf{I_c}$ whilst $\mathbf{I_b}$ varies, it is usual to have a tuning control loop around the cavity and amplitude feedback around the system and this is shown in Fig. 4.3.

Referring back to Fig. 4.2 and the mechanism of phase stability (section 4.2), and relating it to Fig. 4.4, one sees that phase oscillations are fluctuations in the phase of $\mathbf{I_b}$ and that these will cause $\mathbf{I_c}$ and hence $\mathbf{V_p}$ to vary also. If the cavity is exactly on tune, $\mathbf{V_p}$ will vary in phase with $\mathbf{I_b}$ and tend to negate the phase stability (depending on the magnitudes of I_b and R). If the cavity is not exactly on tune, the phase stability will be augmented or further diminished according to the direction of detuning. Cavities must therefore be operated off tune in the correct direction. This is the concept of Robinson stability[5] and analysis of the vector diagram, Fig.4.5, reveals the following numerical criteria:

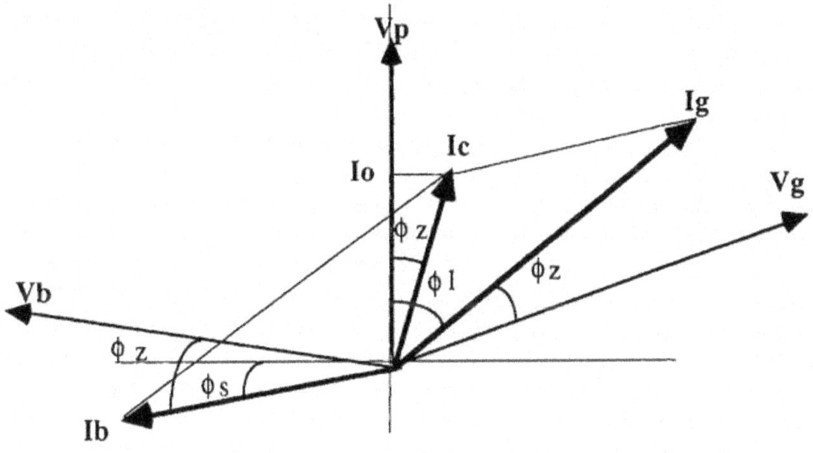

Fig. 4.5 Simple r.f. system - vector diagram.

if the beam loading ratio Y is defined as the ratio of the real part of the beam current to that of the cavity current

$$Y = \frac{I_b}{I_c} \tag{4.18}$$

then for stability

$$Y \frac{\sin 2\phi_z}{\sin \phi_s} < 2 \tag{4.19}$$

and

$$\tan\phi_s < 0 \tag{4.20}$$

where ϕ_s is the beam phase angle as in Fig. 4.2, and ϕ_z is the detuning angle - the phase difference between the cavity voltage and current.

Equation 4.20 shows that to be stable the cavity must be detuned to a lower frequency than the generator frequency; and equation 4.19 shows that the required detuning is proportional to the beam current.

For the generator to be matched at all times, it would need to see a constant impedance $\mathbf{Z_g}$ given by

$$Z_g = \frac{V_g}{I_g} = \frac{V_p}{I_g} \tag{4.21}$$

However, $\mathbf{I_g} = \mathbf{I_c} - \mathbf{I_b}$ and hence varies with $\mathbf{V_p}$, ϕ_z and $\mathbf{I_b}$. $\mathbf{V_p}$ and ϕ_z may be varied during ramping (if injection is not at full energy) and $\mathbf{I_b}$ will vary during injection and with the beam decay (as may ϕ_z). So there will inevitably be reflected power under certain conditions.

The generator can be matched to a given load by adjusting the coupling (see section 4.6.7). The cavity coupling coefficient β is defined as the ratio of the cavity input impedance and the characteristic impedance of the feeder system. However, with beam loading the required coupling factor becomes,

$$\beta = 1 + \frac{P_b}{P_c} \tag{4.22}$$

(where P_b and P_c are the powers transferred to the beam and cavity respectively)[6] and is set according to equation 4.22 so as to give zero reflected power when the maximum beam power is required and the cavity detuning is at the design value for that condition. This means that reflected power will only occur when the beam loading is less than the design maximum and/or the cavities are detuned differently. Calculations must be carried out to check whether the reflected power under such conditions is ever likely to result in a total power demand greater than that when maximum beam loading is present, and if so it must be allowed for in the specification of the power source.

4.5.5 Feeder losses

A transmission line, usually including an isolation device, and also including power splitters if more than one cavity is being fed from one amplifier, is required to transmit the r.f. power from the source to the cavity, and this system will not be loss-free. This will be further discussed in section 4.7 below, but typically the losses in the feeder system should not exceed 10% of the transmitted power. These losses will be known from the specifications of the components when the feeder system has been designed, and can be added to the other power requirements.

4.6 Cavities

4.6.1 Introduction

Electrically, the accelerating gap constitutes a capacitance, and to energise this economically it must be resonated with an inductance. The practical way to do this in a storage ring is to make the gap part of a resonant cavity. A cavity in this context is a hollow metallic vessel, with a hole at each end for the electron beam, an accelerating gap across which the electrons pass, an arrangement for feeding in r.f. power, and a means of tuning.

The geometry must be such that a high electric field is produced across the gap in the direction of travel of the beam, with minimum power dissipation in the walls. This is achieved by using either the E_{010} mode, in a high frequency cavity, or the TEM mode in a low frequency cavity (see sections 4.6.3, 4.6.4 below).

The cavity format has many advantages. There are no external fields, and it can be inserted directly between metallic flanges in a straight section of the storage ring; it has low resistive losses; it can be vacuum-tight without the need for any external vacuum chamber; and apart from a ceramic window for the r.f. power input, it can be wholly metallic and extremely clean.

Cavities can be of various forms. The shape of the cavity (or cavities) will depend on the frequency chosen, on the space available and on an assessment of such matters as power requirements, beam loading, higher order mode effects, etc. To avoid confusion, the term "cavity" will be used here to describe one physical unit which may consist of N cells in line electromagnetically coupled together internally. There may also be n cavities around the ring.

The aim when designing cavities is to make it possible to establish the required total peak gap voltage V_p (around the ring) with minimum total power input P_{rf}, taking into account a number of constraints, of which the principal ones are initial capital cost and space. To do this it is necessary to maximise the transit time corrected shunt impedance per cell Z_c.

If the voltage per cell is v_c and the power dissipation per cell is p_c, then

$$v_c = V_p / Nn \;;\; p_c = V_p^2 / 2N^2n^2Z_c \tag{4.23}$$

the power dissipation per cavity

$$Np_c = V_p^2 / 2Nn^2Z_c \tag{4.24}$$

and the total rf power required,

$$P_{rf} = P_b + V_p^2 / 2NnZ_c \tag{4.25}$$

Power costs reduce as Z_c, n and N increase, but, as will be explained below, for a given beam aperture there is a clear limit to Z_c, and increasing Z_c above about 80% of this maximum may increase cavity construction costs disproportionately due to manufacturing complexity. Increasing n or N increases initial costs and requires straight section space. Space is usually quantised in units of straight sections and the storage ring designer will try to restrict the r.f. requirement to just one, or at most two straights, to leave maximum space for insertion devices. These considerations, taken together with the voltage requirement (see section 4.3 above) and the choice of frequency (see section 4.4 above) quickly dictate the number, length and approximate power ratings of the cavities; the complexities come in the detailed optimisation and subsequent detailed design.

The remainder of this section will discuss the various different types of cavity and the tools for designing them, together with the secondary considerations such as higher-order modes and practical details including cooling, power input coupling and tuning devices.

4.6.2 *Superconducting cavities*

The main requirement of a cavity is to have minimum resistive wall losses and a superconducting cavity is clearly optimum in this respect. Moreover because the wall losses are very small (though not zero for r.f. fields) the shape can be chosen to optimise other features, in particular to minimise higher order modes (HOMs). The field at the gap can be as high as allowed by breakdown and field emission effects and not limited by power or cooling implications; this means that a higher energy gain per metre can be achieved, leaving more space for insertion devices. Nevertheless, superconducting cavities have never yet been chosen for a synchrotron radiation source (though they have recently been proposed for the planned French machine SOLEIL).[7]

This is for two reasons - cost and complexity, and beam loading. A synchrotron radiation source has a relatively high beam current (hundreds of mA) and the power transferred to the beam is considerable - this is the power emitted as synchrotron radiation and one aim of the overall machine design is that it should be large. Because of this, one must have a substantial r.f. power source whether or not the cavities are superconducting.

Moreover, if most of the total power input to the cavity is the beam power, as would be the case with superconducting cavities, any fluctuation in beam current, either in amplitude (e.g. during filling or because of a partial beam loss) or in phase (e.g. due to a longitudinal oscillation) would require very precise and rapid reaction from the r.f. power amplifier to prevent the gap voltage changing in proportion, with consequent risk of instability due effectively to positive feedback. Put another way, Robinson stability (see section 4.5.4) would require a large detuning, and the phase and amplitude feedback loops (as in Fig. 4.3) would require to have a high bandwidth and loop gain. As was shown in the case of Robinson stability, the smaller the beam power in comparison to the cavity dissipation, the more stable is the r.f. amplifier/cavity/beam system.

Superconducting cavities, and also their cryogenic systems, are complicated and expensive to construct and to operate. Expertise is required which is not always readily available. For these reasons superconducting cavities have not so far been adopted for a synchrotron radiation source, and will not be discussed further here. However, superconducting cavities are well-established in the field of accelerators and storage rings for nuclear and high energy particle physics, and should such a cavity be being considered, for instance because of its superior HOM performance, the extensive literature relating to such machines should be consulted,[8] though remembering that different conditions of beam loading, and different criteria for optimisation of the design may well apply.

4.6.3 Low frequency cavities

Most of the storage rings constructed have employed high frequency rf systems, however, a significant number are <250 MHz. In this case the cavities will be constructed from a coaxial line short circuited at one end and capacitvely loaded at the other. Figure 4.6 is a schematic diagram of such a cavity. If the loading capacitance tends to zero, then the geometric length, l, of the cavity will approach $\lambda/4$, and it is consequently known as a $\lambda/4$ cavity.

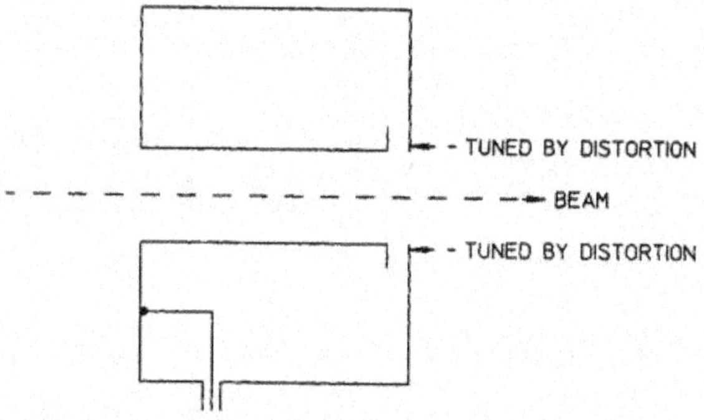

Fig. 4.6 Low frequency single-gap cavity.

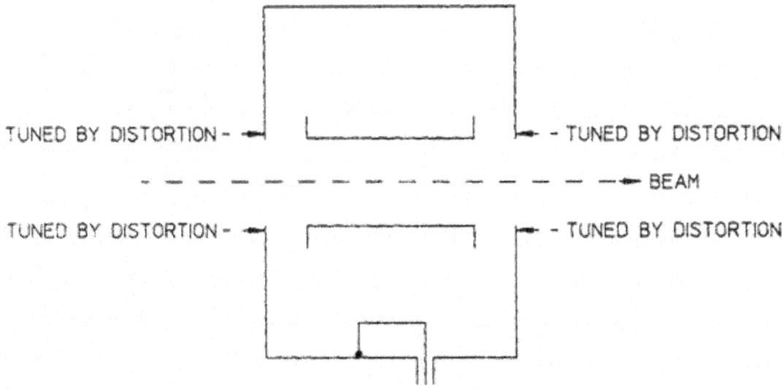

Fig. 4.7 Low frequency double-gap cavity.

As space in the straight sections is usually limited, the amount of capacitive loading is usually a compromise, small to give a high value of shunt impedance, but high to have a small physical size.

It is possible to have 2 gaps in one cavity, as in figure 4.7, and in this case the cavity is known as a $\lambda/2$ cavity.

Both the VUV and the X-Ray storage rings at NSLS, Brookhaven[9] employ $\lambda/4$ coaxial cavities. The compact synchrotron radiation source Helios-2[10] is being constructed using a high capacitvely loaded $\lambda/4$ cavity.

Physical size, (not only allowable length, but beam pipe radius, input coupling port, tuner port and vacuum pump port), construction material, complexity of manufacture, and cost of manufacture are important considerations.

Low frequency $\lambda/4$ cavities can have a Zc of > 1.5 MΩ, but generally 1 MΩ or less is acceptable, as in Helios-2. For high frequency cavities, such as Helios-1, SRS and KEK Photon Factory cavities Zc in the order of 8 MΩ are achieved.

4.6.4 High frequency cavities

By this title is meant cavities resonant at 250 MHz or more, where a basically pill-box shape resonant in the E_{010} mode is of similar cross-sectional dimensions to other storage ring components and where cavities can be cascaded into multi-cell units within one straight section - i.e. the necessary half wavelength spacing is, say, less than 1m. The most commonly used frequency is around 500 MHz, for which a good choice of power source is available and the cavity dimensions match well to the space available and to the beam pipe size in typical machines. For the highest energy rings, 350 MHz has been chosen to take advantage of the very high power klystrons available around that frequency.

The simplest form of cavity is the single cylindrical pill-box (Fig 4.8), made of copper to maximise its shunt impedance. It is resonant in the E_{010} mode, which means that the electric field is everywhere parallel to the beam direction and its amplitude is a maximum on the beam axis. It will require an input power coupling device (see section 4.6.6 below), a tuner (see section 4.6.7), a hole at each end for the beam, and cooling tubes soldered or brazed on the outside. Such a cavity may well be used for a low energy, low current machine. However, its transit time factor (TTF) is bad, its shunt impedance is not the maximum possible and becomes worse if the cavity is shortened to improve the TTF, and it is not particularly free of HOMs.

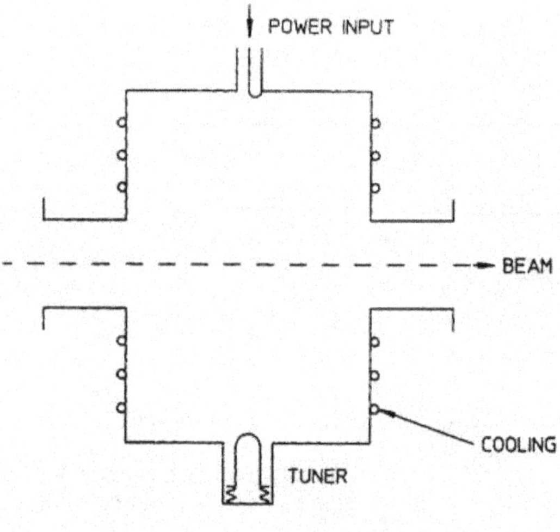

Fig. 4.8 High frequency cylindrical (E_{010} mode) pill-box cavity.

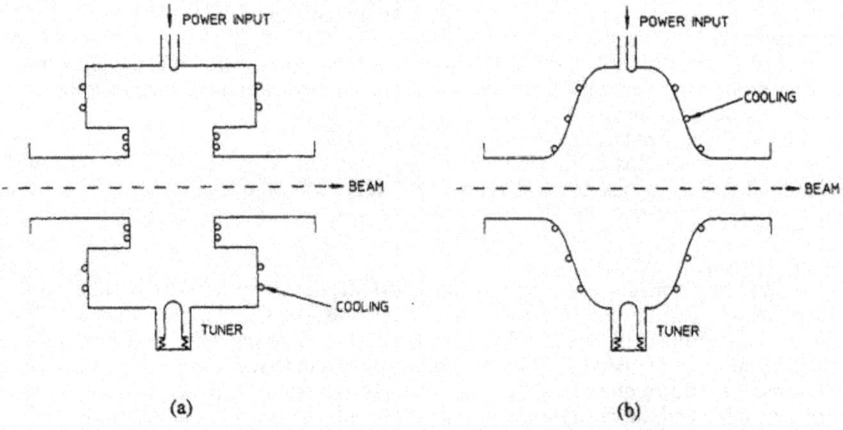

Figs. 4.9a & b Re-entrant E_{010} cavity (i.e. with "nose cones"). Higher shunt impedance E_{010} cavity.

Two improvements can be made, singly or together. To improve the TTF the cavity can be made re-entrant by adding nose-cones (Fig. 4.9a). The improvement in TTF outweighs additional losses which are due to high current densities in the nose-cones, but removing the heat generated by these current densities may not be easy. To improve the shunt impedance, the longitudinal cross-section of the cavity can be made much more rounded (Fig 4.9b). A simple way of looking at this change is that the ratio of wall area to cavity volume has thus been minimised, but full computer field calculations (see section 4.6.5 below) are needed to optimise the design in this way. This shape has the added

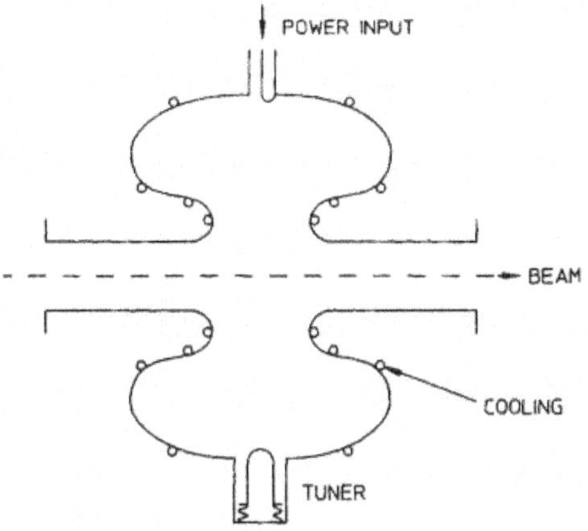

(c)

Fig. 4.9c E_{010} cavity with nose-cones and optimised shunt impedance.

advantage that it tends to have fewer HOMs, but it is more expensive to manufacture. The two improvements (nose-cones and rounded corners) can be combined (Fig 4.9c) to give an overall optimised TTF-corrected shunt impedance. The only reason for not always using this shape is that the increased cost of the cavities, due to their complexity (including the cooling tubes or channels) may not be justified by the saving in the power required.

For a low energy ring, one single-cell cavity may be adequate, but for rings needing a higher total gap voltage more cells may be necessary, and these can be provided by single cell cavities separately energised (from one or several amplifiers - see section 4.7 below) or by multicell cavities. A multicell cavity consists of several cavities in line, electromagnetically coupled by holes in the common walls so that only one power feed is needed (Fig 4.10). In its simplest pill-box form, the central hole provided for the beam may provide adequate inter-cell coupling. This arrangement tends to be cheaper than using separate cavities. However, it introduces further complications. Such an assembly of n cells has n times as many resonant modes as a single cell. Normally the π-mode is used, in which the fields in successive cells are 180° out of phase and the distance between gap centres is a half-wavelength, so that the particle always sees the maximum accelerating field. The coupling is adequate if, taking into account the cavity Q-values and the effects of the dimensional tolerances, the frequency adjustments that might be necessary, the effects of detuning and so on, there is no simultaneous excitation of a neighbouring mode (i.e. the Q-curves do not overlap). If nose cones are used to increase the shunt impedance, this will greatly reduce the central coupling and slots have to be added. There is then a further compromise between the dimensions of the nose-cones and of the slots. Provision of cooling channels becomes further complicated.

Another disadvantage of the multicell design is that there are also n times as many HOMs, and so a much greater likelihood that some will be excited. However, this is a complex issue as the impedances of the HOMs will be different, and in general tend to be lower, than for single-cell cavities.

Examples of single-cell cavities with nose cones may be found at the SRS,[11] and the Photon Factory,[12] a single-cell cavity without nose cones is used at Elettra,[13] and multicell cavities are used at ESRF.[14]

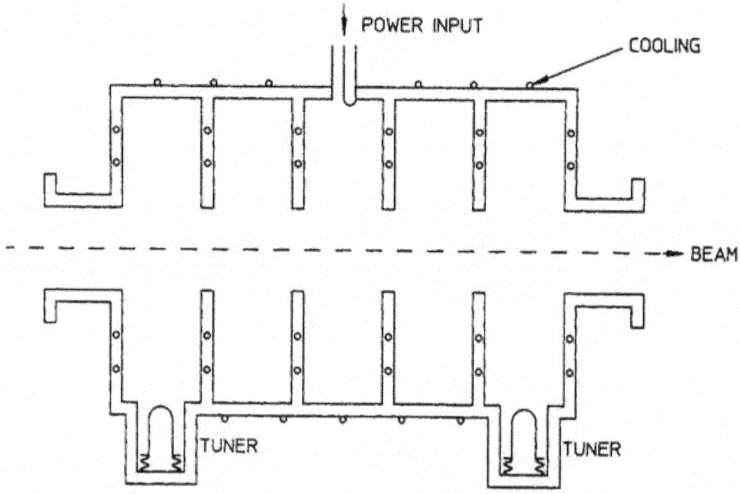

Fig. 4.10 High frequency multi-cell cavity.

Finally, it may be necessary, for a high current machine, to try to reduce or remove the HOMs. Methods for doing this are discussed in section 4.6.8 below.

Fig. 4.11 shows a single-cell 353 MHz cavity for the APS.

Typical values of shunt impedance for single-cell cavities with nose cones are 7 to 8MΩ depending on the size of the beam aperture. Without nose cones this falls 2.5 to 3.5MΩ. A 5-cell cavity such as at ESRF has a shunt impedance of about 28MΩ. [N.B. some designs use a version of equation 4.11 which omits the factor of two, and therefore quote twice the above values for shunt impedance.]

4.6.5 Computer codes for cavity design

Cavities must be designed to have the correct resonant frequency to a high degree of accuracy and to have the maximum practical transit time corrected shunt impedance. These quantities depend on the precise distribution of the electromagnetic fields and the wall currents. It is also desirable to know the frequencies and shunt impedances of the higher order modes.

For most practical cavities the geometry is too complicated for analytical solution of the electromagnetic field distribution, and although model cavities can be constructed and exhaustively measured, it is a costly and time consuming exercise. Also some of the higher order modes may be missed.

Computer codes have been developed and improved upon since the early 1960's. For example, the now obsolete code LALA was used to design the SRS cavities. The codes naturally split into two main categories:-

i) two dimensional codes for cavities with axisymmetric or cylindrical symmetry,

ii) fully three dimensional codes.

Examples of two dimensional codes are SUPERFISH,[15] used at the KEK Photon Factory to design the accelerating cavities. OSCAR2D[16] has been used at Trieste for ELETTRA, and URMEL-T[17] has been used successfully at many Laboratories. A new code developed at Novosibirsk SUPERLANS is also available.

The main three dimensional codes to be used are MAFIA[18] and PRIAM.[19]

Fig. 4.11 353 MHz cavity for APS.

The Los Alamos group has compiled a compendium[20] of the many computer codes now available for modelling the electromagnetic fields in cavities.

Despite the obvious limitations, (no cavities are truly axisymetric when tuners and input couplers are included), designers will use the simpler 2D codes so far as possible because of their relative simplicity and shorter computing time. 3D codes tend to be used as a final check and to explore particular specific effects, except where the designer is particularlyexpert in their use and has ample computing power available.

4.6.6 *Input coupling*

Generally the input coupler is also used as the vacuum device separating the ultra high vacuum of the cavity from the air filled feeder system.

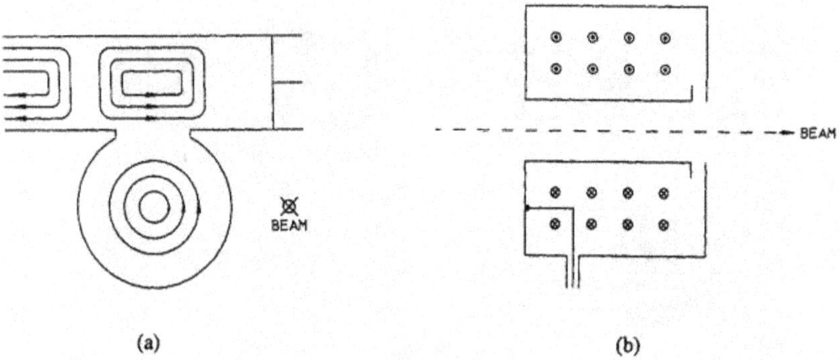

Figs.4.12a & b Aperture coupling from waveguide to cavity. Loop coupling into a cavity.

Only two methods are employed to couple power from the feeder system into the cavity:-

i) aperture coupling, which can be used at high frequencies when waveguide is used as the feeder. A window is formed in the common wall between the waveguide and the cavity to give magnetic coupling, see figure 4.11. Only the SRS, ALS and Helios-1 employ this method of coupling.

ii) loop coupling, again coupling is achieved using the loop to couple the correct magnetic mode, as in figure 4.12. This method can be used at all frequencies.

The windows, whether aperture or loop coupling, also usually separate the ultra high vacuum of the cavity from the atmospheric pressure (or greater) of the feeder system. The aperture window will be a disc of material, typically alumina, but sometimes beryllia, mounted in a frame which will attach the window to the cavity and provide some cooling. The loop window is more complicated, needing the centre electrode of the loop to pass through the ceramic, usually alumina, making a good vacuum seal. Both the mounting frame and the loop are usually cooled.

There can be breakdown problems[21] with the windows as the fields tend to be in the multipactor regions, and several coating techniques to overcome this problem have been tried. Most windows are now coated with titanium nitride,[22] and the SRS has successfully used copper black.[23]

4.6.7 Cavity tuning

Because of its high Q and the need to avoid reflected power or phase errors due to mistuning, the cavity's resonant frequency must be matched to the required frequency to within about 1 part in 10^5.

It is impossible, even with the latest computer codes and extensive model measurements, to guarantee this in a manufactured cavity, and temperature variations are likely in any case to cause errors of this magnitude. Changes in beam loading and amplifier coupling can change the resonant frequency or the offset required for Robinson stability (or for best capture of electrons during injection). For all these reasons it is necessary to tune the cavity continuously.

For high frequency cavities it is usual to tune the cavity by movable pistons or paddles, as shown in figure 4.9 earlier. These pistons vary the magnetic or electrical energy, and thus the frequency of the cavity. These devices are lossy, and need to be adequately cooled, but allow a tuning range of up to 3%. Plungers are used at the SRS, ALS, KEK Photon Factory and other places.

Axial squeezing is also used, primarily on low frequency cavities, but also on superconducting cavities. The deformation must be small to avoid fatigue fracture and damage to vacuum seals. This method is employed at Siberia-1 and Siberia-2 at the Kurchatov Institute in Moscow, and was initially used at the NSLS.

The NSLS have also used the temperature of the cavity coolant to tune the cavity.

More recently at the NSLS electronic tuning has been tried. A ferrite loaded inductance was coupled to a loop inside the cavity, the resonant frequency of the cavity was varied by varying the ferrite bias current.

It is important to remember that tuning the fundamental mode of the cavity in any of the above ways will also tune any HOM's, although usually with less effect than on the fundamental.

4.6.8 Higher order mode damping

Higher order modes are only a problem if the beam can excite them, therefore it is important that all the modes in a cavity are measured, and then compared with the probable beam spectra. At the NSLS the many modes in the cavities were damped by coupling the power in each mode out of the cavity with a tuned electric probe. However, this then made tuning the fundamental difficult as squeezing the cavity changed the frequency of the modes also.

At the SRS, advantage was taken of using aperture coupling to use waveguide filters and a special cavity matching unit to filter all frequencies from the cavity apart from the fundamental. Attention is also given to the cavity temperature to make sure all possible modes are not tuned to dangerous frequencies by the tuneing plunger.

Developments of this technique are taking place at Trieste and Berkeley, where orthogonally mounted waveguides with aperture coupling to the cavity are employed to damp all modes.

So called modeless cavities are being designed, these are basically spherical cavities with low shunt impedance. These designs are generally for superconducting cavities.

4.6.9 Power input limitations

The cavity power can be limited by either voltage breakdown caused by incorrect design or poor vacuum conditions, or by insufficient cooling. The same is true of the input coupling device. Currently loop couplers of up to 250 kW power throughput are available commercially. Aperture windows are available up to 100 kW.

At the ESRF the five cell cavity is fed by two input loops in parallel to handle up to 350 kW per cavity including beam loading.

4.6.10 Cavity construction techniques

There are many ways of constructing cavities, all have to be compatible with ultra high vacuum techniques. The paper by I Wilson, Cavity construction techniques[24] in the proceedings of the CERN Accelerator School, RF Engineering for Particle Accelerators is a full and clear explanation.

4.7 Power supplies

4.7.1 Introduction

The most commonly used power devices are power tetrodes for low frequency operation, i.e. below about 300 MHz, and klystrons for high frequency applications, i.e. above 300 MHz. Recently developed devices such as inductive output tubes, IOT's, may find favour in the future as it is predicted that they will operate from 100 MHz up to 3 GHz, with output powers of up to 1 MW.

4.7.2 Low frequency: Tetrodes

Figure 4.13 is a schematic diagram of a tetrode, and figure 4.14 shows typical characteristic curves.

The anode and screen grid are at a positive potential and the control grid a negative potential with respect to the cathode. The number of electrons emitted from the cathode is controlled by the control grid, while the screen grid, which is maintained at r.f ground, prevents capacitive feedback from anode to control grid.

The r.f. tetrode has a coaxial construction, with the cathode on the inside and anode on the outside, allowing effective cooling of the anode. Output power is limited by the maximum current density available from the cathode and maximum power density that can be dissipated from the anode. The dimensions and spacing of the electrodes are limited to avoid:

(i) variation in signal levels - the length of the anode must be much less than the free-space wavelength of the signal to be amplified.

(ii) higher order modes between the anode and the screen grid - the perimeter of the anode must be much less than the free-space wavelength

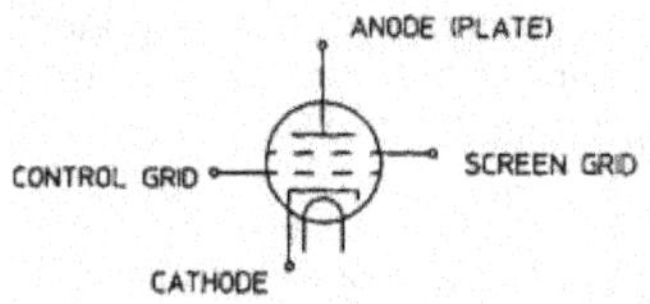

Fig. 4.13

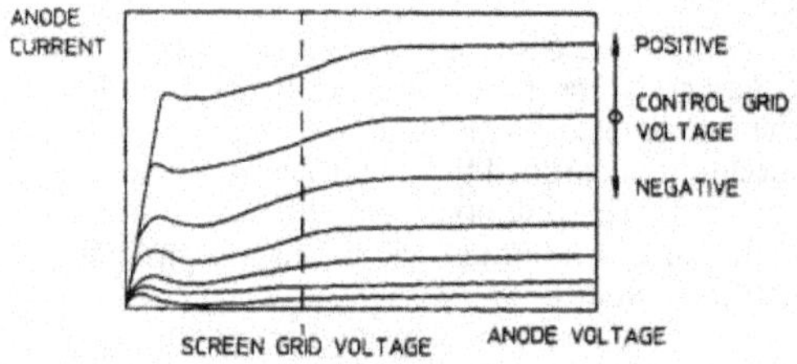

Fig. 4.14

Figs. 4.13 & 4.14 Schematic diagram of a high frequency tetrode. Typical characteristic curves of a tetrode.

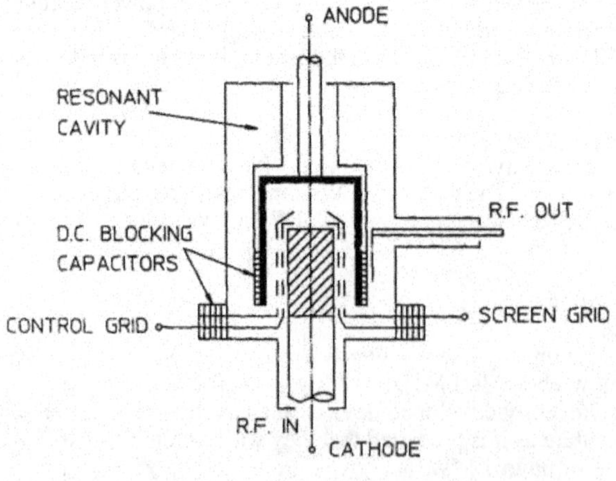

Fig. 4.15 Circuit arrangement for a high frequency tetrode amplifier.

(iii) transit time problems - spacing between the anode and cathode must be much less than the r.f period. Increasing the anode potential to overcome this problem may result in 'flashover' between the electrodes.

The coaxial construction of the tetrode allows effective construction of input and output circuits, figure 4.15 shows a typical arrangement. The most commonly used arrangement is grounded control grid. As both grids are at r.f ground, there is good isolation between the input and output circuits. However as the anode current flows in the input circuit both the input impedance and the gain are lower than if a grounded cathode

The efficiency and harmonic content of the output signal of the power amplifier depend on the class of operation employed. In class A operation, the tetrode conducts throughout the whole r.f cycle, the signal harmonic content is low, but the efficiency is only 50%. On the other hand when the tetrode only conducts during the positive half of the r.f cycle, class B operation, the signal harmonic content is higher, but the efficiency is almost 80%. Most operation is a mixture of these two classes, giving a compromise between low harmonic content and high efficiency. If high efficiency is important then the tube can be operated in class C, where the tetrode operates for less than half of the r.f cycle. Efficiencies of more than 90% can be achieved, but not only is the harmonic content high, the amplifier gain is reduced.

The maximum power output from a single tetrode amplifier is between 125 kW and 150 kW, however many ingenious methods of paralleling several tetrodes in one amplifier circuit and paralleling several amplifiers can be employed.

4.7.3 High Frequency: klystrons

Figure 4.16 is a schematic diagram of a two cavity klystron. Typically klystrons used as amplifiers in storage rings will contain four or more cavities, to achieve the high gain and high efficiencies necessary. A klystron has four main parts;

(i) an electron gun - to produce the high current electron beam, (not shown in the figure)

(ii) interaction region - containing the r.f cavities,

(iii) an axial magnetic field, usually supplied by an electro-magnet,for klystrons of these frequencies,- to confine the electron beam, (not shown in the figure) and

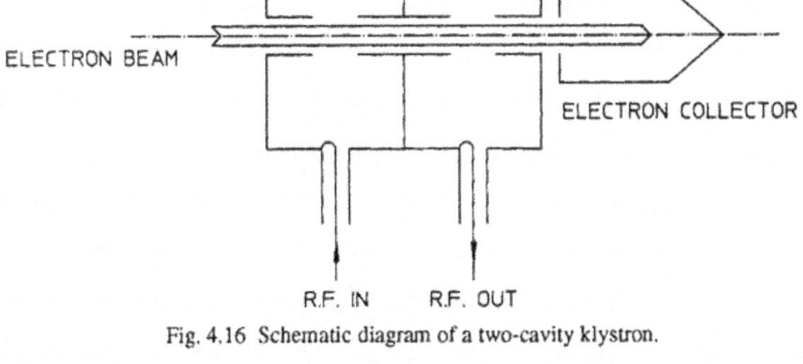

Fig. 4.16 Schematic diagram of a two-cavity klystron.

(iv) the collector - where the 'spent' electrons are collected.

The cavities may be integral with the beam tube (and under vacuum) or external to the beam tube allowing easier tuning.

The klystron extracts power from a bunched electron beam, but the bunches are formed by velocity modulation, not by switching the electrons on or off or controlling the number of electrons leaving the cathode by a control grid,

The electron beam passes through a cavity resonator excited in the TM_{010} mode, giving an axial electric field across the gap in the drift tube. According to its phase, the r.f field in the cavity accelerates or retards the electrons as they cross it. On leaving the cavity the electron beam is velocity modulated but not current modulated.

Electrons crossing the gap when the field is zero, pass without any change in their velocity (appearing as straight lines in figure 4.17). Other electrons are accelerated or retarded as shown, and they form electron bunches until space-charge limits the process.

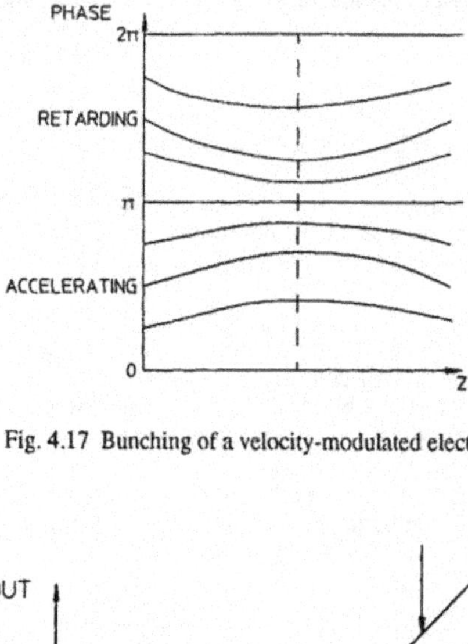

Fig. 4.17 Bunching of a velocity-modulated electron beam.

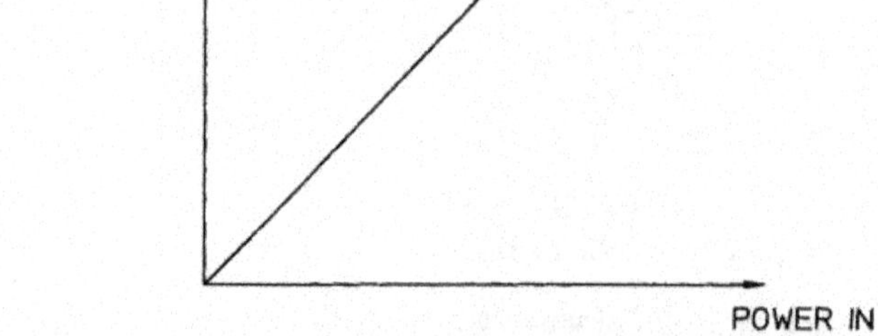

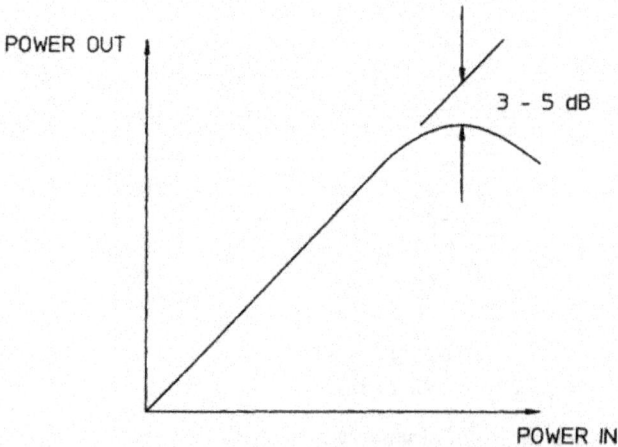

Fig. 4.18 Typical transfer characteristics of a klystron.

At this point the r.f current modulation is a maximum and velocity modulation is zero. If a cavity is placed at this point, the electron beam would induce an r.f current in it. As electrons move down the tube they disperse, then form again.

If the cavity at the plane of maximum bunching is tuned to be resonant at the signal frequency, the induced field is in phase with the existing beam current and so the cavity field is maximum retarding as the centre of the bunch crosses the gap, thus increasing the bunching of the beam. Adding more tuned cavities increases the gain. The last cavity is used to extract the r.f. power.

A 10% efficiency increase can be achieved by tuning the penultimate cavity to a higher frequency than the signal frequency. The induced current now sees an inductive impedance and the resultant phase shift brings more electrons into the bunch.

A further 10% increase in efficiency can be achieved by the incorporation of a second harmonic cavity, which shortens the bunches, increasing the peak component of the current.

As stated earlier, tightness of the bunches formed by velocity modulation is limited by space-charge effects. It is easier to obtain high efficiency with an electron beam which has a low current density and a high velocity, i.e. a low perveance.

The bunching length is independent of input signal, until the drive level is very high, when it causes a reduction in the bunching length. Driving the klystron harder causes less bunching. The output characteristics are shown in figure 4.18. Operation is at or near saturation to achieve the highest possible efficiency. Typical efficiencies for the klystrons presently used are between 60% -70%, at power levels up to 1.3 MW.

4.7.4 Wide band: inductive output tubes

Figure 4.19 is a schematic diagram of an inductive output tube. The IOT has four main parts:

(i) an electron gun - to produce the high current electron beam,
(ii) interaction region - containing the r.f cavity,
(iii) an axial magnetic field, usually supplied by an electro-magnet - to confine the electron beam, and
(iv) the collector - where the 'spent' electrons are collected.

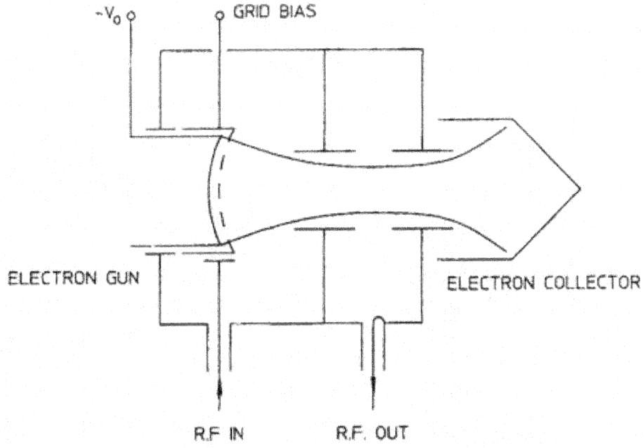

Fig. 4.19 Schematic diagram of an inductive output tube.

The electron beam is formed by an electron gun, which is biased so that current flows only during the positive half of the input r.f. cycle. The bunched electron beam induces an r.f. current in the cavity. The spent electrons are then collected by the collector. The input operation is similar to a tetrode, and the output operation is similar to a klystron. One manufacturer calls the device a 'klystrode'.

Tubes in the frequency range of 470 MHz to 860 MHz, with output powers of 60 kW and efficiencies of greater than 50% are presently available. However, it is possible that IOT's will be useful from 100 MHz to 3 GHz with output powers of up to 1 MW.

4.7.5 Choice of number of power sources

Arguments can be made for multiple or single power sources. For instance at the Pohang Light Source, PLS, at 2 GeV the radiation loss per turn is 300 keV and the peak accelerating voltage is 1.8 MV, and stored currents of 250 mA are intended to be achieved. Four 60 kW klystrons, each feeding a separate single cell cavity are to be used. The arguments given for this approach are:
(i) no crosstalk between cavities,
(ii) the phases of the accelerating voltages are controlled by low level phase shifters in the drive to the klystrons,
(iii) many TV transmitters available that can be modified without difficulty
(iv) better redundancy
(v) coaxial feeder system (6 1/8") can be used instead of waveguide.

The 2GeV SRS at Daresbury has a radiation loss per turn of 280 keV, and the required peak r.f. voltage is 2.1 MV. Stored currents of 30 mA are achieved. A single 300 kW klystron was chosen to feed four single cell cavities. The arguments for this choice were:
(i) a fully developed klystron at this level was available
(ii) a waveguide feeder system could be used which was simpler and more flexible than a coaxial system
(iii) simpler control of cavity voltage
(iv) the relative phase of the accelerating voltage is fixed by the waveguide system
(v) a single high power klystron system was significantly cheaper than the multiple low power klystron option.

Similar arguments can be made for the use of single tetrode amplifiers or a multi-tetrode amplifier.

4.7.6 Transmission

Many of the devices used in the power amplifiers need to be 'isolated' from the load that they feed. For example a typical klystron must feed power into a load, with a voltage standing wave ratio (VSWR) of 1.2:1 and a beam loaded cavity cannot be guaranteed to provide such a load under all conditions (see section 4.5.4 above).

For this reason non reciprocal devices are used between the power source and the cavity, these use the interaction of the ferrimagnetic resonance of ferrites with the circularly polarised r.f. magnetic fields. An isolator allows waves to travel in one direction with little interaction, while those travelling in the opposite direction are absorbed . As the ferrite has to dissipate all this power, the isolator is now rarely used. The three port isolator overcomes the dissipation problem by using the ferrimagnetic resonance to divert the r.f. power into a matched load. In figure 4.20 power fed into port 1 always comes out of port 2, power fed back into port 2 will be directed to the matched load on port 3.

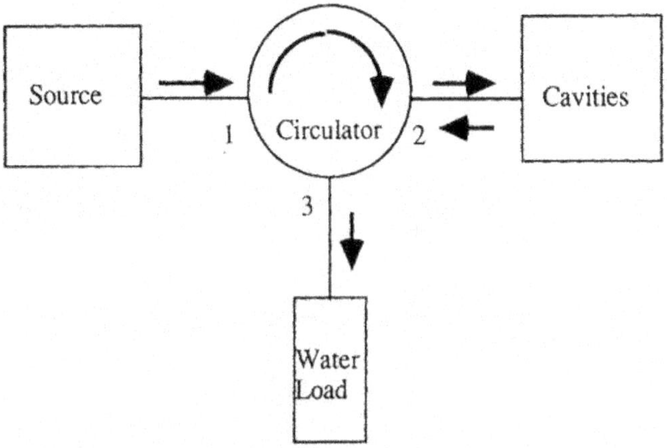

Fig. 4.20 Three port isolator, or circulator.

There are only two practical choices of transmission line, either coaxial line or rectangular waveguide. The coaxial line has no cut off frequency in its normal TEM mode and can be used down to dc. Figure 4.21a shows the power handling of various sizes of coaxial line has a function of frequency, while b shows the attenuation. So coaxial line can be used for high frequency systems where the power handling requirement is low.

Waveguide has a lower frequency cut off which is dependent on the waveguide dimensions. The lower the frequency the larger the waveguide. Figure 4.22a and b show the power handling and attenuation of various sizes of waveguide as a function of frequency. It is obvious that waveguide becomes too cumbersome at lower frequencies.

4.7.7 Voltage control

The voltage requirement is discussed in section 4.3. For a storage ring that has full energy injection the required voltage is set by the control system, and amplitude feedback is employed to keep the required voltage constant whilst allowing the cavity power to vary as the beam loading varies. The feedback also allows the required voltage to stay constant even if the cavity tuning is changed. With multiple cavity systems, if one cavity was completely detuned, the amplifier power would automatically increase to keep the integrated voltage constant.

For a storage ring not employing full energy injection the required voltage during injection, energy ramp and at full energy are programmed by the control system, but the amplitude feedback ensures that the voltage tracks the set values accurately.

4.7.8 Phase control

In a storage ring with multiple cavities, the phase of the r.f. input to each cavity needs to be in synchroism with the phase of the beam. For example, the orbit length between cavities 1 and 2 in the SRS is 19 3/4 λ, and the phase difference between the r.f. inputs is -90°. This phase difference is fixed by the waveguide components, but high power waveguide variable phase shifters have been installed. The phase of the storage ring r.f. is set by a low power phase shifter on the input to the power amplifier, which has to be matched to the phase of the injection accelerator r.f.

The output phase of the power amplifier can change with varying operating conditions. For instance for a 500 MHz klystron with a normal beam voltage of 100 kV the phase change is almost - 6°/kV and approximately 0.5°/°C. Depending on the specification of the sevices to the amplifier, a phase feedback system around the amplifier may be employed.

Beam instabilities may occur in the longitudinal plane which appear as phase oscillations in the cavity. This can be overcome by employing phase feedback from the cavities to the input of the amplifier. Sometimes a beam pickup device will be used instead of the cavities for the feedback signal.

4.7.9 Frequency control

The choice of frequency is discussed in section 4.4. For efficient injection it is usual to have the frequency of the injection accelerator and the storage ring harmonically related. The exact operating frequency is chosen so that the beam horizontal closed orbit is

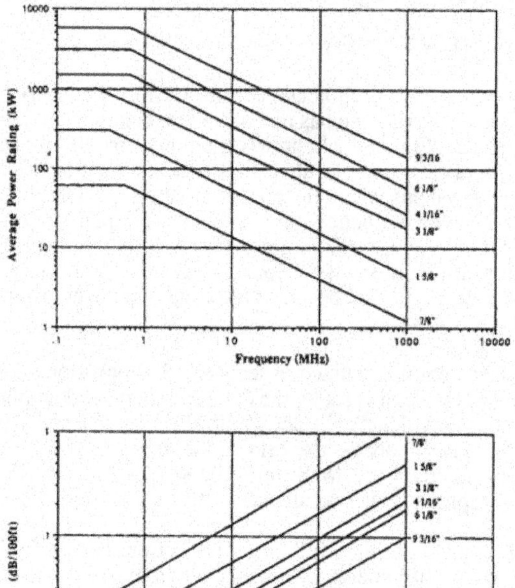

(a)

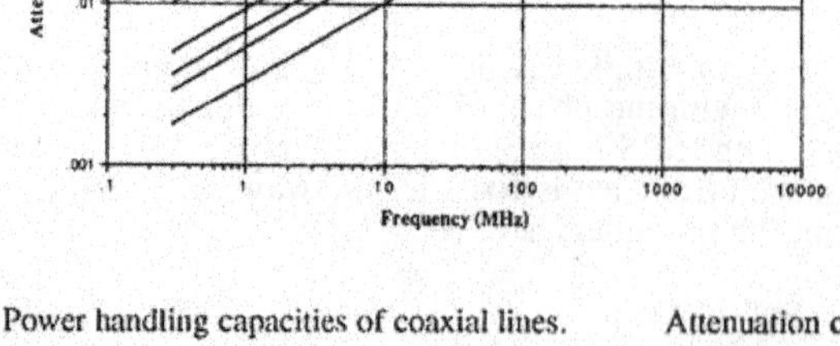

(b)

Figs. 4.21a & b Power handling capacities of coaxial lines. Attenuation curves for coaxial lines

central in the vacuum vessel. Magnetic length errors in the dipoles or positional errors in the quadrupoles will cause the orbit to be offset, this can be adjusted by changes to the frequency.

Beam orbit stability is a critical area to users of synchrotron radiation sources, and if the orbit stability needs to be better than 10 μm, the stability of the oscillator needs to be better than at least 1 part in 10^8 for a machine with a mean radius of 15 m.

4.7.10 DC power supplies

The r.f. power output and the losses in the amplifiers are provided by DC power supplies. For example a 35 kW, 200 MHz tetrode amplifier would need a 10 kV 5 A power supply for the anode supply, up to 1 kV, low cuurent for the screen grid, 200 V at 150 mA for the control grid, and 6.3 V at 160 A for the heater supply.

Similarly a 350 kW 500 MHz klystron would need a cathode power supply capable of supplying 50 kV at 13 A. A typical focus supply is 150 V at 10 A, with a heater supply of 25 V 25 A.

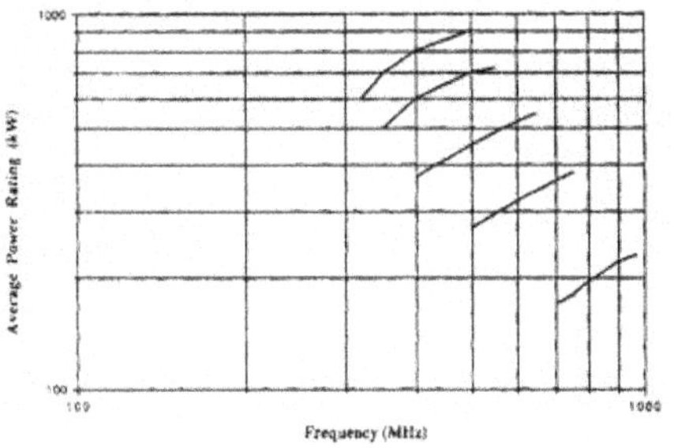

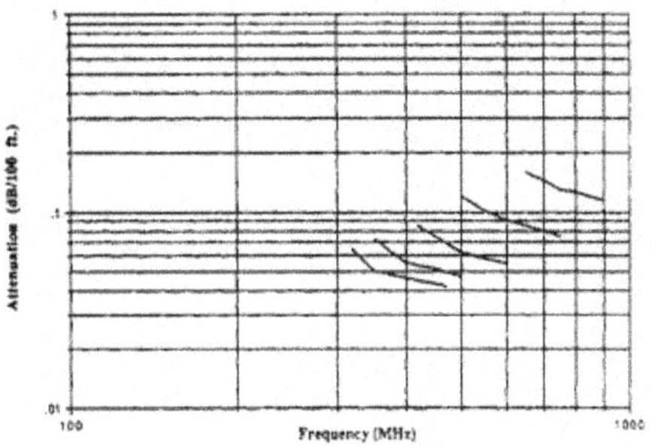

(a) (b)

Figs. 4.22a & b Power handling capacities of waveguides. Attenuation curves for waveguides

Along with other protection circuits, such as spark detectors, on the ceramic components, circuits to detect sudden increase in DC current, detect if the focus power supply is outside its limits in the case of a klystron, whether the heaters are at the correct level, cooling water flow is sufficient and overtemperature monitors are included in the DC supply, and will "crow bar" the supply, removing the DC voltage in ~ 10 μsec.

4.7.11 RF protection

The power to the cavities must be interrupted by a further protection system if any of a number of events occur. These triggering events include:

(a) the total power reflected to the source becoming greater than the limit set by the amplifier manufacturer, or in some cases the SWR being greater than a set value. For example at the SRS there is a trip level of 4 kW maximum reverse power to the klystron, but at high power there is a SWR trip set at 1.2:1.

(b) the cavity vacuum becoming greater than the set limit, typically 1 x 10^{-7} torr.

(c) isolator or circulator, cavity, window, or tuner cooling water flow failing.

(d) the temperature of isolator or circulator, cavity, window or tuner going outside set limits.

A fast pin-diode switch in the drive to the power amplifier is usually used to remove the r.f. power. This operates within 1 μsec.

References

1. M.Sands, SLC Report SLAC-121, Nov. 1970, pp 78-90
2. Ibid., pp 90-97
3. G.Saxon and T.E.Swain, Daresbury Laboratory Report DL/SRF/R6, 1975, p 2
4. "European Synchrotron Radiation Facility: Supplement II - The Machine", p 41 published by European Science Foundation, Stasburg, May 1979
5. K.W.Robinson, Cambridge Electron Accelerator Internal Report CEAL-1010, Feb. 1964
6. Design Study for a Dedicated Source of Synchrotron Radiation, Daresbury Laboratory Report DL/SRF/R2, 1975, p 30
7. M.P.Level et al, Proc. Particle Accelerator Conference 1993, pp 1465-7
8. W.Weingarten, CERN Report CERN 92-03, June 1992, p 318
9. K.Batchelor, J.Galayda and R,Hawrylak, IEEE Trans. Nuc. Sci. NS-28 (1981) pp 2839-2841
10. Private Communication R.J.Anderson, Oxford Instruments
11. Design Study for a Dedicated Source of Synchrotron Radiation, Daresbury Laboratory Report DL/SRF/R2, 1975, p 32-33
12. K.Batchelor and Y.Kamiya KEK Report KEK 79-25 (1975)
13. A.Massarotti et al Proc. 2nd. European Accelerator Conference, Nice, June 1990, pp 919-921
14. ESRF Foundation Phase Report, February 1987, p CII-175
15. K.Halbach and R.F.Holsinger, Part. Accel. Vol 7 (1976) pp 213-222
16. P.Fernandes and R.Parodi, IEEE Trans. Mag.-21 Vol. 6 (1985) p 2246
17. T.Weiland, Part. Accel. Vol 15 (1984) pp 245-292
18. T.Weiland, Proc. of the conf. on computer codes and the linear accelerator community, LA-11857-C (1990)
19. R.K.Cooper et al, Nucl. Instrum. Methods B40/41 (1989) pp 959-964
20. H.S.Deaven and K.C.D.Chan (Eds) LA-UR -90-1766 (1990)
21. D.M.Dykes et al, IEE Trans. Nuc. Sci. NS-32 (1985) pp 2800 2802
22. K.M.Welch, SLAC Report SLAC-PUB-1472, August 1974
23. S.Thomas and E.B.Pattinson J. Phys. D: Appl. Phys., Vol. 3, 1970 pp 1469-1474
24. I.Wilson, CERN Report CERN 92-03, June 1992, pp375-395

Bibliography

1. CAS CERN Accelerator School - RF Engineering for Particle Accelerators. Edited S.Turner. 2 vols. CERN 92-03, June 1992.

2. CAS CERN Accelerator School - Fifth General Accelerator Physics Course. Edited S.Turner. 2 vols. CERN 94-01, January 1994

3. Design Studies and Conceptual Design Reports for Synchrotron Radiation Sources (e.g. SRS, ALS, ELETTRA, APS, ESRF, SPring-8, CEA, SOLEIL, DELTA, Pohang Light Source, BESSY-II, etc)

4. Numerous papers and the Proceedings of various Accelerator Conferences, notably: The US Particle Accelerator Conference series (alternate years since 1967) and the European Particle Accelerator Conference series (alternate years from 1988).

CHAPTER 5: MAGNET DESIGN

NEIL MARKS
Daresbury Laboratory,
Warrington WA4 4AD, U.K.

5.1 Introduction

The first part of this Chapter introduces the basic types of electromagnets in the synchrotron radiation source, and discusses their properties and roles in a qualitative way intended for the non-specialist. Electromagnets are present in all particle accelerators but there are design requirements and constraints peculiar to synchrotron sources; these are explained both in terms of the physics of the components and the engineering constraints in practical circumstances. Economic factors have significance both in resolving certain features of the magnet design and determining the direction of the project; such criteria that will be addressed by both the magnet designer and the project manager are presented.

Non-specialists should be aware that all the magnets dealt with in this Chapter are of the 'electromagnet' type, using current carrying coils to generate the required magnetic fields. Permanent magnets are used in synchrotron sources as Wiggler and Undulator Insertion magnets, which are described in Chapter 14.

5.1.1 Dipole Magnets

This is the most fundamental type of magnet which, as its name suggests, has two poles which generate a uniform magnetic field, usually in the vertical direction, across the path of the electron beam. This causes the particles to travel along the arc of a circle and, with many dipoles present, to traverse the required complete circular path.

The cross section of a typical dipole is shown in the left-hand diagram of Fig. 5.1; the particle beam is located in a gap between the two poles formed by the 'C core', which is assembled from high quality magnetic steel. The required magnetic field is generated by electric currents flowing in the two coils around the poles; total currents (referred to as Ampere-turns) of a few tens of thousands of Amperes are usually required. Variations on this basic geometry are possible and are found in accelerators intended for particle physics. However, the open structure of this design is ideal for synchrotron sources, as it provides space radially outwards from the electron beam for the emerging radiation.

In addition to bending the electron beam, dipole magnets in a synchrotron radiation source have the additional capability of generating beams of radiation (see Chapter 1). In earlier sources, the dipoles were the sole radiation sources present, but in later projects wiggler and undulator insertion devices (see Chapter 14) have become the principal generators of radiation. However, radiation from dipole magnets is caused by fundamental processes and cannot be prevented even though the consequential energy loss requires a high power radio frequency system to maintain the energy of the circulating particle beam (see Chapter 4). It is therefore sensible to use this radiation to supplement the more intense and selective radiation emanating from the insertion devices.

The specification of the dipole magnets for a synchrotron source will be strongly influenced by these features. It is essential to choose a field amplitude that will provide the

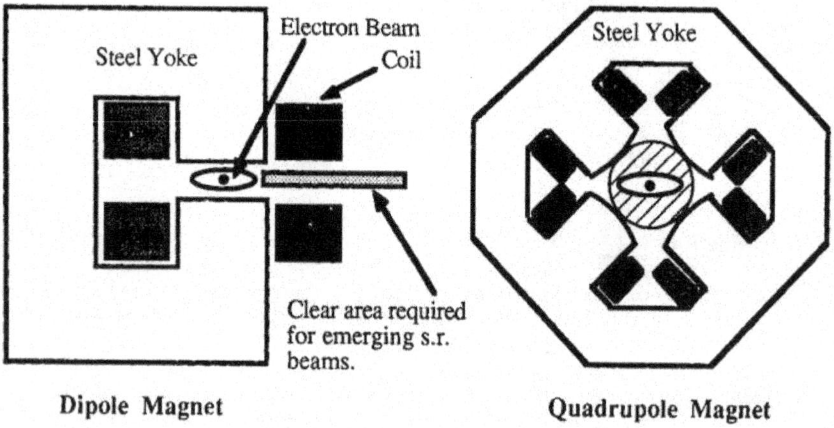

Fig. 5.1. Basic geometry of dipole and quadrupole magnets.

required radiation spectrum from the bending magnets, whilst, at the same time giving an acceptable power rating for the radio frequency system, for both the spectrum and the r.f. power are determined by the dipole field strength and the energy of the circulating electron. This usually results in a choice of modest field strengths, typically less than 1 Tesla (10 kilo Gauss). Consequently, the engineering of dipole magnets in synchrotron sources is usually significantly less demanding than that in high energy proton machines, where the absence of any significant amount of synchrotron radiation and the need to limit the size of the total installation result in much higher flux densities being used.

The choice of the number of dipoles in a synchrotron source is also important, as this influences the lowest emittance possible in the lattice (see Chapter 2) and hence the source brilliance. The lowest achievable emittance varies as the cube of the bending angle of each individual dipole magnet, so a good design will segment the dipoles into a large number of short units. This could lead to a significant increase in the project budget, as the total ring circumference may have to increase, due to the inclusion of more straights between the dipoles; the magnet costs may also alter, as a larger number of shorter units will also change the magnet economics. The choice of number of dipoles is therefore a decision that is fundamental to the source specification and budget.

Recently, there have been indications of interest in using the dipole radiation for specialised applications by designing for a range of fields in the source's bending magnets. In the European Synchrotron Radiation Facility [1] (ESRF, Grenoble, France) and in the SIBERIA-2 facility [2] (Budker Institute of Nuclear Physics, Russia) special low field sections were incorporated into the dipole magnets to provide 'softer' radiation at longer wavelengths down some beam-lines. In two recent studies for new sources, the use of superconducting dipoles at various positions around the magnet ring is proposed. These dipoles will produce 'hard' radiation, which can be used for experiments requiring high energy photons. They will therefore supplement or replace the radiation emerging from the specially installed wiggler 'wavelength shifters'. Such concepts are part of the ongoing development and optimisation of the source design to meet the user's needs.

5.1.2 Quadrupole Magnets

The quadrupole magnets provide the essential focusing to the electron beam that limits its physical dimension in both the horizontal and vertical planes. The cross section of a typical quadrupole magnet is shown in the right-hand diagram of Fig. 5.1. As in the dipole magnet, the electron beam travels through a region between poles that define the nature of the magnetic field; the hatched circle in the diagram does not correspond to any physical object, but defines the inscribed radius, which, like the gap in the dipole, sets the overall dimensions of the magnet. In the quadrupole there are four poles, each with an associated energising coil. The magnetic fluxes generated by these poles completely cancel out at the centre of the magnet, where there is therefore zero magnetic field; off-centre in either plane the transverse field increases linearly with increasing distance from the centre. Particles that are on the centre axis see no magnetic field and are undeviated; off-centre particles, however, are deflected in a way similar to the refraction of light when it passes through a lens. The quadrupole is therefore a gradient magnet and its strength is specified in terms of the gradient of the magnetic flux density between the poles (Tesla per metre).

Due to the basic laws of physics, it is not possible to generate a quadrupole field in a single magnet that will focus in both horizontal and vertical planes. If the magnet is configured to focus horizontally, it will defocus vertically, and *vice versa*. However, combinations of horizontally focusing and vertically focusing magnets at different positions along the circumference of the accelerator can result in overall focusing in both planes. Thus, most circular particle accelerators require a minimum of two different types of quadrupole magnet, whilst in the most advanced synchrotron sources many different 'families' of quadrupoles are required to sculpture the shape of the beam at different parts of the ring. This is partially to meet the different requirements of the various insertion devices, for it is essential to control either the size of the electron beam or its angular divergence to optimise the characteristics of the radiation given off by the wigglers and undulators. The designer of a state-of-the-art synchrotron source will therefore specify six to eight different quadrupole families that will require different strengths and polarities and which also may need different apertures for the circulating beam.

Unlike the dipoles, the design of the quadrupoles for a modern synchrotron source has a number of critical features and constraints. It has been explained that, in the ideal case, a quadrupole magnet has zero magnetic field at its physical centre. However, small manufacturing errors and inhomogeneities in the magnet steel will move the position of this zero field point a small distance; in technical terms, the magnetic centre will be different from the physical centre of the magnet. This is a problem in any particle accelerator, and the magnet engineer will take pains to make the discrepancy as small as possible. However, in the case of synchrotron sources, this is a critical issue. This is partially due to the beam optics, as the variations in beam size required for the optimisation of the insertion device radiation, outlined above, cause the circulating electron beam to be highly sensitive to magnetic centre error in the quadrupoles where the beam dimensions are large. The other factor relates to the steering of the synchrotron radiation beam down the beam-lines to the experiments. Errors in the quadrupole centres will result in beam movements when there is a small change in the quadrupole strength; these are unacceptable for the high precision experiments that are now carried out using the very small cross section beams. The movements can be corrected, but are best minimised at the point of generation.

A further complicating factor in the engineering design of quadrupoles in synchrotron sources is the need for space for emerging radiation beam-lines in the region immediately adjacent to the circulating electron beam (see the corresponding area in the diagram of the dipole magnet). This means that the cross section of the quadrupoles will almost certainly differ from the simple design shown on the right-hand side of Fig. 5.1, and would have one or both outer vertical sections of the steel removed. This produces engineering problems relating to the magnet's mechanical stability, which critically controls the deviation of the magnetic centre. These various requirements therefore combine to present the designer with unusual challenges which call for high quality mechanical engineering, which will be reflected in the budget for the quadrupole magnets.

5.1.3 Sextupole and other Lattice Magnets

The complete assembly of electromagnets around a circular particle accelerator is referred to as the magnet lattice (see Chapter 2) and will usually contain other types of magnets in addition to the dipoles and quadrupoles.

Sextupoles will be needed in a synchrotron source to control the electron beam's chromaticity, the variation of focusing with electron momentum. This magnet is similar to the quadrupole, but has six poles rather than four. As in the case of a quadrupole, the magnetic field in the sextupole is theoretically zero at the magnet's centre but off centre varies as the square of the distance from the centre. There will also be multiple families of sextupoles with different strengths and polarities required, depending on the position of the sextupole in the lattice. The various constraints described for the quadrupoles also apply to the sextupole magnets; the magnetic centre must be accurately located, and the outer vertical yoke limb may be omitted to provide space for emerging beam-lines (the magnetic nature of the sextupole makes it essential to retain the other, inner vertical yoke limb). The specification of the sextupole magnets will reflect these requirements, which may, however, be a little less critical and onerous than those in the quadrupole.

Other 'multipole' magnets, such as eight pole (octupole) magnets may also be needed, for these can improve the stability of the stored electron beam by providing a damping mechanism known as 'Landau damping'. However, the other essential magnetic elements of a synchrotron source lattice will be dipole correctors. These small magnets have the same type of magnetic field distribution as the main dipole magnets, but produce lower amplitude fields that are capable of independent adjustment in order to provide local steering control of the electron beam position (and hence of the radiation emerging from the source). Static corrector dipoles that are powered by direct current are used to correct the closed orbit around the source, but it has now become accepted that dynamic electron beam position control is essential for a state-of-the-art synchrotron source. Such systems use data from both the photon beam position monitors in the beam lines and the electron beam position monitors in the storage ring, and feed-back technology then generates the necessary correction signals, which are applied to small specialised dipoles. These magnets need a frequency response matching the specification of the feed-back system and are required at many points in the lattice. The stringent control of beam position during the many hours of lifetime of a stored beam in a source has now become an important issue to experimenters using synchrotron radiation, and increasingly effective technical solutions to this problem are expected to emerge in the near future; see Chapter 13 for a fuller description of beam stabilisation and steering systems.

Circular accelerators also require high frequency switching magnets for beam injection and, where a booster synchrotron is used as injector, to extract beam from this accelerator for subsequent injection into the storage ring source. The magnets and power supplies for the injection and extraction systems have a very specialised technology and are therefore separately dealt with in Chapter 3.

5.1.4 A.C. and D.C. Magnets

The principal accelerator in a synchrotron radiation source will be an electron storage ring in which the magnets are powered for many hours by constant direct current.

In some cases electrons may be injected into the storage ring at an energy that is less than the full operational value and, under such circumstances, the storage ring magnets will need to be 'ramped' to raise the electron energy to the required value. This process is usually slow, taking at least a minute, and significantly longer in some storage rings. Providing the magnets are assembled from laminations[1], such a slow process will not call for any basic modifications in the magnet design, which can be treated as a direct current exercise. However, if significant changes in magnetic field occur in seconds or less, the magnet design must reflect this dynamic requirement. Such situations may occur in a storage ring where a very short ramp time is required, but will principally be met with in a booster synchrotron that is used to pre-accelerate electrons for subsequent injection into the main ring. These injectors often have cycle frequencies of the order of 10 Hz. Under such circumstances the magnet designer must consider the power losses that will occur due to the oscillating magnetic fields and must also ensure that the magnet steel has good magnetic properties over the range of flux densities that will occur during the acceleration cycle. Hence, the design of booster synchrotron magnets will be more demanding in these respects, resembling the technology used in high quality power transformers.

5.1.5 Design and Fabrication Considerations

The design of the magnets will be based on the specification of the field quality over the horizontal and vertical region occupied by the circulating electron beam. As the source will be a storage ring in which the circulating beam is to be held for many hours, the specification of field quality in all three types of main magnets (dipoles, quadrupoles and sextupoles) will be stringent; it is usual for accuracies of the order of a few parts in ten thousand to be required in the dipole. The designer then has the task of determining the correct shape of the magnet poles to provide the necessary quality of field, as it is the geometry of the iron in the pole regions that chiefly determines the field distribution in the gap. Exact calculations are possible in some circumstances, but complex pole shapes and the non-linear behaviour of the iron present major difficulties. This work is therefore usually carried out using computer codes which predict the field distribution for a specific magnet geometry that is defined by the designer. The calculation is iterative, with small changes to the pole shape being applied, until the required distribution is achieved. In the past, two-dimensional codes were used to determine the flux distributions in the transverse plane of the magnet; as the fall off in field in the azimuthal direction at the magnet ends is important in determining the full interaction with the electron beam, separate calculations in this third plane could also be carried out. Now, three-dimensional programs are available and can be used very effectively to model the field distribution of a magnet.

Specific techniques have evolved for the construction of accelerator magnets. Very high mechanical tolerances will be required for the magnet yoke, accuracies of the order of ± 25 μm usually being specified for the dimensions of the steel yoke in the pole region. The yokes will usually be assembled from laminations that have been stamped from commercially available electrical steel. Laminations are normally only required in a.c. power applications, such as transformers, and as the magnets in the synchrotron source will operate with direct current in the coils, the use of laminated material may appear to be inappropriate. However, it is important to preserve a high degree of uniformity in both the magnetic properties and the shape and physical dimensions of the pole between the dipoles around the ring and the different quadrupole and sextupole magnets within a family. This is achieved by stamping the complete set of laminations for all the magnets in a group before assembly commences, and then randomising the laminations throughout all the magnets in that particular family; steel from different "batches" produced at the steel mill is

---

[1] For a fuller discussion of the advantages of laminated assembly see Sec. 5.5.1.

therefore present in each magnet. This is referred to as "shuffling" and is regarded as an essential procedure for any successful accelerator project.

After assembly, the magnets must be installed with very high positional accuracy; the important topic of supporting and aligning the magnets is covered in Chapter 11.

5.1.6 Economic Issues

In the closing section of this non-technical resume, the principal economic factors that will be encountered during the design of the synchrotron source magnets are discussed. The most significant criterion relates to the choice of dipole field, which has been mentioned above in Sec. 5.1.1. The conflicting issues are, on the one hand, the capital cost of the power supplies and radio frequency system, together with the dipole and r.f. running costs, which increase with larger dipole field, and, on the other hand, the capital costs of the buildings and infrastructure, which decrease with increasing dipole field and smaller ring circumference. The optimisation of these factors will be be modified by the specification of the radiation spectrum that is required from the dipoles. A similar situation arises with the choice of the number of dipoles, as Sec. 5.1.1 also explained how this determines the minimum possible electron beam emittance and influences the dipole magnet costs. The choice of these parameter will have major consequences on the total project budget and are usually determined early in the course of a design.

The physical and good field aperture required for the beam also has a major bearing on the economics of the magnet system. Critical parameters are the gap height in a dipole magnet, and the inscribed radius of the poles in quadrupoles and sextupoles. These have a very strong influence on the capital cost of the magnet systems, to such an extent that project managers will often refer to the dipole's economics in terms of the price per millimetre of gap. These dimensions also influence the running costs of the systems, as the energising Ampere-turns in the coil increase linearly with gap in a dipole and as the square of the radius in a quadrupole, leading to higher losses and power bills. Designers will, therefore, wish to minimise the aperture of the magnets.

Pole width will also influence capital costs, but will have negligible effect on operating costs in a d.c. magnet. In dipoles, quadrupoles and sextupoles, the required physical extent of the good field region will be a principal parameter in determining the pole breadths. In the case of the dipole, this parameter must be chosen not only to provide the required horizontal good field region, but, if straight magnets are used, also to accommodate the sagitta of the beam. This is the horizontal movement of the beam away from the centre line of the magnet, given by the displacement of the circular beam arc from the chord joining the two magnet ends. The magnitude of this displacement depends on the number of magnets around the circumference and on the bending radius, and can vary from a few millimetres in large accelerators to some tens of millimetres in straight magnets in smaller sources. The need to provide additional horizontal aperture to accommodate the sagitta can be eliminated by using curved dipole magnets, which follow the path of the electron beam as it traverses the dipole. The assembly of such curved magnets is more complex, but the increase in fabrication cost is compensated for by the reduction in aperture, and can lead to cost saving in all but the largest accelerators.

For each type of magnet present in the lattice, the total current in the coils (in Ampere-turns) is fixed by the required magnetic field or gradient strength, but the designer has a choice of the total cross section of conductor that is to be used to fabricate the coil. Whilst there is a technical lower limit to the amount of conductor (usually copper) present, this being set by the practical problem of cooling the coil, it is usual to use conductor cross sections that are much greater than this theoretical minimum. The economically optimum cross section is then determined by considering the total capital cost of the magnets, which will decrease as less conductor is used, and offsetting this against the total cost of powering the magnets over some time period, up to the life of the source, as this will increase if smaller conductors are used. Thus, this optimisation involves an estimate of the operational life of the new source, and the cost of electricity during that time. Managers responsible for a project may also wish to demonstrate that the design is ecologically sympathetic, in which case a large conductor cross section, that will use significantly less power during operation, will be chosen. With the total power budget of a large source running into many megawatts, such considerations will become increasingly important.

5.2 Fundamentals of Magnet Design

Many features of the design of the magnet systems for a synchrotron source will be similar to those applicable to particle accelerators in general. This section therefore briefly summarises these fundamentals as a basis for the introduction of practical considerations and the more specialised treatment of radiation source magnets later in the Chapter.

5.2.1 Field Distributions in Free Space

The space occupied by the beam between the magnet poles is free of current carrying conductors and in such a region the magnetic scalar potential can be defined and used to determine the allowed field distributions. A full presentation of this theory is given in a number of text books[3]; a brief summary is given below·

Maxwell' s equations for magneto - statics are :

$$\nabla.\mathbf{B} = 0\,; \tag{5.1}$$

$$\nabla \times \mathbf{H} = \mathbf{j}\,; \tag{5.2}$$

in the absence of currents :

$$\mathbf{j} = 0\,; \tag{5.3}$$

then we can put

$$\mathbf{B} = -\nabla\phi \tag{5.4}$$

where ϕ is the scalar potential. Substituting in Eq. (5.1) gives Laplace' s equation :

$$\nabla^2\phi = 0 \tag{5.5}$$

In two dimensions and for finite applications this has solutions :

$$\phi = \sum_{n=1}^{\infty}(J_n r^n \cos n\theta + K_n r^n \sin n\theta); \tag{5.6}$$

with solutions for the induction in cylindrical coordinates :

$$B_r = \sum_{n=1}^{\infty}(nJ_n r^{n-1}\cos n\theta + nK_n r^{n-1}\sin n\theta) \tag{5.7}$$

$$B_\theta = \sum_{n=1}^{\infty}(-nJ_n r^{n-1}\sin n\theta + nK_n r^{n-1}\cos n\theta) \tag{5.8}$$

where J_n and K_n are geometric constants.

These equations describe all possible physical distributions of flux density in two dimensions within any radius r that is free of iron and coils; they are also valid for the integral of the three dimensional fields along an axis through the magnet. It can be seen that for a particular value of n, there are two degrees of freedom given by the magnitudes of the corresponding values of J and K; in general these connect the distributions of the flux density in the two planes. Hence, with the values of these two constants determined, the distributions in both planes are also defined; distributions in the vertical and horizontal directions are not independent of each other.

Each value of the integer n in the two-dimensional magnetostatic equations corresponds to a different ideal flux distribution as generated by different standard magnet geometries. The three lowest values, n=1, 2, and 3 correspond to dipole, quadrupole and sextupoles flux density distributions respectively. This is illustrated in Fig. 5.2, where lines of constant scalar potential are shown for the cases of dipole, quadrupole and sextupole distributions. These are all for the situation where the J_n are zero; that is for the case for 'right' as opposed to 'skew' magnets, which would have similar distributions, but rotated by $\pi/2n$. Fig. 5.2 also gives a qualitative indication of the direction and amplitude of the magnetic induction B on the axes.

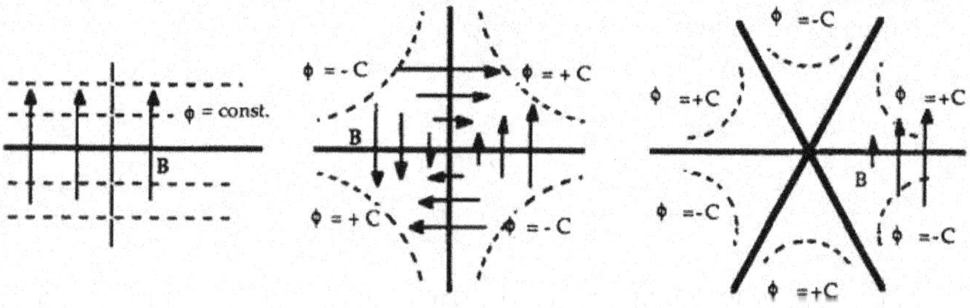

Fig. 5.2. Lines of scalar potential ϕ and magnetic induction B for dipole, quadrupole and sextupole distributions; the lines of B are schematic and indicate amplitude and direction on axes only.

The lines of constant scalar potential, ϕ, define the shapes of the ideal infinite permeability poles; this follows from the definition of scalar potential (5.4) and the well known theorem that **B** is normal to a ferromagnetic surface with infinite permeability. The transformation of the cylindrical Eq. (5.6) for ϕ into Cartesian coordinates, for the cases of n=1, 2, and 3 gives the equations for these 'right' ideal poles:

Dipole $$y = \pm g \tag{5.9}$$

Quadrupole $$2xy = \pm R^2 \tag{5.10}$$

Sextupole $$3xy^2 - y^3 = \pm R^3 \tag{5.11}$$

where g is the half gap in the dipole and R is the inscribed radius in the quadrupole or sextupole.

To design perfect magnets with field distributions corresponding to a single value of n, infinitely wide poles of the correct form, made from infinite permeability steel with currents of the correct magnitude and polarity located at infinity are sufficient; in practical situations they are neither possible nor necessary. Adjustments to the ideal pole shape overcome the necessary finite sizes of practical poles and the proximity of conductor close to the beam aperture and appropriate solutions will be described below. Before examining such specific situations, it is worthwhile considering the theoretical consequences of the symmetry of the pole cut-off points and the yoke geometry.

5.2.2 Symmetry Constraints

The geometric symmetry present in the standard magnet design imposes significant constraints on the allowed values of J_n and K_n and this is illustrated in Table 5.1. This is a simplified approach to the more general theory originally described by Halbach[4], and the Table applies only to 'right' as opposed to 'skew' magnets; it does, however, illustrate the restrictions conventional symmetries impose on the allowed harmonics. It can be seen that depending on the symmetry that is present, the allowed error harmonics of the required field are significantly reduced, irrespective of the pole shape, provided it maintains the correct symmetry. For example, a magnet with pure dipole symmetry can only have sextupole, decapole, etc. error fields, whilst the allowed harmonics in a fully symmetric quadrupole are limited to twelve pole, twenty pole, etc. Given these limitations on the possible error fields that can be present in a magnet, the magnet designer has additional techniques that can be used to reduce further the errors in the distribution; these usually take the form of small adjustments to the pole profile close to the cut-off points. The symmetry constraints described in the Table have special significance for some of the design problems encountered in synchrotron sources, and will be referred to later in the Chapter.

It should be appreciated that these symmetry constraints apply to the perfect magnet design geometries. Construction should closely follow the design but small tolerance errors will always be present in the magnet when it is finally assembled and these will break the symmetries. Thus, a physical magnet will have non-zero values of all J and K coefficients. The magnet designer should then predict the distortions resulting from manufacturing and assembly tolerances and this information becomes the basis for the specification covering the magnet manufacture, so ensuring that the completed magnet will meet its design criteria.

Table 5.1. Symmetry constraints in right dipole, quadrupole and sextupole geometries.

Magnet	Symmetry	Constraint
Dipole	$\phi(\theta) = -\phi(2\pi-\theta)$ $\phi(\theta) = \phi(\pi-\theta)$	All $J_n = 0$ K_n non - zero only for n = 1,3,5, etc.
Quadrupole	$\phi(\theta) = -\phi(2\pi-\theta)$ $\phi(\theta) = \phi(\pi/2-\theta)$	All $J_n = 0$. K_n non - zero only for n = 2, 6, 10, etc.
Sextupole	$\phi(\theta) = -\phi(2\pi-\theta)$ $\phi(\theta) = \phi(\pi/3-\theta)$	All $J_n = 0$. K_n non - zero only for n = 3, 9, 15, etc.

5.3 Design of the Practical Poles and Yokes

This section will review the techniques for designing yoke and pole configurations for practical applications. The discussion, generally, is not particular to synchrotron sources, though any features that require particular emphasise will be highlighted. As in the rest of the Chapter, most of the contents of this section will be devoted to normal i.e. non-superconducting magnets. However, there are now a number of feasibility studies that are studying the possibility of using superconducting dipoles as part of the magnet lattice, and a sub-section will therefore discuss these specialised devices.

Each type of magnet will be treated in detail, with a discussion of the most suitable yoke and pole configuration. However, it is possible to make some general comments on the design procedures necessary to determine the correct pole shape to produce the required field distribution. This is normally carried out with the aid of computer codes which predict a field distribution for a given geometry. It is then the task of the magnet designer to generate a suitable model and iterate the field calculations, modifying the input geometry until the required distributions are obtained. Knowledge of the allowed and forbidden error harmonics helps this process, for the designer can choose the most suitable basic symmetry and then optimise the remaining field components. This process does not usually eliminate any one of these higher order harmonics, but adjusts their amplitude and sign so that over the good field region the error harmonics subtract from each other to give an acceptable spatial distribution. Following the determination of the optimum pole shape, the designer will then ensure that over the specified range of operation of the magnet, the flux densities in the yoke do not cause unacceptable non-linearities. Finally, models that include the asymmetric errors introduced by the finite mechanical tolerances that will occur during manufacture are used as the input data sets; the 'forbidden' harmonics will then be present, and from their predicted magnitude, an assessment of the necessary specification of the tolerances can be made.

A large number of different codes are available for the pole and yoke field calculations. Early codes were two-dimensional magneto-static programs written in accelerator laboratories; MAGNET and POISSON are typical examples. More recently, three dimensional codes with magneto-dynamic capabilities have been written and are made available through a number of commercial organisations. Notwithstanding, it is possible, in the case of dipoles, to use the two dimensional codes to predict performance in all three planes. The design procedures in the plane transverse to the beam is now treated in detail for the different types of magnets, whilst at the end of this section, a general discussion on the use of two dimensional codes to determine the behaviour of a dipole in the third, azimuthal plane, is given.

5.3.1 Dipoles

A number of different dipole yoke designs are used in accelerators and these are shown in cross section in Fig. 5.3. They include the 'H' magnet (centre), in which the fully symmetric yoke has vertical return limbs on either side of the poles and hence the particle beam. A development of this design, the window-frame magnet (right), has the exciting coils (shown cross hatched) located in the median plane, and this has the advantage of providing highly homogeneous magnetic fields with simple parallel poles. The H design has been used for the magnets in the Booster synchrotron at the ESRF, Grenoble[5], but it is generally accepted that the 'H' and the window-frame designs are inappropriate for the main ring of a synchrotron source, as the presence of steel radially outwards from the electron beam represents a major obstruction to the emerging radiation beam-lines. It is, therefore, usual to make use of the 'C core' design for a radiation source.

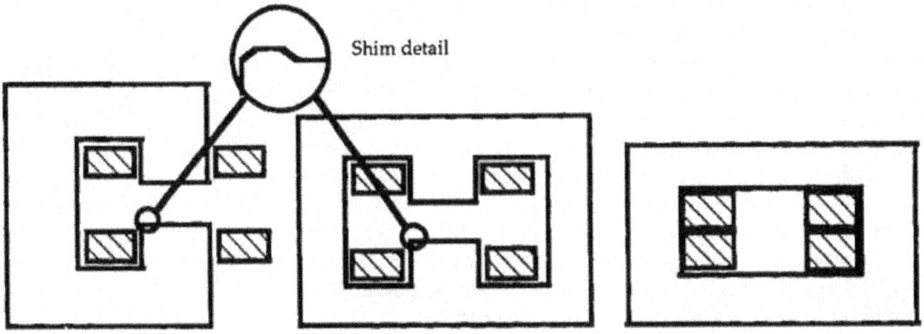

Fig. 5.3. Cross section geometry of 'C cored', 'H' and window frame dipole magnets.

The poles of both the C and H cored magnet require the addition of shims at the pole edges to compensate for their finite width and such a shim is shown in diagrammatic form in Fig. 5.3. The area and shape of the shim are optimised by iterating with codes, as previously described, and in accelerators with high field dipoles this is an exacting process; the flux density in the shim is significantly higher than elsewhere on the pole and any non-linear behaviour of the steel will then cause a change of field distribution with excitation level. However, in most synchrotron sources, the dipoles are not high field magnets and a simple shim having a shape similar to that shown in the diagram is adequate. Note that the shim has bevelled sides to avoid the local saturation that would occur at a right angle corner. As a guide to shimming a simple dipole, the pole geometry used for the ESRF storage ring[6] is shown in Fig. 5.4. This pole gave a field quality better than ± 2 in 10^4 over a horizontal good field region of 70 mm (total), indicating that the poor field extended inwards from the pole edge by approximately 75% of the gap height; this is a useful 'rule of thumb'.

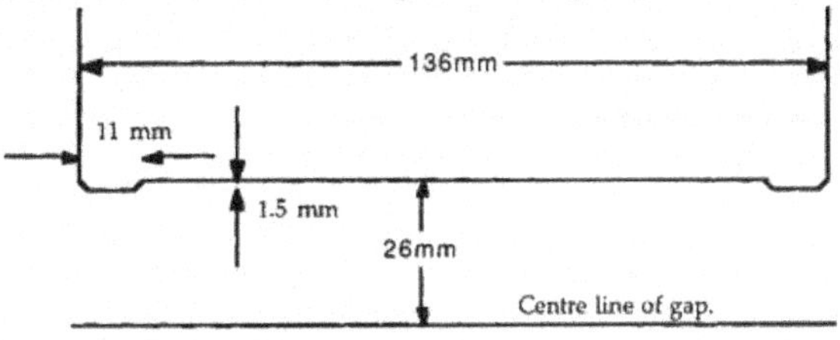

Fig. 5.4. Pole geometry used in the dipole of the ESRF storage ring.

Note that it is usual when determining the optimum pole geometry for a dipole to examine the variation of vertical flux density from the dipole centre as a function of horizontal position in the dipole; i.e.:

$$\frac{B_y(x) - B_0}{B_0} \qquad (5.12)$$

Examining the 'C core' geometry shown in Fig. 5.3, it is clear that the steel does not have full dipole symmetry as defined in Table 5.1. Hence harmonic errors with even values of n (quadrupole, octupole etc.) are possible together with the standard dipole field errors of sextupole, ten pole, etc. However, the pole area is fully symmetrical, so the only possible source of even value n errors is non-linear behaviour of the yoke steel caused by low permeability. With an operating field of 0.8 T in the gap of the ESRF dipole, this problem did not arise. In higher field magnets with wide poles, however, a small gradient across the good field region may be predicted, with the higher field on the side of the return yoke. Usually this can be corrected by the use of shims of slightly different size on the two sides of the pole; this solution is, however, amplitude dependent. The mechanical deflection of the C cored yoke under the stress of magnetic forces should also be considered. This will result in a small narrowing of the gap, increasing the dipole field. The gap will be smaller on the side away from the return yoke, generating a quadrupole field with opposite polarity to that caused by local saturation in the yoke. Hence, the two effects will partially cancel. However, they are both undesirable, and designers should include sufficient steel in the top, bottom and back legs to prevent either effects exceeding one to two parts in ten thousand. This presents little technical problem in a dipole, and the resulting increase in cost, which is a small percentage of the total magnet capital expenditure, is well worth paying to give a mechanically and magnetically stable design.

In some designs, such as the ALS (Berkeley)[7], SRRC (Taiwan)[8] and ELETTRA (Trieste)[9], gradients are built into the dipole, making these combined function dipoles. The gradient is specified by the 'n' value of the magnet, which should not be confused with the integer n used to define the magnet harmonics. The field index n in a gradient magnet is defined as:

$$n = -\left(\frac{\rho}{B_0}\right)\left(\frac{\partial B}{\partial x}\right), \tag{5.13}$$

where ρ is the radius of curvature of the beam and B_0 is the central dipole field.

Note that n can be positive or negative depending on the gradient of the field and is a dimensionless unit. Such a magnet is identical to a quadrupole with the physical centre shifted from the magnetic centre. The revised pole equation is determined for a central field of B_0, with the magnetic centre shifted by X_0.

Then
$$B_0 = \left(\frac{\partial B}{\partial x}\right)X_0;$$

therefore
$$X_0 = -\rho/n;$$

using Eq. (5.10) the pole equation becomes

$$y = \pm\frac{R^2}{2}\frac{n}{\rho}\left(1-\frac{nx}{\rho}\right)^{-1}$$

or
$$y = \pm g\left(1-\frac{nx}{\rho}\right)^{-1} \tag{5.14}$$

where g is half the vertical gap at the centre of the magnet.

The magnet poles will therefore be a section of a hyperbola, but if ρ is very large compared to nx, the pole shape approximates to a linear wedge. Note that to compensate for the pole cutoffs, dissimilar shims will be necessary at each end of the poles, with the largest area shims being needed at the open end, where the gap is largest.

One unusual feature of the dipole designed for the ESRF, Grenoble, should be mentioned here. During the design phase of the project, there was a request from the experimental community for soft dipole radiation beams generated by a field equal to one-half of the main dipole flux density, i.e., by the 6 GeV electron beam in a field of 0.4 T. This required a short section of each dipole to produce half amplitude field with the same uniformity as in the full amplitude section. Doubling the gap height would have destroyed the field quality, so layers of non-ferromagnetic material were introduced, magnetically isolating two short sections of pole both above and below the beam from the main yoke. These were called 'floating poles',[1] and one such isolated segment is illustrated in Fig. 5.5. The non-magnetic material introduces an additional reluctance into the yoke at the magnet end, and will result in a plateau region of half field providing the total thickness of the non-ferromagnetic inserts equals the gap height; the correct pole shape at the gap ensures the resulting field homogeneity. More recently, a similar feature has been included in the Russian radiation source, SIBERIA-2.[2]

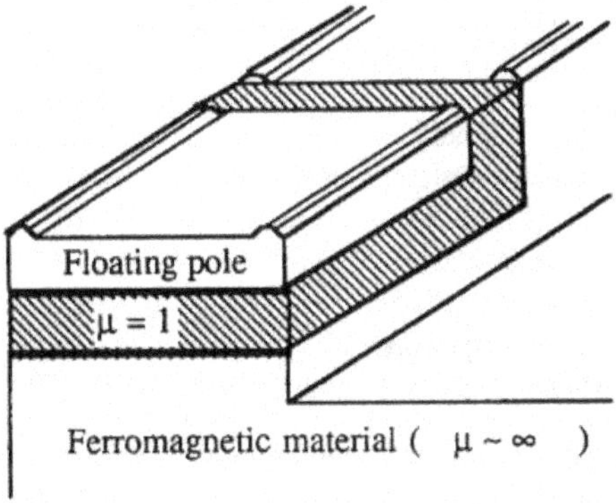

Fig. 5.5. Floating pole used in a short section of the ESRF dipole.

As a guide to the typical configuration of the dipole magnets in a state-of-the-art synchrotron source, Table 5.2 gives the principal dipole parameters for what are, currently, the three largest sources. The ESRF (Grenoble)[10] is complete and operational, whilst the APS (Argonne Laboratory)[11] and SPring-8 (Jaeri-Riken)[12] are presently in an advanced state of construction.

Table 5.2. Dipole magnet parameters of current high energy synchrotron sources.

Radiation Source:	**ESRF**	**APS**	**SPring-8**
Max beam energy (GeV)	6.0	7.0	8.0
Dipole field (T)	0.802	0.599	0.679
Number of Dipoles	64	80	88
Dipole length (m)	2.450	3.06	2.804
Magnet gap (mm)	54.0	60.0	64.0
Total good field (hor. x ver.)(mm^2)	70 x 32	60 x 40	60 x 30
Good field quality (p to p)	5.0×10^{-4}	2.5×10^{-4}	5.0×10^{-4}

For comparison, Table 5.3 presents the basic dipole parameter for three medium energy sources, the ALS at Berkeley[13], the SRRC, Taiwan[14], and ELETTRA, Trieste[9, 15]. Unlike the three high energy rings described above, which all have constant field dipoles, the three medium energy sources have combined function, gradient dipoles; this is far from being a universal feature of such lower energy rings but the data serve as examples of this type of magnet. Other significant differences between the different energy rings are apparent; the medium energy sources use smaller numbers of much shorter dipoles, that have central orbit fields that are of the order of twice that in the large rings. Note that the

Table 5.3. Dipole magnet parameters of three medium energy sources.

Radiation Source:	**ALS**	**SRRC**	**ELETTRA**
Type of dipole	Gradient	Gradient	Gradient
Max beam energy (GeV)	1.9	1.3	2.0
Dipole field at orbit (T)	1.30 (*)	1.24	1.212
Gradient T/m	-5.19	-1.71	-2.86
Number of Dipoles	36	18	24
Dipole length (m)	0.854	1.22	1.44
Magnet gap at orbit (mm)	50.0 (*)	52.0	70.0
Total good field (hor. x ver.) (mm^2)	60 x 40	60 x 24	50 (radially)
Good field quality (p to p)	4.4×10^{-4}	4.0×10^{-4}	2.0×10^{-4}

(* the ALS has straight dipoles, so these parameters are the typical values on axis)

combined function magnets all have horizontally defocussing, vertically focussing, gradients. In the case of the ALS, this is particularly strong as the bending magnet provides most of the vertical focusing and this is reflected in the large vertical aperture needed in the dipole.

5.3.2 Superconducting Dipoles

At the time of writing, the use of superconducting magnets in synchrotron sources has been limited to high field wavelength shifting insertion magnets. However, two new design studies (DIAMOND at Daresbury[16] and SLS at the Paul Scherrer Institut[17]) have now presented outline proposals for using superconducting dipoles at certain positions within the magnet lattice. The intentic n is to include a limited number of high field dipoles, which in total would not generate unacceptable radio frequency power requirements, to produce extra hard radiation down a limited number of beam-lines without introducing the specialised insertion devices that would normally be necessary.

It is not the intention to discuss the design of superconducting magnets in detail; this is a major topic in accelerator engineering and has been refined with great expertise at the high energy physics laboratories around the world. The large proton accelerators used at such laboratories generally require the maximum possible dipole field and a generic magnet design has evolved. As the accelerators have circumferences measured in kilometres, the dipoles are usually completely straight, with fields up to 7 T generated by coils, with a cosine theta current distribution, which are located close to the beam vacuum vessel. A cylindrical iron yoke is placed outside the coils to provide some reduction in the required Ampere-turns, to limit the stray flux, and to add to the essential mechanical constraint of the superconductor. Many hundreds of dipoles are generally required, so the design is adapted to mass-production, with assembly costs minimised.

The requirements for superconducting dipoles in a synchrotron source are radically different. Whilst fields that are uneconomic for a conventional magnet are specified, the first priority will not be to obtain the maximum achievable flux density. For example, the Swiss light source design[17] proposes inductions of between 2.2 T and 4.7 T and these figures are now regarded as modest for a superconducting dipole. However, other constraints, such as the position and angle of emergence of the photon beams at the magnet ends, the need for a warm bore vacuum vessel to dissipate the incident synchrotron radiation flux generated in the dipole and the necessity of an appreciable radius of curvature in the magnet, make different demands.

Both designs considering the use of superconducting dipoles cite the work carried out at the Budker Institute, Novosibirsk[18]. This has resulted in a magnet named after its inventor, Dr. P. Vobly. The design has a number of unique features. Iron is used close to the beam, but the coil configuration prevents any appreciable saturation; this significantly reduces the total reluctance along the flux path and hence the exciting Amp-turns. The coil sees flux densities appreciably lower than those at the beam, so the superconductor can be operated at a higher current density. The design has a warm bore and utilises a number of different coils distributed at different vertical positions in the dipole. This last feature eliminates any saddle coils located in the plane of either the electron beam or the emerging synchrotron radiation and therefore helps to simplify the end geometry of the magnet.

It is understood that the design of such magnets is still evolving and it is likely that the two cited projects, which are currently in the design phase, would make use of superconducting dipoles that are further developments of the outline designs currently described in their published studies.

5.3.3 Quadrupole Magnets

As with dipole magnets, a number of different configurations are feasible for quadrupole magnets. The basic cross section referred to in Sec. 5.1.2 and shown on the right-hand side of Fig. 5.1 can be regarded as the standard quadrupole design. It is fully symmetrical, with high mechanical stability, and can be viewed as the equivalent of the H-dipole (see centre diagram in Fig. 5.3). For high gradient applications, the design shown on the left-hand-side of Fig. 5.6 is more suitable. Saturation in a quadrupole usually first appears at the root of the pole, and this design maximises the steel in this area. Furthermore, it places the exciting Ampere-turns immediately adjacent to the gap, thus producing good field gradient to within close proximity of the coil; note the similarity to the window-frame dipole, that has the same advantage. Unfortunately, this design limits the access to the vacuum vessel and is therefore unsuitable for a synchrotron source, as there is no space available on the outer side of the electron beam for emerging synchrotron radiation beam-lines.

The most common arrangement to provide such space is shown in the right-hand diagram of Fig. 5.6. The principal feature of this geometry is the split in the magnetic yoke in the median plane,

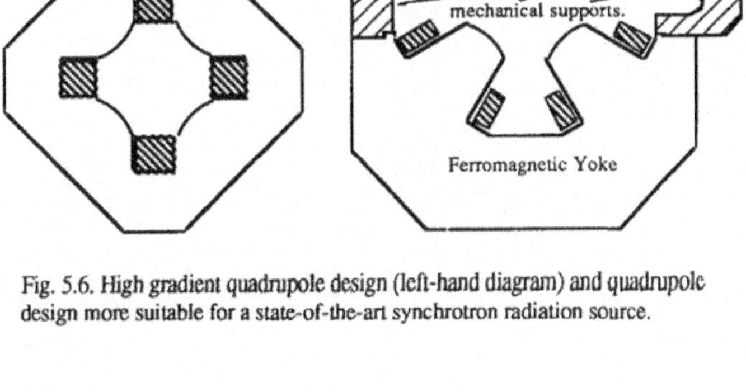

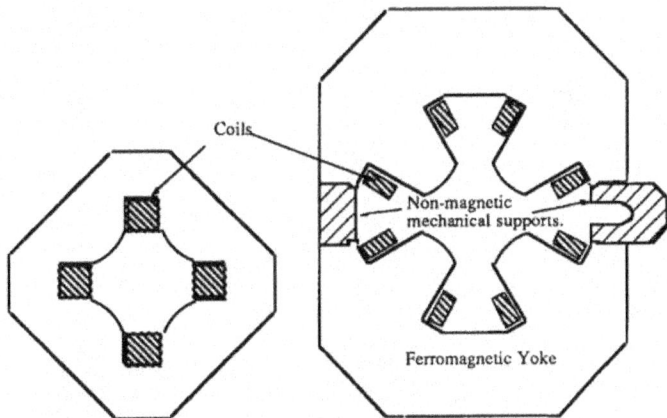

Fig. 5.6. High gradient quadrupole design (left-hand diagram) and quadrupole design more suitable for a state-of-the-art synchrotron radiation source.

and the use of non-magnetic mechanical spacers to support the upper half of the quadrupole; these spacers are usually constructed from stainless steel or cast aluminium. It is then possible to have a range of different support pieces designed to accommodate the widely varying beam-line geometry that will be encountered close-in to the quadrupole on the outside of the electron beam; these can be freely interchanged to build up magnets for the different azimuthal positions in the straight sections, with the certain knowledge that there will be no consequential change to the magnetic performance of the quadrupole. Note also that the coils have been moved back up the pole to provide clear space for the vacuum vessels of the emerging beams and the poles are broadened towards their roots, to reduce the flux density at his position. The necessity of this last feature will depend on the level of peak gradient that is specified in the quadrupole.

It will be noted that accommodation of radiation beam-lines is only required on the outside of the accelerator ring, so the mechanical arrangements described above are only necessary at this position, between the two outer poles. However, magnetic symmetry considerations dictate the necessity of making all poles with the same geometry and with full symmetry about the pole centre line. If the space along the $x=0$ axis were to be used to accommodate coils or provide greater cross section of steel at the pole root, the quadrupole symmetry described in Table 5.1 would be broken and 8 pole, 16 pole etc. fields would be present. In fact, the yoke geometry of this design also breaks the fourfold symmetry but, as in the case of the 'C cored' dipole, the additional error fields would mainly be due to gross non-linearities in the steel; these should be avoided so that the broken symmetry of the backlegs can be more easily tolerated than pole and coil asymmetries close to the good gradient region. An assessment of the eight-pole field generated by the reluctance of the top and bottom yokes was studied during the ESRF design study[1] and found to be less than 1 part in 10^4 in field over the useful aperture.

The arrangement of fully splitting the magnetic yoke on the horizontal centre line of the magnet has been used by a number of recent major projects. The storage rings of the ESRF, (Grenoble)[10] and the APS project (Argonne National Laboratory)[11] have this feature, as does the medium energy ALS ring (Lawrence Berkeley Lab)[13]. A similar solution has now been proposed for the SLS project (Paul Scherrer Institut)[17]. However, it is fully acceptable to use the non-magnetic spacer only in the outer vertical limb of the quadrupole (i.e., the right-hand vertical yoke component in Fig. 5.6) and have ferromagnetic steel completing the vertical yoke on the inner limb. If the poles and coils are fully symmetrical, no flux will flow in this section of the yoke, so its use is superfluous. Furthermore, it could be argued that this arrangement adds a further degree of asymmetry to the magnet. However, it has the appreciable advantage of providing a shunt for the vertical component of the earth's field and other stray magnetic fluxes around the beam. Notwithstanding, no difficulties are reported from the ESRF that are due to the inclusion of non-magnetic spacers on both sides of the yoke[19].

A further variation is proposed for the SPring-8 project (Jaeri-Riken)[12]; the storage ring quadrupoles in this radiation source will have magnetic steel traversing the horizontal median plane on both sides of the electron beam, with an extension in the radially outwards direction to accommodate the emerging photon beams. Two different designs are proposed and these will provide sufficient flexibility to accommodate all possible beam-line geometries.

It should be pointed out that the two cross sections shown in Fig. 5.6 have identical configurations for inscribed radius, pole shape, coil cross section and thickness of steel in the back-leg (in the case of the left-hand diagram the top and bottom yokes are magnetically in parallel with the sides, so only half the thickness is required in each separate limb). Furthermore, the gradient distribution in the left-hand design will be superior to that in the right-hand diagram, as the proximity of the coils will sustain the good field over a wider aperture, though this good field extension will be influenced by the coil properties. Hence, the engineering changes necessary to allow for the correct positioning of radiation beam-lines result in substantial increases in the size and cost of the magnet.

In a third generation light source, there will be an appreciable number of different quadrupole families, with different requirements for maximum gradient. It is likely that the families with the highest gradient requirements will need a design similar to that shown in the right-hand diagram of Fig. 5.6, with broadening poles to limit pole root saturation. Usually, the lengths of these strongest quadrupoles will correspond to the maximum achievable gradient over the specified aperture. Weaker families will, however, probably have their specification determined by the need to have their magnet length much greater than the inscribed diameter, so that end effects do not dominate. It is then probable that the gradient will be sufficiently small to permit the use of parallel sided poles. There will also be appreciable variation in the aperture requirements at different parts of the lattice, as the size of the circulating electron beam is proportional to the square root of the beta functions, which vary periodically around the ring. The quadrupole designer must therefore decide on the number of different cross sections needed to meet the requirement of all the families. A small number will minimise the tooling costs of the fabrication exercise, but will be wasteful on materials and power. A large number of different cross sections will significantly reduce the total steel and conductor requirements and decrease the total quadrupole power consumption, but result in more jigs and fixtures needed for assembly, etc. The optimum is determined by careful assessment of the total manufacturing costs, and discussions with potential suppliers can be very valuable in resolving this issue. Significant variations in the solution to these problems are evident between the existing and proposed third generation sources. The APS and the BESSY II (Berlin)[20] design studies propose a single quadrupole cross section, whilst ESRF and SPring-8 use two different cross sections; different length magnets with the same cross section profile are, however, usual. The major parameters of the quadrupole families in the three main high energy sources, ESRF, APS and SPring-8 are presented in Table 5.4. For comparison, the parameters of a typical lower energy source, the Berkeley ALS, are also presented.

Table 5.4. Parameters of quadrupole families in three large state-of-the-art synchrotron sources and a lower energy source.

Radiation Source:	ALS	ESRF	APS	SPring-8
Max beam energy (GeV)	1.9	6.0	7.0	8.0
Total no. of quadrupoles	72	320	400	480
Number of families	3	8	5	10
No. of diff. cross sections	1(*)	2	1	2
Inscribed radius (mm)	32.5	36.0	40.0	42.5
Max quadrupole length (m)	0.445	0.9	0.8	0.97
Max strength (T/m)	17.0	20.1	19.1	17.6
Gradient quality (p to p)	3.0×10^{-4}	5.0×10^{-4}	3.0×10^{-4}	5.0×10^{-4}
Good gradient dia.(mm)	60.0	54.0	50.0	70.0

(*) One family of ALS quadrupoles has a different coil configuration to the other two.

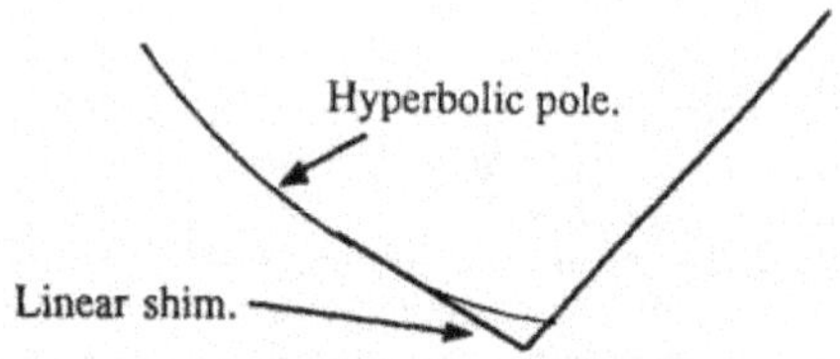

Fig. 5.7. Details of standard shim used in a quadrupole.

The pole in the design shown on the right-hand side of Fig. 5.6 requires pole shims in order to optimise the region of good gradient. The standard shim used in a quadrupole to improve the extent of the good gradient region is shown in Fig. 5.7. The standard hyperbolic pole is terminated in a linear region that is tangential to the curve at a point a little way in from the pole edge. This point is determined using magneto-static codes to examine the resulting quality and extent of the good gradient region. The position of the tangent along the hyperbolic pole is varied until the optimum quadrupole field is obtained.

It should be emphasised that whilst in a dipole it is adequate to plot the variation of vertical field against horizontal position, in the case of the quadrupole the examination of flux density is not sufficiently precise to provide an adequate assessment of the suitability of the magnet. It is vital to judge the quadrupole design in terms of the gradient, i.e., by examining the first derivative of the normal components of flux density on the axes:

$$\frac{\partial B_y(x)}{\partial x} \quad \text{or} \quad \frac{\partial B_x(y)}{\partial y}. \tag{5.15}$$

In many modern codes, the facility for presenting the first differential is provided as a standard package, whilst those using earlier programs must calculate first differences from the printed output of flux density as a function of horizontal position.

5.3.4 Sextupole Magnets

Providing the differences in symmetry are taken into account, many of the comments made concerning the quadrupole magnet can also be applied to the sextupole magnets. Space for the emerging radiation must be provided on the outside of the magnet, and this is best achieved by using a non-magnetic mechanical component, similar to that used in the quadrupole. However, the sextupole symmetry makes it essential to have a magnetic return limb on the side horizontally opposite to this (i.e., at the 180° position). In principle it is then not necessary to have magnetic steel at the 120° and 240° positions, for there will be no flux associated with the main sextupole excitation at these locations. Most projects have joined the steel to form a ring, open at one place only, giving the yoke geometry shown in the left-hand diagram of Fig. 5.8 In principle it is possible to introduced gaps at the positions in the yoke where steel is not essential (right-hand diagram in Fig. 5.8), though this precludes the use of the sextupoles to generate correction dipole field. This was overcome at the ESRF by using small 1 mm gaps. These decoupled the three sections magnetically in the case of

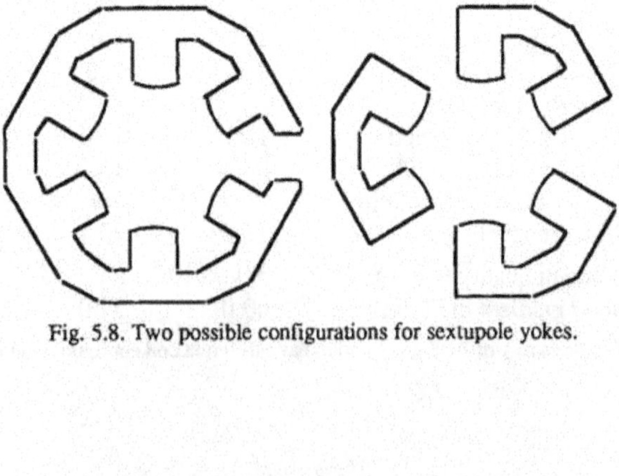
Fig. 5.8. Two possible configurations for sextupole yokes.

sextupole excitation, whilst still preserving a low yoke reluctance for correction dipole excitation, as described below. Thus, error fields associated with small asymmetries in the sextupole circuit will be strongly attenuated, whilst only a small increase in the excitation in the coils placed on the sextupole for dipole correcting fields will be necessary.

The nature of the harmonic errors introduced by these various configurations can be estimated by consideration of the symmetry conditions of Table 5.1. Whilst both geometries shown break the full six-fold symmetry of the ideal magnet, any error field present in either design will be due to steel non-linearities. The three fold split shown in the right-hand diagram of Fig. 5.8 can have the 12, 24 etc. pole series present as well as the normal 18, 30 pole etc. errors. It also has the problem of not providing a path for the vertical component of the earth's magnetic field in parallel with the gap, as well as not being able to support dipole correction field (see Sec. 5.3.5). Such steering fields can be generated in the sextupole magnet shown in the left-hand diagram, for dipole fields are allowed by the symmetry, as are quadrupole, octupole etc. These would only be present as error fields with appreciable amplitudes if there were gross mis-positioning of coils or poles, and, as such, represent no problem.

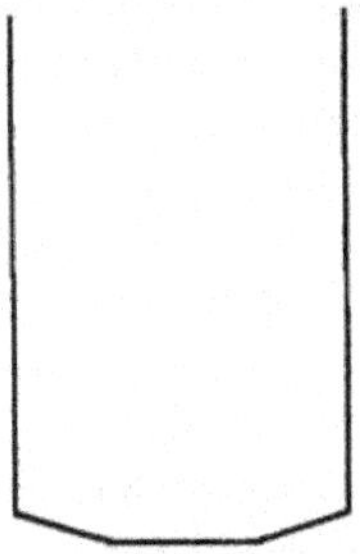

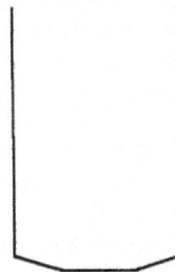

Fig. 5.9. Suitable pole contour for a sextupole magnet.

Pole contouring in a sextupole is less critical than in a dipole or quadrupole. The use of the correct third-order curve is unnecessary and a rectangular pole, with corner chamfers is quite acceptable, as shown in Fig. 5.9. The standard codes are used to obtain this profile; again, it is important not to judge the quality of the magnet on the flux density but to examine the second derivative of the vertical field on the horizontal axis; i.e.:

$$\frac{\partial^2 B_y(x)}{\partial x^2}. \tag{5.16}$$

When specifying the amplitude of the sextupole field, designers should ensure that there is no confusion between the value of the second differential and the coefficient of x^2 in the expansion of the vertical field on the horizontal axis. Both have units of T/m^2, but there is, of course, a factor of two difference between the values.

Modern magnetic computational codes usually provide the values of the field differentials, whilst the designer using the older programs will need to take second differences of the vertical flux density predicted by the code. Note that whilst many codes will give a harmonic analysis of the field, usually in terms of the coefficients of the cylindrical components given in Eqs. (5.7) and (5.8), the adjudication of field or gradient quality from a single harmonic value is dangerous; the sum of all appreciable harmonics present must be considered. In any magnet with a finite aperture, harmonic errors will always be present, but will have opposing amplitudes and polarities in the good field region of a well designed magnet; the only valid criteria are the actual values of flux density, for a dipole, or its differentials, for quadrupoles and sextupoles.

The basic parameters of the sextupole magnets of the current world's largest sources are given in Table 5.5; again the parameters of the ALS sextupoles are included for comparison. Note that design reports for both APS and SPring-8 specify sextupole strength as the second differential of field, whilst the values in the Table are for sextupole coefficient, which is half that differential. It should also be observed that in all cases, the inscribed radius in the sextupole is greater than that in the quadrupole magnet. This is not because higher field gradient qualities are required in the sextupole, but reflects the difficulty of locating the pole on the 30° axis without interfering with the emerging radiation.

Table 5.5 Parameters of sextupole magnets in the ALS, ESRF, APS and SPring-8 sources.

Radiation Source:	**ALS**	**ESRF**	**APS**	**SPring-8**
Beam energy (GeV)	1.9	6.0	7.0	8.0
Total number of sextupoles	48	224	280	336
Number of families	2	7	4	7
Number of diff. cross sections	1	1	1	2
Inscribed radius (mm)	35.0	45.0	50.0	46.0
Max sextupole length (m)	0.2	0.4	0.24	0.50
Max field coefficient (T/m^2) (*)	500.0	165.0	245.0	210
Sextupole grad. quality (p to p)	–	120×10^{-4}	25×10^{-4}	30×10^{-4}
Good gradient diameter (mm)	60.0	52.0	50.0	70.0

(* note that these values are sextupole coefficient, which is half the field second derivative).

5.3.5 Correction Magnets

Correction magnets are vital in a modern synchrotron source. Of prime importance are dipole correction fields, for these will be required to control the closed orbit, to give optimum accelerator performance, and also to provide the fine steering for optimising the transmission of radiation down the beam-lines and onto the experiments; this latter feature has become a matter of great importance for successful experimental programmes. In addition to the dipole correctors, higher order fields for adjustment purposes may also be required. Skew quadrupoles, which generate a quadrupole field that has an angle of $\pi/4$ to the standard focusing fields, can be used to adjust the coupling between vertical and horizontal emitances and higher order fields, such as octupole, can improve beam damping.

In order to save space and to introduce correction fields at the most effective points in the lattice, much effort has been applied in trying to include correction coils in the main magnets; these are most suitable for generating the d.c. closed orbit correction fields as opposed to the dynamic beam steering correctors. To generate vertical magnetic correction fields, an obvious solution is to add low current windings to the yoke of the dipole magnets, and control them from individual power supplies. However, there are problems with this proposal; in booster synchrotrons there will be oscillating magnetic fields, which may result in induced voltages of appreciable amplitudes in the auxiliary coils. There will be no such problems with the storage ring dipoles, but in many cases the horizontal β function is low in the region of the bending magnets, so dipole fields intended to control and correct the closed orbit will not be very effective when located in the main dipoles. The quadrupoles therefore appear to be a better option, for the β values are maximum in the appropriate quadrupole magnet, and the standard quadrupole symmetry allows vertical and horizontal dipole field with equal ease (See Chap. 2 for a discussion of β values in the lattice). Pairs of coils can be mounted around sets of poles and connected so as to generate dipole field across the magnet. A diagram showing this arrangement to generate vertical dipole field in a quadrupole is shown in the left-hand diagram of Fig. 5.10. Note however

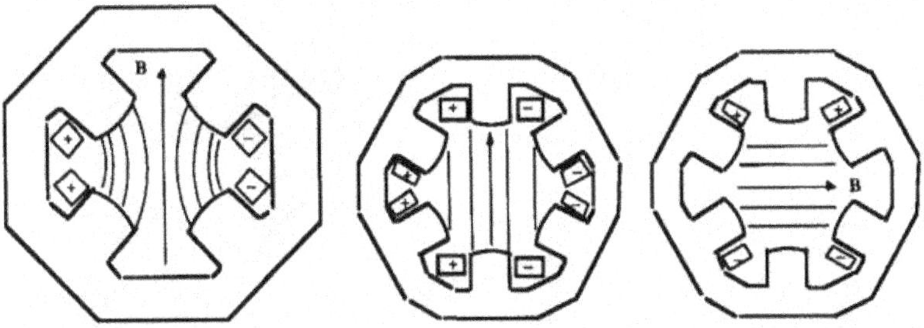

Fig. 5.10. Configuration and polarity of dipole correction coils on a quadrupole and sextupole magnet.

that the split yoke design, described in Sec. 5.2.3 above does not provide a return path for vertical dipole field so the corrections fields are limited to the horizontal plane if this configuration is used. The most serious problem, however, is the very poor quality of dipole field that is produced from this arrangement. The poles are much closer together at the edge of the aperture and hence the dipole field amplitude varies by about 30% over the inscribed diameter. This corresponds to a sextupole error on the dipole field; in the case of the ESRF design, it was estimated that if full dipole correction were applied on a quadrupole, the resulting sextupole field would be significant compared to the weakest lattice sextupole. Hence, the application of dipole steering would change the chromaticity. This possibility was therefore abandoned. Notwithstanding, a number of projects, including SUPERACO (Orsay Laboratory) and SLS (Paul Scherrer Institute)[17] make use, or propose to make use of dipole windings on the quadrupole magnet.

The sextupole magnets are much more suitable for the generation of vertical and horizontal dipole correction fields. The configuration shown in the centre diagram of Fig. 5.10 uses two separate pairs of coils to adjust the Ampere-turns coupling the different poles, and, by using a cosine theta distribution, vertical correction fields with quality of the order of ±2% can be generated. The six-pole configuration is not quite so suitable for generating horizontal correction fields. The required current distribution is shown in the right-hand diagram of Fig. 5.10 and field calculations show that good quality dipole field is generated over a very limited range close to the x axis. Depending on the extent of the required vertical good field region, this may be adequate.

The ESRF has both horizontal and vertical dipole correction capability incorporated in the sextupole magnet; the fields are generated by individually powered backleg as opposed to pole mounted windings, as these provide the flexibility and economy of space when the two field orientations are generated on the same yoke. In the case of the APS, a six-pole magnet that can accommodate the the vacuum chamber and the beam-lines has been used for the generation of both horizontal and vertical dipole correction fields.

Irrespective of whether auxiliary coils are added to the quadrupole and sextupole magnets, it is usually necessary to add small magnets, specifically designed to provide steering fields, into the lattice. All major projects, including the ESRF, the APS and SPring-8, have incorporated such trim dipoles into the lattice and these can be used as part of both the dynamic beam position control system, that is vital in a modern synchrotron source, and the closed orbit correction. These small correctors usually take the form of 'U' shaped steel yokes, with coils mounted on the limbs of the 'U', which are then placed in either the vertical or horizontal plane, depending on the orientation that is required for the magnetic field. However, in the case of the SPring-8 design the arrangement shown in

Fig. 5.11 is proposed. Three sets of coils are wound onto the rectangular yoke; the two coils 'A' provide vertical field across the electron beam, whilst the four coils 'B' give horizontal flux. With separate power supplies for these windings, full steering is obtained from a single magnet. Because of the importance of the dynamic beam position control, this topic is separately treated in Chapter 13.

This section has concentrated mainly on dipole correctors, but higher order fields, such as skew quadrupole and octupole are likely to be required and there may even be a need for individually controllable distributed quadrupole fields, as these provide a diagnostic method of measuring the variation of beta around the lattice. A magnet capable of developing all fields from dipole to ten-pole, with any angular orientation, was developed and used in the SRS (Daresbury Laboratory)[21,22]. The magnet consisted of twelve poles mounted on a circular yoke, with twelve individually powered backleg windings to provide the flexibility described. The magnet is currently in use to provide the main chromaticity sextupole fields in this second generation synchrotron source, with horizontal and vertical dipole used for steering, skew quadrupole for coupling control, and octupole for Landau damping during injection and energy ramping. The magnet provides great efficiency in use of lattice space but has the disadvantage of developing sextupole field that is much poorer in quality than that produced from the classical six pole design.

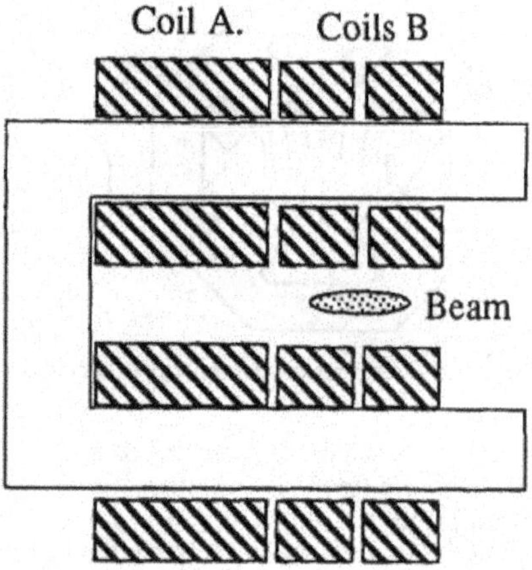

Fig. 5.11. Dipole correction magnet providing both horizontal and vertical field in a single design.

5.3.6 Magnet Ends

It will be appreciated that it is the integrated field through the magnet that determines the interaction with the electron beam and hence the behaviour of the fields at the ends of the magnet is significant.

The designer will wish to terminate the magnet in such a way as to determine the magnetic length, and control the quality of transverse field as the beam passes out into the field free straight. In an a.c. magnet, as would be used in a booster synchrotron injector of a radiation source, there is also the additional requirement of ensuring that flux does not enter the end of the magnet with a component that is normal to the end laminations, thus producing unacceptable eddy losses in this region. The standard solution is to 'roll off' the flux at the end of the magnet by increasing the gap over a limited region. This gives a control of the end field profile, limits the flux penetrating into the straight and prevents the local saturation of the steel that would result from using a square end. A number of standard roll-off geometries are available, but for low to medium field magnets, the main technical criterion will be to provide as big a roll-off as possible in a short space without introducing discontinuities of gradient that would lead to non-linearities. Of crucial importance is the control of field quality by altering the profile of the magnet as the gap increases. Where there are a number of different families of quadrupoles or sextupoles having different lengths, it is important to ensure that *the ends do not contribute* appreciably to the integrated harmonics in the magnet; the length of the different magnets

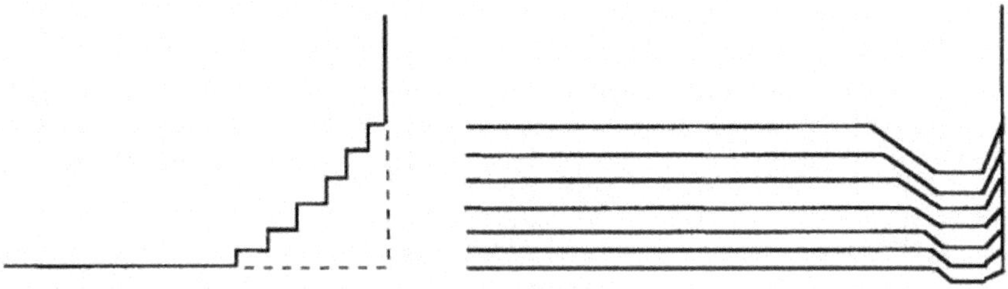

Fig. 5.12. End termination in a dipole; the increase in magnet gap seen azimuthally in the roll-off region (left-hand diagram) and the exaggerated shims in the transverse pane required to correct the field with the larger gaps (right-hand diagram).

will then be irrelevant when considering the full effect of magnetic errors on the beam behaviour.

An example of the enlarged shims that can be used in a dipole in the end roll-off is given in Fig. 5.12. As the gap increases in the azimuthal plane (left-hand diagram) it is necessary to enhance the size of the shim in the transverse plane in an attempt to maintain acceptable field distributions (right-hand diagram). Note that as the pole width in the dipole will have been determined by establishing the minimum dimension that provided an acceptable field distribution, no shim will be found that gives the specified field quality with the larger gap; however, the contribution of this end region to the total integrated magnetic length will be quite small, so a sub-standard local field quality can be accepted.

A similar technique can be adopted for a quadrupole, with a number of increasing inscribed radii used in the roll-off regions. As the radius is increased, the pole hyperbola should be adjusted. It is usually acceptable to approximate this to an arc of a circle in the end regions.

The standard two-dimensional computer codes can be used to determine the field distributions in the three-dimensional situations occurring at the magnet ends of a dipole. Field quality for each non-standard geometry can be examined as a two-dimensional slice through the magnet, and the profile of each roll-off section adjusted accordingly. The code can then be used to predict the variation of field in the azimuthal direction by taking a section in this third plane on the beam centre line. Such a model would take a short section with the normal gap, include the complete roll-off and extend into the straight at the end of the magnet which is nominally field free; such a cross section is shown in Fig. 5.13. The coil at the right-hand side of the diagram corresponds to the coil at the end of the magnet; the position of this coil strongly determines the flux penetration into the straight section and hence it is important to reflect the physical geometry in this region. However, the coil and the return yoke on the left are

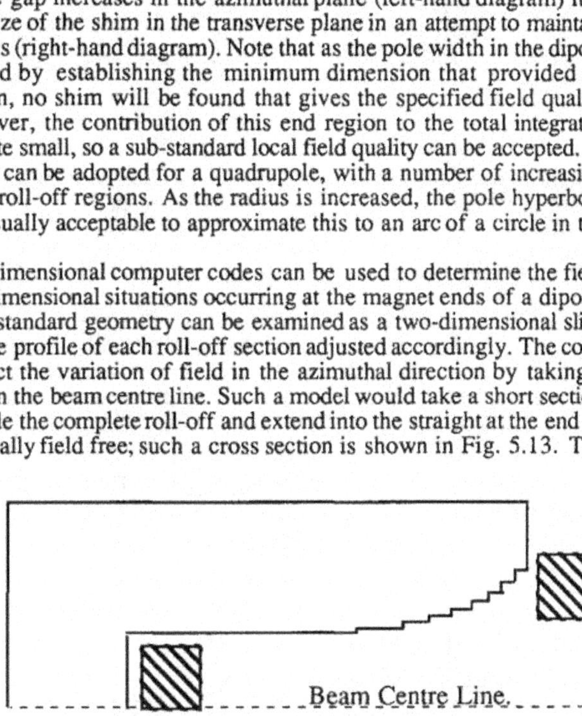

Fig. 5.13. Azimuthal cross section through magnet end with pesudo return yoke, used to calculate end field distribution.

artificial and are added to the model to provide a current and flux return path respectively. The two sets of calculations can then be combined numerically to give an estimate of the field length of the magnet at different radial positions. It should be recognised that this procedure is an approximation as it assumes zero rate of change in the azimuthal direction of the transverse components of flux density for each transverse slice that is taken. Under most circumstances, where the end fields contribute less than 10% to the total strength of the magnet, the procedure gives acceptable results. If the correct computation of the end field distributions is vital, three-dimensional codes should be used.

5.4 Coils

The major section above reviewed the implications of the magnet geometry. Whilst the coils necessary to generate the magnetic fields were shown in the various diagrams, the principal topic of discussion concerned the profiling of the steel poles and yokes. In this section, therefore, the more theoretical aspects of the other vital component of any electromagnet, the coils, will be considered.

5.4.1 Excitation

The excitation required for a given magnetic configuration is based on the application of Stoke's Theorem to Eqs. (5.2); this gives:

$$\oint \mathbf{H}.\mathbf{ds} = \sum NI \qquad (5.17)$$

i.e., the line integral of magnetic field around a closed loop equals the total current cutting the loop; this will be recognised as Ampere's law. In air:

$$\mathbf{B} = \mu_0 \mathbf{H} \qquad (5.18)$$

and by substituting Eqs. (5.4) and (5.18) into (5.17) and integrating over a limited region in air:

$$NI = \frac{1}{\mu_0}(\phi_1 - \phi_0) \qquad (5.19)$$

where ϕ_1 and ϕ_0 are the values of scalar potential along two equipotential lines and NI is the Ampere-turns, generated by remote currents, between the two lines.

Eqs. (5.6) for non-skew symmetrical magnets (i.e., where all J_n=0) is then applied along the pole centre line. By considering the difference in scalar potential between the magnet centre and the pole surface along this centre line, the general equation for Ampere-turns at the pole surface is obtained for any n:

$$NI = \frac{1}{n}\frac{1}{\mu_0}\left(\frac{B_y}{x^{(n-1)}}\right)r^n \qquad (5.20)$$

where r is the inscribed radius of the magnet (or half gap height in a dipole) and the term in the bracket is the magnet strength in $T/m^{(n-1)}$.

If this is applied to dipoles, quadrupoles and sextupoles, expressions for the Ampere-turns per pole, assuming infinite permeability in the steel, are obtained.

Ampere-turns per pole in a dipole ($\mu=\infty$):

$$NI = \frac{Bg}{\mu_0} \qquad (5.21)$$

where g is <u>half</u> the magnet gap. Ampere-turns per pole in a quadrupole ($\mu=\infty$):

$$NI = \frac{GR^2}{2\mu_0} \tag{5.22}$$

where G is the quadrupole gradient in T/m. Ampere-turns per pole in a sextupole ($\mu=\infty$):

$$NI = \frac{G_S R^3}{3\mu_0} \tag{5.23}$$

where G_S is the sextupole gradient (coefficient not derivative) in T/m^2.

In practical circumstances, the loss of magneto-motive force in the yoke due to the finite permeability of the steel will necessitate small adjustments to these expressions. For the dipole, a simple approximation is possible and the expression becomes (μ=finite):

$$NI = \frac{B(g + \ell/\mu)}{\mu_0} \tag{5.24}$$

where ℓ is the path length of the flux through <u>half</u> the yoke; the term ℓ/μ is known as the 'reluctance' of the steel. The expressions for the steel reluctance in quadrupoles and sextupoles are more complex and depend on a number of parameters, including the pole breadth and length.

5.4.2 Choice of Current Density

This topic was briefly discussed in the opening section addressed to non-technical readers. The problem is easily defined; the optimum current density is determined by minimising the sum of magnet capital and operational costs over the life of the synchrotron source. However, estimating the individual components of these costs is more difficult. The capital sum needed for a particular magnet can be estimated in consultation with manufacturers and data on how the material, tooling, handling costs etc. will vary as the current density, and hence the coil cross section, is changed should be sought. After capital estimates have been obtained, their 'amortisation' over the life of the installation is calculated. Not only the capital sum is considered, but the nominal interest that would have to be paid if the capital were to be borrowed is included; exclusion of this figure gives a false optimum. Different funding authorities will have varying attitudes to this topic, and designers should seek appropriate advice before making a final choice of current density.

5.5 Materials and Fabrication

In this final section, the materials that play the most crucial role in the magnet — the steel from which the yoke is constructed and the conductor used in the coils — will be reviewed. Fabrication methods for laminated magnets will also be summarised.

5.5.1 Magnet Steel

Whilst many magnets built for d.c. applications, such as cyclotron yokes and large experimental spectrometer magnets, are fabricated from forged low carbon solid steel, it is usual to assemble the yokes of synchrotron and storage ring magnets from laminated material. This has a number of advantages:

A wide range of steel with different magnet properties is available, and the designer can choose the material most suitable for the particular accelerator application.

If the magnet is required to change field during operation, to ramp the energy of the beam for example, eddy current effects will be present and even rates of change as small as a fraction of a Tesla per minute would cause appreciable disturbance to the beam in a magnet with a solid steel yoke. Even in magnets with steady d.c. excitation, the switch-on transients will last several minutes in a solid yoke.

After stamping, the location of laminations in the completed magnets can be randomised by means of a shuffling process; this ensures that each magnet contains roughly the same mix of magnetic properties and physical dimensions and provides far greater homogeneity between magnets of the same family within a lattice.

It is then important to choose a lamination thickness that is suitable for the magnet design and which provides best economy during the construction process. In general laminations of up to 5 mm are suitable for d.c. and slowly varying applications and the reduction in the number of individual laminations that have to be handled during the manufacturing process would lead to significant savings in the construction budget. However, few manufacturers offer such thick material and the stamping of the laminations would require very heavy duty hydraulic presses; steel of 1 mm to 1.5 mm is therefore often used for such applications. Many booster synchrotrons operate at repetition rates of the order of 5 Hz to 10 Hz; 1.5 mm is then the maximum suitable thickness if excessive losses are to be avoided. Finally, at 50 Hz material of thickness 0.35 mm is regarded as the power industry norm.

The magnetic and loss characteristics of electrical steel laminations is controlled by the heat treatment applied during the rolling at the mill and also by the chemical composition of the material. Of particular importance is the exclusion of carbon and the addition of carefully controlled quantities of silicon. High silicon steel gives very low a.c. losses, low coercive force (which is valuable for preventing excessive residual fields at a low injection energy) and high permeability at low fields. These desirable characteristics are obtained with the penalty of poor permeability at high inductions. As the quantity of silicon in the steel is reduced, the high field properties improve whilst the a.c losses increase and the low field behaviour worsens.

Examples of three different types of laminated steel are illustrated in Table 5.6.

Table 5.6. Loss and magnetic parameters for three different grades of commercially available steel.

Type of steel	**Non-oriented Low silicon**	**Non-oriented High silicon**	**Oriented High silicon (*)**
Silicon content	~1%	~3%	~3%
Lam. thickness	0.65 mm	0.35 mm	0.27 mm
a.c. loss (50 Hz, 1.5 T peak field)	6.9 W/kg	2.25 W/kg	0.79 W/kg
Permeability:			
at B=0.05 T	995	4,420	not quoted
at B=1.5 T	1680	990	> 10,000
at B=1.8 T	184	122	3,100

(*) Data for oriented steel is for properties parallel to the rolling direction.

The table shows loss and magnetic parameters for three different grades of commercially available steel. The first two, described as 'non-oriented', demonstrate the substantial increase in low field permeability and the reduction in a.c. losses that result from the inclusion of silicon at about 3% concentration (and in the case of the a.c. loss, the availability and use of thinner laminations); however, it can also be seen that permeability at high inductions is nearly halved. The steel detailed in the final column is of the 'grain oriented' type; this material shows high anisotropy, produced by the method used for rolling the material at the steel mill. In the direction of the grain, the magnetic properties are excellent, with very low losses and improved permeability at all values of magnetic induction; low coercivity is also obtained. At right angles to the rolling direction, all properties are considerably worse than those found in non-oriented material. Because the flux will normally change direction in a lamination used for an accelerator application, it is difficult to take advantage of the superior properties of grain orientated material and it is seldom used. Generally, then, a non-oriented high silicon grade electrical steel would be used for any application in which there are appreciable oscillating fields. Such material would be needed in a booster synchrotron, or in a servo-controlled dipole steering magnets in a storage ring where the feedback loop is required to have a frequency response that exceeds a few Hertz. Low silicon grades, containing about 1% silicon, would be used for the d.c. magnets in a storage ring to give the best possible permeability in the critical shim areas on the pole, where inductions of up to 1.5 times the flux density at the beam can occur. Note, however, that some silicon is required to mechanically stabilise the steel and exclude the need for a further annealing process after the laminations have been punched.

In alternating field situations, it is important to have adequate inter-lamination insulation to prevent excessive eddy current loss. For many years, coatings of inorganic oxide on one side of the laminations were used. These coatings had good heat and radiation resistance but suffered from the disadvantage of having thicknesses up to 50 μm and hence reducing the packing factor of the stack of laminations. More recently, steel manufacturers have introduced a range of inter-lamination insulations that can be as little as a few micro-meters thick; both organic and inorganic materials are available, the latter being more suitable for accelerator applications on account of their heat and radiation stability.

In all matters relating to the steel's magnetic and loss properties and the choice of inter-lamination insulation, designers are advised to work closely with the potential steel suppliers. When designing a.c. magnets, data can be obtained that provides the specific loss (usually in Watts per kilogramme of steel) as a function of field oscillation amplitude, often at a number of different frequencies, 50 Hz and 60 Hz being common. Accelerator applications usually employ magnetic cycles which are significantly different to the operating norms of the power industry. The magnet designer may therefore need to interpolate or extrapolate from the curves provided in order to obtain a realistic estimate of the expected loss in the a.c. magnet. The loss data supplied by the manufacturer will include eddy loss, which varies as the square of both the a.c. frequency and amplitude, and hysteresis loss, which is linear with frequency but shows a more complex variation with amplitude. Bear in mind also that eddy loss per kilogramme will vary as the square of the lamination thickness, but, for material with identical chemical and rolling history, the hysteresis loss will be independent of lamination thickness. The designer can then make use of these relationships for numerical calculations. In some circumstances, the designer may be able to persuade the potential supplier to carry out loss and magnetic measurements for the particular application and to consider special treatments to give the required low losses and good magnetic properties. In such circumstances, the designer is relying very much on the good will of the steel company, for, in all but the largest projects, the amount of steel that is to be purchased is a very small fraction of the monthly output from the mill.

5.5.2 Conductor Material

Copper conductor is used in the great majority of accelerator magnet coils. However, a small number of projects have made use of aluminium for the conductor material, so it is worthwhile examining the economic and technical issues surrounding the choice between these two conductor materials.

Aluminium has 65% the conductivity of pure copper, so an aluminium coil would need 1.54 times the cross section to give the same resistance; the total coil would have a volume somewhat greater than this factor on account of the increased length of conductor at the semi-circular ends. Aluminium is a light metal, having only 30% of the density of copper, so this enlarged coil would still be only approximately 50% of the weight of the copper windings. The price of bulk metals quoted on the commodities market fluctuates; at the time of writing the price per tonne of aluminium is about 63% that of copper. Therefore the conductor material for the aluminium coil is only about 30% of the cost of the copper required to give the same resistance.

Why, then, are all accelerator magnets not constructed using aluminium conductor? The reason is that the cost of the conductor material is only a small fraction of the cost of the total magnet and whilst the use of aluminium gives lower coil conductor raw material cost, the larger volume of conductor leads to increased cost in nearly all other items. The manufacture of coils is labour intensive; the larger coils will result in most operations (winding, wrapping with outer-ground insulation and release film, etc.) taking longer. The tools required for the coil fabrication (winding mandrill, potting mould etc.) will be larger. The magnet yoke will also be larger, with increased steel, tooling and handling costs. The consequence of these changes is an equation with many components, the values of which can really only be determined by an experienced manufacturer. Experience, however, indicates that in electrical engineering situations where high precision and extensive manual intervention is required, as with an accelerator magnet, copper appears to be the optimum material. In simple situations, such as distribution transformers, where the coil winding and insulation is straightforward, aluminium can sometimes be the more economic solution.

There are technical differences between copper and aluminium that should also be considered. Joining copper sections, to make continuous pieces of conductor that are longer than those delivered from the copper producer, can be carried out by brazing in air; aluminium must be welded in an inert atmosphere, so this increases the technical complexity of producing aluminium coils. The other significant difference favours aluminium, for this metal does not work-harden when bent and mildly distorted, events that will occur during coil winding. Copper does become more brittle when worked in this way and a manufacturer who makes an error during coil winding cannot recover the material without re-annealing before further use.

5.5.3 Construction

The construction of accelerator magnets is a highly specialised manufacturing process and a number of firms in Europe, USA and Japan specialise in this field. Given below is a brief resume of the current methods being employed.

Construction of laminated yokes commences with the manufacture of a stamping tool that will be needed to press the individual laminations out of the packs of steel delivered from the rolling mill. Lamination stamping is an established technique in the electrical engineering industry, and many workshops are capable of producing the necessary tooling. In the case of accelerator applications however, the highest standards of workmanship are required, with pole profile tolerances of ±25 μm often being called for. A large hydraulic press is required to perform the stamping operation. Lamination dimensions should be checked regularly (of the order of every 5,000 laminations) during the stamping run, and the tool should be resharpened by skimming the head whenever any deterioration in lamination tolerances are observed.

After all laminations have been stamped, they should be shuffled, as described in Sec. 5.5.1 above. The yokes are then assembled on a precision fixture which locates the individual laminations in preparation for permanent bonding. It is essential to control the amount of steel that is assembled for each magnet in order that the magnetic lengths will be the same to a high tolerance. It is therefore usual to weigh the laminations as they are introduced into the stacking fixture, and keep to a standard weight to within one lamination.

Prior to bonding, the stack of laminations is compressed on a hydraulic ram; in the past, gluing using epoxy resin was the standard method for securing the laminations, but this has now been superseded by machine welding at the corners and mid outer sides of the yoke to produce a rigid block. To avoid shorted turns between the laminations no welding should occur on the inner faces of the magnet. Hence, a thick, rigid end plate (of the order of 30 mm thickness, but greater for large cross sections) is required to prevent the poles spreading, and this plate is often manufactured by gluing end laminations together.

Coil manufacture is dominated by the need for scrupulous cleanliness. Prior to winding the conductor onto the coil former it is usual to carry out some form of abrasive cleaning to remove all dirt and grease. This is done immediately before wrapping the conductor material with glass cloth. Winding under tension onto a former then produces the required coil geometry. The glass cloth introduced during this stage is intended to provide the inter-turn insulation only, so on completion of winding, the thicker outer ground class cloth is wrapped around the coil. The coils are finally wrapped in release tape, a substance that will not bond to the resin that will be used for impregnation, and the complete assembly inserted into a mould.

A number of moulds containing coils are then introduced into a vacuum oven, as the next batch process stage is most efficient when carried out on as many units as can be assembled into the oven. The coils in vacuum are then heated, usually significantly above 100° C, to dry the assemblies. The baking continues until it is known, either from experience or gas analysis, that all water has been expelled. The moulds are then flooded with liquid resin and the vacuum tank immediately let up to atmosphere so that the resin is forced into the insulation between the layers of the coil. The resin is then cured according to the instructions given by the resin supplier. The curing cycle will depend on the temperature at which the coils are expected to operate; in general, high operating temperatures require high curing temperatures, to give the finished coils the necessary thermal stability.

The choice of resin for the impregnation of the coil, is important, for these are organic materials and can therefore degrade rapidly in high radiation environments. The storage rings in light sources do not generally have this problem; beam loss during injection will lead to high energy radiation being absorbed by the magnet components, but loss during stored beam will, due to the very nature of the facility, be minimal. The booster injector will have a similar radiation history if only used for filling the main ring, but may suffer from radiation damage if used for other purposes between fills. Notwithstanding, it is highly desirable to use a high radiation compatible resin for the potting of the coils on all the accelerator magnets, for there are no great drawbacks in using such high quality material. Proprietary resins are available with guaranteed radiation tolerances; these were developed for high energy proton synchrotron applications and therefore should give long reliable operating life in a synchrotron light source.

Inert fillers can be used to reduce the the amount of resin needed for impregnation, but this economy is not recommended, as it can be used to obscure poor winding and impregnation techniques. Providing such fillers have not been used, the impregnated glass cloth in the completed coil should be fully transparent, and it should be possible to see the conductor clearly. Any white or opaque areas indicate that the coil was not fully impregnated, and this represents a source of both mechanical and electrical weakness. The completed coil should also be fully water tight, and high voltage tests are often carried out with the body of the coil fully immersed in a water bath, with only the terminals clear. Such rigorous testing is recommended to provide reliable and durable coils.

5.6 References

1. N. Marks and M. Lieuvin, *IEEE Trans. on Magnetics* **24,2** (1988) 741.
2. V. Korchuganov, E. Levichev and A. Philipchenko, *Proc. of 1993 Particle Accelerator Conference,* (1993) 2793.
3. W.K.H. Panofsky and M. Philips, *Classical Electricity and Magnetism,* Addison-Wesley, 65.
4. K. Halbach, *Nuclear Inst. and Meth.,* **74** (1969) 147.
5. J-M. Filhol, *Proc. of 1992 Eur. Par. Accel. Conf.,* (1992) 471.
6. M. Lieuvin, A. Ropert, L. Farvacque, *Proc. of 1992 Eur. Par. Accel. Conf.,* (1992) 1347.
7. R. Keller, *Proc. of 1993 Particle Accelerator Conference,* (1993) 2811.
8. C.H. Chang, H.C. Liu and G.J. Hwang, *Proc. of 1993 Particle Accelerator Conference,* (1993) 2886.
9. F. Gnidica et al, *Proc. of 1992 Eur. Par. Accel. Conf.,* (1992) 1358.
10. European Synchrotron Radiation Facility, Grenoble, France, *Foundation Phase Report,* (1987).
11. Argonne National Laboratory, *7 GeV Advanced Photon Source Conceptual Design Report,* **ANL-87-15**, (1987).
12. Jaeri-Riken SPring-8 Project Team, *SPring-8 Facility Design (Revised),* (1991).
13. Lawrence Berkeley Laboratory, *1-2 GeV Synchrotron Radiation Source,* (1986) 63.
14. C.H. Chang et al, *Proc of 11th Int. Conf. Mag. Tech.,* (1989).
15. Sincrotrone Trieste, *ELETTRA Conceptual Design Report,* (1989).
16. M.W. Poole et al, *Proc. of 1993 Particle Accelerator Conference,* (1993) 1494.
17. Paul Scherrer Institut, Switzerland, *Conceptual Design of the Swiss Synchrotron Light Source,* (1993).
18. G.N. Kulipanov, *Rev. Sci. Instrum.,* **63** (1992).
19. A. Ropert, *Proc. of 1993 Particle Accelerator Conference,* (1993) 1512.
20. BESSY Laboratory, Berlin, *The BESSY II Parameter List, version 1,* (1993).
21. N. Marks, *Proc 6th Int. conf. Mag. Tech* (1977).
22. R.P Walker, *Proc 7th Int. conf. Mag. Tech* (1981).

CHAPTER 6: MAGNET POWER SUPPLIES

Rudolf Richter
Sincrotrone Trieste
Padriciano 99
34012 Trieste, Italy

6.1 Introduction

Magnet power supplies are the components of a synchrotron radiation source that supply the current to energise the various magnets which build up the ring. The different excitation levels and range for the current settings are defined by the magnet design, the optics and the different operation energies of the accelerator. Generally speaking a power supply converts the 3 phase input power into a DC- current or a well defined current function for a group of magnets of an accelerator.

From the applications point of view the supplies can be divided into two major groups:

a) static converters with low dynamic requirements for storage rings and transfer lines being highly stabilized DC current sources

b) supplies with high dynamic requirements for cyclic accelerators such as booster synchrotrons used as injectors, providing a well defined current function which could be trapezoid-like or a type of sine wave function, respecting tight specifications related to that function.

Within these groups it is quite common to subdivide the supplies into groups according to the load and power rating:

i) bending magnet supplies are in the range of several hundred kilowatts to several megawatts,

ii) quadrupole and sextupole magnet supplies are in the range of several ten kilowatts to several hundred kilowatts,

iii) corrector magnet supplies are in the range of several hundred watts to several kilowatts.

Since the reliability of a facility is very important the power supplies have to be built in a straightforward and simple manner, with particular attention to:

reliability
avoiding electro-mechanical components wherever possible
ease of repair
simple structure
high degree of standardisation of the components of the supplies
identical standards of the interface to the control system

Figure 6.1 shows the block diagram of a power supply representing the main parts of that system: the basic power supply consisting of the power part with its regulation and interlock, the interface unit to the control system containing a micro processor, DAC (digital to analog converter), ADC (analog to digital converter) and the interface for commands and status in order to drive and control the supply.

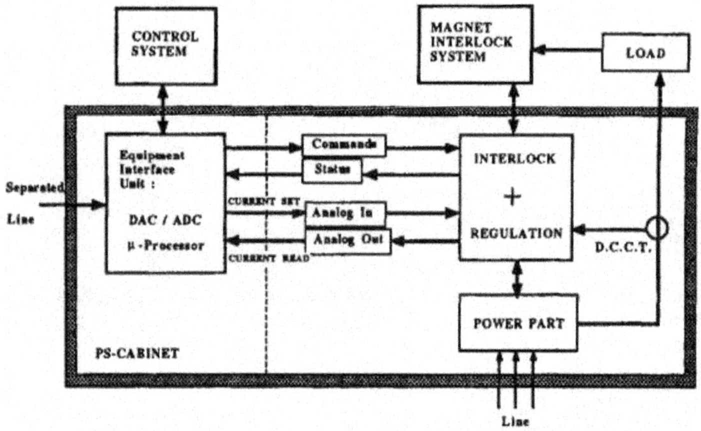

Fig.6.1. Block diagram of a power supply.

6. 2 DC-current Supplies

6.2.1 Power Part

6.2.1.1 The principle of rectifying

As mentioned previously a power supply has to convert the AC power from the three phase system, to which it is connected, into a DC or AC power for a single phase system supplying a magnet or a chain of magnets. This can be provided by a device which is able to rectify and control the AC input voltage.

Applying an AC voltage on a diode and a resistor as a load (see Fig. 6.2a), the diode is conducting during the positive half sine wave. The current through the resistor follows exactly the driving voltage. Using a bridge structure (see Fig. 6.2b), the negative half sine wave will also contribute to the current through the resistor.

In a three phase system the phases are shifted by 120 degrees relative to each other. Using three diodes, one for each phase, one has an array as shown in Fig. 6.2c. Each diode is conducting at the positive half sine wave, the commutation from one phase to the other happens at that moment when the voltage of the following phase becomes higher than the voltage at the conducting diode, and the next diode is taking over the conduction of the current.

To utilise also the negative part of the sine wave one has to choose a bridge structure, for example, the one shown in Fig. 6.2d. The commutation of the current in the bridge works as follows: starting with phase R+, the diodes are conducting in the following sequence: R+ with S-, R+ with T-, S+ with T-, S+ with R-, T+ with R-, R+ with S-. In all these examples the load is a resistor, therefore the current follows exactly the variation in time of the voltage in accordance with Ohm's law. Fig. 6.2d shows also the ideal output voltage V_{di} of a 6 pulse diode bridge.

Substituting the diodes by thyristors the moment of turning on of the conducting path can be controlled. A thyristor is turned off when the current through the device goes below the minimum conducting current, which is close to zero. Triggering the thyristors at the same moment as the commutation of the current in a diode bridge, also called the

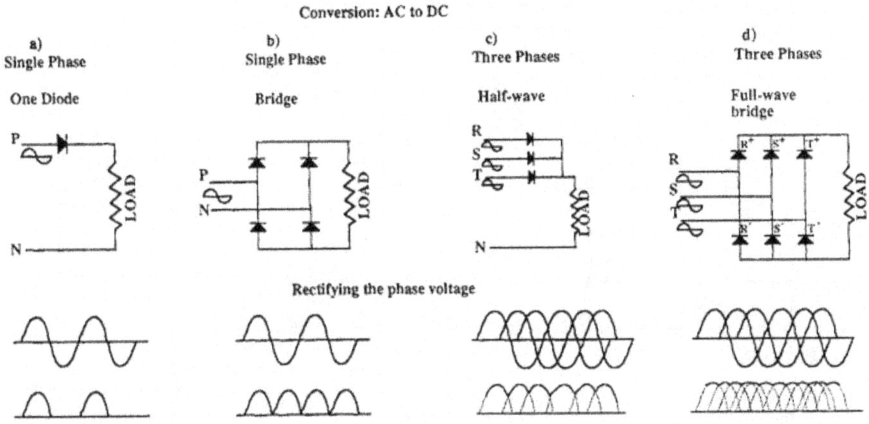

Fig. 6.2. Basic rectifiers.

moment of natural commutation, it has identical output voltage. By delaying the trigger against the moment of natural commutation the output voltage is controlled. The ideal output voltage $V_{di\alpha}$ of a bridge is given by [1]:

$$V_{di\alpha} = s\,(q/\pi)\sqrt{2}\,V_{ph}\,\sin(\pi/q)\,\cos\alpha = V_{di}\,\cos\alpha \qquad (1)$$

α: control angle (phase between natural commutation and trigger of thyristor)
V_{ph}: rms voltage of one phase
q: number of commutations during one period of the phase
s: $s = 2$ for bridges, $s = 1$ for a single phase arrangement

The characteristic curve of this bridge as a function of the firing angle α is shown in Fig. 6.3a. Curve a shows the characteristic curve for a resistive load, and curve b shows the characteristic curve for an inductive load.

The behaviour changes if, instead of a resistor, an inductance is connected as a load. The inductance as an energy storing device keeps the current going even at negative voltages, see Fig. 6.3b, which means that the energy stored in the inductance will be fed back into the grid. If feedback into the line should be avoided, a "freewheeling diode" is used in parallel to bypass the bridge, the energy being dissipated by that diode and the resistive part of the load.

On an inductive resistive load one 6 pulse bridge can operate with positive voltage and positive current as well as with negative voltage and positive current, which is called a 2 quadrant operation; by combining two 6 pulse bridges in parallel with opposite output voltage a 4 quadrant converter can be built up:

first quadrant:	positive voltage, positive current
second quadrant:	negative voltage, positive current
third quadrant:	negative voltage, negative current
fourth quadrant:	positive voltage, negative current

Due to the large power capacity of the grid the mains voltage remains stable if a load, which does not exceed the power capacity, is connected to it. Therefore a converter connected to the grid has to be considered as a voltage source. By measuring the output current and by controlling the output voltage according to the desired current the converter becomes a current source.

6.2.1.2 Reactive power and higher harmonics of the line current

In electro technics the power factor $\cos\phi$ is related to the fundamental line frequency and is defined in the following way [1]:

real power: $P = V_{ph}\, I_{ph} \cos\phi$ (2)

reactive power: $Q = V_{ph}\, I_{ph} \sin\phi$ (3)

with: $S = \sqrt{P^2 Q^2}$ (4)

V_{ph} phase voltage
I_{ph} current drawn of that phase

Considering the idealised theory of functioning of the bridge, which means pure sine wave line voltage and ideal continuous DC current, the reactive power depends on the control angle α as follows: by the control angle α (the phase between natural commutation and trigger of thyristor) the mean output voltage is controlled, see section 6.2.1.1, resulting also in a phase shift on the primary side of the rectifier between the grid voltage and the current drawn by the supply in each phase (see Fig. 6.3c). The current is delayed against the voltage by α. The real power drawn by the supply, expressed with the output voltage V_{di} and the output current I_d is given by [2]:

$$P\alpha := V_{di}\, I_d \cos\alpha = V_{di}\, I_d \cos\phi. \tag{5}$$

It can be shown that under this ideal condition the cosine of the control angle α is equal to the power factor.[2] The reactive power is given by:

$$Q\alpha := V_{di}\, I_d \sin\alpha = V_{di}\, I_d \sin\phi. \tag{6}$$

This means that a thyristor controlled converter produces reactive power due to its function of controlling the output voltage; this is true even for a purely resistive load. Figure 6.3d shows the reactive power of a converter for a fixed output current as a function of $\cos\alpha$; the effect of a freewheeling diode[3] is also shown.

If the converter is conducting an ideal DC current, the current drawn at $\alpha = 0$ out of a single phase is a square wave function. The Fourier analysis of this current function

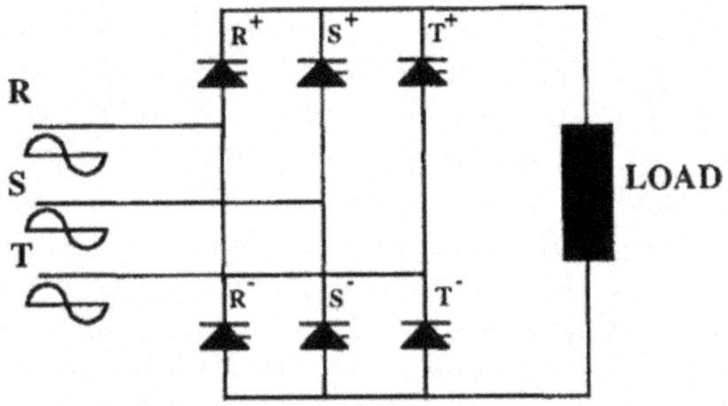

Commutation:

R^+ with S^- → T^-

S^+ with T^- → R^-

T^+ with R^- → S^-

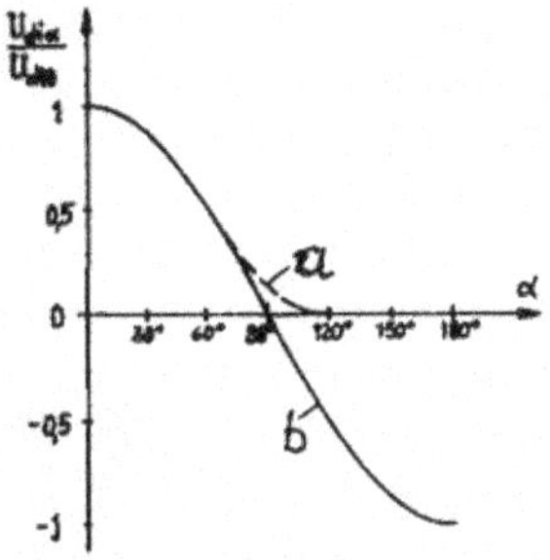

a: resistive load
b: inductive load
Fig. 6.3a. Characteristic curve.

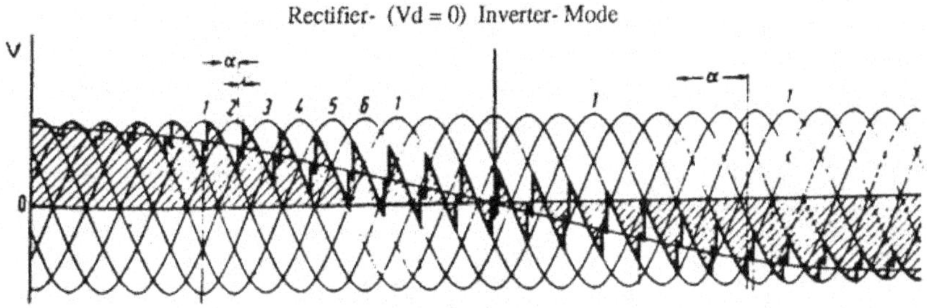

Fig. 6.3b. Output voltage $V_{di\alpha}$.

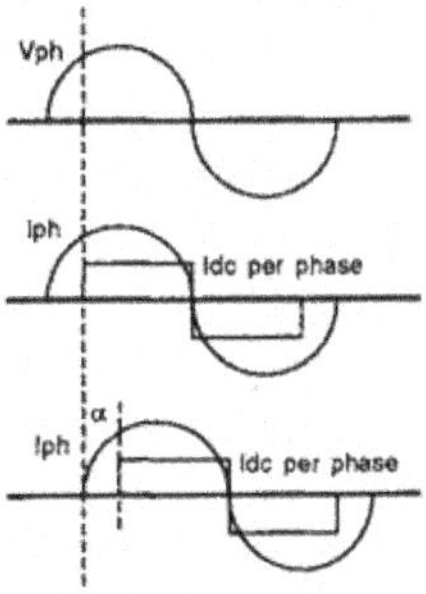

Fig. 6.3c. Phase shift by α.

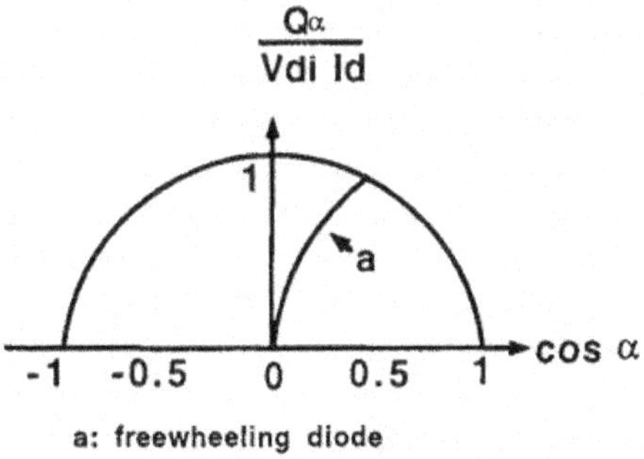

Fig. 6.3d. Reactive power

Fig. 6.3. 6-pulse bridge.

leads to the higher harmonic content shown in Table 6.1.[1] The dependence of the harmonics in the primary current with α can be found in the literature.[1,2,4] Depending on the power capacity of the grid, the reactive power and the higher harmonics of the line current produced by a thyristor controlled converter can create significant distortions of the line voltage, which can harm other loads connected to the same grid.

Table 6.1. Higher harmonics of output voltage and primary current.[1]

P	3		6		12	
v	V_{vi}/V_{di} [%]	I_{vi}/I_{phi} [%]	V_{vi}/V_{di} [%]	I_{vi}/I_{phi} [%]	V_{vi}/V_{di} [%]	I_{vi}/I_{phi} [%]
2	-----	50.00	-----	-----	-----	-----
3	17.68	-----	-----	-----	-----	-----
4	-----	25.00	-----	-----	-----	-----
5	-----	20.00	-----	20.00	-----	-----
6	4.04	-----	4.04	-----	-----	-----
7	-----	14.29	-----	14.29	-----	-----
8	-----	12.50	-----	-----	-----	-----
9	1.77	-----	-----	-----	-----	-----
10	-----	10.00	-----	-----	-----	-----
11	-----	9.09	-----	9.09	-----	9.09
12	0.99	-----	0.99	-----	0.99	-----
13	-----	7.69	-----	7.69	-----	7.69
14	-----	7.14	-----	-----	-----	-----
15	0.63	-----	-----	-----	-----	-----
16	-----	6.25	-----	-----	-----	-----
17	-----	5.88	-----	5.88	-----	-----
18	0.44	-----	0.44	-----	-----	-----
19	-----	5.26	-----	5.26	-----	-----
20	-----	5.00	-----	-----	-----	-----
21	0.32	-----	-----	-----	-----	-----
22	-----	4.55	-----	-----	-----	-----
23	-----	4.35	-----	4.35	-----	4.35
24	0.25	-----	0.25	-----	0.25	-----
25	-----	4.00	-----	4.00	-----	4.00

P: number of pulses per period
v: harmonic number
V_{vi}: output ripple voltage (rms) of the vth harmonic
V_{di}: ideal output voltage
I_{vi}: input ripple current (rms) of the vth harmonic
I_{phi}: input current of the phase

6.2.1.3 Voltage ripple and its suppression

As demonstrated in Fig. 6.2 the ripple depends on the number of pulses per period of the phase, see also Table 6.1. Taking into consideration the attenuation of the vacuum chamber and the maximum allowed error in the field the maximum allowed current ripple could be defined. Taking into account the impedance of the magnets the allowed voltage ripple is then defined.

The standard way of reducing the ripple is a passive low pass filter. A damped filter with only one capacitor path and the damping resistor in that path can be used; a filter with one damped and one undamped capacitor path gives a steeper attenuation with frequency. The design of a passive filter is basically done by considering the higher harmonics produced by the rectifier with reference to the line frequency.[1,5]

For magnets with a time constant of the order of 1 sec a passive filter may be sufficient for the attenuation of the voltage ripple, but the line variations, which have to be considered as very low frequency ripple, can pass through without any attenuation. In this case an active filter is required.

Magnets with time constants of the order of 0.2 to 0.5 sec, typical time constant of a quadrupole or sextupole, usually need an active filter for sufficient ripple attenuation.

A typical solution for large power supplies is an active filter consisting of a small power supply acting on a choke or transformer. The reference signal is the output voltage of the supply. This system works as a voltage feedback independent of the main regulation (see Fig. 6.4a). This active filter limits the dynamic behaviour of the supply for ramping procedures (see section 6.5.2). Another possibility is a voltage source in parallel to the main rectifier as described in section 6.5.2.[6]

The classical linear regulator (see Fig. 6.4b) gives good dynamic behaviour for ramping, but for high current supplies, due to the number of components, its reliability is less than a converter configuration shown in Fig. 6.4a. The normal use of a linear regulator is for low power applications in monopolar or bipolar versions, such as for corrector magnets.

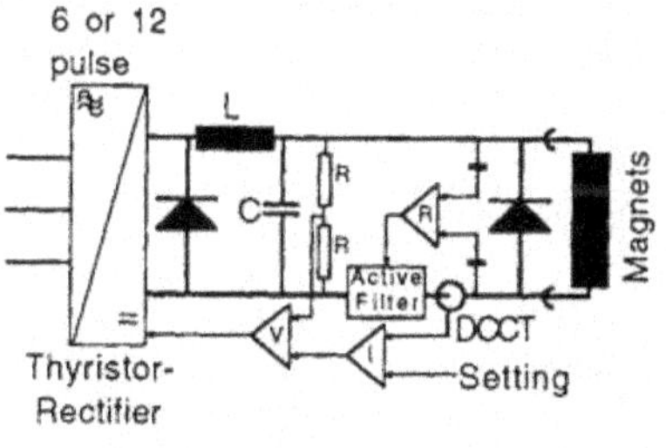

Fig. 6.4a. Thyristor rectifier.

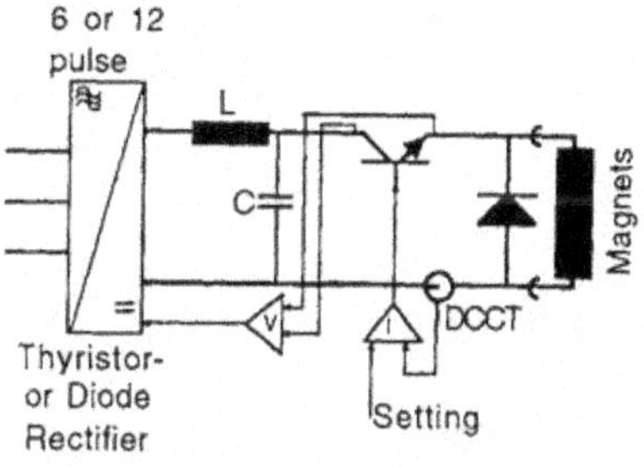

Fig. 6.4b. Linear regulator.

Fig. 6.4. Classical converter configurations.

6.2.1.4. *Switch mode technique*

This technique is based on turning on and off the voltage across the load. By varying the time of the pulse trains the output current is controlled (see Fig. 6.5). There are two possible methods[1]:

i) keeping the pulse duration T_p of the output voltage, which is equal to the current conducting time, constant and changing the repetition period T of the pulses (see Fig. 6.5a).

ii) keeping the repetition period T of the pulses constant and changing the pulse duration T_p of the output voltage. This is known as pulse width modulation (PWM) (see Fig. 6.5b).

The basic structure of a switch mode supply consists of an input rectifier with a filter as input stage providing an intermediate DC voltage, a high frequency inverter or chopper, and an output rectifier with filter (see Fig. 6.5c). The output voltage V_0 depends on the intermediate DC voltage V_{DC} as follows:

$$V_0 = V_{DC}\, T_p/T. \tag{7}$$

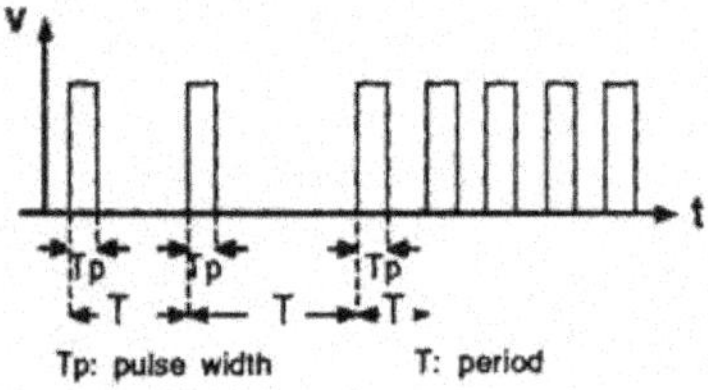

Fig. 6.5a. Period variation.

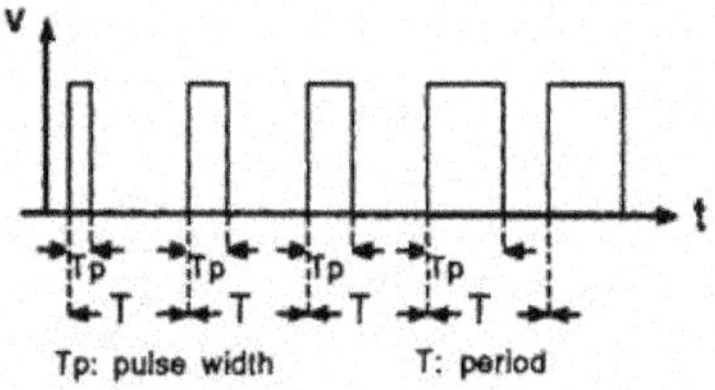

Fig. 6.5b. Puls width modulation. (PWM)

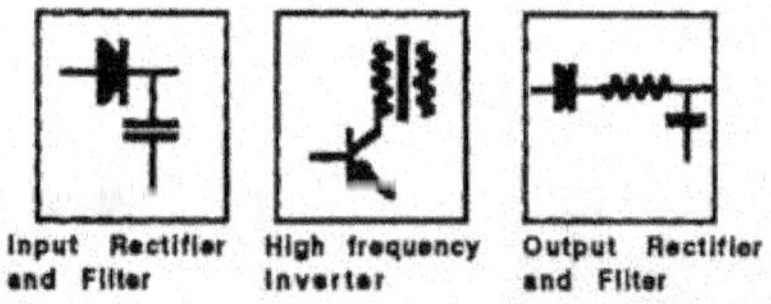

Fig. 6.5c. Principal PS structure.
Fig. 6.5. Switch Mode Principle.

The characteristics of a switch mode supply can be summed up as follows[7]:

high bandwidth,
power factor of 0.95 (for 3 phase input)
small residual ripple at output
light weight, reduced volume

The input stage is a normal rectifier supplying a single switching stage or a DC bus to which DC to DC converters in this technique are connected. The switch stage can be built up in different ways:

i) DC chopper, see Fig.6.6
ii) resonant converter, see Fig. 6.7
iii) bridge converters, see Fig. 6.8

Examples of a DC chopper are the quadrupole and sextupole supplies of the Advanced Photon Source,[8] see also Fig. 6.6. The supplies consist of a DC bus system to which DC to DC converters are connected. The converters work with a switching frequency of 20 kHz with the following ratings:

Quadrupole: 0 to 460 A, stability: $\pm 6 \cdot 10^{-5}$
Sextupole: 0 to 200 A, stability: $\pm 3 \cdot 10^{-4}$

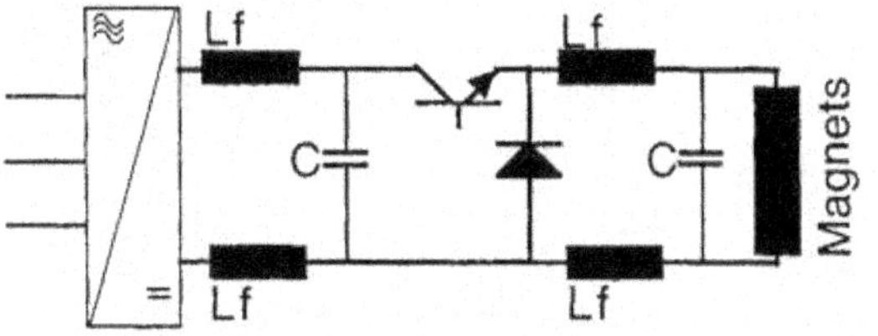

Fig. 6.6. DC chopper.

For LEP a double resonant converter of 37.5 kW with a switching frequency of 12.5 kHz is used[9] (see Fig. 6.7), providing the following ratings[7]:

125 V	300 A	stability: $\pm 2 \cdot 10^{-4}$
188 V	200 A	stability: $\pm 2 \cdot 10^{-4}$
250 V	150 A	stability: $\pm 2 \cdot 10^{-4}$

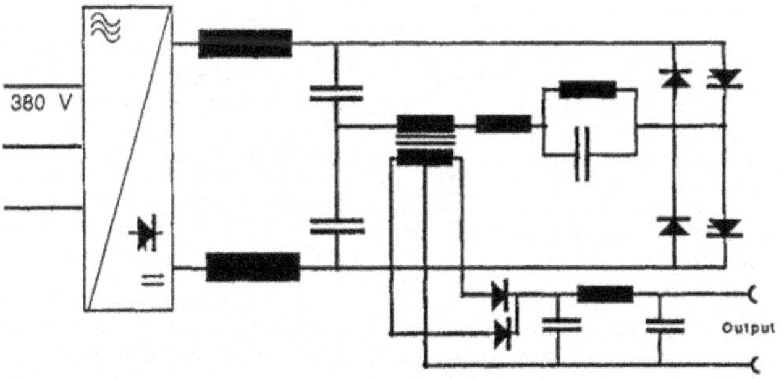

Fig.6.7. Resonant inverter.

The corrector supplies for LEP[10] are examples of a half bridge inverter working with a switching frequency of 50 kHz followed by an inverter bridge for bipolar operation (see Fig. 6.8). The rating is:

675 W ± 135 V ± 5 A stability: $\pm 1 \cdot 10^{-5}$

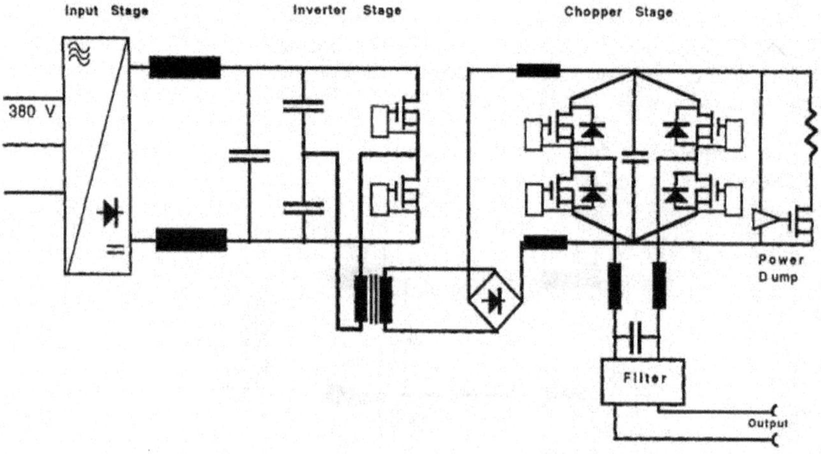

Fig. 6.8. Bridge inverter.

6.3 Correction Schemes

6.3.1 Additional correction with a main magnet

The correction schemes are not only related to corrector magnets. For example, for tune feedback systems and for overcoming the effects of insertion devices there is the need for varying the current of single or small groups of quadrupole magnets of one magnet family. This problem can be solved in different ways:

i) providing correction windings in the magnet and powering them by an additional supply,

ii) connecting an additional supply directly to a single magnet of the family. In these two solutions the main supply providing the main current and the additional supply taking care of the current of a specific magnet of the family "see" each other by the ratio of the impedance of the whole family to the inductance of the single magnet. The transformer ratio of the main windings and the correction windings of the single magnet, if there are correction windings, have also to be taken into account. The supplies communicate to each other any variation of their output voltage, which are the ripple, variation of the output voltage due to regulation actions, turn on and off and, last but not least, emergency turn off. The feasibility of these solutions depends on the requested accuracy, the amount of the correction of the main current, the needed output voltage of the supplies, which is defined by the impedance of the magnets and the transformation ratio between the inductance of the main magnets and the correction coil or the single magnet powered by an additional supply.

iii) splitting up the whole family into independent subgroups. This provides the highest degree of flexibility and avoids all problems of cross talk between power supplies feeding the same magnet; however, care has to be taken for the calibration of the supplies to maintain the correct currents with respect to the overall family.

6.3.2 Additional correction by the same supply

If the correction to be made corresponds to a small fraction (order of a few percent) of the nominal current of the magnets, the requested correction can easily be added to the main setting provided that the sum of the maximum main setting and the maximum correction corresponds to the maximum current of the supply. The correction can be added either in the digital part or in the analog part of the regulation. In the case of dynamic correction the bandwidth is limited by the bandwidth of the system power supply and magnet, as well as the maximum available voltage of the system.

An application for dynamic correction is a tune feedback system acting on a quadrupole magnet supply in order to maintain the tune constant during energy ramping.

Another example is the feedback systems for the closed orbit or photon beam steering, see chapter 13. Adding the correction signal to the DC setting of the correction magnets the same power supply and magnet can be used for the feedback, provided that there is sufficient reserve in the output voltage of the supply to achieve the desired bandwidth for the feedback system.[11]

6.4 Regulation and Control

6.4.1 The principles of the regulation

In order to achieve the specified requirements of the current precision of the power supply, the basic guideline is to provide two regulation loops. The main one is a slow current loop guaranteeing the overall current stability. The second one, which is underneath the first one, is a fast voltage loop acting against fast variations. The design of this part of the regulation may even contain other loops depending on the chosen technology for the power part.

Due to the quality of the available components for the analog regulation overall stabilities of the regulation loops of the order of 10^{-6} are achievable. The determining parts for the stability and reproducibility of the supply are therefore the reference and the current reading device. The reference, usually a DAC, has to be chosen carefully, taking into account the required resolution, the thermal drift of this component and its long term stability. DAC's are available commercially with 16 bit resolution, 1/2 LSB (least significant bit) of linearity and a temperature coefficient of 1 ppm/°C. The choice of the current reading device is important. For small current a shunt may be used, while for higher currents, typically > 100 A, a zero flux current transducer (DCCT) is used. The DCCT is based on the detection of the zero flux in a magnetic core, which is placed around the bar conducting the output current of a supply, by compensating the induced field in the core by the output current of the supply. The current needed for the compensating field is converted via a very precise resistor (burden) into a voltage signal. DCCT's are available commercially with a resolution of 0.05 ppm, a linearity better than 5 ppm, a temperature coefficient of 1 ppm/°C and a typical bandwidth in the order of 10 kHz.

6.4.2 Strategy for the current calibration

The task of the calibration is to define precisely the output current of a power supply. The correlation between the bit pattern defining the requested current and the actual output current, as well as the correlation between the actual output current and the bit pattern representing the reading of the output current of the power supply has to be defined. The function indicated in Fig. 6.9 describes the overall functioning of a power supply as follows:

f: the transfer function between the I_{set} [digit] and the I_{out} [A], i.e. the power supply output current corresponding to the digital setting, describing the whole system: the digital to analog conversion, all regulation and the power part.

f*: the conversion function between the I_{set} [A] (imposed by the operator) and the I_{set} [digit], being the inverse function of f.

g: the transfer function between the I_{out} [A] and the I_{read} [digit], describing the whole current reading system: the DCCT, the electronics and the analog to digital conversion.

g*: the conversion function between the I_{read} [digit] and the I_{read} [A] seen by the operator, being the inverse function of g.

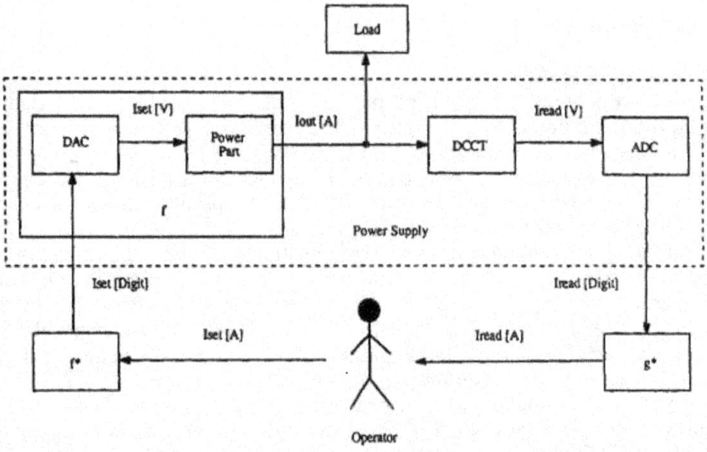

Fig.6.9. General system description.

The calibration of the device is to determine by measurement the functions f and g with reference to the digits imposed on the DAC. As a DAC is more stable than an ADC the calibration should be referred to the setting of the device.

Assuming that the adjustment of the regulation has been done and having adjusted properly the offset and gain of the DAC (max. digit =10 V), the function f can be measured by varying the supply from its minimum to maximum current, measuring the output current with a reference current measuring system as a function of the digital setting. For function g the procedure is similar; offset and gain of the ADC have to be adjusted to guarantee no underflow at zero current and no overflow at maximum current. The ADC output (digits) are measured as a function of the output current defined by the reference current measuring system.

Functions f and g are determined by making a polynomial fit to the measured data. The specification for the linearity is normally very tight, however a linear function usually fits the measured data sufficiently well. A second order fit may be considered in order to improve the reproducibility in the range of a few ppm. The functions f* and g* are the inverse functions of f and g respectively. In this way the system is completely described.

Following this procedure the nominal current I_{nom} of the power supply is defined by the fitted value, which corresponds to the maximum digits, of function f. The relative errors have to be referred to function f.

6.4.3 Regulation requirements

The tolerances requested of a power supply are determined by the quality of the machine optics to be achieved. These requirements have to be transferred into parameters related to the current or voltage of a supply, and can be summarised as:
absolute accuracy, linearity, following error between setting and actual current during dynamic operation, overshoot of the current changing from dynamic into static operation, resolution of the current setting, current ripple, short and long term stability.

6.4.3.1 Absolute accuracy

The absolute accuracy is defined by the quality of the chosen current reading device being the reference for the calibration measurements. Using a high precision shunt or a DCCT as reference the absolute accuracy of the supply can be found by comparison of the output current measurement by this reference with the measurement by the internal current reading device. The errors of a power supply are referred to the nominal current. For the measurement of absolute accuracy there are two possibilities:

i) determine the calibration factor at the maximum excitation, the reading of the current reference gives the nominal current. The reading of the internal current reading device has to stay within the requested limits. A straight line taken through that measured point and zero defines the reference straight line for the other deviations;

ii) to fit the set of measured points as indicated in section 6.3.1, defining by this fit the reference line for the other deviations. The nominal current of the supply is given by the fitted value of this function at maximum imposed current.

6.4.3.2 Linearity

The linearity can be defined as the deviation of the actual current from the reference line as described above, normalized to the nominal current.

6.4.3.3 "Following error "

The "following error" describes the error in the current during a current change and can be defined as the difference between the actual current and the imposed current by the DAC at any moment of the change. The difference should be normalised to the nominal current. This error also has to be referred to the requested speed of the current change.

6.4.3.4 Overshoot

In case of abrupt changes of the setting the overshoot has to be less than the specified value:

$$\Delta I_i/\Delta I_S < \text{specified value}$$

where

ΔI_i: peak to peak amplitude after first reaching the new setting;

ΔI_S: step change of the setting value.

6.4.3.5 Resolution

The resolution is determined by the maximum number of bits of the DAC, for example:

16 bits: $\Delta I / Inom = 1.5 \cdot 10^{-5}$

14 bits: $\Delta I / Inom = 6.1 \cdot 10^{-5}$

12 bits: $\Delta I / Inom = 2.4 \cdot 10^{-4}$

6.4.3.6 Higher harmonic content of the output current

As the higher harmonic content of the output current is determined by that of the output voltage and the impedance of the load, it is quite usual to define the maximum allowed ripple of the output voltage normalized to the nominal output voltage and rely on the calculation, taking into consideration the impedance of the load. Measurement of the harmonic content is also usually made in this way since direct measurement of the current ripple is difficult due to the poor signal-to-noise ratio.

6.4.3.7 Stability against line variations

In designing the power supply one has to take care that the supply has sufficient voltage reserve to remain stable against short and long term line variations. The situation regarding in-house produced variations and the general stability of the grid must always be investigated in any given case.

The short term stability can be defined as the reaction of the supply to fast variations of the mains of the order of 1 to 2% of the nominal line voltage which are induced by turning on and off other loads connected to the mains grid (instantaneous steps).

The long term stability describes the stability of the supply with respect to the slow variations of the mains of the order of ± 10% of the nominal line voltage.

6.4.3.8 Short and long term current stability

The short and long term stability of the output current are not precisely distinguishable. For short term stability one may consider the stability over one hour, whereas the long term stability covers a period of at least 24 hours, taking into account the thermal drift of the supply itself and the variation of the ambient temperature.

6.4.3.9 Dynamic requirements

The requirements for the dynamic behaviour of a supply are determined by the operation modes of the accelerator, for example, cycling procedures for overcoming hysteresis effects and energy ramping of an accelerator. Cycling magnets for hysteresis can normally be done without taking care of the precision; during the ramp the speed of the ramp itself is determined by the maximum output voltage and the time constant of the magnet. By applying a current function with varying ramping speed the overshoot of the current, when changing from dynamic into static operation, can be avoided. The cycling of each set of magnets is independent of the other.

Energy ramping of an accelerator beam, however, requires much more care in the precision because of the required synchronism between various supplies. In order to keep the optics of the machine constant at all current settings the different magnets have to be changed simultaneously, maintaining the ratio between them constant, i.e. they have to "track" together. The tracking error is influenced by the ripple during the ramp, the linearity and the following errors of the individual supplies. As long as the velocity of the ramp is slow (order of several minutes) any normally designed DC power supply for a storage ring should be sufficiently precise for energy ramping. The situation changes as soon as a fast ramp (one minute or less) is required; as the linearity is dominated by the quality of the DAC, the specification of the ripple and the following error during the ramp heavily influence the design of the supply.[6,13] Fast ramping is usually only a requirement for synchrotron supplies.

6.5 Booster Supplies

6.5.1 General

A booster synchrotron requires a continuous energising and denergising of the magnets at higher dI/dt than is normally applied during energy ramping of storage rings. Two different ways of powering the magnets have been realised, one is the ramped DC power supply designed under the constraints of the required dynamic behaviour; the other is resonant excitation, also known as the White circuit. With ramping supplies care has to be taken for the dynamic reactive power due to the inductance of the magnets; the electrical power needed for the energising and denergising of the magnets increases with the repetition frequency of the accelerator. The load to the grid and the consequential distortions of the line voltage have to be seriously considered in designing such a power supply system, so as not to disturb other users of the grid and to respect the demands by the local electricity supplier. The costs for these systems are not negligible. Apart from technical and accelerator physics considerations, the choice between ramping supplies and resonant excitation depends also on economic considerations: the cost of a converter for ramp excitation and the dynamic reactive power compensation versus the cost of the resonant circuit and the excitation power sources.

6.5.2 Ramping Supplies

The advantage of the ramp excitation is the flexibility in generating different waveforms for the current variation at the expense of a low repetition rate: up to now supplies with a repetition rate ranging from less than one up to a few Hz have been realised. Ramping power supplies are designed on the basis of static power converters, taking special care of the dynamic behaviour during the ramp. The design considerations have been described in the sections above.

An example is the power converter for the heavy ion synchrotron SIS at GSI Darmstadt [6] for a rise time of 117 msec from minimum to maximum. The design consists of a 12-pulse thyristor fully controlled bridge acting as a current source providing the main power to the magnets. A fast voltage source is connected in parallel providing the voltage necessary at the load in order to guarantee the desired current performance in dynamic and static operation (see Fig.6.10); the parameters are listed in Table 6.2.

Table 6.2 Ratings of the power supplies of SIS.[6]

Power rating [MW]	Imax [kA]	dI/dt [kA/sec]
26	3.5	± 19
1.6	1.76	± 10
0.51	0.82	± 6

current range 2.2% to 100% of I_n
accuracy and ripple of current normalised to the instantaneous current of the cycle less than $\pm 2.5 \cdot 10^{-4}$.

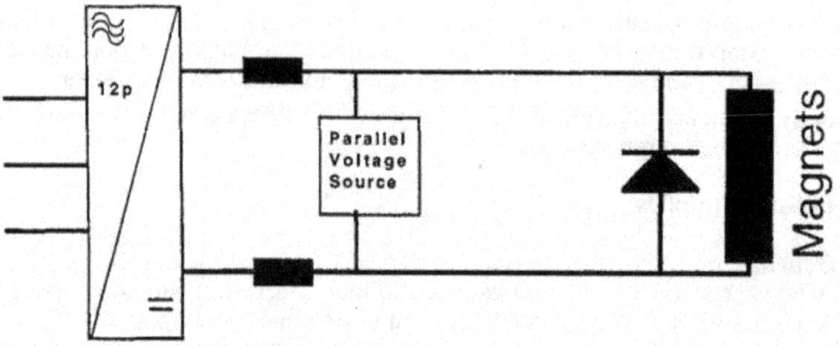

Fig. 6.10. Fast ramping supply.

Other examples are the power supplies for the booster bending magnets of DELTA[13] at the University of Dortmund for a rise time of 300 msec from minimum to maximum. The design consists of a supply with a DC interlink followed by a PWM inverter built up by GTO (gate turn off) thyristors (see Fig. 6.11) with a dynamic current range from 20% to 100% of I_n in 0.3 sec, a stability and a ripple of $\leq 1 \cdot 10^{-4}$.

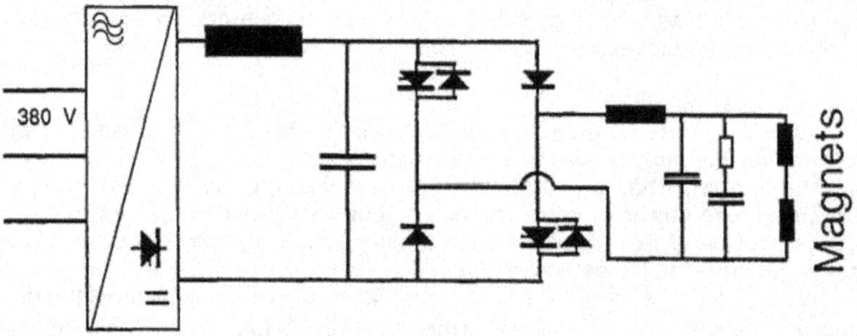

Fig. 6.11. PS scheme for DELTA.

6.5.3 *The White Circuit*

6.5.3.1 *Introduction*

The resonant excitation of synchrotron magnets is based on a parallel resonant circuit. Since only unidirectional field is useful it is convenient to add a DC-bias current to the sine-wave form. Therefore the magnet current has the following function:

$$I_m = I_{dc} - I_{ac} \cos\omega t \tag{8}$$

The latest construction of booster synchrotrons for synchrotron light sources have shown that it is possible to utilize a single parallel resonant circuit for the magnets without having too high a voltage across the magnets with respect to the ground. This can be achieved by keeping the inductance of the magnets sufficiently low, and working with repetition rates of 10 Hz. The complex structures which have been used for the large fast cycling (50 Hz) synchrotrons such as DESY or NINA[14] can thus be avoided. The problems[15] related to those large structures will not be discussed, as they are not relevant for the simpler structures. Figure 6.12 shows two possibilities of arranging a simple structure. All magnets are connected in series and in parallel to the C2, the resonant circuit built up by the transformer and C1 matches the AC power source to the resonant circuit of the magnets.

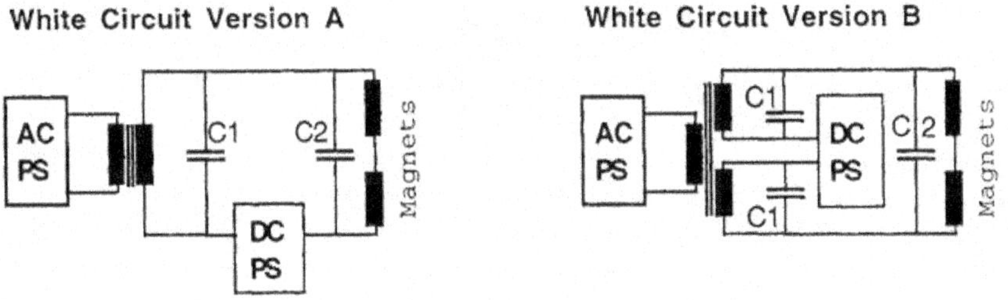

Fig. 6.12: Basic white circuits.

6.5.3.2 DC power sources

The DC power source for the bias current in both the cases are shown in Fig. 6.12 connected to the network at the point where the AC voltage is minimum. Thus its equipment design is the same as any static power converter, for example, a 6-pulse or 12-pulse thyristor converter followed by a low pass filter circuit, its capacitor supporting the separation of the AC voltage from the DC source. In the case of an active filter in the DC source an LC series resonant circuit (case one) or an additional capacitor (case two) may be useful.

6.5.3.3 AC power sources

The AC power sources can be divided into two groups, those with and without DC interlink.

An example of a supply system with AC sources but without DC interlink is the plant for DESY II[16] (see Fig. 6.13). The bending magnet circuit is driven by a square quadrant cycloconverter[4] followed by a Steinmetz circuit. As the cycloconverter is a system that converts three phase 50 Hz AC into three phase 10 Hz this avoids the difficulties caused by the fluctuating power. The matching of the White circuit representing a single phase load to the inverter is done by the Steinmetz circuit, which needs careful tuning.[17] The quadrupole magnet circuit is driven by a modulated rectifier operating as a two quadrant converter supplying a unidirectional current to the load. Care has to be taken with the feedback of reactive power to the line. If there is no independent grid available with loads connected to that grid which are insensitive to large power variation, the maximum power rating of this converter depends on the power capacity of that grid.

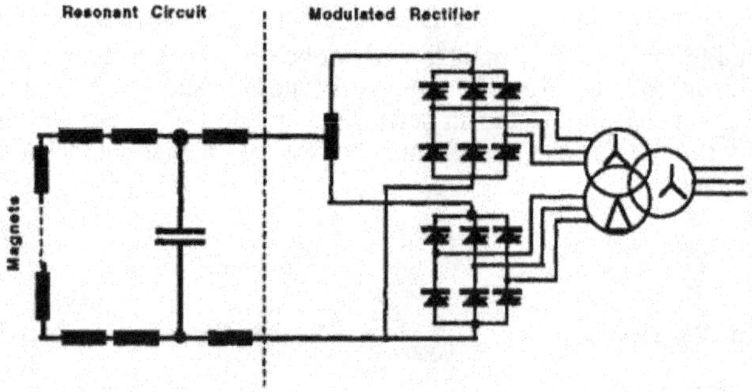

Modulated Rectifier

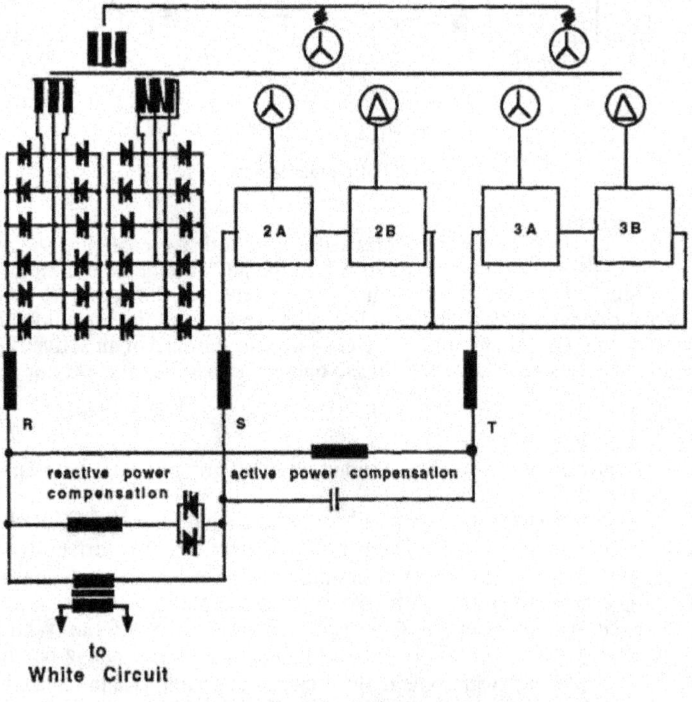

Cycloconverter with 3-phase output and Steinmetz Compensation

Fig. 6.13. Principles of DESY II supplies.

For various White circuits a series resonant inverter[18] has been used. The principal scheme is shown in Fig. 6.14. A 6-pulse or 12-pulse fully controlled bridge

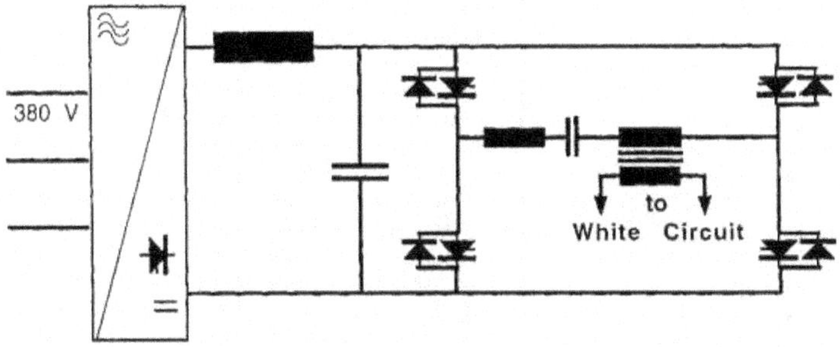

Figure 6.14: Series Resonant Inverter

followed by a low pass filter acts as a voltage source on a one phase inverter with a series resonant circuit as load. This supply works with a DC interlink using thyristors for the inverter and it is a load commutated type.[1] A sufficient capacitive phase shift between output voltage and current has to be maintained to guarantee a safe turn off of the thyristors. This is done by driving the inverter sufficiently below the resonanant frequency of its load. Due to the wave form this inverter produces odd harmonics. The White circuit is coupled via a transformer in series with the series resonant circuit acting also as a filter for the harmonics. If the suppresion of the harmonics is not sufficient a series resonant filter connected in parallel to the primary winding of the transformer of the White circuit may be added. This filter has to be tuned to the frequency of the harmonic to be suppressed. The AC amplitude of the magnets is tuned by the DC voltage of the inverter.

Another supply wth DC interlink is the PWM inverter[19] which is based essentially on the same scheme as the series resonant inverter, using forced commutated thyristors or GTOs it is able to interrupt the output current at any moment. The inverter works in a four quadrant output operation as a current source to the White circuit representing a parallel resonant load. Choosing an appropriate pulse pattern a good suppression of the harmonics can be achieved.

In the latest designs of booster synchrotrons separated function lattices have been chosen. The bending and quadrupole magnets have therefore to be powered independently, but have to track together. More than one White circuit is therefore used, their regulation guaranteeing the tracking between these circuits. As an example, Figure 6.15 shows the control and regulation scheme for the booster of BESSY.[20] By detecting the AC peak field an AC feedback loop has been implemented for each circuit, providing a stability better than $\pm$ 1x 10^{-4}. The phase lock of the different White circuits has been achieved by controlling the time delays between the triggers for the inverters of the quadrupole circuits (slave) and the master trigger for the bending magnets. The zero crossing of the AC amplitudes of the quadrupole circuits is detected relative to the zero crossing of the AC amplitude of the bending circuit. From these measurements the corrections for the triggers of the inverters of the quadrupole circuits are calculated.

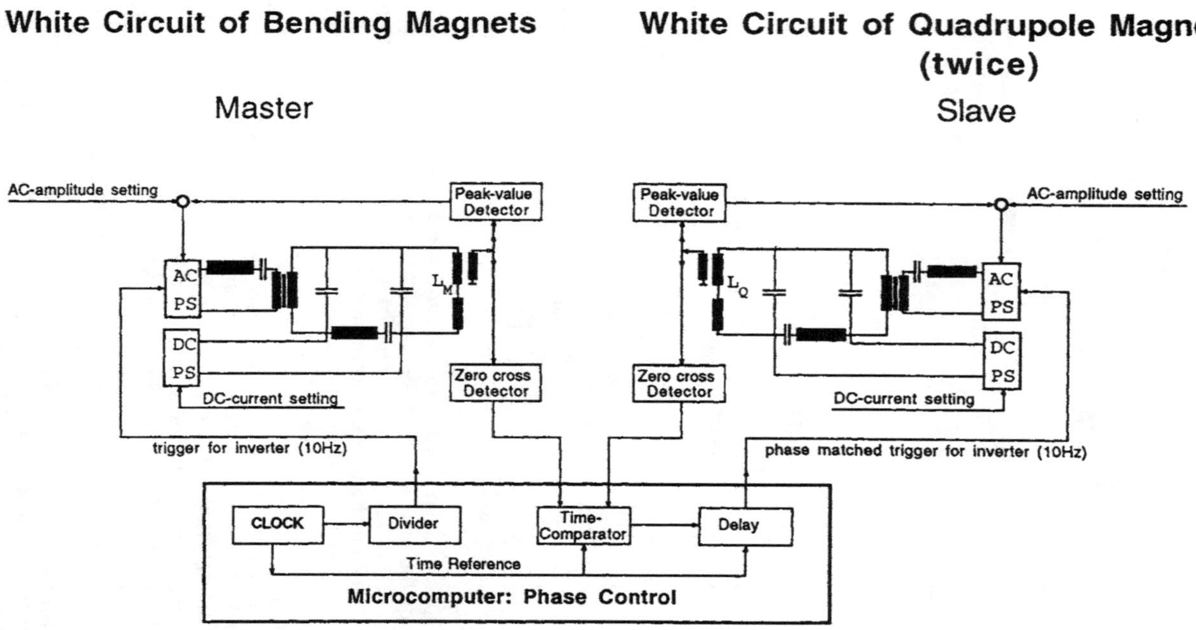

Fig.6.15. Control of white circuit

References

1. Hütte, *Elektrische Energietechnik, Band 2 Geräte* (Springer Verlag, 1978).

2. G. Möltgen, *Netzgeführte Stromrichter mit Thyristoren* (Siemens AG, Berlin - München, 2. Auflage 1970).

3. M. Häusler, "Three-phase Bridge Rectifiers with Freewheeling Diodes", Elektrotechnische Zeitschrift **94** (1973) 230-234.

4. B. R. Pelly, *Thyristor Phase-Controlled Converters and Cycloconverters,* (John Wiley, New York, 1971).

5. P. Proudlock, "Achieving High Performance", CAS Cern Accelerator School, Power Converters For Particle Accelerators, Cern 90-07 (1990) 55-79.

6. R. Fink and R. Wagnitz, AEG Berlin; G. Breitenberger and H. Ramakers, GSI Darmstadt;*New Principles For Power Supplies For Synchrotron Magnets Without Tracking Error,* EPAC (1990).

7. P. Proudlock, "Switch mode Power Converters", CAS Cern Accelerator School, Power Converters For Particle Accelerators, Cern 90-07 (1990) 103-120.

8. D. G. McGhee, *Circuit Description of Unipolar DC-to-DC Converters for APS Storage Ring Quadrupoles and Sextupoles,* Proc. of the 1993 IEEE Particle Accelerator Conference, Washington, D.C. (1993) 1271-1273.

9. B. Hennevin, S. Volut and A. Duapquier, *Study and realisation of a 37.5 kW double resonance converter using GTO thyristors switching at 12.5 kHz,* EPE Grenoble (1987).

10. R. Forest and C. D. M. Oates, *The design of a high stability bipolar 700 W continous variable output switched mode power converter,* EPE Grenoble (1987).

11. R. Richter and R. Visintini, *The ELETTRA Storage Ring Steerer Power Supplies,* EPAC (1994).

12. R. Richter and R. Visintini, "Basic considerations on power supply calibration", Sincrotrone Trieste, internal note ST/M-TN-93/4.

13. W. Bothe, Delta Group, University of Dortmund, *Magnet Circuits For Delta,* EPAC 90 (1990).

14. J. A. Fox, *Resonant Magnet Network and Power Supply for the 4 GeV Electron Synchrotron NINA,* Proc. IEE 112 (1965) 1107-1126.

15. N. Marks, *The Exitation and Elimination of Delay Line Modes of Resonances in the Magnet Power Supply System of a Fast Cycling Electron Synchrotron,* proc. Int. Conf on Magnet Technology, Oxford (1967) 409-420.

16. W. Bothe and P. Pillat, *Magnet Exitation Circuits for DESY II,* IEEE Trans. Nucl. Sci. (1985) 3752-3754.

17. W. Bothe, "Resonant Exitation Of Synchrotron Magnets", CAS Cern Accelerator School, Power Converters For Particle Accelerators, Cern 90-07 (1990) 271-303.

18. W. Bothe, "Aufbau der Stromversorgung für die Führungsmagnete des DESY", Elektrotechnische Zeitschrift **A 84** (1963) 231-235.

19. J.M. Filhol, P. Berkvens and J. F. Bouteille, *General description of the ESRF injector system,* EPAC (1988).

20. G. v. Egan-Krieger, D. Einfeld, W.-D. Klotz, H.Lehr, R.Maier, G. Mülhaupt, R.Richter and E. Weihreter, *Performance Of The 800-MeV Injector For The BESSY Storage Ring,* IEEE Trans. Nucl. Sci. **30** (1983) 3103-3105.

CHAPTER 7: MAGNETIC MEASUREMENTS

RICHARD P. WALKER
Sincrotrone Trieste
Padriciano 99
34012 Trieste, Italy

7.1 Introduction

Magnetic measurements are an essential part of the component tests that must be carried out to guarantee the correct functioning of a storage ring. This is particularly true for modern low emittance storage rings where there are strict tolerances on the field quality of both lattice magnets and insertion devices. The importance of magnetic measurements to the construction of accelerators in general is illustrated by the fact that regular International Magnet Measurement Workshops are held, and a special school on Magnetic Measurement and Alignment was recently organized.[1] The particular problems associated with measurement of insertion devices inspired a recent international workshop specifically on this topic.[2]

Considering firstly the **lattice magnets** - dipoles, quadrupoles, sextupoles, and corrector magnets (see Chapter 5) - magnetic measurements play several roles. Prototype magnets are usually constructed in order to verify the magnet design and magnetic measurements are very important at this stage to allow the need for any changes in the pole profile or end design to be determined and if necessary tested. Measurement of the series production magnets are equally important. Although dimensional checks are usually carried out as part of quality control procedures these are not always as easily interpretable or as accurate as a direct magnetic field measurement. Ideally measurements should be carried out in parallel with the production in order to spot any changes in the machining or assembly as quickly as possible. In the case that the magnets do not conform with the specification, corrective measures, for example by shimming, are usually determined by trial and error on the basis of magnetic measurement data. Even if the magnets do conform with the specification it is common to employ some sorting technique based on the measurement results to minimize unwanted effects on the electron beam.

Modern low emittance rings also result in strict tolerances for magnet positioning, particularly of the quadrupoles. To achieve this means in general not relying on the mechanical references but positioning the magnet directly in accordance with the magnetic centre. Magnetic measurements therefore play another role in either determining the offset between physical and magnetic centres, or in allowing reference targets to be positioned with respect to the magnetic centre on the measurement bench.

In order to carry out the magnetic measurements it is desirable to set up a dedicated area with the following features :

Fax: +39-40-226338 - E-mail: R.P.Walker@ELETTRA.Trieste.it

• adequate access and lifting arrangements for routine installation of a large series of magnets.

• quiet and clean surroundings, not subject to ground vibrations or interruptions in electrical supply.

• air conditioning to maintain a reasonably constant temperature, of the order ± 1-2°C.

• demineralized water circuit(s) with easily adjustable pressure, and monitoring of pressure and temperature at the inlet and outlet of each magnet, together with the flow rate.

• general purpose power supply capable of operating with single magnets of all magnet types. Alternatively, one or more of the power supplies that will eventually be used to power the magnets in the ring can be used.

• calibration facility for magnetic field probes, consisting of a dipole magnet with a sufficiently large volume of homogeneous field and an NMR system (Sec. 7.2.3).

In the case of **insertion devices** (see Chapter 14), magnetic measurements are not only essential to guarantee minimum effect on the ring operation, but also to obtain the best quality of the emitted radiation. A separate location for these measurements is preferred since they are usually of a less routine nature and often take a much longer period of time per device. Since most devices are usually constructed using permanent magnets, a power supply and water cooling system are not generally required. A temperature controlled environment is however very important, within ± 1°C or better, because of the variation of the strength of the permanent magnet material (NdFeB) with temperature.

7.2 Magnetic Measurement Techniques

Various physical principles have been applied to the task of magnetic field measurement. For the magnitude of field produced in storage ring magnets, the two most popular techniques are based on the principles of magnetic induction and the Hall effect, however several others have also been used. General reviews of the various techniques have been published[3,4] and several reviews of magnetic measurement systems have appeared in the proceedings of the International Magnet Technology Conferences.[5,6,7]

7.2.1 Magnetic Induction

In this method a change of magnetic flux (Φ) linked by a circuit (i.e. coil) induces a voltage (V) according to Faraday's Law :

$$V = -\frac{d\Phi}{dt} \tag{7.1}$$

where the flux is defined as the integral of the normal field component over the surface of the coil, multiplied by the number of turns (N) :

$$\Phi = N \oint_S B.dS \, . \tag{7.2}$$

It follows that the integrated voltage gives the change in the average field value over the area of the coil (A) between the initial and final conditions :

$$\int V dt = \Phi_1 - \Phi_2 = N A \, (\bar{B}_1 - \bar{B}_2) \,. \tag{7.3}$$

Using S.I. units, Volt-seconds corresponds directly to flux in Weber (Wb) or equivalently, Tesla-metre^2.

The change of flux may be induced by various means : translating, flipping, rotating or vibrating the coil in a static field, or changing the magnetic field with a static coil (e.g. measurement of a.c. or pulsed magnets). The choice of coil geometry, and the method of introducing the flux change, allows great versatility in the type of measurement that can be made. Point measurements of the field can be made with a coil of small cross-section ("search-coil"), or alternatively line integrals can be measured with a long, thin coil shaped to the desired path. Field gradients can be measured using two coils at a fixed separation connected in series opposition. A major use of the technique is the rotating or "harmonic coil" method[8] for determining the multipole field content of a magnet, which will be discussed in detail in Sec. 7.4.1.

A further variant of the induction method is the "moving wire" technique in which only one arm of the coil within the field is moved, the return arm remaining fixed and usually passing outside the magnet. The voltage in this case is the rate of change of the flux normal to the direction of movement of the wire, i.e.

$$V = N B_n \, l \, v \tag{7.4}$$

where N is the number of turns in the wire bundle, l its length and v its velocity.

Depending on the application, the measured quantity can either be the voltage as a function of time using a fast voltmeter, or the integrated voltage, using an electronic integrator. Two different types of integrator are commonly used.[9] The analog integrator, or "fluxmeter", is usually based on the Miller integrator circuit and provides an output voltage proportional to the integral of the input voltage, which can then be sampled by a voltmeter; several analog integrators are available commercially.[10,11,12] More recently digital integrators have been developed using a voltage-to-frequency converter and digital counter. A version developed at CERN is available commercially.[13] One commercial digital voltmeter can also function as an integrator.[14]

The induction method is particularly attractive since the response is completely linear with the field strength. The only uncertainty is the effective area of the coil, which can be determined by measuring the response in a region of homogenous field with known strength, but in many cases it is sufficient to simply use the measured dimensions of the coil.

7.2.2 *Hall Effect*

The second main technique employs the effect discovered by E.H. Hall (1879), whereby a voltage (V_H) is created in a thin slab of material (usually a semiconductor) in which a current (I) flows perpendicular to the direction of a magnetic field. The voltage is orthogonal to both field and current directions (see Fig. 7.1) and to a first approximation varies linearly with both field and current.[15] Usually a stabilized d.c. current source is used, however a.c. excitation is also a possibility. Commercially available Hall plates have a sensitivity in the range 5-20 μV/Gauss at the nominal operating current, usually in the range 50-400 mA. The output is therefore sufficient to be read directly by a sensitive voltmeter; alternatively an amplifier circuit may be used. The method results in a point measurement of the field since the active area of the plate is usually quite small, typically between 0.5 x 1.0 mm and 2.0 x 5.0 mm.

Hall plates can reach high accuracy, of the order of 0.5 Gauss (or 1 in 10^4 at higher field levels), provided a number of factors are taken into consideration. The main disadvantage of the Hall plate for field measurement applications is that each individual plate must be calibrated against a reference field (e.g. NMR, see Sec. 7.2.3) since non-linearities are significant : even with an appropriate load resistor in the output circuit the linearity is typically in the range 0.2-2 %, which is not sufficient for this application. The magnet measurement laboratory should therefore be equipped with a suitable electromagnet and NMR system to enable calibrations to be carried out. Depending on the application, the calibration needs to include both positive and negative field directions since the calibration curve is not symmetric. It may also be necessary to repeat the calibration several times during the life of the probe due to ageing effects. This is a poorly documented and not very well understood phenomena. However, it appears that changes in the Hall voltage zero-offset and sensitivity may arise from either diffusion of doping impurities in the semiconductor or effects related to mechanical stress introduced by ageing of the glue which binds the semiconductor to its substrate, which can even result in the formation of micro-cracks. Hall plates should therefore be treated with care and not subjected to mechanical shocks or large temperature fluctuations.

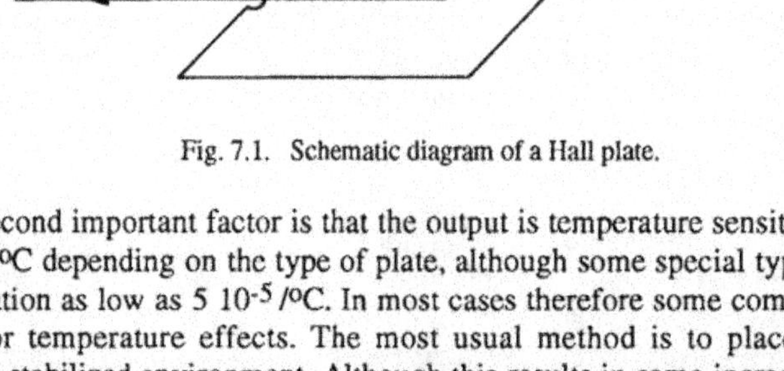

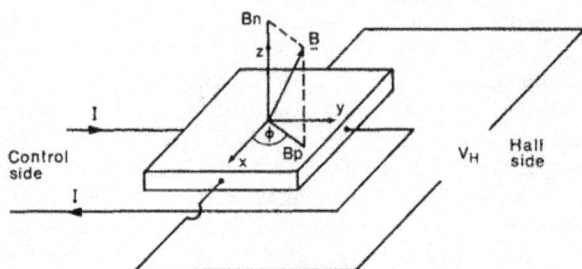

Fig. 7.1. Schematic diagram of a Hall plate.

A second important factor is that the output is temperature sensitive, in the range 0.04-0.1% /°C depending on the type of plate, although some special types are available with a variation as low as 5 10^{-5} /°C. In most cases therefore some compensation has to be made for temperature effects. The most usual method is to place the plate in a temperature stabilized environment. Although this results in some increase in the size of the probe, this is to be preferred over other methods involving either measurement of the temperature and numerical correction, or compensation using an electrical circuit, since a constant temperature helps also to reduce thermal e.m.f.'s introduced by temperature differences between the output terminals on the plate; the contact between the copper electrode and InAs plate material forms a thermocouple source of about 300 µV/°C. For this reason the plate should be mounted so as to have a good thermal contact between the connections using heat conductive paste. A low thermal emf solder should also be used for the extarnal connections.

A further important factor in some applications is the so-called planar Hall effect.[16] This arises due to the difference in magneto-resistance perpendicular and parallel to the magnetic field. As a result an additional voltage is developed proportional to the square of the field in the plane of the plate (B_p) that depends on the angle between the directions of the field and the current flow (see Fig. 7.1) :

$$V_p = \beta B_p^2 I \sin 2\Phi \, . \tag{7.5}$$

The planar Hall effect coefficient, β, is different for each type of plate and also varies from one plate to another of the same type. The maximum error that can arise (V_p/V_H) is typically in the range 7 10^{-4} to 3 10^{-2} times B_p^2/B_n where B_n is the field normal to the plate.[6] In many cases the symmetry of the measurement geometry can lead to there being no significant planar Hall effect term, e.g., measurement of the vertical field component in the median plane of a conventional accelerator magnet. In the general case it must be eliminated, for example by measurement of all 3 field components, and then applying a correction based on the measured planar coefficient. Alternatively, the field can be inverted, or the plate flipped through 180°, and the measurement repeated; since the planar term depends on B_p^2 it remains constant while the main Hall voltage reverses in sign.

Various commercial instruments are available employing Hall plate probes that are ready for use in a field measuring system, for example with temperature stabilization, built-in calibration curve, computer interface etc.[17,18,19,20]

Hall probes can be used in making measurements of superconducting magnets at cryogenic temperatures, however in strong magnetic fields another electron transport phenomena occurs that results in oscillations in the relationship between the Hall voltage and field : the Shubnikov-de Haas effect.[15] Since the oscillations have an amplitude of about 0.1 % the probe calibration must be carried out at cryogenic temperatures. To reach the 10^{-4} level of accuracy usually required for field mapping one has also to take into account the fact that the calibration can differ after each thermal cycle, for example by performing an in-situ calibration technique.

7.2.3 Nuclear Magnetic Resonance

The Nuclear Magnetic Resonance (NMR) technique is the primary absolute standard for calibration purposes. Particles with a spin (usually either protons for the range of fields being considered here, or deuterium for fields above 2 T) exhibit a resonant absorption at a frequency that is directly proportional to the magnetic field, the constant of proportionality being the particle's gyromagnetic ratio. The field is thus obtained directly from a very precise and accurate measurement of frequency, independent of temperature. Commercial instruments are available that cover the range from 0.043 T to 13 T with a series of probes.[13] The main disadvantages of the technique are the need for a homogeneous field (of the order 0.1% /cm or better), and the lower field limit. Its application to magnet measurements is therefore very limited : only the field in the central region of dipole magnets can be measured, i.e. excluding the fringe field. Its use is therefore usually restricted to calibration work.

7.2.4 Force on a current carrying conductor

When a current carrying circuit is placed in a magnetic field it is subjected to forces that tends to increase the flux that links it. If the flux would increase by a translation in the x-direction then the force on the coil is given by :

$$F_x = I \, {d\Phi}/{dx} \tag{7.6}$$

with I in Amps and Φ in Webers (T m^2) the force is given directly in Newtons. Measurement of the force therefore provides an absolute measurement of the field. various "current balance" techniques have been developed in the past for accurate measurement of generally large and homogeneous fields.[4] An application of the method that has been used widely in the past for accelerator magnets is the "floating wire" method.[21] A long flexible wire passes through the magnet aperture along the whole length of the magnet. The return part of the circuit is completed outside the magnet. One end of the wire is fixed while the other is stretched over a pulley to maintain a constant tension, but is free to orientate itself. When a current is passed through the wire it tends to curve, the force tending to increase the flux in the circuit being balanced by the tension : $T = I\,B\,\rho$, where ρ is the radius of curvature. It follows that if the ratio of tension to current (T/I) is chosen to equal the magnetic rigidity of a charged particle ($B\rho$) then the wire will follow the same trajectory as a particle of that energy. This method has now almost entirely been replaced by field mapping techniques (see Sec. 7.3.1).

A new application of this method that has been applied to the measurement of insertion devices consists in sending a pulse of current along the stretched wire and detecting the resulting transverse oscillation of the wire[22] (see Sec. 7.5.3).

7.2.5 Magneto-optical Effect

The magneto-optical or Faraday effect, by which the plane of polarization of light is rotated by an amount that is proportional to the magnetic field and the path length in the active medium, has been applied as a technique for visualizing field patterns, particularly for the centre location of multipole magnets[23] (see Sec. 7.4.3).

7.3 Storage Ring Dipole Magnets

7.3.1 Point-by-point Field Measurements

The most common technique for obtaining detailed information about the field in dipole magnets is to make a series of point measurements with a probe consisting of a single or a group of Hall plates, although point coil and integrator techniques are sometimes also used. In this way information can be obtained both about the field quality in the bulk of the magnet, allowing a direct comparison with the results of a 2D magnetic field calculation, as well as the 3D effects due to the ends of the magnet. In addition, field maps can be used to calculate the actual particle trajectories and field variation perpendicular to it allowing a better modelling of the magnet behaviour. This is a clear advantage over the field integral measurement schemes (see Sec. 7.3.2); the disadvantage however is the complexity of the measuring equipment and the long time that is usually needed to map each magnet. A solution sometimes adopted is to measure the prototype in this way and then only carry out integral measurements of the series production.

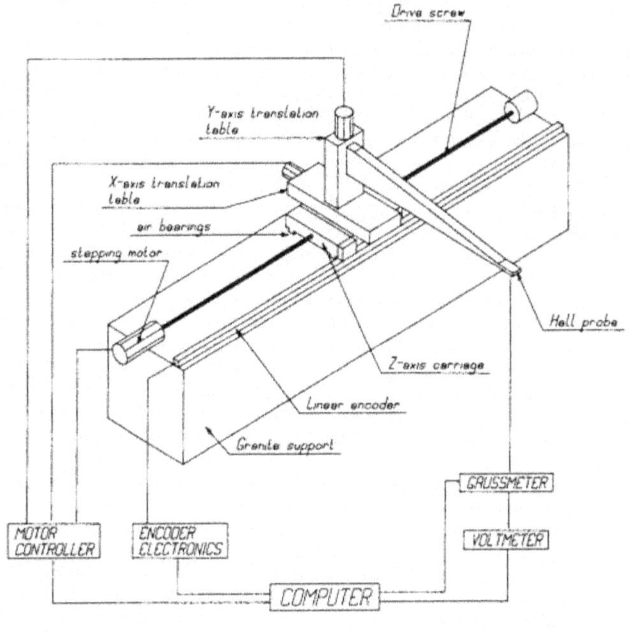

Fig. 7.2. Schematic diagram showing the main components of a general purpose 3-axis field measurement bench.

In the case of the usual open-sided, or "C-frame", magnet design the probe is usually mounted at the end of a long slender arm that can enter the magnet gap, and which is connected to a general-purpose 2-axis (x,z) or 3-axis (x,y,z) positioning machine (see Fig. 7.2). A flexible control system is needed to carry out automatic measurement sequences, either along straight or curved trajectories, or carry out a "field mapping" over a 2D area. An "H-frame" construction is, however, often chosen for transfer line and synchrotron magnets, or dipoles with a large bending angle, and in this case the probe cannot enter from the side. Alternative techniques are to use a rotary stage with a curved probe arm that can enter from the end of the magnet, or a "mouse" system in which the probe is carried along a circular track. Some systems are available commercially.[24]

Measurements are usually only carried out in the median plane, where by symmetry only the vertical field component exists. The simplest kind of measurement consists of a linear scan in the radial direction perpendicular to the nominal trajectory, at the magnet centre or at any point along it. A plot of $[B(r)-B(0)]/B(0)$ in the case of a dipole magnet, or $[B(r)-B(0)-(dB(r)/dr|_{r=0})r]/B(0)$ in the case of a combined function magnet gives the transverse field homogeneity for direct comparison with 2D magnetic field computations. The data can also be fitted with a polynomial to extract information about the individual multipole field harmonics :

$$B_y = B_1 + B_2 r + B_3 r^2 + B_4 r^3 \ldots \tag{7.7}$$

where B_1 = dipole, B_2 = quadrupole, B_3 = sextupole, etc. However, some care has to be taken with this type of analysis since the polynomial terms are not orthogonal, which means that the value obtained for any particular term depends on which terms are included in the series.

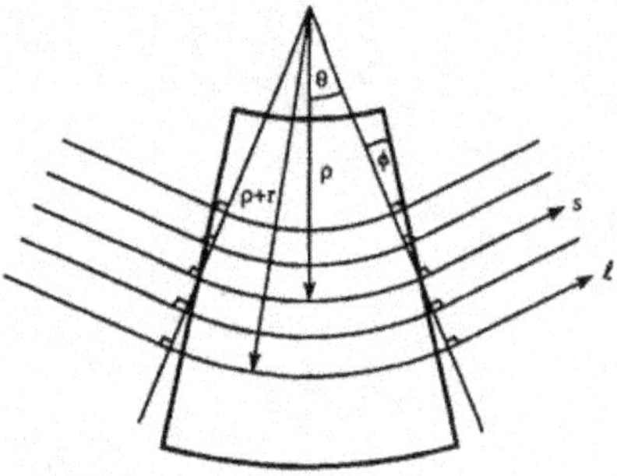

Fig. 7.3. Measurement of a dipole magnet by means of a series of curved arcs and straight lines.

The integrated field properties can be determined by scanning the probe along the arc of a circle of the nominal bending radius. Outside the nominal magnet boundary the scan may be continued along a straight line in order to include the fringe field region (see Fig. 7.3). The first quantity of interest that can be determined in this way is the effective magnetic length, defined as the field integral along the nominal trajectory (circular arc plus straight sections at the end) divided by the field at the magnet centre :

$$L_{eff.} = \frac{\int B\,ds}{B_o} . \tag{7.8}$$

Information about the integrated field quality can be obtained from scans performed along a series of parallel (i.e. concentric) curves, as shown in Fig. 7.3.

The data can be analysed in different ways, and some care has to be taken over the interpretation of the results. One method consists of evaluating the field integrals along each of the curves $\int B(r)\,dl$ and fitting to a polynomial : $\int B(r)\,dl = b_1 + b_2 r + b_3 r^2$ ...etc. The resulting coefficients however do not represent the integrated quadrupole and sextupole fields acting on the beam because of the curvature of the trajectory. We can express the integral above as an integral with respect to the nominal trajectory (i.e. along s rather than l) as follows :

$$\int B(r)\, dl = \int [B_1(s) + B_2(s)\, r + B_3(s)\, r^2 \ldots]\, (1 + r/\rho)\, ds \tag{7.9}$$

where $B_1(s)$, $B_2(s)$, etc. are the field and its derivatives at any s and we have used the fact that the path length elements are related as follows : $dl = (1 + r/\rho)\, ds$, where ρ is the local radius of curvature. The following correspondence can then be made between the coefficients of the fit of the field integrals and the integrals of the local field derivatives :

$$\begin{aligned} b_1 &= \int B_1\, ds \\ b_2 &= \int B_2\, ds + \int B_1/\rho\, ds \quad . \\ b_3 &= \int B_3\, ds + \int B_2/\rho\, ds \end{aligned} \tag{7.10}$$

Using the measurement data it is often required to express the results in terms of a "hard-edge model" that can then be used to model the magnet in beam dynamics calculations. This can be done as follows :

$$\begin{aligned} b_1 &= B_1\, L_{eff.} \\ b_2 &= \tilde{B}_2\, L_{eff.} + B_1\, (\theta - \tan\phi) \quad . \\ b_3 &= \tilde{B}_3\, L_{eff.} + \tilde{B}_2\, (\theta - \tan\phi) \end{aligned} \tag{7.11}$$

In the above $L_{eff.}$ is the effective length of the dipole field defined in Eq. (7.8) and θ is the half-bend angle. If ϕ is the nominal end-angle then $\tilde{B}_2$ is the average field gradient; alternatively $\tilde{B}_2$ can be set to the measured gradient at the centre of the magnet, then ϕ becomes the effective end-angle. $\tilde{B}_3$is the effective sextupole field.

An alternative to scanning along parallel curves is to use curves of constant curvature displaced in the x-direction, simulating the measurement that can be made with an integral coil. Provided the bend angle of the magnet is small the results are equivalent to the method above. For example, for total bend angles less than 20°, the absolute error in determining the effective gradient is less than 0.5%.

A more sophisticated analysis can be made by determining the real electron trajectory in the magnet. This can be done by direct numerical integration of the equation of motion, which in the median plane is given as follows (Cartesian co-ordinates) :

$$x'' = -\frac{e}{\gamma mc}\left[(1 + x'^2)^{3/2} B_y\right] . \tag{7.12}$$

General purpose routines for integration of ordinary differential equations of this kind are widely available.[25] Alternatively a tracking algorithm can be used which approximates the trajectory as a circular arc between points which are sufficiently close together that the field is constant between them.[26] For this type of analysis measurements are usually carried out on a rectangular grid of points, with a higher density of points at the edge of the magnet where the field is changing most rapidly. Interpolation of the field at any other point can conveniently be made using a bicubic spline.[25] The calculation is repeated with electrons of different energy until the correct bending angle is obtained. Having obtained the real electron trajectory, the field can be interpolated along parallel curves to determine the real integrated gradient, sextupole etc. as above.

Another possibility is to calculate the trajectories of displaced particles in order to determine directly the values of the matrix elements that best describe the linear beam optics properties of the magnet.[27] For example, from particles displaced by an amount Δx_i or with a difference in angle $\Delta x'_i$ at the entrance and giving changes Δx_f and $\Delta x'_f$ at the exit of the magnet, the matrix elements can be calculated as follows : $m_{11} = \Delta x_f/\Delta x_i$, $m_{12} = \Delta x_f/\Delta x'_i$, $m_{21} = \Delta x'_f/\Delta x_i$, $m_{22} = \Delta x'_f/\Delta x'_i$. The same can be done in the vertical plane, by integration of the equation :

$$y'' = \frac{e}{\gamma mc}(1 + x'^2)^{1/2}(B_x - x' B_z) \tag{7.13}$$

where the components B_x and B_z are derived from the relationship curl $\underline{B} = 0$: $B_x = (dB_y/dx)\, y$ and $B_z = (dB_z/dx)\, y$.

7.3.2 Integrated Field Measurements

Integral field measurements can be made with long coils that cover the entire region of the magnet, including the fringe field. Such a technique, although providing less information than a complete field mapping is well adapted to the measurement of a large series of magnets, because of it speed, sensitivity and reproducibility. Measurement of field integrals along a straight line, as for quadrupoles and higher order multipole magnets (Sec. 7.4), would be the simplest approach both from the coil construction and measurement point of view. For example, the HERA proton ring superconducting dipoles were measured using a harmonic coil similar to that used for the quadrupoles.[28] For the types of magnet used in synchrotron radiation sources however it is generally preferred to measure the field integrals along the nominal electron trajectory, and parallel to it, in which case the coil must have the same radius of curvature as the magnet. Flipping or rotation of the coil is then obviously no longer possible, which places some restrictions on the measurement technique.

Measurement of the field integral can be made most simply by integrating the induced voltage while increasing the magnet current from zero to the desired operating level. In this case however the remanent field in the magnet is not taken into account. A better method is to start with the coil outside the magnet, preferably in a zero-field chamber to eliminate the earth's field. The coil is then inserted in the magnet and the current increased while integrating the voltage. Elimination of linear integrator drift can be made by removing the coil and replacing in the zero-field chamber.

The field homogeneity is usually determined by a translation of the coil in the horizontal plane. The induced voltage depends only on the changes in flux and so is insensitive to the dipole field. In this way a greater sensitivity to the higher order harmonics is obtained, compared to the method above. By placing two coils in the magnet, connected in series opposition, and translating one coil, a further gain can be made by eliminating errors due to power supply fluctuations.

In general the best results are obtained by a null technique, measuring the difference between each series magnet and a fixed reference magnet, which are powered in series. For example, in the APS dipole measurement system[29] the integrated voltage from a series magnet bucked by the signal from a fixed coil in a reference magnet is automatically recorded during a current ramp. Horizontal translation of the test coil gives field homogeneity data. A relative accuracy and reproducibility of better than 10^{-4} has been achieved. For the measurement of the ESRF series dipole magnets a special "azimuthally traversing coil" bench was built.[30] In this system an assembly of 5 coils is translated between the magnet under test and a reference magnet, which are situated sufficiently far apart that the field is essentially zero between them. Thus, only differences of the field integrals with respect to the reference magnet are measured, along the nominal trajectory and at 2 radial positions on either side of it. A relative accuracy of about 10^{-4} of the field integral was achieved in this case also, with a sensitivity of $5\ 10^{-5}$.

The construction of suitable long curved coils for these types of measurement requires some special care and attention.[31] For example, the construction of a 1.8 m long, 4 mm wide, coil with 72 turns and a 4 m radius of curvature for measurement of the ALS booster dipoles is described in Ref. 32.

7.4 Quadrupoles, Sextupoles and Higher Order Multipole Magnets

7.4.1 Harmonic Coil Method

7.4.1.1 Basic theory

The standard technique for measuring these kinds of magnets is the harmonic coil method, by which a coil loop is rotated in a magnet and the induced voltage analysed as a function of rotation angle, the Fourier components giving directly the harmonic, or multipole, content of the magnet. Figure 7.4 shows the geometry of the most common "radial coil" arrangement. The alternative "tangential coil" arrangement is considered in Sec. 7.4.1.6. A detailed theory and error analysis has recently been published[33] while a more basic introduction can be found in Ref. 34.

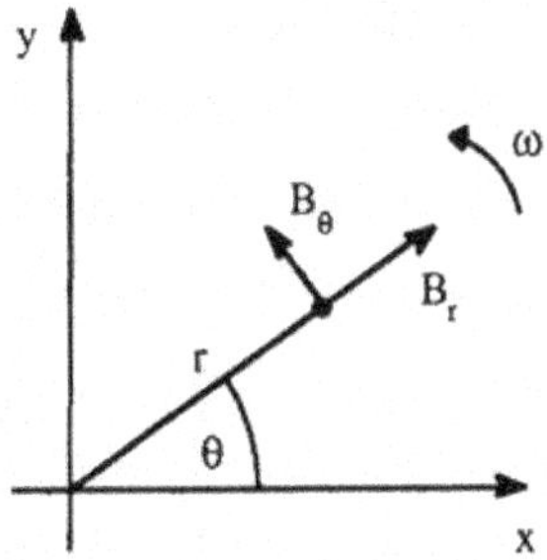

Fig. 7.4. Radial harmonic coil geometry.

The analysis is most naturally carried out using a polar coordinate system, for which the radial (B_r) and azimuthal (B_θ) field components in a 2D magnet can be expressed as a sum of multipole components with coefficients A_n, B_n as follows :

$$B_r = \sum_n (A_n \cos n\theta + B_n \sin n\theta)\, r^{n-1}$$

$$B_\theta = \sum_n (-A_n \sin n\theta + B_n \cos n\theta)\, r^{n-1} . \tag{7.14}$$

The corresponding expressions for the horizontal (B_x) and vertical (B_y) field components can be written as follows :

$$B_x = \sum_n (A_n \cos (n-1)\theta + B_n \sin (n-1)\theta)\, r^{n-1}$$

$$B_y = \sum_n (-A_n \sin (n-1)\theta + B_n \cos (n-1)\theta)\, r^{n-1} . \tag{7.15}$$

The above can be expressed in a more convenient form using complex notation, which will be of use later when displacement and rotation errors are considered :

$$B_y + iB_x = (B_\theta + iB_r)e^{-i\theta} = \sum_n (B_n + iA_n)\, \tilde{z}^{n-1} \tag{7.16}$$

where $\tilde{z} = x + iy = re^{i\theta}$. In the above, A_n refer to "skew" field components while B_n refer to "normal" field components. The index n determines the order of the field : $n = 1$ corresponds to dipole, $n = 2$ quadrupole, $n = 3$ sextupole and so on.

Consider a single wire rotating around the central axis of a magnet with a radius r and angular velocity ω (Fig. 7.4). The voltage induced in a length l of the wire is, using Eq. (7.4) :

$$V = \omega\, r\, B_r\, l . \tag{7.17}$$

The wire can also be considered part of a coil, whose return loop passes through the centre of rotation and therefore has no influence on the voltage. In this case the induced voltage is the rate of change of normal component of flux integrated over the coil surface:

$$V = -\omega \frac{d}{d\theta} \int_0^r B_\theta l \, dr . \tag{7.18}$$

The two expressions are equivalent since in the air-gap of a magnet div $\underline{B} = 0$, and hence in polar coordinates $\partial B_\theta / \partial\theta = -\partial(rB_r)/\partial r$. For a coil with radius R_{coil}, length L, number of turns N, the voltage is therefore given in terms of the multipole coefficients, using Eq. (7.14), as follows :

$$V = N\omega L \sum_n (A_n \cos n\theta + B_n \sin n\theta)\, R^n_{coil} . \tag{7.19}$$

If the measured voltage is Fourier analysed it yields components :

$$V = \sum_n (V_{n,cos} \cos n\theta + V_{n,sin} \sin n\theta) \tag{7.20}$$

from which the multipole terms can be obtained directly as follows :

$$A_n L = \frac{V_{n,cos}}{N\, \omega\, R^n_{coil}} \qquad B_n L = \frac{V_{n,sin}}{N\, \omega\, R^n_{coil}} . \tag{7.21}$$

Although the definition of the multipole terms, Eq. (7.14), refers to the 2-dimensional case, it may easily be shown that it is also valid for the integral of a 3-dimensional field, provided the integration starts and ends in regions of uniform field. Thus if the coil is sufficiently long that it extends into a region of near-zero field on either side of the magnet, A_nL and B_nL in the above refer to the integrated multipole components of the magnet, independent of the actual coil length. Alternatively, a short coil of known length can be used to determine the harmonics A_n and B_n in the central region of a multipole magnet.

It is common to express the error terms as a ratio of the field amplitude component to that of the main field component at a given reference radius, $R_{ref.}$, which may be different from the coil radius. For the normal field components in a quadrupole, for example, this ratio is obtained from the measured voltage components as follows :

$$\text{relative field error} = \frac{(B_n L)\, R_{ref.}^{n-1}}{(B_2 L)\, R_{ref.}} = \frac{V_{n,\sin}}{V_{2,\sin}} \left(\frac{R_{ref.}}{R_{coil}} \right)^{n-2} \tag{7.22}$$

and similarly for the skew components. Sometimes results are expressed in terms of the field gradient, rather than the field; in this case the contribution of the n'th harmonic to the integrated gradient, relative to the main component, at the reference radius is given by

$$\text{relative gradient error} = (n-1)\, \frac{V_{n,\sin}}{V_{2,\sin}} \left(\frac{R_{ref.}}{R_{coil}} \right)^{n-2} . \tag{7.23}$$

Care must therefore be taken in comparing magnetic measurement results from different laboratories that the results refer to the same radius (or radius relative to the magnet inscribed radius) and the same quantity, field or gradient.

7.4.1.2 Practical considerations

The first application of the harmonic coil technique used a continuously rotating coil, at a frequency of several tens of Hz. The induced voltage was transmitted to a spectrum analyser, which gave directly the harmonic content of the magnet. The measurement accuracy, however, was limited by rotational errors and difficulties with slip rings or commutator brushes. Although rotating coil systems continue to be used, with the availability of modern integrators, digital voltmeters and rotational encoders the method of choice for the most accurate measurements is generally a single slow rotation. The two most common techniques are :

i) measure the induced voltage as a function of time during continuous rotation. In this case the velocity of rotation must be kept constant for the equations presented in 7.4.1.1 to be valid.

ii) measure the integrated voltage as a function of angular position. In this case we have :

$$\int V\, dt = NL \sum_n \left(-A_n \sin n\theta + B_n \cos n\theta \right) \frac{R_{coil}^n}{n} . \tag{7.24}$$

In this case the result no longer depends on the angular velocity. The first applications of this method using analog integrators performed step-by-step measurements, reading and zeroing the integrator after each angular interval. More recently digital integrators employing a voltage-to-frequency converter (VFC) allow measurements to be made during continuous rotation; for example, the trigger from the angular encoder can be used to switch the output of the VFC between two counters so that there is no loss of counts while one is being read-out. A commercial instrument is available of this type.[13]

Typically between 128 and 512 measurements of voltage or integrated voltage are made per revolution, which is adequate for the number of harmonics usually considered to be significant, up to $n = 10$, i.e. 20-pole. The subsequent analysis is carried out using the Fast Fourier Transform (FFT) algorithm. It is a useful technique to remeasure the data on the counter-rotation as a test of reproducibility. In the case of integrated voltage measurements correction must be made for linear integrator drift, which would introduce spurious skew components. This can be achieved, for example, by making an extra (2^n+1)th measurement or using the data from the counter-rotation.

Fig. 7.5. **Photograph of the Danfysik Model 692 Multipole Magnet Measurement System (reprinted with permission from Danfysik A/S).**

For each of the measurement methods, it is clear from Eqs. (7.19) and (7.24) that the sensitivity for measuring the nth harmonic increases as R^n_{coil}, thus the measuring coil should have the largest possible radius within the magnet aperture. Taking "typical" parameters for the measurement of a quadrupole magnet : $B_2 = 20$ T/m, $R = 30$ mm, $L = 0.5$ m, $N = 10$, $\omega = 2\pi/10$ s^{-1}, we see that the quadrupole component gives a peak induced voltage of 57 mV. A sensitivity of 10^{-4} for the nth harmonic therefore implies a voltage sensitivity of 5 μV, which is readily achievable. If the voltage is integrated, the peak signal for the main component is 0.045 Vs. In this case the sensitivity for the higher harmonics depends on n; 10^{-4} sensitivity for the n=10 harmonic implies an integrated voltage signal of 0.9 μVs, which is within the capabilities of modern integrators.

The absolute accuracy of determining the integrated field strength depends on the accuracy with which the coil radius is known. For example, 15 μm uncertainty in a nominal 30 mm radius coil leads to an uncertainty of 0.1% in the integrated strength. However, provided all magnets of the same type, e.g. all quadrupoles, are measured with the same coil, the effect of such an error is minimal.

Several other factors have to be taken into account to achieve high accuracy. The rotation of the coil must be very precise, of the order of 10 μm , and so air bearings are often used. The angular resolution is also important; encoders with a resolution of 0.01°, i.e. 15 bit accuracy, are commonly used. A great deal of effort usually goes into the coil construction to achieve sufficient rigidity and stability to avoid sagging.[31] The connection between the drive motor, coil and encoder is also a critical item, and should be of the universal joint type to prevent bending moments on the coil or encoder if the axes are not precisely aligned.

A useful feature is a separate short coil (e.g. of the tangential type) of calibrated length, or a Hall plate (Sec. 7.4.2), mounted on the rotating coil former or on a special-purpose support to allow the field strength and quality in the magnet centre to be determined. This allows the magnetic length to be determined, as well as a comparison with the results of 2D field calculations to optimize the pole profile. The possibility to be able to move the coil or probe along the magnet axis is a further useful option since it allows errors in the pole profile to be localized and the effect of the magnet end geometry on the multipole content to be studied.

Commercial systems are available[24,35] one of which is shown in Fig. 7.5. The measurement coil, which rotates in air bearings, is clearly seen. This system allows an automatic positioning of the magnet on the moveable table underneath the coil, after which the reference targets on the top of the magnet can be positioned with respect to the axis of a laser beam, which is parallel and vertically displaced with respect to the coil (magnetic) axis.

7.4.1.3 *Determination of the magnetic centre and magnetic axis*

The effect of an offset between the centre of the magnet and the measuring coil can be calculated most efficiently using the complex notation, Eq. (7.16). Each winding is displaced by an amount $\Delta z = \Delta x + i\Delta y$. The field components at the shifted position therefore become :

$$(B_\theta + iB_r)e^{-i\theta} = \sum_n (B_n + iA_n)\,(\tilde{z} + \Delta\tilde{z})^{n-1}\ . \tag{7.25}$$

Note that since both the coil winding and its centre of rotation are displaced by the same amount, the angle it makes with the x-axis (θ) is not altered in the expression above. Use of a general expression for the voltage for an arbitrary $\tilde{z}$ would give an incorrect result. Assuming $\Delta\tilde{z}$ is small, and therefore expanding the above to first order gives the following :

$$(B_\theta + iB_r)e^{-i\theta} = \sum_n (B_n + iA_n)\,[\tilde{z}^{n-1} + (n-1)\,\Delta\tilde{z}\,\tilde{z}^{n-2}]\,. \tag{7.26}$$

The induced voltage therefore becomes :

$$V = N\omega L\ \text{Im}\ \sum_n (B_n + iA_n)\,[R^n_{\text{coil}}e^{in\theta} + (n-1)\Delta\tilde{z}\ R^{n-1}_{\text{coil}}e^{i(n-1)\theta}] \tag{7.27}$$

where Im represents the imaginary part. In the case of a pure multipole magnet of order n therefore :

$$V = N\omega L\,B_n\left[R^n_{\text{coil}}\sin(n\theta) + (n-1)\ \Delta x\,R^{n-1}_{\text{coil}}\sin((n-1)\theta) + (n-1)\ \Delta y\,R^{n-1}_{\text{coil}}\cos((n-1)\theta)\right]. \tag{7.28}$$

We may thus calculate the horizontal and vertical offsets from the relative values of the(n-1)th harmonics of the induced voltage as follows :

$$\Delta x = \frac{R_{coil}}{(n-1)} \frac{V_{n-1,sin}}{V_{n,sin}} \; ; \quad \Delta y = \frac{R_{coil}}{(n-1)} \frac{V_{n-1,cos}}{V_{n,sin}} \; . \tag{7.29}$$

For example, a sensitivity of 10^{-4} of the voltage induced by the main harmonic leads to a sensitivity for measuring the magnetic axis offset of 3 μm in the case of a 30 mm radius coil.

If a multipole magnet of order n is rotated by an angle α with respect to the measurement coil the induced voltage can be written as follows :

$$V = N\,\omega\, B_n L\, R^n_{coil} \sin(n(\theta-\alpha)) = N\,\omega\, B_n L\, R^n_{coil}\, [\sin(n\theta)\cos(n\alpha) - \cos(n\theta)\sin(n\alpha)] \; .$$

The angle of rotation (roll error) can therefore be deduced from the magnitude of the ration of the skew and normal components of the main harmonic as follows :

$$\alpha = \frac{1}{n} \tan^{-1}\left(\frac{-V_{n,cos}}{V_{n,sin}} \right) . \tag{7.30}$$

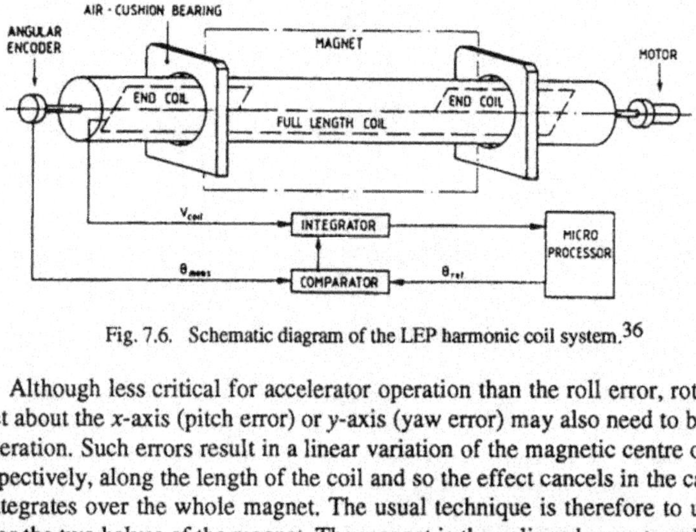

Fig. 7.6. Schematic diagram of the LEP harmonic coil system.[36]

Although less critical for accelerator operation than the roll error, rotation of the magnet about the x-axis (pitch error) or y-axis (yaw error) may also need to be taken into consideration. Such errors result in a linear variation of the magnetic centre offset, Δy or Δx respectively, along the length of the coil and so the effect cancels in the case of a coil that integrates over the whole magnet. The usual technique is therefore to use separate coils for the two halves of the magnet. The magnet is then aligned so as to null Δx and Δy offsets for both coils. A coil assembly including separate end-coils as well as a long integral coil is described in Ref. 36 (see Fig. 7.6). Another technique is to divide the long coil into two halves, whose output can then be measured individually or in series.

7.4.1.4 Rotation errors

A limit to the accuracy of the location of the magnetic centre and the determination of the multipole content comes from imperfect rotation due, for example, to imperfect bearings or bending of the coil support structure due to gravity. Both of these types of error lead to a displacement of the coil axis with respect to the magnetic centre which varies with the angle of rotation, θ. Equation (7.27) can be used for the induced voltage, except that in this case the shift in position is a variable $\Delta\tilde{z}(\theta) = \Delta x(\theta) + i\Delta y(\theta)$. Making a Fourier analysis of the displacement :

$$\Delta\tilde{z}(\theta) = \sum_{m=-\infty}^{+\infty} \Delta\tilde{z}_m \, e^{im\theta} \tag{7.31}$$

results in the following expression for the induced voltage in the case of a pure *n*th order multipole magnet :

$$V = N\omega L\, B_n \left[R_{\text{coil}}^n \sin(n\theta) + \text{Im}\ \{(n-1)\, R_{\text{coil}}^{n-1} \sum_{m=-\infty}^{+\infty} \Delta\tilde{z}_m e^{i(m+n-1)\theta}\} \right]. \tag{7.32}$$

Thus, additional spurious harmonics are introduced of order n-1$\pm m$, where m is the Fourier component of the displacement error, and which scale with radius as the (n-1)th harmonic. To have a better understanding of the relation between the type of rotation error and the resulting effect on the voltage, it is necessary to express $\Delta\tilde{z}$ in terms of its component parts, Δx and Δy :

$$\Delta x(\theta) = \sum_{m>0} a_m \cos(m\theta) + b_m \sin(m\theta)\ ;\quad \Delta y(\theta) = \sum_{m>0} c_m \cos(m\theta) + d_m \sin(m\theta)\ . \tag{7.33}$$

Relating the $\Delta\tilde{z}_m$ terms to the a_m, b_m, c_m, d_m then gives the following :

$$V = N\omega L\, B_n \left[R_{\text{coil}}^n \sin(n\theta) + (n-1)\, R_{\text{coil}}^{n-1} \begin{Bmatrix} \frac{a_m+d_m}{2} \sin(n-1+m)\theta + \frac{a_m-d_m}{2} \sin(n-1-m)\theta \\ + \frac{c_m-b_m}{2} \cos(n-1+m)\theta + \frac{c_m+b_m}{2} \cos(n-1-m)\theta \end{Bmatrix} \right]. \tag{7.34}$$

Thus, a_m and d_m terms produce normal components while b_m and c_m terms produce skew components, in each case of the harmonics (n-1$\pm m$).

An example of such a rotational error is the effect of a sag of the coil due to gravity. This occurs predominantly when the plane of the coil is horizontal, and so during one rotation the coil is displaced in the vertical plane with a periodicity of π, i.e. in terms of the analysis above there is ac_2 term, which in a quadrupole magnet therefore introduces spurious skew-dipole and skew-sextupole terms. The magnitude of the spurious field error, relative to that of the main component, is in general $(n-1)\delta / 2R_{\text{coil}}$, where δ is the amplitude of the rotational error. For example, to reach an accuracy of 10^{-4} in measuring the relative harmonic components at R_{coil} = 30 mm in a quadrupole magnet therefore requires rotational errors of less than ± 6 µm.

The fact that a spurious dipole term can be created in a quadrupole, or in general an (n-1) term when $m = 2(n-1)$, gives rise to an error in determining the magnetic centre given by $\Delta x = -(a_m - d_m)/2$ and $\Delta y = (b_m + c_m)/2$. Thus, rotational errors should be minimized for accurate determination of both multipole errors and magnetic centre location. Fortunately however there is a way of reducing the sensitivity to such errors, by means of bucking coils.

7.4.1.5 Bucking coils

A bucking coil is an additional coil mounted in the same plane as the main coil, in general with different outer and inner radii and number of turns, connected in series opposition in such a way as to reject the signal from particular field harmonics. Combinations of coils, which may also be located in symmetric planes, may also be used. One use of the technique is to reject the fundamental (*n*th) component, in order to reduce the dynamic range required of the measuring instrument and so improve the resolution of the higher harmonics. Rejection ratios of 1000 or more can be achieved, depending on the accuracy of the coil construction. A second use is to null the (*n*-1)th harmonic, in order to reduce the sensitivity to rotational errors.

Since the total voltage is simply the sum of the voltages generated by each wire, the previous expressions are easily generalized to the case of more than one contributing coil. For example, the voltage corresponding to the main harmonic in an *n*th order multipole magnet is given as follows :

$$V = \omega B_n L \sin(n\theta) \sum_{j=1}^{N_{coil}} R_j^n N_j \tag{7.35}$$

where N_j is the number of wires in the *j*th bundle. Note that N_j can be negative, to indicate the direction in which the wires are connected, and that if all wires pass within the magnet aperture, and start and return at the same end then $\sum N_j = 0$. A dipole field may therefore be suppressed with $\sum R_j N_j = 0$, a quadrupole term with $\sum R_j^2 N_j = 0$ and so on. Many configurations are possible; some typical schemes are given in Table 7.1. Each consists of two coils, *a* and *b*, with relative number of turns N_a and N_b. The relative radii for each coil are as specified, the maximum value in each case being 1.0. The coils are connected in series opposition, i.e. in the order $R_{a1}, R_{a2}, R_{b2}, R_{b1}$ so that N_j is positive for R_{a1} and R_{b2}, negative for the others.

Table 7.1. Bucking coil arrangements for the radial harmonic coil method.

Suppressed Components	N_a	N_b	R_{a1}	R_{a2}	R_{b1}	R_{b2}	Ref.
1	1	1	-0.333	0.333	0.333	1.0	34
2	1	1	0.0	0.707	0.707	1.0	36
1, 2	1	2	0.0	1.0	0.25	0.75	37
1, 2	1	2	-0.5	1.0	-0.125	0.625	38
2, 3	2	7	-0.5	1.0	-0.4224	0.6266	29

It follows from the above, and by inspection of Eq. (7.32), that suppressing the (*n*-1)th harmonic eliminates the term due to the rotation errors. Two schemes which provide for cancellation of both quadrupole and dipole terms, i.e. with $\sum R_j^2 N_j = \sum R_j N_j = 0$ are given in the Table, as well as one that compensates quadrupole and sextupole terms for sextupole magnet measurements. The general disadvantage of such schemes is that there is some reduction in the absolute signal level induced by the higher harmonics, however this can be minimized by a careful selection of parameters. For example, for the first scheme in the Table which removes both dipole and quadrupole components[37] the sensitivity for the sextupole component is $\sum R_j^3 N_j = 0.188$, as opposed to 1.0 for a single coil with the same outer radius and number of turns. With the second scheme[38] however, the sensitivity is increased to 0.633.

7.4.1.6 Tangential coils

Another coil geometry that has advantages for certain kinds of measurements is the "tangential coil" arrangement, in which the coil windings are located at a common radius, but different angular positions, on the surface of the winding frame (see Fig. 7.7). For each coil there are now two contributing terms; in the case of a coil with angular separation ϕ we have :

$$V = N\,\omega\,L\,R\left[B_r(\theta+\phi/2) - B_r(\theta-\phi/2)\right]\,. \qquad (7.36)$$

Inserting the expression for B_r and expanding we obtain :

$$V = 2N\,\omega\,L\sum_n R^n \sin(n\,\phi/2)\left[-A_n\sin(n\theta) + B_n\cos(n\theta)\right]\,. \qquad (7.37)$$

Compared to the result for the radial coil geometry a factor of 2 has appeared because of the fact that both parts of the coil contribute to the signal, which is now modulated by a term involving the angular separation, ϕ. A case of special interest is $\phi = \pi$ for which there is sensitivity only to the odd harmonics n = 1, 3, 5, 7, ..., etc., and so this arrangement is often referred to as a "dipole coil". The coils at $\pi/2$ and $3\pi/2$ when subtracted yield only n = 2, 6, 10, 14, ..., etc. i.e. a "quadrupole coil". These are particular examples of windings that form the basis of the Morgan coil, which contains a number of windings each of which is sensitive to a different harmonic.[39]

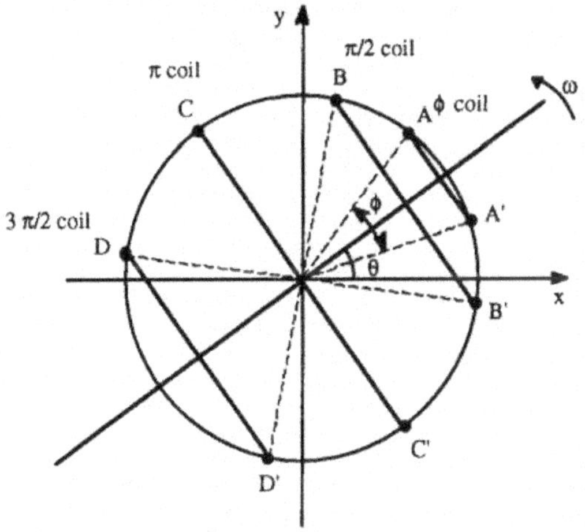

Fig. 7.7. Tangential harmonic coil geometry.

The above configurations can be used to buck-out unwanted harmonics. With N_ϕ, N_D and N_Q turns in the ϕ, dipole and quadrupole coils respectively the conditions for bucking out the dipole term becomes : $N_\phi \sin(\phi/2) - N_D = 0$. Bucking of the quadrupole occurs when : $N_\phi \sin(\phi) - 2N_Q = 0$. One solution[40] which gives good sensitivities up to n = 15 is obtained with ϕ = 19.47° and $N_D = N_Q = 1$, $N_\phi = 6$. As with the radial configuration, bucking the (n-1)th harmonic in an nth order magnet effectively removes the disturbing effect of rotation errors, however there are other errors specific to the case of the tangential coil.[33] A combination of both radial and tangential coils is used in the coil assembly developed at Argonne for the measurement of the APS quadrupoles and sextupoles.[29]

7.4.2 *Hall Plate Methods*

Although much less convenient and accurate than a harmonic coil system, multipole magnet measurements can also be carried out with Hall plates. The main accuracy limitation is due to the fact that both probe calibration errors and the planar Hall effect can introduce spurious harmonics. For example, for a rotating Hall plate aligned perpendicular to the radial B_r field component, the planar Hall effect due to the B_θ component is to introduce a spurious skew-octupole error.

General purpose scanning systems such as the one described in Sec. 7.3.1 can also be used for multipole measurements. If the plate can be made to follow a circular trajectory, then the multipole terms can be recovered directly from a Fourier analysis of a single field component, for example $B_y(\theta)$, according to Eq. (7.15). Once again however the planar Hall effect, depending on the orientation of the plate in x-z plane, could introduce spurious dipole and sextupole harmonics.

Alternatively, measurements can be made in the median plane (y=0) for which :

$$\begin{aligned} B_y &= B_1 + B_2 x + B_3 x^2 + B_4 x^3 \ldots \\ B_x &= A_1 + A_2 x + A_3 x^2 + A_4 x^3 \ldots \; . \end{aligned} \tag{7.38}$$

Thus, in principle by measurement of both field components in the median plane both normal and skew multipole terms can be obtained. However, in practice the skew-quadrupole term (A_2) is strongly influenced by the angle of the B_x plate, and the skew-sextupole term (A_3) contains a contribution from the planar Hall effect, so that measurements of this kind are not usually carried out. A further problem, as mentioned in connection with dipole measurements (Sec. 7.3.1), is the difficulty in extracting information about the individual harmonics. Instead, the usual method of analysis is to consider an overall field or gradient error. Scans along the length of the magnet at various x-positions (y=0) allows the integrated field uniformity and the magnetic length to be determined.

7.4.3 *Other Techniques*

A stretched wire system was used at DESY for the measurement of the integrated strength, and determination of the magnetic axis, of the superconducting quadrupole magnets for HERA.[28] A single CuBe wire is used with return loop outside the magnet. The field was measured by making horizontal and vertical movements and integrating the induced voltage. A special bench has been developed at SLAC that uses the same stretched wire for determination of both the magnetic centre of a quadrupole and the electrical centre of an attached beam position monitor.[41] In this case the wire is vibrated and the induced voltage monitored using a spectrum analyzer. The magnetic centre is located to an accuracy of a few microns by nulling the signal at the drive frequency. The same system allows the wire to be scanned to allow measurement of the multipole content.[42]

Earlier techniques that have been used for determination of the magnetic axis of a quadrupole, when rotating coil systems were not sufficiently accurate, were the floating wire method and an optical technique.[23] In the former case, a wire is stretched through the magnet, and moved by trial and error until a position is obtained such that no deflection is observed when a current is passed through the wire. Accuracies of 50-100 μm could be obtained. A method that was used considerably was based on the scattering of polarized light on aligned particles in a colloidal solution of ferrous oxide. Accuracies of the order of 25 μm were achieved. The method has continued to be developed,[43] most recently for application to the centre location of long superconducting magnets.[44,45]

7.5 Insertion Devices

Compared to the more conventional dipole and quadrupole magnets, insertion devices have particular requirements for magnetic measurements that in general have led to different approaches being taken. The magnetic field distribution must be optimized not only to minimize the effect of the device on the electron beam, but also (in the case of undulators) to maximize the degree of constructive interference that determines the intensities of the radiated harmonics. Measurement of both field distribution and integrated properties are therefore needed, of both transverse field components, with high accuracy. The magnet geometry, up to 5 m in length with small gap, 20 mm or less, also restricts the kind of measurement that can be performed. A sufficiently accurate determination of the field integrals using the same point-by-point method needed for the calculation of electron trajectory and radiation spectrum is difficult to achieve. It is advisable therefore to use more than one technique and to compare the results.

7.5.1 Point-by-point Field Measurements

The usual method consists of a Hall plate scanning technique, somewhat similar to that employed for dipole magnets (Sec. 7.3.1). A high density of points is generally used, between 20 and 100 points per period, for sufficient accuracy in the trajectory and radiation calculation and also to reduce statistical noise. Since measurements are usually carried out at a range of transverse (x) positions, and for different magnet gaps, it is an advantage to reduce the time as much as possible by taking data "on-the-fly" rather than "step-by-step" as is more common in dipole field mapping. Speeds of between 20 and 50 mm/s are commonly used. Triggering of the measurements can be carried out in different ways : at the computer level by reading the encoder and checking if the desired position has been reached, or at the hardware level by counting either stepping motor or encoder pulses. A fast voltmeter is required to measure the Hall plate voltage - either directly, or using the analog output from a commercial system. In either case, choice of integration time is very important, since it determines the resolution, noise level, and of course maximum measurement rate.

Several factors have to be taken into consideration if accurate field integral values are to be obtained, for example the positioning accuracy of the probe along the z-axis. A simple analysis shows that random errors with an rms value of σ_z give rise to an rms field integral error given by :

$$\sigma_I = 2\pi \frac{N}{\sqrt{2N_{pts}}} B_o \, \sigma_z \tag{7.39}$$

where N is the number of periods of the insertion device, N_{pts} the number of data points and B_0 the field amplitude. For example, 1000 readings in a 50-period magnet with peak field 0.5 T with positioning error of 10 μm results in a random field integral error of 0.35 Gm, which is not insignificant compared to typical accuracy requirements. An accurate position measuring device (linear encoder, or interferometer) and precise triggering are therefore essential to obtaining reliable data. Statistical uncertainty in the Hall plate reading due to resolution and noise can also be important in some cases. An rms field error σ_B gives an rms field integral error given by :

$$\sigma_I = \sqrt{N_{pts}} \, \Delta z \, \sigma_B \tag{7.40}$$

where Δz is the spacing of the data. Using the example above, with Δz = 2.5 mm, a random error of 0.2 G gives an uncertainty in the integral of 0.016 Gm. A larger effect however is introduced by uncertainty in the zero-field offset which most probes have, due to the resolution of the system and/or drift due to temperature fluctuations. An uncertainty of 0.1 G in the present case leads to an integral error of 0.25 Gm. Finally, care must be taken to accurately calibrate the Hall plate, in particular the even terms which affect directly the field integral.

A measurement of both transverse field components is generally necessary, since the field errors are of similar magnitude in both planes. Also, certain kinds of device have a non-zero B_x component by design, for example to produce elliptical or helical electron trajectories. However, some difficulty arises in the measurement of a small B_x component in the presence of a large orthogonal B_y component, due to both angular positioning errors and the planar Hall effect. The former introduces a spurious oscillating component proportional to B_y but this is of no great importance since it does not contribute to the net field integral. The planar Hall effect on the other hand is proportional to $B_y{}^2$ and can therefore significantly affect the field integral value. One possibility to remove the effect is to carry out a series of measurements with the B_x plate rotated in different positions around the x-axis to locate the null orientation. Alternatively, a numerical correction can be applied based on the measured planar Hall coefficient. Rather than make specific calibration measurements, requiring equipment to accurately position and rotate the plate in a suitable magnet, the coefficient (β) can be determined in-situ from measured $B_x(z)$ and $B_y(z)$ undulator fields by fitting : $B_x = \alpha B_y + \beta B_y^2$; the first term accounts also for an angular misalignment. Alternatively a Fourier analysis can be made of the central periodic part of the undulator : $B_x = \alpha \cos(kz) + \beta \cos(2kz)$. All of the above correction methods are valid in the case that there is no B_z component and are therefore not valid at large vertical displacements from the median plane.

Determination of the electron trajectory from the measurement data is simpler than in the dipole magnet case : since the maximum angular deflection is always much smaller than unity, Eq. (7.12) can be approximated as follows :

$$x'(z) = \frac{e}{\gamma mc} \int_{-\infty}^{z} B_y(z)\, dz \,. \tag{7.41}$$

A second integration yields the displacement $x(z)$. Replacing B_y with B_x gives the vertical trajectory. Further analysis would generally include calculation of the radiation spectrum emitted by the device. For an accurate calculation up to the 5th harmonic using simple trapezoidal rule integration at least 20 measurement points per period are required. To extend the calculation to higher harmonics either more points are needed, or spline interpolation may be used.

7.5.2 *Integrated Field Measurements*

Insertion device field quality specifications usually place very small limits on the first field integrals ($I_{x,y}$), typically in the range 0.25-1.0 Gm, and on the integrated multipole content, corresponding to 0.25-0.5 Gm at a radius of 5 mm. The requirements for the measurement system are therefore approximately an order to magnitude smaller than this. Since it is very difficult to obtain this kind of accuracy from point measurements, a number of other techniques have been developed.

The "flipping coil" system developed at the ESRF[46] uses a long integral coil, typically 10 mm wide and up to a few metres in length, formed from stretched wires. Litz wires are generally used with the individual strands connected in such a way as to form a coil of between 20 and 50 turns. Flipping the coil and integrating the voltage over 4 steps of 90° allows both field integrals to be obtained, automatically compensating linear drift. Measurement over the return rotation allows a check of the accuracy. The alignment of the coil to the magnet is not critical. Due to variations in the width of the coil the integral varies periodically with the longitudinal (z) position of the coil with respect to the undulator.[47] The system should therefore allow for the possibility to scan automatically in the z-direction, in order to determine the true (mean) integral value. With a 20 turn coil, and 10 mm separation for example, the sensitivity is 0.025 Gm per µVs for a 180° coil rotation. Integrators are available with a sensitivity of 0.1 µVs. The typical reproducibility of such a system is 0.01 Gm (r.m.s.) with an absolute accuracy of about 0.1 Gm. By carrying out a series of measurements at different x-positions the integrated multipole field errors can be determined from the variations $I_y(x)$ and $I_x(x)$, as in Eq. (7.38).

An alternative to the flipping coil technique could be to use the moving wire method, as used for quadrupole measurements (Sec. 7.4.3), with the advantage of eliminating the effect of coil width variation.

A one-period long coil moving along the axis in an ideal periodic field has zero net flux, and hence gives a zero voltage signal. It can be used therefore as a very sensitive method for locating field errors in the magnet, and measuring the total field integral. An advantage of this approach is that it can quite easily be incorporated into the Hall plate scanning bench to cross-check the results, as in the systems set up at HASYLAB[48] and Argonne.[49] A digital integrator can be used which stores the data for later readout by the computer, or alternatively the output from an analog integrator can be read by a voltmeter. As with any integrator method, care has to be taken to remove the effects of linear drift.

7.5.3 Other Techniques

A novel "pulsed wire" technique has been developed recently for measurement of both single permanent magnet blocks and complete devices.[22] In this method a short (~20 μs) current pulse propagates along a stretched wire and excites a transverse wave, the velocity given to the wire at any point being proportional to the magnetic field. The wave propagates along the wire and the displacement, proportional to the field integral, is detected using an optical interrupter. Accuracies of about 0.5% in the peak field amplitude have been achieved. The method is very rapid and was developed particularly as a convenient way of checking undulators installed in free-electron laser experiments.

A floating wire method has also been used successfully on one occasion to measure the effects of transverse (B_x) field errors.[50]

7.6 A.C. and Pulsed Magnets

Alternating current magnets used in booster synchrotrons are most often measured only with a d.c. excitation. However, sometimes it is desired to check the performance with a.c. excitation to assess the effects of eddy currents, particularly to include the effect of the vacuum chamber. The most common technique is that of magnetic induction in a static coil - either a point coil or long curved coil. Using an analog integrator the coil output corresponds directly to the average field over the coil surface area. A convenient technique is to use a digital oscilloscope to store the resulting waveforms. The magnetic length can be obtained by dividing the result from the integral coil with that of a point coil at the magnet centre. Field homogeneity can be studied by means of two matched coils in series opposition, one fixed at the origin and another translated to different transverse positions. Such a technique was used to test the ALS booster dipole magnet prototype both with a 2 Hz sawtooth and 10 Hz sinewave excitation.[51] The use of Hall probes in this context is not very common, however a Hall probe with a fast sampling voltmeter was used to make 50 Hz measurements of the prototype dipole for the TRIUMF KAON Factory.[52]

In the case of pulsed magnets, such as injection and extraction kicker and septum magnets, field measurements are most usually carried out with coils, although the frequency response of the Hall plate would not preclude its use. Since the times involved are very short - typical pulse lengths range from a few μs for kicker magnets to a few hundred μs for septa - a simple RC integrator can be used.[5] The same techniques as for a.c. measurements can be used to perform measurements of field homogeneity.

7.7 Magnetic Materials Testing

The quality of the magnetic field of a device depends not only on the geometric precision with which it is machined and assembled but also on the quality of the magnetic materials used in its construction. For the main storage ring magnets a careful selection of the steel material is usually carried out. Various instruments that are used to check that the desired properties have been met are described in Ref. 53.

Most insertion devices are constructed with permanent magnet material, and the resulting field quality is largely determined by the detailed properties of the individual blocks. It is common therefore to characterize the blocks magnetically in some way, in order to determine the optimum configuration to use for the assembly. The standard techniques consist of flipping[54], or spinning[55], the magnets inside a pair of Helmholtz coils and measuring the induced voltage in order to determine the magnetization components. Point-by-point Hall plate and integral coil measurement systems have also been applied to the task of individual block measurements in order to take into account the significant inhomogeneity of the magnetization.[47]

7.8 References

1. *Proc. CERN Accelerator School, Magnetic Measurement and Alignment*, CERN 92-05, Sept. 1992.
2. *Proc. Int. Workshop on Magnetic Measurements of Insertion Devices*, ANL/APS/TM-13, Oct. 1993.
3. J.L. Symonds, *Rep. Prog. Phys.* **18** (1955) 83.
4. C. Germain, *Nucl. Instr. Meth.* **21** (1963) 17.
5. P.G. Watson and R.F. DiGregorio, *Proc. Int. Symposium on Magnet Technology*, Stanford (1965), p. 393.
6. O. Runolfsson, *Proc. 6th Int. Conf. Magnet Technology*, Bratislava (1977), p. 802.
7. K.N. Henrichsen, *Proc. 8th Int. Conf. Magnet Technology*, Grenoble (1983); *J. de Physique C1* **45** (1984) 937.
8. W.C. Elmore and M.W. Garrett, *Rev. Sci. Instr.* **25** (1954) 480.
9. M.I. Green, Ref. 1, p. 206.
10. Walker Scientific Inc., Rockdale St., Worcester, MA 01606, USA.
11. Dowty RFL Industries Inc., Instrumentation Division, Powerville Rd., Boonton, NJ 07005-0239, USA.
12. Magnet-Physik Dr. Steingrover GmbH, Emil-Hoffmann Str. 3, D-5000 Köln 50, Germany.
13. Metrolab Insruments SA, 110 ch. du Pont-du-Centenaire, CH-1228 Geneva, Switzerland.
14. Solartron Instruments, Victoria Rd., Farnborough, Hampshire, GU14 7PW, England.
15. B. Berkes, Ref. 1, p. 167.
16. C. Goldberg and R.E. Davis, *Phys. Rev.* **94** (1954) 1121.
17. F.W. Bell, 6120 Hanging Moss Rd., Orlando, Florida 32807, USA.
18. Bruker Analytische Messtechnik GmbH, Wikingerstr. 13, D-76189 Karlsruhe 21, Germany.
19. Group 3 Technology Ltd., P.O. Box 71-111, Rosebank, Auckland 7, New Zealand.

20. Lake Shore Cryogenics Inc., 64 East Walnut St., Westerville, Ohio, USA.
21. L.G. Ratner and R.J.Lari, *Proc. Int. Symposium on Magnet Technology*, Stanford (1965), p. 497.
22. R.W. Warren, *Nucl. Instr. Meth.* **A272** (1988) 257.
23. J.C. Cobb and J.J. Muray, *SLAC-39* (Revised), November 1965.
24. Field Effects, 6 Eastern Rd., Acton, MA 01720, USA.
25. See for example, *"Numerical Recipes : The Art of Scientific Computing"*, W.H. Press *et al.* (Cambridge University Press, 1989).
26. P.J. Bryant, CERN 84-04, 1984.
27. J.N. Galayda *et. al., IEEE Trans. Nucl. Sci.* **NS-28** (1981) 2593.
28. P. Schmüser, Ref. 1, p. 240.
29. S.H. Kim, *et al., Proc. 1993 US Particle Accelerator Conference*, p. 2802.
30. M. Lieuvin, in *Proc. Int. Magnet Measurement Workshop (IMMW-7)*, GSI Darmstadt, June 1991.
31. M.I. Green, Ref. 1, p. 103.
32. M.I. Green *et al., Proc. 1989 US Particle Accelerator Conference*, p. 1972.
33. W.G. Davies, *Nucl. Instr. Meth. Phys. Res.* **A311** (1992) 399.
34. L. Walckiers, Ref. 1, p. 151.
35. Danfysik A/S, DK-4040 Jyllinge, Denmark.
36. L. Walckiers, *IEEE Trans. Magn.* **MAG-17** (1981) 1872.
37. O. Pagano *et al., J. de Physique C1* **45** (1984) 949.
38. K. Halbach, PEP Note 208, Feb. 1976.
39. G.H. Morgan, *Proc. 4th Int. Conference Magnet Technology*, Brookhaven National Laboratory (1972), p. 787.
40. S.H. Kim, Argonne National Laboratory, LS-167.
41. G.E. Fischer *et al., Proc. 3rd European Accelerator Conf.*, Berlin, 1992, p. 138.
42. P. Tenenbaum *et al., Proc. 1993 US Particle Accelerator Conference*, p. 2838.
43. R. Sughara *et al.*, KEK Report 89-9, Sept. 1989.
44. J. Le Bars and F. Kircher, *Proc. 3rd European Accelerator Conference*, Berlin, 1992, p. 1403.
45. M.A. Goldman *et al., Proc. 1993 US Particle Accelerator Conference*, p. 2916.
46. J. Chavanne, ESRF-SR/ID-89-27, September 1989.
47. D. Zangrando and R.P. Walker, *Proc. 3rd European Accelerator Conference*, Berlin, 1992, p. 1355.
48. J. Pflüger, *Rev. Sci. Instr.* **63** (1992) 295.
49. L. Burkel *et al.*, ANL/APS/TB-12, Argonne National Laboratory, March 1993.
50. D.C. Quimby, *J. Appl. Phys.* **53** (1982) 6613.
51. M.I. Greeen *et al., Proc. 1989 US Particle Accelerator Conference*, p. 1969.
52. A.J. Otter and P.A. Reeve, *Proc. 1991 US Particle Accelerator Conference*, p. 2363.
53. J. Billan, Ref. 1, p. 17.
54. D.H. Nelson *et al., Proc. 9th Int. ConferenceMagnet Technololgy*, Zurich (1985).
55. A. Luccio *et al., Nucl. Instr. Meth.* **219** (1983) 213.

CHAPTER 8: VACUUM SYSTEMS

JOHN NOONAN and **DEAN WALTERS**

Advanced Photon Source, Accelerator Systems Division,
Argonne National Laboratory, 9700 South Cass Avenue,
Argonne, Illinois 60439, USA

8.1 Introduction

8.1.1 Background

Even with a "perfectly" designed lattice of electron optics, the accelerator would be useless without containing the particle beam in a vacuum. The scattering cross section of a high energy particle is so high that the beam would propagate only a few meters in atmosphere. In a storage ring the density of residual gases in the vacuum chamber is usually the limiting factor in sustaining a stored beam of high energy particles. Obviously, since the goal for experimenters is the longest possible time for data acquisition, the storage ring should have the longest beam lifetime. So a casual review of the need for vacuum systems would support designing a system that produces the lowest pressure. Unfortunately, as will be developed throughout this chapter, optimal design of the vacuum system and optimal particle beam optics are often in conflict. The most efficient placement of pumps should be near the major gas sources. Unfortunately, one of the bending, focusing, or steering magnets often needs to occupy the same space. Innovative and creative designs have been developed to resolve the conflicts between the lattice components and vacuum requirements. This chapter will provide a condensation of lessons learned over the last 40 years. As a result of innovative vacuum system design, accelerators routinely achieve ultra-high vacuum pressures (10^{-9} Torr) even though the power density of the beam has increased enormously. "Third-generation" synchrotron light sources that are being built and commissioned today have photon energy densities approaching 750 watts/mm^2.[1-4]

A conflict also arises between optimizing vacuum pumping and minimizing the beam coupling impedance to make the high current beam stable. To minimize the coupling impedance, the vacuum chamber should have a large cross-section, be as smooth as possible, and be made of high conductivity metals. The best vacuum system design has large pump-out ports as close to the gas source as possible. The pump ports break up the smooth cross-section needed for low beam impedance. In addition, valves, flanges, bellows, etc. must be specially designed to expose smooth interfaces.

There are a number of key concepts that are used in design and operation of accelerator vacuum systems. This chapter will describe vacuum measurements, and how they relate to the particle beams, efficient creation of vacuum, and sources of gas and their control. Special requirements of vacuum systems for rf cavities, windows, and insertion devices will be discussed.

John Noonan e-mail: noonanjr@aps.anl.gov; fax: (708) 252-5948
Dean Walters e-mail: drw@oxygen.aps.anl.gov; fax: (708) 252-5948

8.1.2 Beam Interactions with Molecules and Beam Lifetime

The particle beam can scatter from residual gas molecule through Coulomb scattering. Hard scattering from the gas causes the particle to lose enough energy or momentum to be removed from the beam being accelerated.[1] Soft multiple scattering causes the beam to broaden and the beam lifetime to decrease. For example, it can be shown that the strong Coulomb scattering cross section depends on the atomic number of the molecule, Z, and the beam energy, γ:

$$\sigma_s = \frac{4\pi\, r_e^2\, Z\,(Z+1) < \beta_y >}{\gamma^2\, A_y} \tag{8.1}$$

where $< \beta_y >$ is the average vertical beta function of the ring, and r_e is the classical electron radius, $\frac{e^2}{mc^2}$;[5] and $A_y = \frac{(g/2)^2}{\beta_y}$, where g is the full chamber vertical aperture. Using the APS storage ring as an example, if the residual gas in the system was nitrogen, and g = 8mm, the beam lifetime due to Coulomb scattering is:

$$\tau_s = \frac{35.5}{P} \qquad (\tau \text{ in hours, P in nTorr}) . \tag{8.2}$$

Particles can be lost from the beam through inelastic collisions with residual gas nuclei (bremsstrahlung). If the energy loss is greater than the rf bucket height, the particle is lost from the beam bunch (see Chapter 4). Again the scattering cross section depends on the atomic number, Z, the beam energy, and the height of the rf bucket. The bremsstrahlung differential cross section and lifetime are:

$$\sigma_\beta = \frac{4\, r_e^2}{137}\, Z\,(Z+\xi)\left(\frac{4}{3}\ \ln\left(\frac{E}{\Delta E_{RF}}\right) - \frac{5}{6}\right)\ln\left(183\, Z^{-1/3}\right) \text{ and}$$

$$\xi = \frac{\ln\left(1440\, Z^{-2/3}\right)}{\ln\left(183\, Z^{-1/3}\right)} \tag{8.3}$$

$$\frac{1}{\tau_\beta} = cN\sigma_\beta \tag{8.4}$$

where c is the speed of light and N is the residual gas atomic density.

If the system residual gas is nitrogen at 300K, the lifetime for the APS storage ring is:

$$\tau_\beta = \frac{53.3}{P} \qquad (\tau_\beta \text{ in hours, P in nTorr}) . \tag{8.5}$$

As illustrated in the example, Coulomb scattering reduces the beam lifetime more than bremsstrahlung.[5]

For completeness, the beam lifetime depends on several loss mechanisms unrelated to scattering from residual gases. There is a "quantum lifetime" that refers to electrons being lost from the phase-stable rf acceptance due to emission of a high-energy, synchrotron-radiation photon (see Chapter 2). The loss mechanism depends on the momentum acceptance of the machine and the rf bucket height (see Chapters 2 and 4). Typically, the lifetime is very long compared to the other loss processes. Touschek scattering is caused by intrabeam collisions that result in energy transfer between particles, and can also result in large energy loss to the particle, causing it to be scattered out of the rf bucket and lost (see Chapter 2).

Touschek scattering is a function of beam energy, current density, and the six-dimensional phase volume of the beam bunch. Touschek scattering is commonly the ultimate limiting lifetime in low energy storage rings, while Coulomb scattering is usually the limiting lifetime in high energy rings. There are exceptions to that statement. For example, the 6-GeV X-ray ring at ESRF operates with beam lifetime approaching 40 hours with a 80-mA beam current and the lifetime is directly proportional to current when the current is distributed in 992 bunches, i.e., multi-bunch mode. However, when the 80 mA is divided into only 16 bunches, the particle density is so high that Touschek scattering drops the beam lifetime to 18 hours and is essentially independent of pressure.[6]

In most cases the beam lifetime is limited by the residual gas pressure in the chamber. However, the problems of accelerator vacuum systems operation is more pernicious. In an elegant perversity of physics, the electron or positron beams are also responsible for generating a major part of the gas load in the chambers.[7] Photons (synchrotron radiation) are generated by the particle beam passing through a bending magnet. The beam-induced photons, especially photons in the visible through soft X-ray spectrum, excite molecules adsorbed on an exposed surface causing them to desorb. Photon-stimulated desorption (PSD) is the largest source of gas in synchrotron light sources. An immediate consequence is that the beam lifetime is inversely proportional to the stored beam current. For example, Figure 8.1 illustrates the number of photons generated per second per mA of stored beam for several light sources due to bending magnet radiation.[8] As a first approximation, the photon intensity (photons/eV/mA/sec), $d^3N/d\varepsilon/dI/dt$, is:

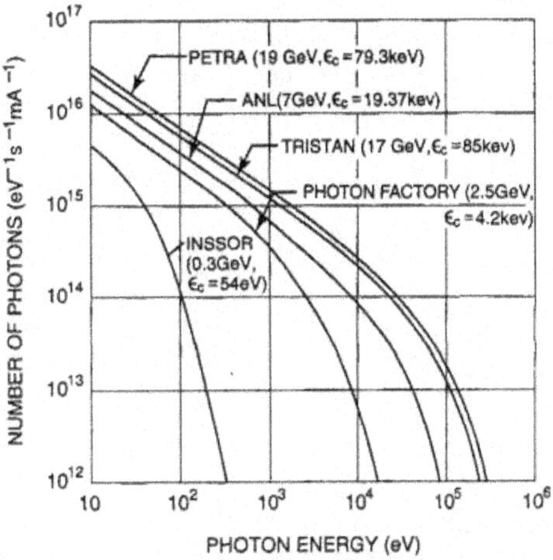

Figure 8.1. Synchrotron radiation spectra of different machines.

$$\frac{d^3N}{d\varepsilon dI dt} = 1.51 \times 10^{14}\left(\frac{\rho}{E^2}\right)\left(\frac{\varepsilon}{\varepsilon_c}\right)^{-2/3} \tag{8.6}$$

where ρ = bending magnet radius (in m)
ε_c = critical energy (in eV)
ε = photon energy (in eV)
and E = particle beam energy (in GeV).

The total photon flux for the APS storage ring at 7 GeV, 300 mA is 1.82×10^{21} photons/second.[1] The photon flux couples to the gas molecules on the surfaces exposed to the beam, excites the molecules, and causes them to desorb. The detailed desorption mechanism is not, however, well known. Measurements have shown that molecular energy levels are excited by photons in the 2- to 100-eV energy spectrum. X-rays are too energetic to couple efficiently to the vibronic molecular levels. However, high energy photons do cause considerable desorption because these photons can stimulate electron emission from atomic cores. The secondary electron yield couples to the vibronic molecular levels to cause electron-stimulated desorption (ESD). Photon- and electron-stimulated desorption coefficients can be calculated for a specific molecule. However, the total desorbed gas load is essentially impossible to calculate for accelerator vacuum systems. A complete analysis of the secondary electron yield, electron fluorescence, and coupling coefficients is required—but not sufficient. The geometric view factor to other exposed surfaces must also be determined since the electron-stimulated desorption is from surfaces exposed to the scattered beam. Then a spatial convolution integral must be calculated. Theoretically it is possible. Practically, an efficient *a prior* computation scheme is not yet available.

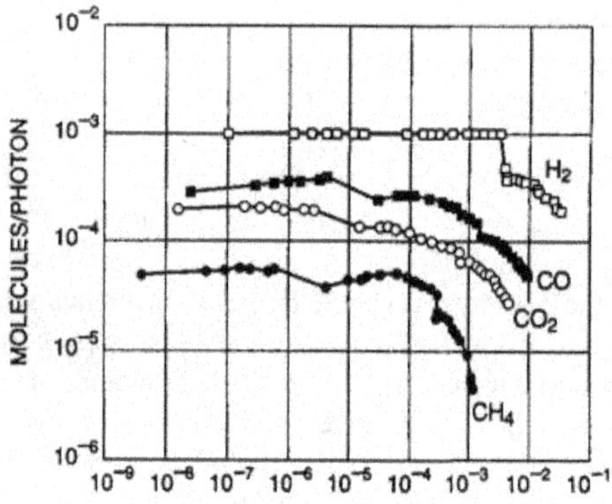

Figure 8.2. Desorption versus photon flux for Stainless Steel.

Engineers at CERN have developed an engineering measurement of the photon-stimulated desorption gas load.[9,10] A chamber that represents the storage ring geometry is installed in an X-ray beamline. Gas flow experiments are made as a function of photon flux. Figures 8.2 through 8.4 illustrate the total gas load as a function of integrated photon flux. The gas load is a convolution of photon and electron-stimulated desorption and the geometric view factor. No attempt is made to separate the various desorption mechanisms.

This engineering measurement has been repeated for a number of storage ring geometries.[1-3,7-9] Although there are major differences in chamber geometries, the integrated gas loads behave similarly. The initial gas loads are enormous. In general, aluminum surfaces are larger sources of gas than

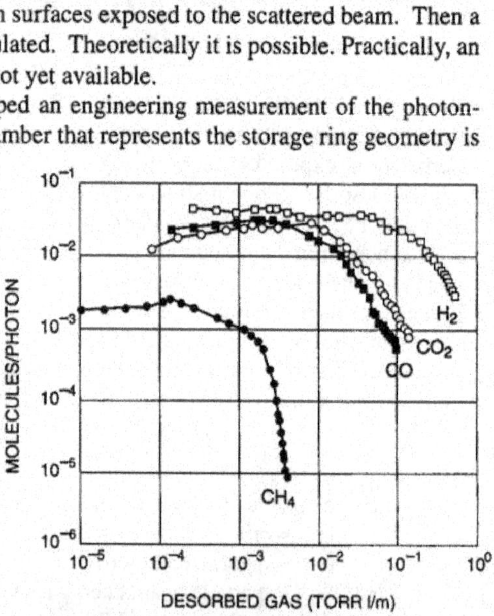

Figure 8.3. Desorption versus photon flux for Aluminum.

copper, and copper is a greater source than stainless steel (see Figures 8.2-8.4).[7,8] In addition, different molecules have different desorption coefficients. As an example, the desorption coefficient for H_2 and CO from oxygen-free, electrolytic (OFE) copper are $7 \times 10^{-5}D^{-2/3}$ and $5.6 \times 10^{-6}D^{-2/3}$ molecules/photon, respectively, where D is the total photon dose. The gas desorption rate, Q_D, can be described as:[10]

$$Q_D = k_v N_P \tag{8.7}$$

where k $= 3.1 \times 10^{-20}$ Torr l/molecule
v = desorption coefficient (molecules/photon)
N_P = photon flux (photon/s)

So the gas load from copper absorbers exposed to X-rays from 8-GeV positrons are 6.5×10^{-5} Torr l/s of H_2 and 4.6×10^{-6} Torr l/s of CO after 100 ampere-hours.[11]

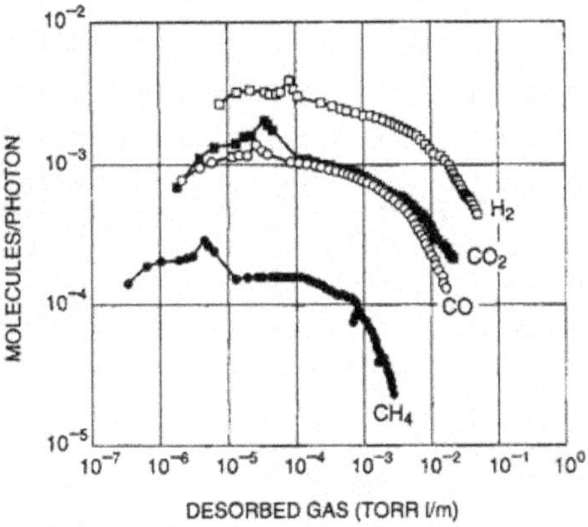

Figure 8.4. Desorption versus photon flux for Copper.

Thermal desorption is the other major gas load. Molecules that are weakly chemisorbed to the vacuum surfaces can couple to the phonon spectrum, generated by thermal excitations in the bulk material. The most notorious example of a gas that can be desorbed because of thermal excitation is water. The first several monolayers of water adhere strongly to metal surfaces. However, a metal surface exposed to atmosphere can have 10's of monolayers adsorbed. The outer layers are weakly bound and thermal excitation can desorb the gas. It is well known that baking the vacuum system improves the base pressure of a vacuum system. The increased temperatures excite more phonons to couple to the adsorbed molecules. The desorption rate can be modeled as an exponential function with an activation energy.[12]

$$\tau = \tau_0 e^{-2.05/kT} \text{ for CO on 316 LN stainless steel.} \tag{8.8}$$

The rf acceleration cavity can produce significant gas loads. In the accelerator, the microwave power not only couples to the beam, but also to adsorbed molecules on the cavity surfaces. Because of the high electric fields and large gas loads, the surfaces can sputter metals atoms into the vacuum system. The metal vapor can deposit on windows and ceramics in the rf cavity. If the coating becomes too thick, insulators can be heated through resistance losses. Catastrophic vacuum failures will result. Special precautions for rf acceleration cavities will be discussed in the next section.

8.2 Vacuum System Design

Although the source of gas in accelerators is unconventional, once the molecules are residual gases in the vacuum system, vacuum technology design and operation are the same for accelerators as for other applications. The consequences of bad vacuum practices are substantially greater than for laboratory scale vacuum systems. One of the special problems of accelerator vacuum systems is the common conductance path through the whole accelerator. Contaminating one chamber will contaminate the whole accelerator. A vacuum leak in one port will worsen the whole system pressure. This section will encapsulate experience in choosing the materials used in the vacuum system, and advantages and disadvantages of vacuum pumps, valves, seals, and gauges.

8.2.1 Vacuum State Equation

There is a simple relationship that governs vacuum systems at steady state:

$$\Sigma Q_i = SP \tag{8.9}$$

where P is the pressure in Torr
S is the effective speed in l/s
and Q_i is the gas flow rate from all sources.

It has been discussed above that gas will be released from normal thermal desorption and by the interaction of photon and electron beams with the system walls. There is also gas flow from bulk diffusion, permeation, and system leaks. Altogether, the total cumulative gas load divided by the effective pumping speed at any location will determine the local pressure. Accelerator storage ring pressures will rise between pumps depending on the physical conductance or reduced pumping speed along the ring. It is a challenge of any accelerator project to simultaneously reduce the gas flow rate and increase the pumping speed. The goal is to lower the pressure which translates into increased beam lifetime.

8.2.2 Vacuum Materials

Judicious selection of materials for use in vacuum chambers and components will minimize the gas load in the accelerator vacuum system. A number of metals have been used in fabricating accelerator vacuum chambers. Stainless steel, copper, aluminum, and Inconel have been used in various accelerators. There is no one metal uniquely suited for accelerator use. Cost, radiation resistance, thermal conductivity, electrical conductivity, fabrication ease, and magnetic properties are factors to be considered in choosing the vacuum chamber materials. The most common metal for third-generation X-ray light sources is aluminum. ALS has chosen 5083 Al; APS, SPring-8, and the SRRC in Taiwan have selected 6063 Al. Aluminum has very good thermal properties, the alloys used have good mechanical properties, and the 6063 Al can be extruded—a significant cost advantage. However, aluminum is difficult to weld repeatably. It can be done, but significant care and attention to detail are required. For example, SPring-8 extrudes the aluminum chambers in a controlled Ar/10% O_2 atmosphere and machines the Al using ethanol as a coolant to minimize oxide thickness. To illustrate how sensitive reliable Al welding can be, the APS devoted significant time and effort to develop an automatic welding system to produce the chambers. Even then, the weld leak rate was 10% until the Al weld rod was extruded in a nitrogen atmosphere. The wire had a thinner oxide, which reduced the oxide inclusions in the weld. This one change reduced the weld failure rate to below 1%. Thick Al oxide layers produce the largest initial gas loads, and the Al chamber will have larger eddy currents in varying magnetic fields.

ESRF selected stainless steel for the vacuum chambers. Stainless steel is easily welded and machined, is strong, and has good magnetic properties. However, stainless steel has poor thermal conductivity. Copper has been selected for one of the storage rings in the B-factory.[13] It has excellent thermal conductivity, good thermal stability and radiation resistance, and has a gas desorption rate intermediate between aluminum and stainless steel. Copper and copper alloys, such as Glid-Cop,[14] are more often used for component fabrication. Copper is commonly used for photon beam stops. The photon beam stop is exposed to high power so it must be able to conduct heat efficiently. It has good mechanical strength and thermal conductivity. One of the advantages of copper is that it can be joined readily to other materials, such as stainless steel.

The APS storage ring absorber illustrates an innovative use of Glid-Cop. As mentioned earlier, the APS stored beam will generate an enormous photon flux. Most of the photons produced in bending magnets will be intercepted by absorbers. An absorber is exposed to over 12 kilowatts of power. Early designs of the absorber involved extremely complex shapes. Recently, a group of mechanical engineers at the APS have developed an elegantly simple design.[15] Figure 8.5a shows the design of the absorber. It is a Glid-Cop body that has slots machined in the front face to disperse the photon beam. The complex water channel geometry is electric-discharge machined (EDM). The absorber is then brazed to a stainless steel support. The support will be welded to a stainless steel ConFlat flange and provide the water cooling for the absorber. Figure 8.5b shows a finite element analysis of the temperatures on the surface of the absorber under irradiation by a 7-GeV, 300-mA positron beam. The temperature would be too great for copper, but is within the mechanical properties of Glid-Cop. Previous absorber designs for high intensity beams were either inadequate thermally or used materials that are difficult to fabricate, such as Be brazed to Cu.

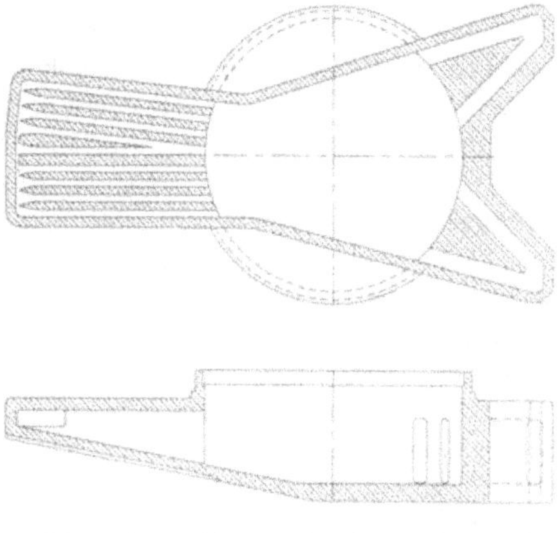

Figure 8.5a. APS absorber design.

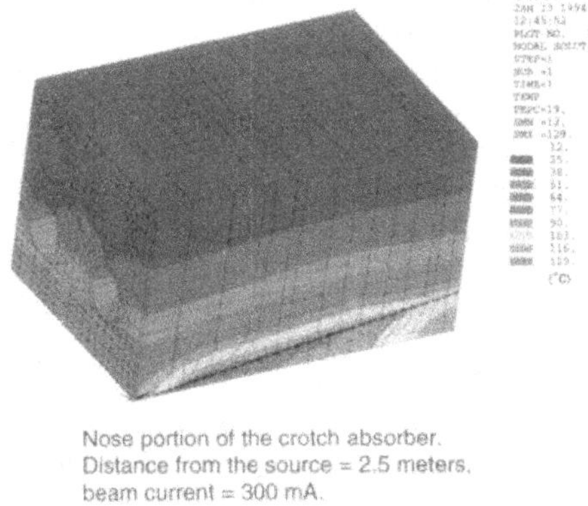

Nose portion of the crotch absorber.
Distance from the source = 2.5 meters,
beam current = 300 mA.

Figure 8.5b. Analysis of thermal gradient on the APS absorber under X-ray beam due to 7-GeV, 300-mA beam.

Careful evaluation of materials for vacuum components, such as electrical breaks, windows, bellows, feedthroughs and valves, is essential for vacuum integrity. Radiation resistance is important for accelerator vacuum components. For example, several machinable glasses decompose under intense irradiation. Most of the metal oxides are not susceptible to radiation damage. Aluminum oxide is the most common insulator material because of high voltage stand-off, reasonable cost, and availability. Borosilicate glasses do darken under irradiation, but are commonly used. Quartz and sapphire are also good window materials in accelerators.

Polymer materials, such as viton, teflon, etc., are vulnerable to radiation damage. There are a number of radiation-resistant polymers available, including Tefzel (a co-polymer of PFTE and ethylene), polysulphonates, polyethylene, and ethylene-propylene, among many others. For example, Teflon has depolymerized by 10^5 rads and viton by 10^7 rads, but Tefzel can withstand 10^9 rads.[16] However, polymer vacuum seals should be avoided, especially in the new high intensity, high-radiation light sources under construction.

The vacuum system design should also include special precautions for bellows and feedthroughs. Radiation in the accelerator tunnel decomposes polymers that insulate power cables. Often these are fluorinated, or chlorinated, polymers. The radiation liberates activated halogen atoms that combine with atmospheric water to form acids. In addition, the radiation produces ozone and activated nitrogen that also react with water to form acids. The acids can corrode the bellows or the braze joints in the metal-to-ceramic breaks on the feedthroughs. Since the bellows and feedthrough vacuum walls are thin, the corrosion can create vacuum leaks. 316L stainless steel and 625 Inconel are corrosion-resistant steels that should be used for accelerator bellows. Ag-Au brazes should be used to fabricate feedthroughs because the Ag-Au alloy is corrosion resistant.

8.2.3 Vacuum Pumps

Because of the unique character of accelerator vacuum systems, capture pumps—vacuum pumps that trap the gas in the pump body—are used predominantly. The most common capture pump is the sputter-ion pump. However, getter pumps are also used to increase the ultra-high vacuum pumping speeds. Capture pumps have significant advantages for accelerators: the pumps have no moving parts; therefore, the maintenance on the pump is minimal and there is no vibration transmitted to the accelerator lattice. The pumps are rugged and radiation resistant—the power supplies are usually out of the radiation environment.

There are three general types of sputter-ion pumps: the diode pump, the noble diode pump, and the triode pump.[17] The diode pump is simple, clean, has the highest pumping speed for the same volume, and highest pumping speed at low pressures of the three types. The pump action is achieved by biasing an anode at high voltage, near a titanium cathode at ground potential in a magnetic field. The magnetic field is sufficiently high to extract electrons from the cathode. The electrons ionize residual gases to produce two pumping processes: ionized gas molecules are accelerated to high energy and implanted into the cathode. In addition, the ion implantation causes the cathode to sputter titanium onto the anode and, since titanium is a very reactive metal, residual gases chemisorb on the titanium layer. The magnetic field increases the electron path to improve the probability of ionizing the residual gases. The ion pump has other advantages besides being vibration free: it uses a simple high voltage cable, it needs no cooling, and it is very reliable with very little maintenance. One long-term problem is feedthrough leaks which are a result of corrosion.[17] Depending on the humidity in the area of the pump high voltage feedthrough it can take many years to occur. As a pump ages, the sputtered titanium will encounter problems adhering to the anode and particulates will form. It is for this reason that ion pumps are usually mounted below the beamline to avoid the possibility of particles collecting in the beam tube. A significant limitation of the diode pump is the low pumping speed for noble gases such as He, Ne, Ar, and Kr. In addition, Ar that is buried can eventually form bubbles under the surface, causing pressure in the bubble to increase sufficiently to rupture the surface and emit a burst of gas from the pump.

Any type of diode pump has the advantage of being a good pressure gauge because it will not be as susceptible to the whisker formations that cause erroneous readings. The noble diode pump uses two materials in the anode to improve the stability of the anode to Ar trapping. The pump loses more speed at low pressure than the simple diode pump. Figure 8.6 illustrates the pumping speed versus pressure for the diode and triode pumps.

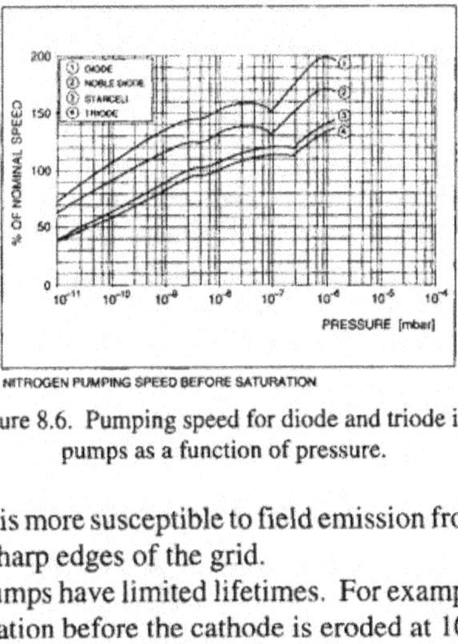

NITROGEN PUMPING SPEED BEFORE SATURATION

Figure 8.6. Pumping speed for diode and triode ion pumps as a function of pressure.

The triode pump is configured such that the titanium cathode is a grid instead of a plate and the pump wall acts as a third electrode. In this case the anode and the pump wall are at ground and the cathode is at high negative voltage. Glow discharge is initiated at higher pressure in triode pumps than in diode pumps. The triode pump has much higher pumping speed and stability for noble gases. However, the pump loses more pumping speed at low pressures than the diode or noble diode pumps. The triode pump is not as good a choice for pressure readings as a diode because it is more susceptible to field emission from whiskers that form in the high fields around the sharp edges of the grid.

Because the pumping action erodes, ion pumps have limited lifetimes. For example a diode pump would have only 400 hours of operation before the cathode is eroded at 10^{-4} Torr pressure, but has over 40,000 hours at 10^{-6} Torr.[18]

Unique sputter ion pumps have been developed for accelerators. A distributed ion pump (DIP) has been developed to operate in bending magnet chambers.[7,19,20] A strip of diode pump is installed in the dipole chambers. The diode pump's magnetic field is from the dipole magnet. Because the magnetic field strength typically is considerably higher than in regular ion pumps, the pumping speed is lower than "lumped" diode ion pumps.[21] Nevertheless, the DIP is effective because the pump is intimately close to the gas source.

There is a dilemma in using ion pumps. The pumps are designed for high vacuum operation, but as the pressure decreases the pumping speed also decreases. To improve system pumping speeds at ultra-high vacuum, getter pumps are used in conjunction with ion pumps. There are two getter configurations: evaporable getters and non-evaporable getters. Evaporable getters have their origins with vacuum tubes. However, most of these early getters are unsuitable for UHV because they contain barium, a metal with high vapor pressure and mobility. Titanium, usually an alloy of titanium and molybdenum, is suitable as an UHV evaporable getter. Bulk titanium is heated until the metal sublimes and the vapor condenses on adjacent surfaces. Clean titanium is extremely reactive—it can be pyrophoric—so chemically active gases that strike the clean titanium layer are captured. Typical pumping speeds for titanium at room temperature are 20 liters/sec/in^2 for H_2 and 60 liters/sec/in^2 for CO.[18] Evaporable getters have several limitations. For a given titanium layer, there are a finite number of active, or pumping, sites. The pumping speed continuously decreases as gas is trapped on the layer. At high pressures, in the 10^{-6} Torr range, the layer will be saturated within a few seconds. The pump can be regenerated by evaporating a new layer of titanium; however, heating the titanium source causes significant pressure rises. The titanium getter is ineffective in pumping noble gases and inert gases, such as methane and higher-order linear organic molecules. Titanium getters share the same particulation problem that ion pumps have. After building up layers of titanium films, particles form and fall into the system.

Non-evaporable getters (NEGs) are reactive metal alloys in which the metal oxides, nitrides, and carbides are soluble.[22] When the NEG alloy is heated, the surface compounds dissolve into the bulk. The surface is reduced to the clean metal which is reactive, similar to the evaporable getter. H_2 is pumped differently. Hydrogen forms a solid solution with the NEG alloy.[23] The hydrogen can diffuse freely through the alloy, forming hydrides. When the alloy is heated, the hydride decomposes, liberating hydrogen into the vacuum system. However, when the heat is removed, the hydrogen reforms the hydride. A variety of NEG alloys are used in UHV systems. The manufacturer of one Al-Zr alloy recommends that the alloy must be heated to ~ 700° C to be fully activated.[22] Alloys based on Zr, V, and Fe are also used, and can be activated at temperatures near 450° C. The NEG alloys should be used with caution because they can be pyrophoric. At room temperature, a surface oxide forms and inhibits further pumping. If there is a large gas load when the alloy is hot, e.g., 250 to 300° C, the surface becomes self-regenerating and release significant heat. NEG pumps have an additional restriction that evaporable getters do not have—there is a limit to the amount of gas that can be dissolved into the bulk of the alloy. Therefore, NEG alloys have a lifetime gas capacity beyond which the alloy will not pump, even after regeneration. Furthermore, there are gases, such as fluorine, that form compounds that do not dissolve into the bulk. The compounds irreversibly contaminate the surface, preventing gettering of residual gases.[23] The formation of the insoluble compounds is called poisoning.

As with distributed ion pumps, many accelerator vacuum systems use distributed NEG pumps. Accelerators at CERN, the AGS at Brookhaven, the APS, and SPring-8 have designed distributed NEG pumps to collect the distributed gas loads in the accelerators.

A novel problem with NEG pumps has recently been reported.[6] ESRF uses distributed NEG pumps in their insertion devices. During operation of the accelerator, the electron beam was dumped when the insertion device magnets were moved. The engineers discovered magnetic particles embedded in the NEG strip, probably from the tooling that crushes the NEG alloy into powder, which form whiskers that protrude into the beam path.

Momentum transfer pumps usually are not used as the primary pumps in current accelerators. Diffusion pumps are no longer used because of backstreaming of organic molecules from the oils. Turbomolecular pumps would be ideal pumps for accelerators except for two problems: until recently these pumps were sensitive to radiation and, more importantly, turbomolecular pumps generate significant vibrations which would be intolerable to accelerator operations. However, turbomolecular pumps are being used routinely in initial phases of the system evacuation as a transition pumps from roughing pump pressures to ion pump pressures. The ALS has installed turbomolecular pumps permanently on the storage ring in order to pump the large gas loads generated in early commissioning operations.[24]

8.2.4 Vacuum Monitoring

There are a variety of instruments to measure the pressure and residual gas mixture of accelerator vacuum systems. Most of the instruments provide an accurate picture of the accelerator vacuum system, if used correctly. However, due to the large number of gauge technologies, gauges, and scientists and engineers, there is a never-ending debate over the optimal vacuum measurement configuration. Rather than continue this argument, the various technologies that are available will be presented.

Total pressure gauges are the most frequently used vacuum monitors. The most common UHV gauge is the "nude" Bayard Alpert ionization gauge tube which heats a filament that produces thermionic electron emission. The electrons are accelerated into a gas ionization chamber, and residual gases are ionized, prevented from escaping from the ionization chamber, and collected on an electrode. By knowing the geometric relationship between the collector and the ionization chamber, the electrode current can be calibrated to the pressure in the vacuum system.[25] The calibration also depends on the gas composition in the chamber. The nude UHV gauge has a large linear dynamic range. The current due to X-ray photoelectron emission from the grids is the limiting feature of a nude UHV gauge; this limit is near 3×10^{-11} Torr. A major difficulty with the gauge is that the calibration factor can change with time.[25]

Another consideration when using ion gauges is their placement with respect to the stored beam. Electrons, whether from the primary or secondary beam, can greatly change the gauge reading. To counteract this, gauges are enclosed in elbows to prevent line-of-sight electrons from entering the gauge. A grounded wire mesh placed in front of the gauge is another successful technique. When mounting a gauge in an elbow, care must be taken to consider the local gas load from the heated walls. Both gauges and elbows must be degassed in order to accurately read the system pressure and not the local elbow pressure.

Although cold cathode gauges have been available for a long time, they have only recently been used in UHV systems and accelerators. The cold cathode gauge generates a high voltage glow discharge in a magnetic field. Since there is no filament, the gauge is very rugged. The glow discharge current is low and, until recently, limited the gauge to pressures above 10^{-8} Torr. However, improved electronic detection of very small currents has extended the cold cathode gauges to use below 10^{-10} Torr.[26] The gauge response to pressure fluctuations at low pressure is slow, and at low pressure the glow discharge can be extinguished. Recent advances in gauge design, i.e., the addition of a quartz lamp to generate a local pressure burst, allow the gauge to be restarted in UHV systems. This type of gauge was always robust, and it now has a large dynamic pressure range. The gauge is most often used as a redundant gauge to hot filament gauges to increase reliability of vacuum control systems.

As with cold cathode gauges, ion pumps operate by generating a glow discharge by field emission of electrons in a magnetic field. The current which sustains the glow discharge is directly related to the pressure in the pump. Therefore, ion pumps can be used as pressure gauges. The pump current is often used as the pressure monitor for the vacuum interlock and control system. The vacuum interlocks monitor the system pressure and activates equipment protection when a large pressure rise is detected. While absolute accuracy is not important for vacuum interlocks, reliability is.

Although total pressure measurements of vacuum systems are essential for safe and reliable operation, analysis of the partial pressures of residual gases in the vacuum system provides vital additional insight. Partial pressure analysis of the vacuum can readily identify existing air leaks, system contamination, or the presence of reactive gases which can poison a process. Recently a a chamber was delivered to the APS that was contaminated with fluorine from a soft solder braze. The symptoms were a very high total pressure and poor pumping speed from a NEG pump. Once a residual gas analyzer was installed, large partial pressures of mass 19 (F) and 20 (HF) were observed—in fact the mass 19 peak was the dominant peak of the RGA spectrum. It became clear that the pump was poisoned, the chamber was significantly contaminated, and, in fact, the whole system had to be disassembled and cleaned. If the RGA had been used earlier, fewer pieces of vacuum hardware would have been contaminated.

There are two principle types of residual gas analyzers: quadrupole energy filters and magnetic sector filters. Magnetic sector RGAs are commonly used in analytical mass analyzers to detect high mass organic molecules. The magnetic sector has very high mass resolution; however, these mass spectrometers are generally large and cumbersome, and require large magnets for operation. The magnetic sector analyzer is also nonlinear in mass. There is one magnetic sector analyzer manufactured and marketed for UHV system operations.[26] It uses computer control to linearize the mass sensitivity and a small magnet, which limits the mass resolution.

The quadrupole electrostatic filter RGA is much more compact. The quadrupole RGA mass analyzes the partial pressure by imposing a DC electric field and a radio-frequency electric field on a quadrupole lens. By properly tuning the voltages, most molecules are bent away from the analyzer axis. Molecules with a selected mass range pass through and are detected. Because they are easy to use, quadrupole RGAs are common in UHV vacuum systems and accelerators and are available from a number of vendors.

Residual gas analyzers are valuable tools for vacuum diagnostics during operations. By monitoring partial pressures of N, N_2, and Ar (masses 14, 28, and 40, respectively), leaks can be identified. RGAs can be used for a special purpose near rf cavities.[27] Large increases in the partial pressure for CO_2 (mass 44) have been observed just before the ceramic window coupling the rf transmission waveguide to the cavities fractures. Since these ceramic windows often fail catastrophically, any precursor signal is vital to protecting the vacuum system.

Leak detectors are special types of residual gas analyzers. These analyzers are tuned to mass 4, helium. However, there is considerably more engineering and design involved, including optimization of gas flow, pumping speed, and sampling efficiency . RGAs can be used for leak detection, but sensitivity to He, used as a probe gas, may be low because geometric view factors are not optimized.

8.2.5 Vacuum System Components

Vacuum components, such as bellows, feedthroughs, valves, photon beam stops, and linear and rotary translators, have special requirements for accelerator use. The stored beam creates a high radiation environment and the high power photon beams from second- and third-generation synchrotron light sources can cause material fatigue and failure. In addition, particulates from moving parts in the accelerator must be minimized.

Bellows and feedthroughs are the most susceptible to radiation-enhanced corrosion. In order to be flexible, the bellows walls are thin, 150-250μ. Unfortunately, corrosion eats through the thin walls very quickly. In order to reduce the effects of corrosion, the bellows should be fabricated with 316L stainless steel or Inconel instead of 304 stainless steel (the most common in the United States). Some 400-series stainless steels are corrosion resistant but are also magnetic, which must be avoided in accelerators.

Unique valves are required in synchrotron storage rings. Modern accelerator vacuum system designs include valves to isolate sections of the rings so that program maintenance and repair can be performed on one sector while the remainder of the accelerator stays under vacuum. The ability to isolate sectors allows the ring to be commissioned faster because only the vented sector will have atmospheric contamination. However, normal valves would create a change of vessel cross-section. As stated earlier, this creates a high coupling impedance to the beam and could cause instabilities. SLAC and VAT have developed a new valve[28] with a lengthened piston which, when the valve seat is raised, draws a body with the chamber cross section into place, minimizing the impedance due to the valve throat. Valves are also used to isolate photon beamlines from the storage ring; these valves encounter special problems in accelerators. The power of the photon beam can be great enough to cut through the valve seat, and valves have literally been cut in half as they are closed. In order to protect a valve, a photon beam stop that shields the valve must be engaged before valves are closed.

All of the third-generation and many of the second-generation synchrotron light sources intercept most, or all, of the photon beam before it strikes the chamber wall. By using a local beam stop, non-uniform heating of the chamber is reduced significantly and thermal distortions are reduced. However, the thermal power striking the absorber can be enormous. Detailed and complex analyses of the photon beam distribution, material properties, and heat flow are required to safely dissipate the power on the absorber. As mentioned earlier, the absorber used downstream of the APS dipole magnets is an elegantly simple design capable of safely dissipating over 12kW of power.[15] Cornell researchers[5] have developed a different absorber design. The Cornell CHESS absorber has a beryllium ring brazed to a water-cooled cylinder.[29] Since X-ray absorption for the low-Z Be is small, the absorber dissipates the X-ray beam through the Be ring body, and the power density is reduced. Beam absorbers for high intensity synchrotron light sources are remarkable examples of state-of-the-art mechanical engineering design.

8.2.6 Vacuum Sealing Methods

Vacuum chamber seals are critical issues in system design. Soldering, brazing, welding, and flange seals are the main techniques used. Proper seal selection involves compromises between ease-of-use and reliability. Flange seals provide the greatest flexibility. Replacing a vacuum component or repairing a leak involves changing a gasket. Welding provides the greatest strength for vacuum seals, and, if done correctly, the greatest ductility. Brazed joints make excellent vacuum seals but they tend to be brittle and are not sufficiently strong. In addition, braze joints have very tight dimensional tolerances in order to achieve high quality bonding. Least desirable are solder joints which typically require use of a flux to make the solder wet to the substrate. Most fluxes leave residues with poor vacuum properties—i.e., they contain zinc, chlorine, fluorine, sulphur—all of which limit the base pressure or form corrosive products. The Stanford Linear Accelerator has developed a handbook of vacuum sealing practices[30] which provides a wealth of practical suggestions for vacuum sealing. The American Welding Society[31] has developed a number of monographs to describe welding procedures for a variety of metals, and the American Society for Metals[32] has published monographs that review brazing of metals. The semiconductor industry has recently been a driving force in developing ultra-clean joining technology. SEMATECH, a company run by a consortium of semiconductor manufacturing companies, has published reports on ultra-clean welding and sealing.[33] The professional society monographs and reports mentioned here are not comprehensive, only representative; a number of companies and other societies also have reports that can help in the design of vacuum seals.

8.2.7 Contamination Control

No matter how well the vacuum system is designed and fabricated, it is useless if extreme care is not taken to prevent contamination of the vacuum chambers. A great deal of effort must go into development of techniques for cleaning vacuum chambers and components. If done properly, the system will have minimal organic contamination, low surface outgassing, and low particulation, all of which affect the operation of accelerator storage rings. Organic contamination usually contains large hydrocarbon molecules. As mentioned in the introduction, the residual gas scattering cross section depends on the atomic number, Z. Recall that Coulomb scattering increases as $Z(Z+1)$. All surfaces exposed to atmosphere are covered with a complex oxide, which acts as a gas source due to thermal and photon-stimulated desorption. Proper cleaning can reduce the oxide thickness, and therefore, the gas load in the system.

SLAC and CERN have devoted considerable research and development effort on finding new methods for cleaning vacuum chambers and components. The SLAC handbook has a number of prescriptions for cleaning stainless steels, copper, aluminum, ceramics, bellows, etc.[30] One of the advantages SLAC has over other accelerator facilities is that they run their own waste treatment plant. Most of the cleaning solutions presented in the SLAC handbook require strong acid or base solutions which typically etch the porous, adventitious oxide, and rebuild a denser, thinner oxide. In addition, these aggressive solutions can etch the base metal, leaving a smoother surface finish. Today most laboratories and state environment regulatory agencies are reluctant to approve these cleaning solutions and considerable research has been devoted to finding an environmentally benign cleaning process—with considerable success. Heated, ultrasonic baths using alkaline detergents can clean aluminum and stainless steel better than many solvents,[34] and ultrasonic baths with acidic detergents can clean copper. Surface analysis of the aluminum, stainless steel, and copper has shown that the native oxide is thin.[34] However, the surface finish using the ultrasonics and detergent cleaning is not as smooth as the finish from the polishing acid baths. Whether or not the smoother acid-bath surface is a lower gas source than the detergent-cleaned surface is still under debate.

The magnitude of accelerator vacuum systems imposes a significant obstacle not encountered with conventional laboratory UHV systems. In order to achieve base pressures in the 10^{-11} Torr range, laboratory systems are baked to 150° C for aluminum, and up to 450° C for stainless steel chambers. This bake-out accelerates the thermal-stimulated desorption. Figure 8.7 illustrates the improvement in vacuum following a bake-out.[35] Although it is often extremely inconvenient and costly to bake an installed accelerator vacuum system, it is not impossible—the ALS, ESRF, APS, SPring-8, and other light sources are designed to be baked *in situ*. In cases where bake-out is not feasible, purging the vacuum system with heated dry nitrogen can improve the pump-down time and base pressure.

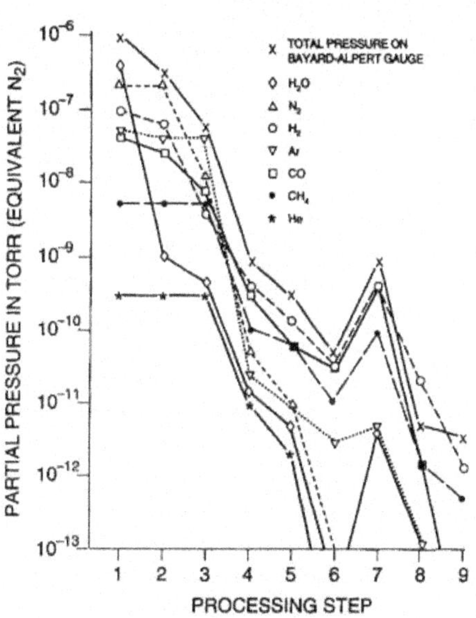

Figure 8.7. Total and partial pressures of a stainless steel UHV system undergoing several process steps.

It is impossible to prevent particles from contaminating accelerator vacuum systems; nevertheless, minimizing particle contamination is vital for accelerator operations. There are two major sources of contamination. During assembly, installation, or venting dust, metal shavings, and fibers can be left on the vacuum components. Turbulence caused by venting the system will then distribute dust throughout the accelerator. It is, therefore, important to develop a slow venting procedure. The second major source of particles are components with moving parts. Metal shavings may be generated when two metals rub against one another. Even worse, if dry lubricants such as MoS_2 are used, lubricant particles are scraped off and, if they are insulating, can become charged. Silver or iridium plating are alternative dry lubricants to MoS_2; unfortunately, the plating layer has a higher coefficient of friction than MoS_2. Semiconductor manufacturers are now developing "particle-free" vacuum systems which are assembled, cleaned, installed, and operated in clean room environments with extremely stringent operating conditions.

8.3 Insertion Device Vacuum Chambers

A principle distinction between second- and third-generation light sources is including insertion devices (IDs) as an integral part of the design and not as an add-on (see Chapter 14). Insertion devices are periodic arrays of magnets which force the electron beam to execute transverse oscillations resulting in the emission of enhanced synchrotron radiation. The X-ray beam power emitted from insertion devices can be extremely intense. For example, insertion devices at the APS will have photon beams with over 12 kW of power.[1] The chambers for insertion devices pose severe demands on the vacuum system design. Small gaps are desirable for insertion device magnets to increase the strength of the magnetic field or the number of periods, both of which result in enhanced synchrotron radiation. However, a small aperture for the beam decreases the beam lifetime due to Coulomb scattering (see Eq. (8.1) and discussion). Thus the design of an insertion device system is a balance between performance and lifetime. The vacuum chamber design significantly affects the trade-off. A vacuum chamber has a finite wall thickness. Conventional chamber designs would have 1.5- to 2-mm wall thickness to sustain the atmospheric force on the chamber wall. The ID magnet gap would be up to 4 mm larger, and the power of the ID is reduced. Typical insertion devices have vacuum chambers with 1-mm walls.[1,2] Thinner wall chambers would deflect significantly, reducing the beam aperture, or even collapse. Insertion device chambers have several additional design problems. First, since the gap is small and the chamber is long, gas conductance from the ID chamber can be limited and can result in a high local pressure. This would result in a decrease in lifetime and an increase in the high energy bremsstrahlung radiation due to the interaction of the beam with the residual gas. A pumping antechamber is included to lower the ID chamber pressure. Strips of NEG getter are installed in the antechamber. Chambers with this cross section and NEG pumps can reach 5×10^{-11} Torr base pressure. Second, the chamber must be shielded from bending magnet radiation, so photon absorbers must be included in the design.

Researchers at the TRISTAN accelerator have developed an insertion device that has the magnets installed inside the vacuum vessel.[36,37] Since there is no vacuum chamber, the magnet gap is the same as the vertical aperture. In order to achieve UHV pressures, the permanent magnets were plated to trap the gases in the magnet. The first encapsulated magnets were nickel plated but more recently TiN coatings have been used. By plating the magnets, the outgassing was reduced enormously.

8.4 Conclusions

Although peripheral to the physics of accelerator storage rings, vacuum is a prerequisite for operation. The primary advantage of improving the vacuum in a storage ring is increased beam lifetime because of reduced bremsstrahlung and Coulomb scattering. There are secondary advantages to lower vacuum. For example, ceramic windows survive longer because the sputtering of rf cavities is reduced due to lower plasma density.

Although accelerator vacuum system design is the same as conventional UHV system design in many respects, e.g., pumping criteria, vacuum diagnostics, and materials selection, accelerator vacuum systems have their own unique design requirements. Major gas sources are caused by photon-stimulated desorption due to synchrotron radiation. The conductance of the accelerator chambers is constricted. A change in chamber cross section worsens the system impedance to the rf power. Photon absorbers tax the ability of present day materials and heat flow designs in order to dissipate the extreme high power densities produced by modern accelerators.

8.5 References

1. 7-GeV Advanced Photon Source Conceptual Design Report, ANL-87-15 (April 1987).
2. "The Red Book, Draft B," European Synchrotron Radiation Facility (January 1987).
3. SPring-8 Project, Part I, Facility Design 1991 (revised) (August 1991).
4. 1-2 GeV Synchrotron Radiation Source Conceptual Design Report, PUB-5172, rev. (July 1986).
5. T. Khoe, "The Effect of the Residual Gas on the Beam Life Time in Electron (Positron) Storage Ring," APS Light Source Note, LS-27 (1985).
6. J. M. LeFebvre, "Vacuum Conditioning," APS-ESRF-SPring-8 First Joint Workshop on Advanced X-ray Light Sources, Grenoble, January 17-19, 1994.
7. E. L. Garwin, "3 BeV Colliding Beam Vacuum System," SLAC Memorandum (August 1963).
8. J. Kouptsidis and A. G. Mathewson, DESY Report: DESY 76.49 (1976).
9. O. Gröbner, A. G. Mathewson, H. Stori, and P. Strubin, "Studies of Photon Induced Gas Desorption Using Synchrotron Radiation," *Vacuum,* **33** (7) (1983), pp. 397-406.
10. A. G. Mathewson, "Vacuum System Design of Synchrotron Light Sources," *AIP Conference Proceedings*, Number 236 (AIP, New York, 1991).
11. S. R. In and S. H. Be, *RIKEN Accel. Prog. Rept.*, **24** (1990), p. 178.
12. A. G. Mathewson, *Int. Workshop on Vacuum Systems for B-Factories and High Energy Synchrotron Light Sources*, Cornell (1992).
13. W. A. Barletta, Design Status of Vacuum System of B-Factory, SLAC (1990).
14. Glid-Cop is a copper, 0.15% Al_2O_3 composite available from SCM Metal Products, Research Triangle Park, N.C.
15. I. C. Sheng, S. Sharma, E. Rotella, and J. Howell, "A Conceptual Design and Thermal Analysis of High Heat Load Crotch Absorber," *Proc. of the 1993 Particle Accelerator Conference* (IEEE, 1993), pp. 1497-1499.
16. P. Beynel, P. Maier, and H. Schönbacher, "Compilation of Radiation Damage Test Data, Part III: Materials used around high-energy accelerators," CERN 82-10 (November 4, 1982).
17. VPT-VacIon Plus Comm., Varian report (Varian Associates, Lexington, MA, 1994).
18. Basic Vacuum Practice (Varian Associates, Lexington, MA, 1992), p. 110.
19. W. Schuurman, "Investigation of a Low Pressure Penning Discharge," *Physica,* **36** (1) (1967), pp. 136-60.

20. M. D. Malev and E. M. Trachtenberg, "Built-in Getter-Ion Pumps," *Vacuum,* **23** (11) (1973), pp. 403-409; "Further Consideration of Built-In Getter-Ion Pumps," *Vacuum,* **25** (5) (1975), p. 211.
21. H. Hartwig and J. Kouptsidis, "A new approach for computing diode sputter-ion pump characteristics (for particle accelerator)," *J. Vac. Sci. Technol.,* **11** (6) (1974), p. 1154-9.
22. T. A. Giorgi, B. Ferrario, and B. Storey, "An Updated Review of Getters and Gettering," *J. Vac. Sci. Technol.,* **A3** (2) (1985), p. 417-23.
23. R. J. Knize, L. C. Emerson, and J. L. Cecchi, "Measurement of the hydrogen recombination coefficient for ZrAl," *J. Vac. Sci. Technol.*, **A5** (4) pt. 4 (1987), pp. 2202-4.
24. Kurt Kennedy, personal communication (1993).
25. M. H. Hablanian, High Vacuum Technology (Marcel Dekker, New York, 1990).
26. Available from Vacuum Technology, Inc., Oak Ridge, TN.
27. R. L. Kustom, personal communication.
28. N. R. Dean, W. R. Fowkes, M. W. Hoyt, H. D. Schwarz, E. F. Tillmann, "SLAC Linear Collider Waveguides Valve," *Proc. of the 1987 IEEE Particle Accelerator Conference*, 87CH2387-9 (IEEE, 1987), pp. 1611–13.
29. D. M. Mills, D. H. Bilderback, and B. W. Batterman, "Thermal Design of Synchrotron Radiation Exit Ports at CESR," *IEEE Trans. on Nucl. Sci.*, **NS26** (3) pt. 2 (1979), pp. 3854-6.
30. SLAC Vacuum Handbook, SLAC-TN-86-6 (1986).
31. American Welding Society, 2501 Northwest 7th Street, Miami, Florida 33125.
32. American Society for Metals, Metals Park, OH 44073.
33. Available from SEMATECH, Austin, TX.
34. R. A. Rosenberg, M. W. McDowell, and J. R. Noonan, "X-ray photoelectron spectroscopy analysis of aluminum and copper cleaning procedures for the Advanced Photon Source," *J. Vac. Sci. Technol.*, **A12** (1994), p. 1755.
35. J. P. Hobson, "Methods of Producing Ultrahigh Vacuums and Measuring Ultralow Pressures," *Surface and Colloid Science,* **11** (1979), p. 187.
36. S. Yamamoto, T. Shioya, M. Hara, H. Kitamura, X. W. Zhang, T. Mochizuki, H. Sugiyama, and M. Ando, "Construction of an in-vacuum type undulator for production of undulator x rays in the 5-25 keV region," *Rev. Sci. Instrum.* **63** (1) (1992), p. 400.
37. H. Kitamura, "Insertion Device Developments," APS-ESRF-SPring-8 First Joint Workshop on Advanced X-ray Light Sources, Grenoble, France, January 17-19, 1994.

CHAPTER 9: ACCELERATOR CONTROLS AND MODELING*

JEFF CORBETT◊ and **CLEMENS WERMELSKIRCHEN**◊

Stanford Synchrotron Radiation Laboratory, Stanford Linear Accelerator Center
Stanford University, Stanford, CA 94309

9.0 Introduction

The first part of this chapter provides an overview of the general requirements for modern synchrotron light source control systems. This description covers different components, architectures, and aspects of the operator interface. In addition, features of the computer infrastructure, on-line communication facilities, and front-end interfaces are described. As the control system is a central part of any accelerator, it interacts with many of the hardware components described in other chapters, and provides exchange, storage, manipulation, and display of data. The control system also contains hardware calibration constants needed to run machine analysis and control programs. The second half of this chapter provides an overview of accelerator models with particular attention to accelerator control. The model is an important part of the control system because it helps physicists adjust the operating conditions. To understand the role of the model, from its inception in the accelerator design phase to its application for machine control, we first consider the components of the model and the interface to the control system. We next derive physical parameters from the model: transport matrices, beta functions, closed orbit perturbations, and the synchrotron integrals. Many of these parameters can be measured experimentally. Procedures to determine calibration constants that make the model agree with experimental measurements are outlined. The final section concentrates on model based machine control.

9.1 Control System Overview

A general function of the accelerator control system is to establish coordination between all hardware components, so that the goal of the accelerator, to control a charged particle beam over a certain period of time, is accomplished. Specifically, the goal of an accelerator may be to produce electrons (or ions) in a beam with short pulse duration, or a beam stored for a long period of time, but the basic requirements of the control system and its components are similar for all accelerators.

Control systems have always been adapted to the specific needs of each individual accelerator. But over the years, the technologies used for the control systems (and so the architecture) have undergone significant changes. In fact, the differences in the control systems are usually driven more by the development of technology and component cost (mainly computer and microprocessor development) than by the scientific goals of the accelerator.

---

* Work supported by the Office of Basic Energy Sciences, Department of Energy contract DE-AC03-76SF00515.

◊ Author e-mail: corbett@ssrl01.slac.stanford.edu and FAX: (415) 926-4055, and
author e-mail: wermelsk@vms.gmd.de and FAX: +49 2241 14 3002.

The following sections will discuss the typical layers of a synchrotron light source control system from the hardware level to the operator interface, and touch on more sophisticated controls. At the same time, this development provides an overview of the historical evolution of accelerator control systems.

9.2 Control System Basics

Other chapters of this book describe the hardware components needed to build and operate a synchrotron light source. Here, we will show how these components are connected through the control system.

From the point of view of the control system, each hardware component consists of a set of control signals, either to control the device or to monitor the device status. In the very simple case, for instance to set the field strength of a magnet, the control system provides a low-level control voltage to the power supply, which in turn drives current through the magnet. At the same time, an electronic sensor detects the current going through that magnet and passes this information back to the control system computer. The first *control systems* consisted of potentiometers (knobs) and ammeters to perform the task of controlling devices like power supplies. As accelerators grew larger, the number of components increased and the components were spread over a larger area, so other control system techniques had to be developed.

The simple example of magnet power supply control shows all the main characteristics of any modern control system. Assume, for instance, the operator wants to adjust certain parameters of the beam. For this, the operator sets the magnet strengths, and the new power supply currents are calculated from calibration constants. Each power supply is adjusted by setting the control voltage through the control system. There is in principle no difference if the control is from local knobs or from the computer interface. The operator then checks the correct operation of the power supply by measuring the supply current. Again, there is no difference if it is through local meters or from values displayed by a computer. The power supply may also have switches and status indicators. This binary information (on/off) is part of the control system, and can be controlled or monitored.

This simple example illustrates the following basic functions required of any accelerator control system:

- it consists of everything between the operator and the hardware that is needed to control and monitor each device
- it gives the operator access to the accelerator from the control room or any office (via computer_network connection)
- it transfers information from all devices, potentially spread over a large physical area, to one or more locations where this information is needed
- it provides analog set point values
- it accesses analog readback values and
- it sets digital or binary values, and processes single bit information

Modern synchrotron radiation sources include more complicated devices than magnet power supplies, but each device is similar from the point of view of the control system. The radio-frequency (RF) drive system, for example, when looked at as a black box, is controlled by exactly the same kind of electrical interface signals. The difference is that the analog set point values control the RF power applied to the cavity. For the control system, it is the same task as before; i.e., provide set point values, monitor power supply values, or set and monitor digital control signals.

The tasks of the control system defined so far are simple, in the sense that the control signals are constant in time (DC). Accelerators also include components that are controlled with time correlations between each other. The injection system, for instance, might include a linac, booster synchrotron, and injection and ejection hardware. In this case, the control system must synchronize time-related signals so that these components operate appropriately during each acceleration cycle.

For these applications, depending on the type of hardware device to be controlled, the control system provides trigger signals for well defined events (linac, septa, kicker), and provides analog set point values that evolve in time (waveforms). For a booster synchrotron, both the RF and magnet currents require time evolving waveforms during each acceleration cycle. Time evolving signals are also acquired to detect beam position, component instabilities, and feedback system signals.

9.3 General Control Systems

During the evolution of accelerators and their control systems over the last 50 years, the control of the accelerators has become more and more complex. In place of knobs, meters, and oscilloscopes, there are now front-end interfaces, computer controls and work stations. The basic task of the control system is still to allow operators to interact with the accelerator, in particular to set and monitor the machine parameters. But in the context of modern control systems with object-oriented databases, the meaning of each machine parameter can now be more general.

Today, each accelerator control system is built around a specific computer architecture which serves as the system infrastructure. The selection of infrastructure has become increasingly important as computer speed increases and costs decrease. The enhanced computing power allows much more sophisticated handling of machine parameters, more complex on-line calculations, more accurate simulations of machine physics, and more user friendly operator interfaces.

Accelerators also have been built to reach higher particle energies. This has spread the hardware components over a larger area. With the evolution of local area computer networks, these high-speed networks can be incorporated into the accelerator control system. One typical function of the local area networks is to connect distributed processors to central processing computers and distributed operator consoles (work stations).

A modern control system is structured into hardware and software layers that manipulate data on different levels of abstraction. The bottom layer interacts with the electrical signals where the processors have to implement real-time control. The top layer provides the human interface where operators can control the accelerator. The layers in between maintain the machine parameter database, and provide data collection, data distribution, networking, and monitoring, and control signal timing.

As the available computing power has increased, more sophisticated application programs can be used for machine control. Theoretical calculations and accelerator models that where formerly used only in the machine design phase are now used for machine commissioning and model calibration.[1] Today's more sophisticated control systems can model the state of the accelerator and the physics of beam-hardware interactions. This begins with calibration factors that convert hardware signals to physical beam parameters. The modern database may also include information for expert systems or artificial intelligence.[2]

9.4 Control System Architectures

Synchrotron light sources differ in size and in requirements for physical layout of the control system. In some cases, such as high-energy rings APS,[3] ESRF,[4] or SPring–8,[5] the control systems must be distributed over a large area. Smaller storage rings can have a more centralized control system. Almost independent of these constraints, the control system hardware components are arranged in a hierarchical structure as indicated in Fig. 1. The actual implementation of the control system will differ depending on the size and operational goals of the machine. As the scope of this book is restricted to synchrotron light-source accelerators, the following discussion will focus on control system architectures for such machines.

The levels of functionality shown in Fig. 1 can be concentrated in a single or a small number of computers. For small accelerators, there might be one computer with no network at all. The more control intelligence (i.e., computers or microprocessors) that is involved, the more the aspects of physically distributed processing capability and computer networking become important. Although the layers shown in the Fig. 1 give the principal overview of the control system building blocks, the meaning or even necessity of the different blocks can vary over a wide range for different control systems.

The control system for a synchrotron light source can also be divided into the control system for the injector and the control system for the light source (storage ring). The degree of separation between these two systems is arbitrary, but there must be some communication between the systems.

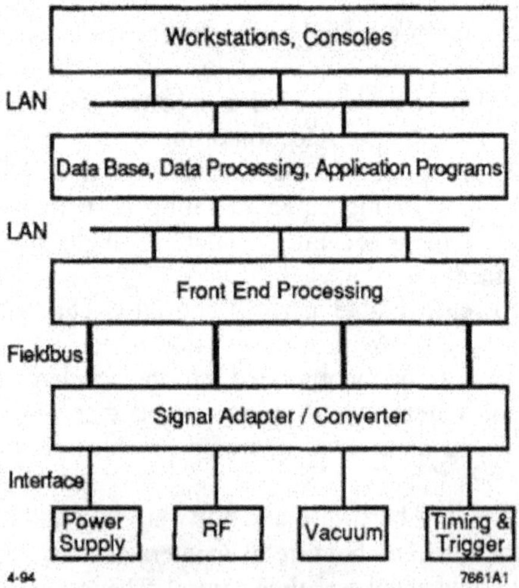

Fig. 1. Schematic of control system hierarchy for hardware components.

In the past, each accelerator had a home-made control system that was adapted to the needs of the individual machine. Components of modern control systems are frequently based on more standardized commercial products. The processors, as well as the networking infrastructure, are often off-the-shelf products, and there are efforts to standardize accelerator control software, in particular, EPICS.[6] There is also commercial software available that can be customized to specific needs.[7,8]

In general, work stations are widely used for the operator interface. They provide high-quality graphic displays, and take over the job of visualizing the machine parameters and handling operator input.

9.4.1 Front-End Systems

The front-end system can be characterized as the *low level* part of the control system. The front-end system interfaces, or connects, physical signals derived from electrical hardware components to the computer. In practice, most signals are not wired directly into the main control room; rather the electrical signals are converted into a digital form that can be transferred via data communication lines. For hardware control, signal values are transferred in digital form and converted into electrical signals. Digital signals have the added benefits that they can be transferred over long distances much easier than analog signals, they are less sensitive to electromagnetic noise, and data transmission errors can be detected. Therefore, the front-end control system components that convert between analog and digital signals (ADC and DAC) can be located close to the hardware components. This also reduces the cost of cables, since a single digital communication line can service many devices.

Some accelerator components, especially pulsed devices like kickers, produce high-energy electromagnetic noise in the surrounding area. In this case, the front-end interface electronics must be designed to isolate the rest of the control system from the induced noise. This can be done by using signal transformers and optical isolation circuits.

For economic and maintenance reasons, front-end systems are set up in a modular design. Standard modules that can be easily arranged and replaced within a housing crate are used to connect the control system to the electrical interfaces of the hardware component. The CAMAC standard, for example, is widely used for this purpose. It has a well established protocol, and modules for many different purposes are commercially available. Traditionally, the CAMAC modules were used primarily for data conversion (buffers, ADCs, DACs), but an increasing number of modules now have their own processing capability and the functionality of front-end systems. Although CAMAC systems originated from the time of low integration level in microelectronics, they have evolved to include microprocessors and minicomputers.

Beginning in the 1980s, when high-power single-chip microprocessors became available, a new front-end standard was created. This system, the VME standard, began with modules that had powerful microprocessors (Motorola 68000) and a data bus structure that was designed for high-speed communication between the modules. By using additional modules that provide data converters and drivers, VME systems are now widely used for front-end purposes in accelerator control systems.[9] Because of the more

CPU–oriented approach of the VME design, a field bus layer was added to the front-end system. This allows each VME processor to handle several hardware components through inexpensive digital communication lines like RS232, GPIB, or MIL–STD–1553B, either connected directly or through an additional interface.

Similar to the VME standard, there are other developments like Multibus and Bitbus. These options provide mechanical and electrical standards that allow the system designer to combine commercial modules, interface modules, and specialized home-made modules.[10-13]

9.4.2 Centralized Control Systems

Centralized control systems are characterized by a database that contains all the machine parameters and the control system software on a single host computer. Since central control systems with a single computer are limited to the resources of the host processor, they are not practical for larger accelerators. Where a centralized control system is adequate, it has the advantage that software management and data organization are easier than with distributed systems.

For small control systems, the central host may be the only processor in the whole control system. In this case, the central host not only has to maintain the database, but also has to perform the task of data acquisition and provide the operator interface.

To increase the performance of a centralized system, the task of data acquisition can be delegated to locally intelligent front-end processors. The central host is then no longer responsible for reading individual data values out of each front-end interface, such as a CAMAC module. Instead, data is transferred in blocks to and from front-end microprocessors, such as intelligent crate controllers or VME crates.

9.4.3 Distributed Control System

Whenever the processor power or data storage capability of the central computer are not sufficient, the control system can be spread over several computers. This may involve distribution of software applications, or splitting the machine parameter database into a distributed database.

A distributed control system consists of a hierarchy of processor layers. They either share a completely distributed database where an application program can be run on any processor, or they are arranged as subsystems where each subsystem is responsible for a certain group of machine components. As in Ref. 14, there can be subsystems for the cryogenics, vacuum, injector, RF, beam transport, and beam switchyard, etc. Each subsystem may also have its own front-end system, communicate with a dedicated front-end through a special subnet, or the subsystems may share the front-end systems by exchanging data through an underlying communication network.

Perhaps the most extreme example of a distributed control system comes from the LEP collider at CERN. This machine has a circumference of 28 kilometers and a huge number of components to control, so the control system is distributed over a large area with a hierarchy of processors, and networks, and a distributed database.[15,16] Similar systems, such as reflective memory techniques for fast orbit feedback, can be used for large, high-energy light sources.

9.5 Networks

The components of accelerator control systems are linked by a network to transmit data. A wide variety of technologies are used for these networks. Depending on the quantity and speed of the data transmission, either coaxial cable or fiber optic connections are used. Specialized links are used on the lower network layers. These must be adapted to the real-time needs of the front-end processors or the communication needs of the control system, especially for timing purposes.

Ethernet, the first high speed local area network standard (10 Megabits/sec) is widely used at this time. Because Ethernet offers CSMA-CD (carrier sense multiple access with collision detect), every station connected to the Ethernet can send data whenever no other station is transmitting. Where collisions occur, data packets must be retransmitted, which can lead to unpredictable transmission delays. Where guaranteed network bandwidth is an issue, token-ring architectures are now in use.

The use of standardized networks within a control system directly couples the control system with the computer network within the whole laboratory. This is often very convenient, but care has to be taken to prevent unauthorized manipulation of the accelerator.

9.6 Application Programs

The control system has to perform a series of tasks to keep the accelerator operational. Each of these tasks is controlled by software packages, commonly referred to as application programs. Examples of application programs include software to monitor beam current, vacuum conditions, power supply settings, and programs for orbit control. The overall design of the control system software is very important since the control system must provide mechanisms for all application programs to access machine parameters. The application programs should also be designed independent of the underlying hardware architecture. This arrangement can be accomplished by providing a well defined software interface through the control system kernel. A well designed kernel makes testing of new application programs much easier, and keeps application programs small, separated, and dedicated to specific purposes. Modern software development techniques and tools can be used to keep the software manageable, and to meet software and quality control standards. This improves overall system reliability.

In the following sections, a selection of basic tasks required of any control system is described. The functions of some higher level application programs will be mentioned. It should be kept in mind that the development and maintenance of application programs is an ongoing task during the lifetime of an accelerator. Research and development in the area of running accelerators under automated control is still progressing.

9.6.1 Machine Parameter Maintenance

The control system maintains a complete database with the status of all accelerator components. Either a centralized or a decentralized database is used to perform this task. The database software can be a commercially available product such as ORACLE,[16] another relational database,[17] or in-house software.

It is often necessary to maintain data for several machine operating conditions, and consequently, the complete set of machine parameters or a subset of parameters has to be saved or restored from external files. The collective set of hardware set points is often referred to the machine configuration. These configurations can be maintained at three parallel levels:

- hardware set points from file archives
- present hardware set points
- present hardware read back values

If the hardware set points are adjusted, the read back values should change to verify the result. Any new configuration can be archived to the computer disk and retrieved at a later date.

9.6.2 Operator Interface

The most visible part of the control system is the operator interface; i.e., the consoles and displays. Most systems use work stations now, but control systems based on personal computers can also be found. For ease of machine control, it is important to have a high-resolution ergonomic display. The X-Windows graphics standard has been adopted, or is being planned for future use, for many synchrotron light source control systems.

Figure 2 shows an example of a graphical presentation for the machine status and operator interaction points for the injector synchrotron at SSRL. This control menu displays all parameters in the transport line between the linac and the booster, and allows the operator to manipulate their values. In practice, a high resolution color control panel gives a user friendly overview of the complex hardware configurations.

9.6.3 Monitoring, Alarms, Logging

Monitoring of the accelerator status is done at several levels of the control system. The front-end system, for example, monitors the correct behavior and function of each component. Monitoring is also done at the higher levels of the control system to check the function of the complete machine; for example, the beam position or beam intensity.

Whenever a device-monitoring application detects a parameter value out of range, an alarm can be generated and transferred throughout the system. These alarms can have different levels of significance. Some may indicate emergencies that lead to automatic shutdown of the machine, while others alert the machine operators of a potential hardware problem.

Alarms and operator actions can also be logged and displayed on consoles. Logging systems provide on-line information about the actual status of the accelerator. For long-term diagnostics, and possibly for accounting purposes, records of significant machine parameters have to be kept in a protocol file. Typical parameters in a protocol file include beam intensity, beam line status, and accelerator startup and shutdown information.

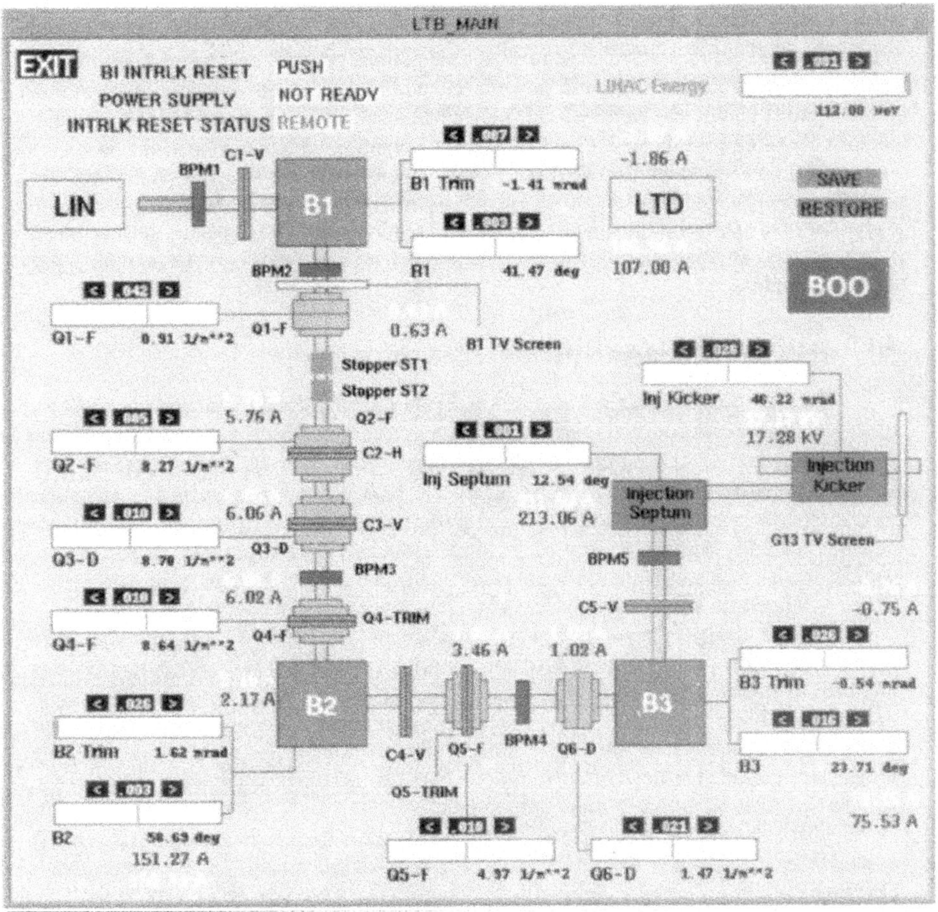

Fig. 2. SSRL injector linac-to-booster transport line.

9.6.4 *Automatic Procedures*

Besides attending to the passive accelerator control tasks described so far, control systems are taking over responsibility for the active operation of the accelerator.

As a first step, the control system does not just passively control machine parameters under operator command, but also executes automated procedures by itself.

This is helpful during a machine startup or shutdown sequence, where all equipment must be turned on or off in a controlled sequence. Correlations between devices and parameters have to be maintained during these procedures to minimize commissioning time, and to protect against equipment damage. This becomes more important as accelerators become increasingly more complicated, to ensure reliable operation with minimum downtime.

Feedback systems (discussed in Chapter 13) are also becoming integrated into the control systems. Feedback systems can be implemented directly through the control system, or they can be implemented as software applications. For feedback systems of all kinds, there is an ongoing effort to increase the performance of the accelerator by using faster equipment.

9.7 Accelerator Modeling Overview

In Matt Sands' landmark paper,[18] he made the remark that with storage rings we must pay "particular concern for their performance as instruments for research... ." The recent explosion in the number of storage rings, with ever increasing performance standards, reinforces this notion. If we adopt this view of the storage ring as a scientific instrument, it becomes apparent that the performance of a synchrotron light source depends critically on the accuracy of the model. Machine control programs, for instance, use the model to manipulate the beam orbit and to adjust the electron beam properties. These requirements alone make the model an important part of any accelerator.

In the design phase, computer simulations are used to construct the *ideal* accelerator model. The goal is to produce a machine with electron beam properties that conform to stringent performance specifications, including photon beam brightness and tolerances to error. As a result, the accelerator design phase is a complicated process that seeks stable solutions to inherently nonlinear constraints. The finished design (the accelerator model) requires many iterations between accelerator physicists, engineers, and the synchrotron radiation user community.

Once the design is complete, the magnets, radio-frequency system, and support systems are constructed and installed in the accelerator tunnel. In the construction process, each component is carefully calibrated (see Chapters. 5, 6, and 7) and the calibration data are entered into an on-line version of the model that will be used for machine control.

However, even with great care in the manufacture of each accelerator component, the electron beam properties often differ from the values predicted by the model. To reconcile the difference, accelerator physicists carefully measure the beam properties and compare the results to the model. Large discrepancies between the measurements and the model often indicate problems that can be fixed at the hardware level. Small errors are accounted for by adjusting calibration factors in the model. The new *calibrated* model is then more suitable for machine control.

9.8 Model Components

Accelerator modeling originated as the art of finding solutions for charged particle trajectories under the influence of electromagnetic fields. It has since evolved into a broad field that relies on sophisticated computer programs. In this chapter, we focus on accelerator models for machine control—in particular, the linear field model. This simplification reduces the basic set of model components to:

- drift sections (no magnets)
- bend and corrector magnets
- quadrupole magnets (focusing)
- beam position monitors (bpm)

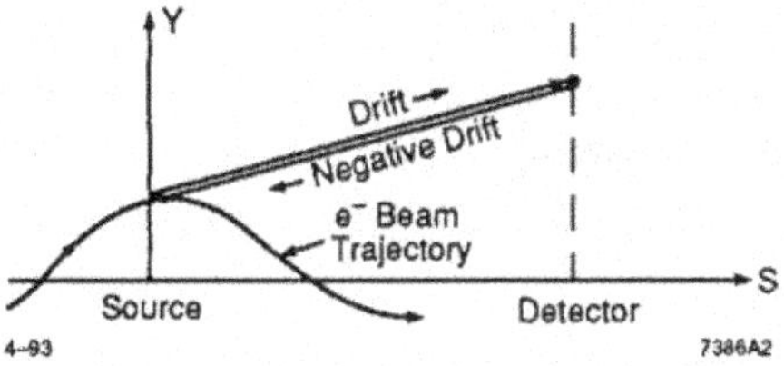

Fig. 3. Construction of a model component for photon beam lines.

The next level of model sophistication usually includes nonlinear elements such as sextupole magnets for chromatic correction, and the RF acceleration system needed to restore energy lost from the beam.

Another important model element particular to synchrotron light sources is the photon beam position monitor. Conventional accelerator codes do not explicitly include photon beam position monitors because the magnetic fields acting on the electron beam do not deflect the photon beam. We can, however, use the element sequence (drift)–(bpm)–(negative-drift) to model the photon beam position.[46] As indicated in Fig. 3, the length of the drift, (and the negative-drift) is the distance from the photon beam source point to the photon bpm. (The *negative-drift* simply returns the beam trajectory to the original location of the photon beam source point.) Using this element, it is possible to include the photon beam position in the accelerator model.

The physical configuration of magnets in a synchrotron light source is often referred to as the lattice. Storage ring lattices are usually built up from a repeating pattern of bend and focusing magnets, collectively referred to as a cell. For many applications, each cell has the special property that the electron beam envelope is the same at both ends of the cell. In a light source lattice, the magnet cells are typically designed to be achromatic (no energy dispersion at the ends of the cells), so that insertion devices installed between the cells are driven by a dispersion-free beam. These devices produce the high quality photon beams characteristic of third-generation synchrotron radiation sources (see Chapter 14).

One of the first achromatic cell configurations, the DBA (double-bend achromat), was developed by R. Chasman and G. K. Green for the NSLS.[19] The DBA contains a strong horizontally focusing quadrupole in the center of the cell to drive the dispersion to zero at the ends. A variant of the DBA, the TBA (triple-bend achromat) is based on similar principles. The TBA configuration forms the unit cell for the ALS (Advanced Light Source).[12] For a more complete discussion of lattice design considerations, see Chapter 2.

Returning now to the individual elements of the model, note that each element can be described with an almost arbitrary degree of mathematical complexity. For instance, both the central field and the fringe fields at the edge of the magnet can be represented by high-order multipole expansions. The most popular magnet model for machine control, however, is the simple linear field approximation, based on the first-order expansion of the equations of motion.[20] For machine control, the linear optics model is fast, and is accurate enough for most applications.

The most widely used format for the accelerator model file presently follows the MAD convention.[21] MAD will make all the standard linear optics calculations and produce a table of the accelerator optical functions suitable for post-processing. The HARMON package in MAD[22] also provides a means to compute sextupole strengths for on-line chromaticity correction. The effect of sextupole fields on the beam envelope and on the position of the beam centroid are often neglected in model-based control applications.

More precise accelerator simulation codes tend to concentrate on higher moments of the magnetic field, edge field effects, geometric corrections to the equations of motion for a small bending radius, and the impact of synchrotron oscillations. As discussed in Chapter 2, much of the emphasis on the design of storage rings for synchrotron radiation (and therefore the modeling codes) has been centered around characterizing the dynamic aperture. Codes like TRANSPORT[23] and Marylie[24] perform a Taylor series expansion of the equations of motion through each element, and simulate beam propagation via higher order transport matrices. These codes can generate a high-order power series expansion around the closed orbit to obtain a *one-turn map* of the phase space for single particle motion. Other programs *slice* each element into thin differential strips, and perform *symplectic integration* to find the particle motion.[25] TRACY and TRACY–II contain a PASCAL compiler with an accelerator physics *library* that allows the user to program an arbitrary sequence of calculations.[26]

Many other modeling programs exist for a wide range of applications. For a compendium of accelerator modeling programs, including a description of their applications, see Ref. 27. Given the available range of high-order programs, it is particularly important to recognize that the choice of program is an integral part of the accelerator model. Depending on the application, one should always bear in mind that the model is an approximation to the actual accelerator hardware, and that very high order models have a limited range of validity for machine control. Even where the range of validity is accurate to higher orders, one may wish to trade model accuracy for computation speed and ease of use.

9.9 Accelerator Model Interface

As described in the previous sections of this chapter, the on-line model is closely connected to the accelerator control system. One function of the on-line model is to monitor and control the electron beam orbit. In this case, the control system converts voltage signals from beam position monitors (bpms) into orbit position data by way of calibration factors. The model then calculates a set of dipole kicks to adjust the orbit, and a different set of calibration factors are used to convert these kicks to power supply voltages. The entire process takes place from a graphical interface connected to the control system.

A closer look illustrates how the calibration factors are integrated into the control system. The example in Section 9.2 outlined the measurement and control of magnet power supplies. For model purposes, a set of calibration factors is used to convert each power supply read back value to a magnetic field strength. This requires at least two signal conversion stages: first, the voltage of a current transducer is converted to a value for amperes flowing through the magnet windings; then, a second set of calibration factors convert the amperes to magnetic field strength. This conversion stage can be complicated when the magnet cores have nonlinear magnetization characteristics. To parameterize the behavior of the iron, a set of polynomial coefficients can be numerically fit to data points measured in the laboratory. These polynomials convert the power supply read back values to field strength. Further details on the magnet measurement and calibration process can be found in Chapter 7.

Once the magnet calibration factors are in place, the model can be used to estimate the beam parameters. For on-line applications, a *skeleton* file containing the lattice geometry and effective length of the magnets forms the basis for the model. The control system loads magnet strengths into this file and the model program calculates the beam parameters. Using a similar procedure, the accelerator model can simulate longitudinal beam dynamics based on read back values from the radio-frequency system.

To provide the experimentalist with more flexibility, it is common to add an extra calibration factor for each lattice component in the database. For instance, the quadrupole strengths can be multiplied by the extra factors before running the on-line model. Proceeding in the other direction, quadrupole strengths calculated by the model are divided by the extra calibration factors before conversion to power supply currents. Although these calibration factors do not strictly conform to the polynomial formalism, they allow the experimentalist to adjust the model to conform with measurements. Section 9.11 of this chapter outlines techniques that can be used to determine these factors experimentally.

9.10 Model-Based Calculations

Model-based machine control relies primarily on calculations of the beam transport matrices and β-functions. The beam transport matrices describe single particle motion in terms of phase-space coordinates as the beam travels through a transport line or storage ring. In this chapter, we restrict the motion to first-order oscillations in the plane transverse to the beam direction. Closely related to the transport matrices are the response matrices that describe the motion of the closed orbit caused by dipole kicks in a storage ring.

The β-functions also describe single particle motion (and closed orbit perturbations), but the β-functions have a much wider range of applications. In particular, the β-functions are used for resonance calculations, to yield statistical properties of the beam, and to parameterize the beam envelope. It is important to note that both the transport matrices and the β-functions are nonlinear functions of the model parameters, and their ability to predict machine behavior depends on the accuracy of the model.

9.10.1 Transport and Response Matrices

Transport matrices are used in accelerator physics to propagate the phase-space coordinates of the beam from one position to the next. In general, the phase space can be six-dimensional, $(x, x', y, y', \Delta p/p, \Delta s)$, where each coordinate is evaluated relative to an ideal reference orbit.

In a synchrotron, the transverse oscillations come about as the result of periodic quadrupole focusing. These are the betatron oscillations. Similar to a harmonic oscillator, betatron oscillations can be represented on a phase-space diagram where the coordinates indicate the displacement and the angle of the motion. The transport matrices map the change in coordinates between any two positions in the accelerator. To first order, we can write

$$(x,x')_2 = \mathbf{R}(x,x')_1 \ , \tag{9.1}$$

where the matrix **R** is a function of the lattice parameters between position 1 and position 2.[28] The elements of **R** have two equivalent representations, either in terms of the lattice parameters or in terms of the β-functions. Accelerator modeling programs often calculate **R** across each element and multiply matrices to find the cumulative effect. To model linear particle dynamics, including synchrotron motion and coupling effects, a 6x6 dimensional transport matrix is used. Nonlinear terms can be added by expanding the equations of motion to higher order, and extending the column vector of phase-space coordinates.[23,28,29] In the accelerator literature, the components of the 2×2 transport matrix are often denoted

$$\mathbf{R} = \begin{pmatrix} R_{11} & R_{12} \\ R_{21} & R_{22} \end{pmatrix} \ . \tag{9.2}$$

The focusing effect of a quadrupole, for example, enters through the R_{21} term.

For a storage ring, we can form a matrix of the R_{12} elements that connect each corrector magnet kick to the orbit displacement at each bpm. This *response matrix* can be either calculated by the model, or measured directly. For precise orbit control applications, such as fast orbit feedback, the direct measurement may be preferred for maximum precision. It is not always convenient, however, to interrupt operation of the machine to measure the response matrix when the machine optics is changed (e.g., insertion device adjustment). To update the response matrix, either a model-based calculation is needed, or the feedback system must have an adaptive feature that can *learn* the new response matrix, following a change in lattice optics.

9.10.2 Beta Functions

Beta functions were first used by D. Kerst to describe transverse particle oscillations in the early betatron machines at General Electric and the University of Wisconsin.[30] Hence, the name *beta functions*. The theory of β-functions was then formalized by E. D. Courant and H. S. Synder for alternating gradient accelerators.[31] Since that time, β-functions have become a standard tool for accelerator analysis. A development of β-function theory, including a wide range of applications to storage rings, was recently compiled by Wiedemann.[20]

The β-functions are extremely useful for accelerator analysis because together with the dispersion function they form a compact representation of the electron beam parameters. They also provide a powerful tool to estimate the effect of field perturbations on the particle beam envelope, the closed orbit, and the impact of resonances on beam dynamics. Mathematically, the β-functions are a solution for single particle motion in the accelerator structure. Physically, the β-functions predict the amplitude and phase of transverse particle oscillations relative to the equilibrium orbit. They also parameterize the phase space ellipse of the particle beam distribution function as the beam circulates around the accelerator. It is important to note, however, that the β-functions are derived *from* the accelerator model.

To see how the β-functions are derived from the model, we first consider the equation of motion for an on-momentum particle oscillating with respect to the closed orbit (see Chapter 2, Section 3):

$$x'' + K(s)x = 0 \ . \tag{9.3}$$

The function K(s) in Eq. (9.3) represents the quadrupole focusing action as a function of position 's' along the ideal orbit. For on-line model applications, K(s) is determined directly from power supply current measurements. Looking ahead to Section 9.11, one of the great challenges of experimental accelerator physics is to determine the exact value of the focusing function K(s) in the synchrotron!

To within a constant phase factor, the two linearly independent solutions to Eq. (9.3) can be combined to yield an expression for the particle position in terms of the β-function, β(s), and the phase advance of the oscillation, φ(s),

$$x(s) = x_o\sqrt{\beta(s)}\ \cos[\phi(s)+\phi_o] \ , \tag{9.4}$$

where x_o and ϕ_o are the initial oscillation amplitude and phase of the oscillation. In this representation, it is clear that the function $\sqrt{\beta(s)}$ gives the amplitude modulation of the motion, and φ(s) is the phase advance for the motion. Furthermore, it is clear that for a system of many particles, each with random oscillation phase ϕ_o, the function $\sqrt{\beta(s)}$ is proportional to the size of the beam envelop at each position 's' (neglecting dispersion effects[20]).

Two parameters closely related to the β-function are $\alpha = -\beta'/2$ (derivative) and $\gamma = (1+\alpha^2)/\beta$. Together, the set of functions {α, β, γ} are often referred to as the Twiss parameters. The quantity $\varepsilon = \gamma x^2 + 2\alpha x x' + \beta x'^2$ (the celebrated Courant-Snyder invariant[31]) is a constant of the motion for lossless betatron oscillations. For a given distribution of betatron oscillation amplitudes, the invariant quantity 'ε' is used to characterize the emittance of the beam. See also Chapter 2 and Section 9.10.5 below.

To calculate the β-functions from the model, we can use the fact that the beam envelope repeats each turn and the invariance of the Courant-Snyder constant to yield a similarity transformation for propagation of the Twiss parameters:

$$\sigma(s_2) = \mathbf{R}\sigma(s_1)\mathbf{R}^T \ . \tag{9.5}$$

where the beam sigma matrix is given by

$$\begin{pmatrix} \sigma_{11} & \sigma_{12} \\ \sigma_{21} & \sigma_{22} \end{pmatrix} = \begin{pmatrix} \beta & -\alpha \\ -\alpha & \gamma \end{pmatrix} \tag{9.6}$$

In this equation, **R** is the same phase-space transport matrix used in Section 9.10.1 to propagate single particle trajectories. Replacing **R** with the transport matrix for one complete revolution around the accelerator, and rearranging terms, we have

$$\{\beta_o, \alpha_o, \gamma_o\} = \mathbf{S}\{\beta_o, \alpha_o, \gamma_o\} \ , \tag{9.7}$$

where the elements of the 3×3 matrix S are functions of the storage ring model parameters (quadrupoles, bends, drifts, etc.).[32] Equation (9.7) can be solved for the initial Twiss parameters, $\{\beta_o, \alpha_o, \gamma_o\}$, at one point in the storage ring. The Twiss parameters in the remainder of the ring are then found by propagating these values via the similarity transformation of Eq. (9.5). The x and y β-functions as calculated by the COMFORT program for SPEAR are indicated in Fig. 4. The phase function,

$$\phi(s) = -\int_o^s \frac{ds}{\beta(s)} \tag{9.8}$$

is found by direct integration, and plotted in Fig. 5 for one-half of the SPEAR ring.

9.10.3 Closed Orbit Perturbations

It is well known that a section of the closed orbit in a storage ring can be analyzed just like a transport line. To find the beam coordinates, we specify the initial conditions (x_o, x_o'), and propagate the trajectory with transport matrices through each element. One difference between transport lines and storage rings is that in a storage ring the coordinates of the closed orbit trajectory (x_o, x_o') always repeat after each revolution. We call this trajectory the closed orbit. A betatron oscillation is motion with respect to the closed orbit that does not repeat each turn (for non-integer tunes).

For a well aligned lattice, the on-momentum beam passes through the magnetic axis of each element. In this case, the initial conditions of the closed orbit are simply $(x_o, x_o')=(0, 0)$. If a dipole field perturbation is introduced, at the point of the kick the new orbit satisfies the closure condition

$$M\begin{pmatrix} x_o \\ x'_o \end{pmatrix} + \begin{pmatrix} 0 \\ \theta \end{pmatrix} = \begin{pmatrix} x_o \\ x'_o \end{pmatrix}, \tag{9.9}$$

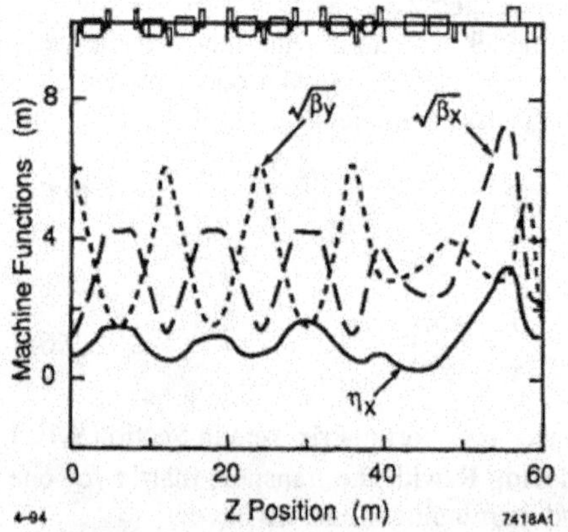

Fig. 4. β-functions for the SPEAR light source lattice.

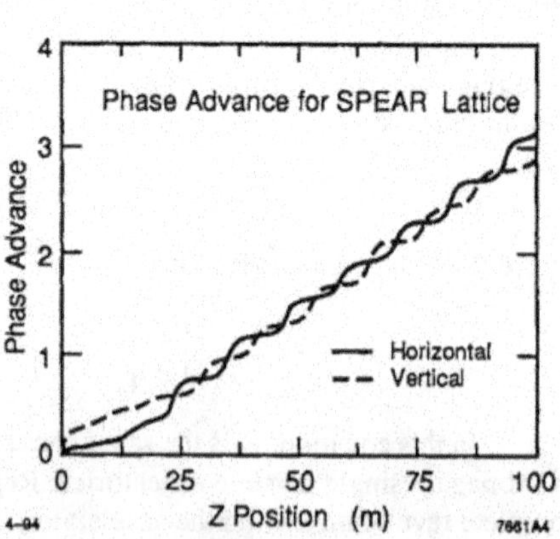

Fig. 5. Phase advance functions for the SPEAR lattice.

where **M** is the one-turn transport matrix, and θ is the magnitude of the dipole kick. The position and angle of the closed orbit at the point of deflection are

$$(x_o, x'_o) = (1-\mathbf{M})^{-1}\,(0,\theta)\ , \tag{9.10}$$

At any other point in the ring, the coordinates of the closed orbit perturbation are found by propagating the orbit via the transport matrix, $(x, x') = \mathbf{R}\,(x_o, x_o')$. If more than one dipole kick is present, the closed orbit is a superposition of orbit perturbations.

As indicated in Fig. 6, the orbit deflection at the location of the corrector magnet in a storage ring cannot be distinguished from a corrector kick in a transport line. In the storage ring, however, the coordinates (x_o, x'_o) of the closed orbit satisfy the closure condition, Eq. (9.9).

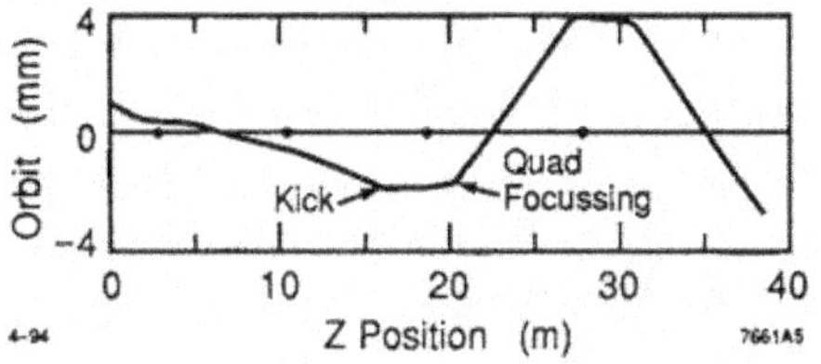

Fig. 6. Local closed orbit deflection from a single dipole kick.

No discussion of closed orbit perturbations is complete without mention of the β-function representation. In this context, the response matrix elements relating a kick $\Delta\theta_i$ at point 'i' to the beam displacement Δx_j at point '*j*' are written

$$\Delta x_j \;=\; \Delta\theta_i\,\frac{\sqrt{\beta_i\beta_j}}{2\sin\pi\nu}\;\bullet\;\cos\nu\,(\pi-|\phi_j-\phi_i|)\ , \tag{9.11}$$

where the phase factors advance from 0 to 2π, and ν is the tune of the accelerator (phase advance per revolution). The absolute value of the phase argument permits the point of observation to be either ahead or behind the kick.

Although the β-function representation of the response matrix is equivalent to the response matrix formulation of Section 9.10.1, it has several features that illuminate the physical behavior of orbit motion. First, from Eq. (9.11), it is clear that dipole kicks located at points where the β-function is large will induce relatively large closed orbit perturbations. For this reason, strong quadrupoles located in regions of *large beta* must be well aligned. Likewise, bpms located at points of large beta are the most efficient way to detect orbit perturbations. Moreover, Eq. (9.11) tells us that integer tunes are unstable, and indicates how to select a corrector with the appropriate phase advance to cancel an orbit perturbation (e.g., correctors to produce a local orbit *bump;* see also Section 9.12.1).

9.10.4 Dispersion

The dispersion function tells us the deviation of the closed orbit for off-momentum particles from the (nominal) closed orbit defined for particles at the synchronous energy (Chapter 2). The dispersion function can either be evaluated from the model or measured directly by varying the radio-frequency. To calculate the dispersion function from the model, we first extend the transport matrix formalism to include momentum dependence, and then solve for the off-momentum closed orbit, often called the η-function. In the horizontal plane, the transport matrix that maps the η-function for one turn around the storage ring becomes,

$$\mathbf{M} = \begin{pmatrix} R_{11} & R_{12} & R_{13} \\ R_{21} & R_{22} & R_{23} \\ 0 & 0 & 1 \end{pmatrix} , \tag{9.12}$$

where the new elements R_{13} and R_{23} give the dependence of horizontal position and angle on energy. Analogous to Eq. (9.9), the initial coordinates (η, η') for the dispersion function satisfy the closure condition

$$\mathbf{M}(\eta,\eta',1) = (\eta,\eta',1) , \tag{9.13}$$

and the dispersion is evaluated at all other points by propagating the initial conditions with the appropriate 3x3 transport matrix computed from the model. Figure 4 includes a plot of the dispersion function for SPEAR. A more general discussion of the dispersion function evaluated in terms of integrals over the β-functions and phase advance is covered in reference.[20] For our purposes, it is important to note that independent of the method of calculation the model must be well calibrated for the predicted dispersion to agree with the measured value.

9.10.5 Synchrotron Integrals

Up to this point, we have demonstrated how the accelerator model can be used to calculate the shape of the electron beam envelope (β-functions, dispersion) and the coordinates of the closed orbit. These calculations do not, of course, constitute a full representation of the electron beam. What is still missing are the particle distribution functions in both the longitudinal and transverse planes.

As described in Chapter 2, photon emission in the presence of dispersion creates a shift in the closed orbit, and consequently excitation of betatron oscillations. Photon emission also stimulates synchrotron oscillations in the longitudinal plane. Both of these effects are remediated by the action of the radio frequency drive system, with the result that the particle distribution functions relax to a stationary equilibrium value. Since the emission of energy quanta (photons) is statistically random, the Central Value Theorem of statistical mechanics tells us that the particle distribution functions are Gaussian.

In the plane transverse to the beam motion, the spread in oscillation amplitudes is characterized by the emittance of the beam. Recalling the Courant-Snyder invariant for betatron oscillations, $\varepsilon = \gamma x^2 + 2\alpha x x' + \beta x'^2$, the transverse particle distribution can be written

$$\rho(x,x') = \frac{1}{\sqrt{2\pi}\ \varepsilon} \exp - \left(\frac{\gamma x^2 + 2\alpha x x' + \beta x'^2}{\varepsilon} \right) . \tag{9.14}$$

In this case, the emittance 'ε' characterizes the distribution of oscillation amplitudes. Neglecting dispersion effects, in a given plane the rms size of the beam cross section is $\sqrt{\varepsilon\beta(s)}$, at point '*s*.' Low emittance (cold) electron beams imply dense distribution functions, leading to high brightness photon beams. The longitudinal distribution is also Gaussian, and can be characterized by an rms energy spread, σ_E.

How are the equilibrium emittance and energy spread determined for stationary stored beams? This question was answered by the synchrotron radiation integrals.[18] In short, the synchrotron radiation integrals summarize the effects of photon emission in the accelerator guide field to yield estimates for the following important parameters:

- energy loss per revolution
- equilibrium beam emittance
- energy spread
- damping times for transverse and longitudinal oscillations
- path length dependence on energy

In the linear approximation, the five primary synchrotron radiation integrals all contain at least one factor of the inverse bending radius (ρ^{-1}) in the argument of the integral (photon emission in dipole fields), with additional dependence on n (field index), η, β, and their derivatives, depending on the integral. To calculate the synchrotron radiation integrals, it has been demonstrated that only the β– and η–function values at the entrance of each magnetic element with finite bending radius are required.[33] This feature makes evaluation of the integrals fast and efficient for on-line modeling applications.

9.11 Model Calibration

In this section, we review experimental procedures that can be used to calibrate the linear optics model. These procedures are based on comparing beam measurements to the values predicted by the model. Discrepancies between the two sets of data are eliminated by adjusting parameters in the model.

Depending on the goals of the experimentalist, several approaches to model calibration are possible. It is easy, for example, to adjust the model so that it predicts the measured tunes. Finding a more accurate model requires more careful measurements, and sometimes complicated numerical analysis. In general, the goal of the model calibration procedure is to first adjust the model to accurately reflect the measured beam parameters, and then adjust the accelerator so that the beam parameters agree with their design values. For a review of diagnostic techniques used for synchrotron light sources, see Chapter 10 and references therein.

9.11.1 Optics Calibration

One common diagnostic is to spectrum analyze transverse oscillations of the beam to determine the tunes of the lattice. If the measured value of the tunes are different from the model value, we can adjust the quadrupole strengths in the model to make the model agree with the measurement. The easiest way to find the new quadrupole strengths is to Taylor-expand the model tunes to first order in terms of the quadrupole strengths. Since the horizontal and vertically focusing quadrupoles dominate ν_x and ν_y, respectively, we can choose one QF and one QD quadrupole family for the Taylor expansion. The choice of quadrupole families is, to some extent, arbitrary. Note that although accelerator model calibration is an inherently nonlinear problem, the equations can often be solved by linear analysis if we make a first order Taylor expansion.

It is also possible to estimate β–function values at discrete locations in the lattice from measurements of the corrector-to-bpm response matrix, or from turn-by-turn oscillation measurements. Each entry of the corrector-to-bpm response matrix, for instance, has a β–function representation (neglecting RF effects)

$$\frac{\sqrt{\beta_i \beta_j}}{2\sin \pi \nu} \cdot \cos \nu(\pi - |\phi_j - \phi_i|) \ , \tag{9.11}$$

where '*i*' and '*j*' indicate the location of the kick and location of the bpm measurement, respectively. Initially, we have a set of model values $\sqrt{\beta_0}$ and ϕ_0 at each corrector and each bpm. These values can be expanded to first order,

$$\sqrt{\beta} = \sqrt{\beta_o} + \Delta\sqrt{\beta} \ ,$$

$$\phi = \phi_0 + \Delta\phi \ , \tag{9.15}$$

and substituted into Eq. (9.11). Retaining only terms to first order, we obtain a linear matrix equation for the perturbed values of $\Delta\sqrt{\beta}$ and $\Delta\phi$. The tunes can also be varied as fitting parameters. Since the problem is inherently nonlinear, the linear fitting procedure may require several iterations for the solution to converge. Although knowledge of the β-functions at discrete locations can be useful for machine control, it is not equivalent to knowing the quadrupole field strengths.

Several algorithms exist for numerically fitting the model parameters (e.g., quadrupole strengths) so that the model predictions agree with the measured response matrix.[34] These algorithms typically compare columns of the model matrix (calculated orbit perturbations) to columns of the measured matrix (measured orbit perturbations). The advantages of using a response matrix for model calibration include the following:

- The beam motion probes the spatial field structure of each quadrupole
- The beam motion tests for bpm linearity
- The corrector kicks can be calibrated
- The large number of measurements provides good redundancy for numerically fitting the model

Moreover, since the response matrix measurements produce difference orbits, the unknown effect of bpm offset errors and dipole field kicks from misaligned quadrupoles are eliminated from the data. Clearly, the measured response matrix contains a wealth of information. To keep the response matrix measurement linear, sextupole magnets, and insertion devices should be turned off whenever possible.

Quadrupole field errors are difficult to analyze, however, because the response matrix has a nonlinear dependence on the field strengths. In this case, we can assign several columns of the measured response matrix as fitting constraints and search for the correct quadrupole strengths for the model with a nonlinear fitting program. The solution is well constrained if orbit perturbations measured in both the 'x' and 'y' planes are used.

As an alternative to nonlinear model calibration methods, it is straightforward to expand the difference between the model response matrix and the measured response matrix to first order in terms of the following parameters:[35]

- quadrupole field strengths
- corrector gain factors
- BPM gain factors

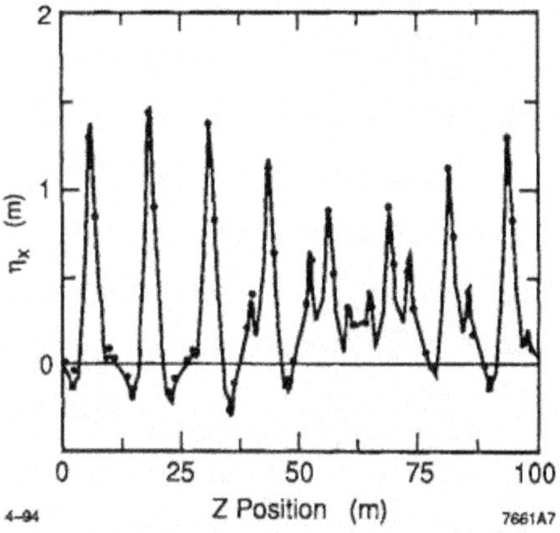

Fig. 7. Model prediction and measurement of dispersion function in NSLS x-ray ring.

If we include contributions from the dispersion orbit caused by the corrector kicks in the horizontal plane,[36] the linearized equations can be written in vector form:

$$\Delta C = \frac{d\Delta C}{d\mathbf{k}} \Delta\mathbf{k} + \frac{d\Delta C}{d\theta} \Delta\theta + \frac{d\Delta C}{dg} \Delta g + \frac{d\Delta C}{d\delta} \Delta\delta , \qquad (9.16)$$

where ΔC is the difference between the model and measured response matrix, and the variables $\{\Delta\mathbf{k}, \Delta\boldsymbol{\theta}, \Delta g, \Delta\boldsymbol{\delta}\}$ are the quadrupole gradient errors, corrector gain factors, bpm gain factors, and momentum shift ($\Delta P/P_0$) produced by each horizontal corrector kick, respectively. Solving for these values by linear matrix inversion helps to bring the model into agreement with the measurements. Before inverting the matrix, error bars can be added for the bpm read back noise to weight the individual equations.

A convincing test of this algorithm was carried out on the NSLS x–ray ring.[36] First, all sextupole magnets were turned off and the insertion devices were retracted to obtain a linear lattice, and the response matrix was measured several times to accumulate statistics. After several iterations of the fitting procedure, the new model agreed with the measured data to an accuracy of about 0.2%, or 2 microns for a 1 mm orbit perturbation. The *calibrated model* also predicted the measured tunes to within experimental accuracy. Hence, the calibrated computer model was consistent with the measured data.

Having established a linear optics model for the x-ray ring without sextupoles, the sextupoles were turned on and the response matrix remeasured. Since a beam passing off-axis through a sextupole magnet experiences a quadrupole field, this time the calibration variables were quadrupole field components at the sextupole magnet locations. The fitting again produced a statistically viable solution, in agreement with the measured response matrix, that indicated the beam was off-center in some of the sextupoles by several mm. This result agreed with bpm measurements of the closed orbit. As a cross check of the optics model with the sextupoles on, the new model was used to predict the dispersion function. As demonstrated in Fig. 7, the predicted dispersion agreed with the measured dispersion, including an error component

The result of the NSLS x-ray ring experiments was to produce a linear optics model that predicted:

- the measured response matrix
- individual quadrupole field strengths
- corrector calibration factors
- bpm calibration factors
- beam position in sextupole magnets
- tune (in both planes)
- the dispersion function

Given this pedigree for the calibrated model, it is reasonable to believe that it will predict the actual β-functions and synchrotron integrals to a high degree of accuracy. The model should also be sufficiently accurate to begin searching for alignment errors.

9.11.2 Beam Based Alignment

Magnet and bpm alignment errors are considerably more difficult to detect than focusing errors. In the first place, prior to searching for alignment errors, the linear optics model should be established. This reduces the number of variables in the problem. The dominant source of orbit perturbations are then dipole field components (e.g., quadrupole misalignments), but the orbit measurement can also be contaminated by bpm read back errors, and can contain a momentum error. Including these terms, the expression for the *measured* orbit can be written in vector notation as

$$\mathbf{x}_{bpm} + \Delta\mathbf{x}_{bpm} = \mathbf{C}\theta + \mathbf{Q}\Delta\mathbf{x}_q + \frac{\eta\Delta P}{P_o}, \tag{9.17}$$

where $\mathbf{C}$ and $\mathbf{Q}$ are the response matrices, as viewed at the bpms, for the corrector kicks ($\boldsymbol{\theta}$) and quadrupole misalignments ($\Delta\mathbf{x}_q$), respectively. The bpm read back errors are contained in the column vector $\Delta\mathbf{x}_{bpm}$. The term $\eta\Delta P/P_o$ is the dispersion component of the orbit that appears if the beam is off-momentum. Note that although the corrector response matrix $\mathbf{C}$ and the dispersion function η can be measured, the response matrix $\mathbf{Q}$ for quadrupole misalignments must be calculated from the model.

In general, the solution for the exact electron beam orbit relative to the design orbit [$\mathbf{x}_{bpm}$ in Eq. (9.17)] is extremely difficult to determine because the number of unknown variables $\{\Delta\mathbf{x}_{bpm}, \Delta\mathbf{x}_q, \Delta P/P_o\}$ is greater than the set of measurements $\{\mathbf{x}_{bpm} + \Delta\mathbf{x}_{bpm}\}$ for a single optical configuration.

The under-constrained data analysis problem makes it difficult to distinguish dipole field errors from bpm offset errors. It is also difficult to distinguish between quadrupole misalignments and bend field errors. Furthermore, unlike the optics calibration procedure, deflecting the orbit does not yield new information (to first order) because dipole fields are spatially homogeneous.

To help solve these problems, the analysis of beam-based alignment data can be broken up into a sequence of piecewise solutions that agree with the model (the GOLD method[34]). An example taken from the SPEAR storage ring in shown in Fig. 8, where, a stray dipole kick from the injection transport line is found by propagating a fitted trajectory in the forward and backward directions past the location of the kick. An isolated bpm error is also indicated in the measured data. The closed orbit perturbation, with the dipole kick included in the model, is shown in Fig. 8c.

Another simple, yet informative, technique for analysis of the absolute orbit is to interpolate the beam position between adjacent bpm readings. In this case, we compute the angle at the first bpm to pass the trajectory through the second bpm. A quick scan of the result graphically illustrates potential bpm or quadrupole alignment errors.

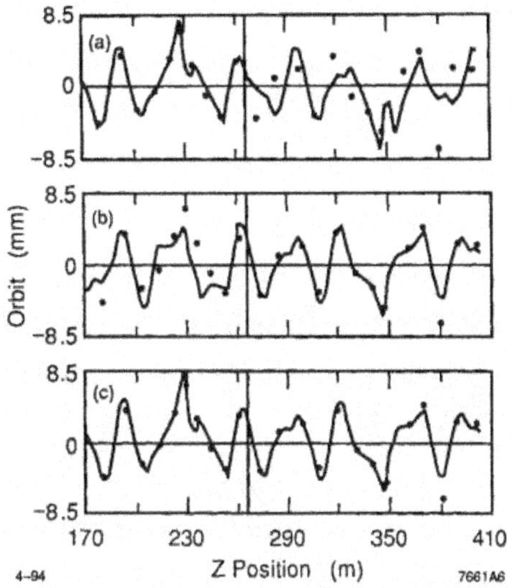

Fig. 8. Closed orbit trajectory, propagated (a) forward and (b) backward, over dipole kick error, and (c) closed orbit with dipole kick.

An algorithm for analyzing sets of three contiguous bpm readings has been tested at BNL.[37] This technique found errors in the AGS Booster orbit that were attributed to improper electrical connections. An extension of the bpm triplet analysis was developed at CERN.[38] This technique relies on an analysis of the χ^2 (residue) generated by least-squares fitting the beam trajectory data for bpm triplets, quadruplets, quintuplets, and so on. The residue method is particularly suitable for large storage rings where the amount of data makes processing difficult—on LEP, for example, a localized field error was found that was corrected by realigning quadrupole magnets.

Alignment errors can also be found by simultaneous analysis of the orbit measured with several different optical configurations.[39] In principle, the bpm and quadrupole offset errors remain constant, independent of the optical configuration. By changing the quadrupole strengths, a set of closed orbit equations [Eq. (9.17)] with different response matrices (**C** and **Q**), and different dispersion functions can be produced. From the expanded set of equations, the quadrupole offsets, bpm offsets, and beam energy offset can be found. The key to this method, of course, is to generate a set of sufficiently different optical configurations to make the solution unique.

9.12 Model-Based Control

The discussion up to now has been directed toward calculation of beam parameters and model calibration. With these tools, we are now in a position to use the model for machine control. The goal, of course, is to guide the electron beam parameters to an optimum state for the photon beam users. This requires model-based calculations with reliable model-to-magnet calibration factors. Once the state of the accelerator is changed, the result should always be verified by appropriate diagnostic measurements.

9.12.1 Orbit Control

As an introduction to model based accelerator control, consider control of the closed orbit. In general, the goal of almost any controlled orbit change will fall into one of the following categories:

- orbit amplitude reduction
- local orbit bumps
- orbit adjustment to a previous reference

Often, the minimum orbit amplitude (as measured at the bpms) is not the optimum orbit for machine operation. For example, it is possible to include dispersion corrections in the orbit control program,[40] to adjust the position and angle of the photon beams, or to ask the orbit correction algorithm to minimize corrector strengths. Local orbit bumps are also used to fine tune the injection rate and to control coupling.

To manipulate the closed orbit, operators use dipole corrector magnets to deflect the beam, and beam position monitors detect the motion (either in the storage ring or on the photon beam lines). To first order, orbit adjustments require finding a solution to the linear matrix equation:

$$\Delta \mathbf{x} = \mathbf{C} \Delta \theta \ , \tag{9.18}$$

where $\mathbf{C}$ is the corrector-to-bpm response matrix, $\Delta\mathbf{x}$ is the desired orbit perturbation, and $\Delta\boldsymbol{\theta}$ is a column vector of the corrector strengths.

Several common methods for 'orbit control' are outlined below. In each case, the mathematical calculation is based on either a computed or measured corrector-to-bpm response matrix. The solution to a set of linear equations (matrix inversion) yields a corrector pattern that will move the orbit toward the desired goal.

9.12.1.1 MICADO[41]

Historically, the MICADO package was one of the first orbit control algoritnms. In progression, MICADO finds the single most effective corrector, the most effective additional corrector, and so on, to implement the orbit correction. The main advantage of MICADO is the ability to find the single most efficient corrector, which may point to the location of an error in the lattice.

9.12.1.2 Harmonic[42]

Every closed orbit perturbation can be decomposed into a Fourier series evaluated on the normalized, periodic phase interval 0 to 2π.[31] For a dipole perturbation, the closed orbit is in fact a periodic solution to a kick term driving the single particle equation of motion, Eq. (9.3). The harmonic spectrum of the orbit contains components in ratio

$$F(n) \propto \frac{1}{n^2 - \nu^2} \ , \tag{9.19}$$

where ν is the betatron tune and n is the harmonic wave number.

Since the spectra of orbit perturbations are dominated by harmonics near the tune, Fourier decomposition on a limited set of harmonics centered about the tune can produce an effective representation of the orbit. Conversely, to compensate components centered about the tune requires only small corrector amplitudes. These features make harmonic orbit correction a useful technique for orbit feedback systems.[42]

Complications with harmonic orbit analysis include:

- non-equal phase interval between bpms,
- Nyquist sampling limits,
- the need for accurate β-function values at the bpm locations

The latter restriction underscores the need for a calibrated optics model.

9.12.1.3 Eigenvector[43]

Given a linear set of equations, $\Delta\mathbf{x} = \mathbf{C}\,\Delta\boldsymbol{\theta}$, we want to find solutions for a corrector pattern ($\Delta\boldsymbol{\theta}$) that will produce the orbit perturbation, $\Delta\mathbf{x}$. One approach to finding $\Delta\boldsymbol{\theta}$ is to solve this problem by the method of eigenvectors. In orbit correction applications, however, the response matrix **C** is not always square. For the general M×N dimensional matrix (M-bpms, N-correctors) there are three possibilities:

M = N (square matrix) In this case, for a full-rank matrix **C**, there exists one unique solution, $\Delta\boldsymbol{\theta} = \mathbf{C}^{-1}\Delta\mathbf{x}$. The inverse response matrix, $\mathbf{C}^{-1}$, can be found by eigenvector decomposition. Whenever a singular (not invertible) matrix is encountered, Singular Value Decomposition[44] can be applied (see below). The action of the eigenvectors is to project the orbit perturbation $\Delta\mathbf{x}$ onto the eigenbasis of the matrix **C**. This projection yields the corrector pattern, $\Delta\boldsymbol{\theta}$. Notice that orbit decomposition in terms of eigenvectors of the response matrix is analogous to a Fourier series decomposition in terms of harmonic basis vectors. Depending on the application, the corrector pattern can be produced from all, or a subset, of the eigenvectors.

M > N (more bpms than correctors). This case is a linear least-squares problem where the number (or choice) of correctors cannot produce all orbit displacements that span the range of orbit perturbations, $\Delta\mathbf{x}$. The standard least squares solution procedure is to multiply Eq. (9.18) by $\mathbf{C}^T$ to make a square matrix: $\mathbf{C}^T\Delta\mathbf{x} = \mathbf{C}^T\,\mathbf{C}\Delta\boldsymbol{\theta}$. The resulting corrector pattern is found by matrix inversion,

$$\Delta\theta = (\mathrm{C}^T\mathrm{C})^{-1}\,\mathrm{C}^T\Delta\mathrm{x}\ . \tag{9.20}$$

In the language of linear algebra, the least-squares procedure projects the orbit perturbation onto a subspace of basis vectors spanned by the correctors (the columns of matrix **C**). The process is analogous to finding the shortest path from a point to a line. Least-squares orbit control can be used to produce a smoothed orbit correction in the presence of bpm noise.

M < N (more correctors than bpms). In this case, the fitting problem is underdetermined, that is, the matrix **C** is not full rank. Under these conditions, SVD will produce the 'pseudo-inverse' of the matrix, $\mathbf{C}^{-1}$. Since SVD minimizes the rms (root-mean-square) value of the solution vector $\Delta\boldsymbol{\theta}$, this is where SVD makes its application to accelerator control particularly attractive.[45] [The minimum rms result is easy to demonstrate. Note that the solution vector, $\Delta\boldsymbol{\theta}$, is a linear combination of the homogeneous solution $\Delta\boldsymbol{\theta}_n$ (null vectors) and the particular solution $\Delta\boldsymbol{\theta}_p$ (in the row space of **C**). Each null vector is a corrector pattern that does not move the beam at the bpms under observation. Since SVD can isolate the particular solution, the magnitude of the corrector kick vector is minimized.]

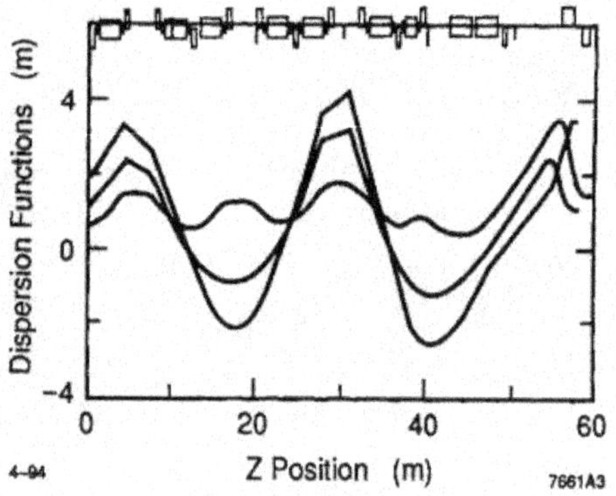

Fig. 9. Model calculation of dispersion function used to guide SPEAR lattice toward a low momentum-compaction lattice.

In the language of control theory, matrix inversion by SVD minimizes the cost function; in this case, the rms current in the corrector magnets. This feature of SVD is routinely applied to SPEAR to reduce the strength of the beam line steering correctors.[46] Here, the position of nine photon beams is held constant (modeled as per Section 9-8), and the SVD algorithm reduces the rms value of up to 30 corrector magnets. The reduced corrector strengths give more overhead for the local beam line steering servos and the global orbit feedback system. For global orbit feedback applications, SVD is useful because one can vary the number of eigenvectors needed for orbit representation (eigenvalue cut-off), and the corrector kicks are minimized at each feedback cycle.[43,47,48]

9.12.2 On-Line Optics Control

Another important function of the accelerator model is to control the optical functions. It is common practice, for instance, to use the model to calculate a new set of quadrupole strengths that will adjust the horizontal and vertical betatron tunes. The operator may also want to change the β-function values of the lattice to achieve different operating conditions. Examples of more complicated lattice modification include keeping the tunes constant while lowering the emittance, or lowering the momentum compaction. In each case, the new quadrupole strengths are calculated from the accelerator model.

The example of controlling the momentum compaction factor in SPEAR illustrates the main features of on-line optics control.[49] For this experiment, the operator wants to lower the momentum compaction, thereby shortening the bunch length, by controlling the dispersion function while keeping the tunes and chromaticities constant. The evolution of the dispersion function through a progression of configuration changes is indicated in Fig. 9.

In order to guide the electron beam optics along a smooth path from the nominal lattice to the new lattice, an incremental series of small modifications is desirable. To change the machine optics in a controlled manner, the operator first needs a model for the present state of the accelerator. In addition, the operator needs accurate calibration factors that will predict the correct power supply settings to implement the desired change in magnet field strengths. Next, similar to the procedure used by off-line lattice designers, a new lattice is calculated, and the result is downloaded onto the machine hardware. At each step, measurements are made to diagnose the new lattice (tune, response matrix, etc.), and the results are compared to the model prediction.

References

1. The following papers contain examples of integrated control system/accelerator modeling packages: L. Catani et al., *Proc. IEEE 1991 PAC* (San Francisco, 1991); D. Dobrott et al., *ibid.*; M. J. Lee, *Proc. 1991 Int. Conf. on Accelerators and Large Experimental Physics Control Systems* (Tsukuba, Japan, 1991); H. P. Chang et al., *Proc. IEEE 1993 PAC* (Washington, D.C., 1993); H. Nishamura, *ibid.*
2. D. P. Weygand et al., *Proc. IEEE 1987 PAC* (Washington, D.C., 1987).
3. *7–GeV Advanced Photon Source: Conceptual Design Report* (Argonne National Laboratory, Illinois 60439, 1987), ANL–87–15.
4. European Sychrotron Radiation Facility, *ESRF Foundation Phase Report, Feb 1987.*
5. JAERI–RIKEN Spring–8 Project Team, *Spring-8 Project* (1991).
6. M. Knot, D. Gurd, S. Lewis, and M. Thout, *Proc. 1993 ICALEPCS* (Berlin, 1993); W. McDowell, M. Knott, and M. Kraimer, *Proc. IEEE 1993 PAC*, (Washington, D.C., 1993).
7. P. Clout et al., NIM **A293**, 456 (1990).
8. B. Ng et al., *Proc. IEEE 1991 PAC* (San Francisco, 1991), p. 1371.
9. J. P. Scott and D. E. Eisert, *Proc. of Europhysics Conf. on Control Systems for Experimental Physics (*Villars-sur-Ollon, Switzerland, 1987), p. 103.
10. N. D. Arnold et al., *Proc. IEEE 1991 PAC* (San Francisco, 1991).
11. G.J. Nawrocki et al., *ibid.*
12. *LBL 1–2 GeV Synchrotron Radiation Source, Conceptual Design Report,* (Lawrence Berkeley Laboratoy, Livermore, 1986); PUB–5172 Rev. (1986).
13. T. Russ and C. Sibley, NIM **A293**, 258 (1990).
14. R. Bork, *Proc. IEEE 1987 PAC* (Washington, D.C., 1987), p. 523.
15. Beetham and the SPS/LEP Accelerator Controls Group, *Proc. IEEE 1987 PAC* (Washington, D.C., 1987), p. 741.
16. P.G. Innocenti et al., NIM **A293,** 1 (1990).
17. C.O. Pak, NIM **A293,** 408 (1990).
18. M. Sands, *The Physics of Electron Storage Rings: An Introduction,* SLAC-121, 1970.

19. R. Chasman, G.K. Green, and E.M. Rowe, IEEE Trans. Nuc. Sci. **NS–22**, 1765 (1975).

20. Helmut Wiedemann, *Particle Accelerator Physics* (Springer Verlag, Berlin, 1993).

21. F. C. Iselin and J. Niederer, CERN/LEP–TH/88–38 (1988).

22. *A User's Guide to the Harmon Program* (CERN, 1982), LEP Note 420.

23. K. L. Brown et al., SLAC–91, Rev. 2, UC–28.

24. A. J. Dragt, et al., IEEE Trans. Nuc. Sci. **NS–32**, No. 5, 2311 (1985). The Lie algebra methods introduced by Dragt are fast becoming a popular tool for accelerator analysis and design. See, for instance, J. Irwin, NIM **A298**, 460 (1990).

25. R. Ruth, IEEE Trans. Nuc. Sci. **NS–30**, No. 4, 2669 (1983).

26. TRACY evolved from the efforts of H. Nishamura and E. Forest, and was rewritten in a very compact and powerful form by J. Bengston. For an original reference, see H. Nishamura, LBL REPORT 25236, ESG–40 (1988).

27. Los Alamos Accelerator Code Group, *Computer Codes for Particle Accelerator Design and Analysis: A Compendium*, second ed., LA–UR–1766 (1990). For linear accelerators, see C.R. Eminhizer, ed., La Jolla Inst., AIP Conf. Proc. **177** (1988).

28. K. L. Brown, SLAC Report–75.

29. D. C. Carey, *The Optics of Charged Particle Beams* (Harwood Academic, New York, 1987).

30. D. W. Kerst and R. Serber, Phys. Rev. **60**, 53 (1941).

31. E. Courant and H. Snyder, Ann. Phys. **3**, 1 (1959).

32. M. D. Woodley et al., SLAC–PUB–3086 (1983).

33. R. H. Helm et al., *Proc. 1973 PAC* (San Francisco, 1973); SLAC–PUB–1193.

34. Examples of beam line error analysis codes based on comparing measured data to the model include: M. Lee et al., *Proc. Europhysics Conf. on Control Systems for Experimental Physics* (Villiars, Switzerland, 1987); SLAC–PUB–4411. M. J. Lee, Y. Zambre, and W. Corbett, *Proc. Int. Conf. Cum Workshop on Current Trends in Data Acquisition and Control of Accelerators* (Calcutta, India, 1991); SLAC–PUB–5701. W. J. Corbett, M. J. Lee, and Y. Zambre, *Proc. 3rd EPACS* (Berlin, 1992); SLAC–PUB–5776.

35. W. J. Corbett, M. J. Lee, and V. Ziemann, *Proc. IEEE 1993 PAC* (Washington, D.C., 1993); SLAC–PUB–6111.

36. J. Safranek and M. Lee, *Proc. of Orbit Correction and Analysis Workshop, Brookhaven Nat'l. Laboratory* (1993).

37. A. Luccio, *Proc. IEEE 1993 PAC* (Washington, D.C., 1993).

38. A. Verdier and J.C. Chappelier, *ibid.*

39. W.J. Corbett and V. Ziemann, ibid.; SLAC–PUB–6112.

40. Examples of simultaneous orbit and dispersion correction used for PEP-I: E. Close et al., IEEE Trans. Nuc. Sci, **NS–26**, No. 3, 3502 (1979). A. Chao et al., PEP–NOTE–318 (1979). M. H. R. Donald et al., *Proc. IEEE 1981 PAC* (Washington, D.C., 1981); SLAC–PUB–2666.

41. B. Autin and G. Marti, CERN–ISR–MA/73–17 (1973).

42. Harmonic representation of the closed orbit is developed in Ref. 31. Fast harmonic orbit control in synchrotron light sources was developed at NSLS: L. H. Yu et al., NIM **A284**, 268 (1984).

43. Eigenvector decomposition is widely used in theoretical and applied physics, and forms the basis for many orbit correction codes. For a recent treatment of applications to storage rings, see A. Friedmann and E. Bozoki, BNL–49527, submitted to NIM.

44. G. Strang, *Linear Algebra and It's Applications*, second ed. (Academic Press, Inc.); see also G. Strang, Amer. Math. Monthly, **100**, 9 (1993).

45. V. Ziemann, *SLAC Single Pass Collider Note* CN–393 (1992); see also E. Bozoki and A. Friedmann, *Proc. IEEE 1993 PAC* (Washington, D.C., 1993).

46. W. J. Corbett, B. Fong, M. J. Lee, and V. Ziemann, *ibid.*; SLAC–PUB–6110.

47. Y. Chung et al., *ibid.*, p. 2263.

48. W. J. Corbett and D. Keeley, *Proc. of Orbit Correction and Analysis Workshop* (Brookhaven National. Laboratory, 1993).

49. P. Tran et al., *Proc. IEEE 1993 PAC* (Washington, D.C., 1993), p. 173.

CHAPTER 10: BEAM DIAGNOSTICS

PETER KUSKE
BESSY
Berliner Elektronenspeicherring-Gesellschaft
Lentzeallee 100
14195 Berlin, FRG

10.1 Introduction

10.1.1 Function and Role of Beam Diagnostics

Beam diagnostics in a synchrotron light source has three important functions: In the early days of the facility it aids in the commissioning of various systems like injectors, transfer lines, and storage rings. During the entire life of the facility beam diagnostics is important for the efficient running of these systems, and last but not least there will always be demands from the users of synchrotron radiation to improve the performance of the source which is unthinkable without beam diagnostics pinpointing existing weaknesses.

From a user's point of view, parameters like beam position, particle distributions, and intensity have to be measured. These parameters are usually linked to user-driven target goals which might be different from one facility to the next. Goals like high spatial stability of an intense high brilliance beam can only be achieved if these parameters not only can be measured, but optimised, monitored, and kept constant either in choosing the right conditions for the operation of the storage ring (passive) or by using feedback in order to stabilise some of these parameters (active).

There might be important users with very special demands, e.g. people doing metrology.[1] They have to know the energy, the magnetic fields, the number of particles, and geometrical factors to very high precision in order to be able to accurately predict the emitted spectrum of synchrotron radiation. The energy of the stored beam must be determined to be better than 10^{-4}. The beam current is needed with high absolute precision and eventually individual particles must be counted in order to use the storage ring as a primary radiation standard with a dynamic range spanning 12 orders of magnitude.

In addition there are parameters more relevant for accelerator physicists: Lattice functions (Twiss parameters) or global lattice parameters like tunes and chromaticities. The measurements of these parameters, the observation of instabilities and the investigation of the non-linear behaviour of the particles usually allows to reach the desired performance of the facility or at least give an insight into the mechanisms which hinder their achievement.

From what was said above, it is obvious that beam diagnostics plays an extremely important role in any synchrotron light source from the very first to the very last day of its operation. On the lowest level of sophistication beam diagnostics has to answer questions like – is there a beam, where is it, how much is it, and how are the particles distributed within the beam. How these questions are tackled will be introduced briefly in the following sections. In case there is no beam, diagnostics has to answer the most difficult question – why the beam was lost. Here the experience of the operators must enter since expert systems are not yet available.

10.1.2 Intensity

The measurement of the intensity of the beam is probably the most important parameter to be monitored in an accelerator facility because it answers the most fundamental questions. The analysis of the recorded data in terms of lifetime and injection efficiency leads to two further key target parameters. The permanent display of the stored beam current together with the lifetime of the beam is the fastest storage ring performance indicator.

At different locations of the chain of accelerators different types of detectors are used.[2] As usual in beam diagnostics one always has to distinguish between destructive and non-destructive monitors. The intensity of the beam can be measured destructively by collecting the charges in a Faraday cup or non-destructively through other physical phenomena[3] like the accompanying image currents measured with a wall current monitor or the magnetic field around moving charges assessed with a current transformer.

The most important current measuring device, however, is the D.C. current transformer (DCCT). In its most advanced form as a parametric current transformer (PCT) it is nowadays state-of-the-art for current measurements in booster synchrotrons and storage rings. The principle of the measurement is based on the precise compensation of the magnetic field created by the beam current.[4] The toroidal sensor is mounted around an insulated piece of vacuum pipe because the wall currents must be interrupted. The high DC sensitivity and extended AC frequency response is achieved by combining an active current transformer and a magnetic parametric amplifier in a common feedback loop.

The speed of this monitor is not sufficient to verify the purity of the single bunch. For some users doing time resolved spectroscopy with synchrotron radiation the single bunch purity is important. With time resolved photon counting a 10^6 dynamic range with a time resolution of 100 ps was achieved at the ESRF.[5] Single bunch purity of better than 10^6 is reached if a cleaning process based on the selective transverse excitation of particles is used.[6]

10.1.3 Beam Position

The position of the beam is in almost all cases measured by means of a set of four button-type pickup electrodes mounted at 45 degrees to the orbital plane in order not to be illuminated by synchrotron radiation. This set typically looks like the one in Fig.10.1 taken from the ALS.[7] Due to the bunched nature of the beam the signals picked up by the electrodes will always contain a strong frequency component at the R.F. frequency which is around a few hundred MHz (see Chapter 4 on RF systems). The amplitude of this fundamental or a component at a multiple of the fundamental frequency of all four electrodes are measured with high absolute accuracy. The beam positions are obtained from a suitable scaled difference-over-sum calculation.[8]

The measurement of beam positions in transfer lines and booster synchrotrons also make use of buttons but the signal treatment is a bit more complicated because of the necessity for timing and synchronisation of the individual measurements. At these locations of the facility fluorescent screens[9] are probably the most useful diagnostic tools even though they are destructive. The screen is inserted at an angle of 45 degrees with respect to the path of the beam and the luminescence due to the impinging beam is viewed with a TV-camera (CCD). The image can be observed directly on a TV monitor giving qualitative positions or a more quantitative analysis can be performed with image data processing.

For the user of synchrotron radiation, who illuminates samples at the end of some ten meters long beam line, the emission angle is usually more important than the position of the source. Due to the long lever arms these measurements can be performed with much higher accuracy by means of photon beam position monitors (see Chapter 13 on beam stabilizing and steering) installed in the beam lines.

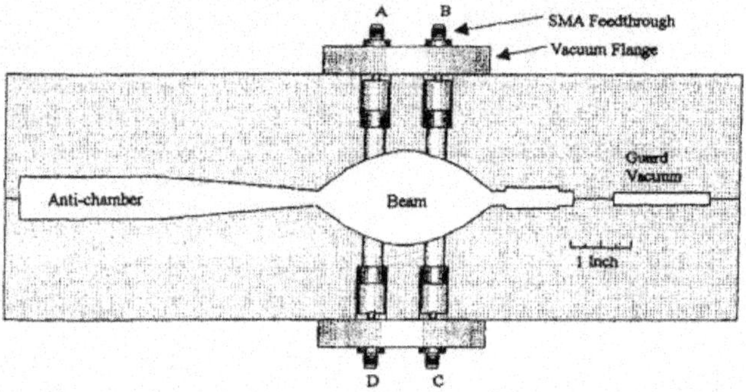

Fig. 10.1: Cross section of the ALS storage ring button installation. The four pickup electrodes are denoted A to D.

10.1.4 Distribution of Particles

In storage rings that are used as synchrotron radiation sources, the small natural emittance is one of the most important design goals (see Chapter 2 on lattices). The width of the Gaussian distribution of particles is closely related to the emittance. In order to verify the achievement of this key parameter the beam profiles have to be determined. Emittances measured and optimised along the chain of accelerators lead to the ultimate performance in terms of injection speed.

The experimental setup for the measurement of the distribution of particles depends again on the location along this chain. As long as the emittance is large and a destructive monitor can be used, the usual approach is based on fluorescent screens.[10] The width of the distribution of particles is measured as a function of the quadrupole strength and the data analysis yields Twiss parameters and the emittance. This technique is applicable in any transfer line, for example, between pre-injector and synchrotron or linac or between booster synchrotron and storage ring.

In storage rings or in synchrotrons the synchrotron radiation is used for emittance measurements.[11] The distribution of synchrotron radiation follows the distribution of radiating particles. The beam profile is obtained from the analysis of the image of the source. From that the emittance can be determined if the Twiss parameters are known. The most severe limitation of this technique is the blurness of the image due to diffraction. The small, wavelength dependent natural opening angle of synchrotron radiation acts like a limiting aperture and the resolution, σ_d, is given by Fraunhofer diffraction[3]: $\sigma_d \approx 0.39(\lambda^2 \rho)^{1/3}$. Here it was assumed that the synchrotron radiation was emitted in a bending magnet with bending radius ρ. The effect is large especially if visible light is used. This is a very convenient spectral region, however, in low emittance storage rings imaging at much shorter wavelengths has to be employed especially if the extremely small vertical emittance is the target.

Similarly, the bunch length can be measured with synchrotron radiation. The temporal structure of the light pulse is the consequence of the longitudinal distribution of particles. The bunch length in low emittance storage rings is very short. Thus, special equipment like a streak camera has to be employed to get down to the desired pico second time resolution.[12] The energy distribution of particles, which is closely related to the temporal structure of the beam, can be inferred from transverse profile measurements at a location where the contribution to these dimension is large (large dispersion).

The most convenient beam diagnostics tool is obviously the one which can answer many questions at the same time. In this respect fluorescent screens even though they are destructive are extremely useful because of their simplicity, reliability, and their direct visualisation of many relevant parameters: position, distribution, and intensity.

10.1.5 Additional Accelerator Parameters

In the storage ring the particles are oscillating in all three dimensions around the closed orbit due to the focussing and defocussing forces of magnets and the action of the cavity. The oscillations occur at characteristic frequencies. The linear tunes, $\nu_{x,y,z}$, are defined as the number of oscillations per revolution. Tunes belong to the class of global parameters. In addition there are local parameters. Their values depend on the location along the lattice structure. Examples are the Twiss parameters and the dispersion. One of the Twiss parameters is the beta function, β. The beta functions are related to the amplitude of the transverse oscillations. An important longitudinal parameter is the dispersion, D. The dispersion describes the orbit of an off-momentum particle. For more details see Chapter 2 on lattices.

Only if the transverse tunes, the shape of the beta functions, and the dispersion are in agreement with the design values we can hope to reach the desired performance. An iterative process is necessary to refine the settings of all the magnetic elements of the lattice structure until the measured parameters agree with the predictions (see Chapter 9 on controls and accelerator modeling).

The chromaticity is a global parameter connecting transverse and longitudinal motions. The chromaticities are the shifts of the transverse tunes with momentum $\xi = \Delta\nu / \Delta p/p$. The natural, negative chromaticity must be compensated by sextupoles in order to reach high beam currents. This stabilises the beam against the head-tail instability (see Chapter 12 on instabilities). In addition the tune spread due to the natural energy spread is reduced and off-momentum particles are no longer lost on transverse resonances. The chromaticity has to be measured in order to verify the correct settings of the chromatic sextupoles.

Sextupoles are elements producing non-linear effects. In many storage rings additional so-called harmonic sextupoles are installed. The intention is to increase the dynamical aperture by reducing amplitude dependent tune shifts and suppressing resonances. For the correct setting of these sextupoles it is highly desirable not only to observe the increase in lifetime or injection efficiency but to measure the tune shifts and the non-linear behaviour directly.

The last important issue of beam diagnostics is the observation of instabilities. Instabilities have a disastrous impact on many of the performance parameters already mentioned. Careful observations in time and in frequency domain using fast oscilloscopes and spectrum analysers are necessary for the detailed analysis and classification of instabilities (see Chapter 12 on instabilities). Cures can only be found if the instability mechanism is understood (see Chapter 13 on beam stabilizing and steering systems).

10.2 Measurement of the Beam Intensity

Different instruments and devices are used for the measurement of the intensity of the beam going from electron gun, linac, microtron over transfer lines to synchrotrons or storage rings.[2] In the following section some intensity monitors for transfer lines and the parametric DC current transformer (PCT) for accurate measurements in synchrotrons and storage rings will be introduced. Finally a few comments will be made about the determination of intensity related parameters like the lifetime of the stored beam and the injection efficiency.

The precise determination of extremely small beam currents in the storage ring which is important for some experiments, either for accelerator studies or metrology, is possible by observing synchrotron radiation with UV sensitive photodiodes.[13] Since intensity is not a problem, single electrons can be detected. This corresponds to current measurements of less than 1 pA.

10.2.1 In Transfer lines

In transfer lines the intensity of the beam can be measured destructively by moving a so-called Faraday cup into the path of the charged particle beam. The charge is collected and the delivered current is used to determine the number of particles. Faraday cups are used after the pre-injector (linac or microtron usually). Properly terminated, the signal contains the time structure of the beam and this can be used, for example, to verify the purity of single bunch operation behind the pre-injector.

A second monitor which is even more useful, is a so-called wall current monitor. This monitor does not destroy the beam, because the image charges flowing in the conducting vacuum beam pipe are measured. In order to do so, a ceramic gap of a few mm is introduced into the beam pipe. This gap is bridged by a number of resistors and the wall current will develop a voltage across these resistors. Like the Faraday cup, the wall current monitor is suitable for time resolved measurements.

Another instrument for the measurement of beam currents in transfer lines and in large rings with only a few bunches circulating, is the current transformer. In its simplest form the sensor is a toroidal core made of high-permeability alloy with n secondary windings. The toroid is mounted around an insulated piece of the beam tube, and the beam, going through the centre of the sensor, can be considered as a one turn primary winding. The concentric magnetic field from the beam current induces a signal in the secondary windings. The transformer differentiates the beam pulse signal with a typical time constant given by the ratio of the inductance of the secondary winding and the load resistor. With the active-passive transformer scheme the response is extended to very low frequencies.[2] This involves a DC coupled high gain amplifier and the DC levels have to be restored by periodic resetting.

The current transformers are commercially available from the French company Bergoz as fast or integrating current transformers (FCT or ICT). The ICT allows the measurement of the total beam charge in pulses shorter than 1 ns.[14]

10.2.2 In Synchrotron and Storage Ring

The current in synchrotrons and storage rings is measured with high absolute precision by the parametric current transformers. This monitor is based on an active-passive transformer but in order to extend the frequency response down to DC at least two additional toroids with primary and secondary windings for a magnetic modulator-demodulator circuit are included in the sensor head. These toroids are driven into saturation by an external modulation current. The magnetic inductions are opposite in the two cores. If both cores are identical then the voltage induced in a common secondary winding will cancel for a perfect compensation of the flux created by the beam and the feedback current. Any imbalance between these currents is detected with the demodulator circuit as a component at even harmonics of the modulation frequency. The error signal is used to adjust the feedback current until the imbalance is cancelled.[2] A description of all the other technical details can be found somewhere else.[4] PCTs are commercially available from Bergoz.

The very large dynamic range (2×10^7), the resolution down to 0.2 μA for an integration time of one second, and the frequency response from DC up to 100 KHz turns the PCT into the ideal absolute current monitor for storage rings and synchrotrons. Its only disadvantages are the temperature dependent zero drift[15] of up to 8 $\mu A/^{o}C$ and sudden changes of up to 2 μA in the current reading every couple of hours.[4]

10.2.3 Determination of Current Related Parameters

Once the intensity of the stored beam is recorded the 1/e decay time can be determined: $I(t) = I_0 \cdot e^{-t/\tau} \simeq I_0 \cdot (1-t/\tau)$. In order to optimise the lifetime it is important to reduce the length of the time interval t needed for a certain precision in the determination of the lifetime. The faster the system responds, the faster the optimum working conditions can be established. If a set of n current measurements with the same rms error taken every Δt are analysed, the relative error of the lifetime would be: $\Delta\tau/\tau \approx n^{-3/2} \cdot \sigma/\Delta t \cdot \tau/I$, where $\sigma/\Delta t$ is the resolution of the current monitor per integration time. The response time (or $n \cdot \Delta t$) will be large when the current, I, is small and the lifetime is long. Plugging in typical numbers for the original ESRF target goals (τ=10 h at I=100 mA) and the parametric current transformer ($\sigma/\Delta t$=0.5 μA/s) n would become 7 for an error of one percent. Therefore, the lifetime measurement should take only a few seconds. A similar lifetime measured at 1 mA of beam current or less would take more than ten minutes. The determination of long lifetimes with a PCT at low beam currents is very time consuming and a different detector like a photodiode has to be chosen.[13] The resolution per integration time could then be as small as a few pA/s and the determination of the lifetime is possible in a very short time interval.

Another intensity related parameter is the injection efficiency. The efficiency should be normalised if the output current of the pre-injector is fluctuating. In the daily operation of the light source lifetime and injection efficiency are key parameters to judge the overall performance.

10.3 Measurement of Beam Positions

As mentioned in the introduction, the position of the beam is usually measured by a set of pickup electrodes, some signal processing electronics, and a computer controlling the measurement and collecting the closed orbit data. These beam position monitor (BPM) systems have the advantage of being non-destructive and they are widely used in linacs, transfer lines, synchrotrons, and storage rings. In the first parts of this section I will describe the electrodes, the various approaches in processing the signals, and I will mention some of the solutions for interfacing the BPMs to the central control system.

The fluorescent screen is the second type of monitor which will be presented in more detail. Fluorescent screens are simple but they destroy the beam. They are used primarily in situations where the positions from single shots have to be measured. This is the case for example in transfer lines and for first turn measurements in synchrotrons and storage rings. In addition to the beam positions, fluorescent screens give information about the distribution and number of particles.

10.3.1 Sensors for Beam Position Monitoring

The position of the beam is sensed with four pickup buttons as shown in Fig.10.1. These circular electrodes can be built and mounted with good precision. They produce a low impedance which is important for the stability of the beam (see Chapter 12 on instabilities). The signal can be simulated easily, is high for short bunches, and can be optimised for high signals and good position sensitivity.

The positioning of the electrodes is imposed by the diamond shaped cross section of the vacuum chamber (see Chapter 8 on vacuum systems) and the desire for a low impedance. The electrodes are either mounted on flanges (ALS[7], APS[16]) or fixed on stainless steel blocks by welding (ESRF,[17] Elettra[18]). The replacement of blocks is easy if they are mounted between two vacuum flanges (Elettra). The desired high absolute accuracy (0.1-0.15 mm) of the measured beam positions depends to a large extent on the correct placement and fixation of the set of electrodes with respect to the magnetic centres of nearby quadrupole magnets. The relative motion of the pickup assemblies and the magnets due to thermal effects induced by synchrotron radiation has to be kept small.[18]

The electrical centre of a set of electrodes is not necessarily identical to their mechanical centre because of small variations in the individual electrodes, the vacuum feedthroughs, and in the cables used for connections. In order to have stable conditions these cables have to withstand the bakeout temperatures (see Chapter 8 on vacuum systems) and the radiation levels close to the accelerator (see Chapter 16 on safety). The electrical centres of the blocks are determined either by the wire technique where an antenna placed in the vacuum chamber simulates the beam or from purely external measurements.[19] The sensitivity of the pickup assemblies to changes of the beam position can only be found with the wire technique.[20]

The number of monitors and their locations along the circumference of the ring depends on the horizontal tune, the type of magnet lattice, and conditions set by the users of the synchrotron radiation, for example independent control and measurement of position and angle of the beam in straight sections.

10.3.1.1 Signals from pickup electrodes

In the ideal case, where the bunchlength, σ, is long compared to the diameter, d, of the circular pickup and the output is properly terminated into R, the signal in the time domain is the derivative of the longitudinal Gaussian particle distribution with respect to time. In the frequency domain the spectrum of a single circulating bunch consists of many lines spaced by the revolution frequency, ω_0. The amplitude envelope of the components is again a gaussian distribution folded with the linear increasing frequency response of the pickup electrodes. The rms output voltage for a single bunch current I at ω is given by[8]:

$$V(\omega) = R\,I\,2^{-5/2}\,\frac{d\,\omega}{b\,c}\,e^{-\omega^2\sigma^2/2}$$

where c is the speed of light and b is the distance between the electrode and the beam. If all h RF buckets are populated equally the spectrum contains only the lines at multiples of the RF frequency, ω_{rf}. The harmonic number h is equal to the ratio ω_0 to ω_{rf}. For the same total current, I, these amplitudes of the lines are increased by a factor h and so is the output voltage. Very often only part of the available buckets are filled and a gap is left for ion clearing.[21] The spectrum will then possess a similar structure as above with some additional lines spaced by ω_0 around the multiples of the RF frequency. These are the kind of filling patterns encountered in synchrotrons and storage rings. Therefore, if the beam positions are determined from signal processing in the frequency domain, the best choice is a multiple of the RF frequency.

The sensitivity of the signal to beam displacements is independent of the observation frequency. For two signals, R and L, of identical pickups facing each other, the difference-over-sum signal is given by[8]:

$$\frac{R-L}{R+L} = \frac{4\,\sin(\phi/2)}{\phi}\,\frac{x}{b} + \text{higher order terms}$$

where ϕ is the angular width of the electrodes given by $\sin(\phi/2) = 0.5 \cdot d/b$ and x is the displacement of the beam.

10.3.2 Signal Processing

Three different signal processing methods are used to derive the normalised, current independent beam positions from the raw pickup signals: Difference-over-sum method, conversion of the differences in amplitude to phase shifts (AM/PM-processing), and the log-ratio technique.[22]

10.3.2.1 Difference-over-sum

The difference-over-sum processing of the signals is the approach taken at nearly all synchrotron radiation facilities because it is the simplest and least expensive. There is still variety in the different designs. At Super-ACO, the signals are multiplexed and analysed in the time domain with a single peak detector.[23] The usual approach is the processing in the frequency domain. The left and right signals can be combined with hybrids in order to create the difference and the sum at the observation frequency. An example is the BPM system at Daresbury.[24] Alternatively, the individual pickup signals are demodulated and the individual amplitudes are detected. This is the most common approach and it can be realised either simultaneously with four matched detector channels[7] (ALS). In all other cases the HF signals are multiplexed and analysed successively with one detector.[25] This has some advantages, since building four matched detectors is difficult and on line calibration procedures are needed.[7] The difference-over-sum calculation is done in the computer after the individual signals were digitised. The exception are the BPM systems in the VUV and X-ray rings in Brookhaven where the sum signal is kept constant and the positions proportional to the difference are obtained with analogue techniques.[26]

Only the ALS BPM system can acquire the beam positions on all 96 pickup stations for 1023 consecutive turns in one shot. This is done with separate broadband signal processing channels (peak detectors) and 8 bit flash ADCs. The Brookhaven system, because it is hard wired and the bandwidth is limited, has no single turn option at all. In most of the other systems provisions have been made to measure the path of the injected beam from four shots. In this case the data has to be accumulated pickup after pickup and turn after turn shifting the sample-and-hold trigger by a corresponding number of revolution times. This "fast" orbit measuring mode is very important for low emittance storage rings, because small alignment errors of the magnets lead to large closed orbit distortions or to no closed orbit at all. Therefore, single turn beam position measurements and first turn orbit corrections are necessary (see Chapter 13 on beam stabilizing and steering systems).

If the signals are processed individually, the major drawback is the large dynamic range which is needed in order to achieve the desired μm resolution. The resolution is limited by the granularity of the analogue to digital conversion (typical 12 bit). To overcome this problem the gain of the system is adjusted close to the maximum input level of the AD-converter. Sometimes even switching from amplification to attenuation is necessary in order to increase the dynamic range (ESRF, ELETTRA).

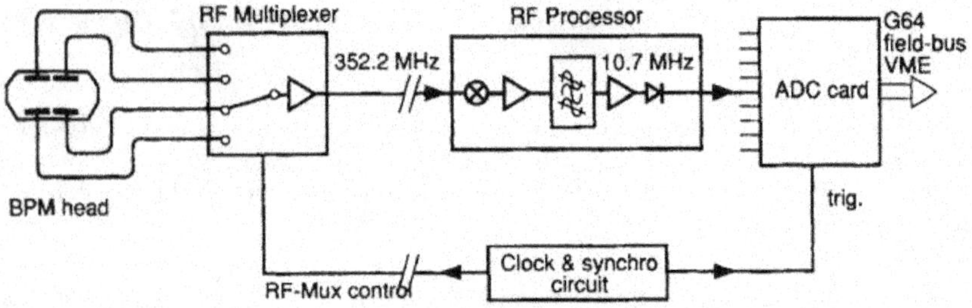

Fig. 10.2: Block diagram of the ESRF BPM system.[5]

Figure 10.2 shows a typical block diagram of a RF multiplexed and tuned receiver detection system. The example is taken from the ESRF where the system is broken up into the front end, the RF processor, and the digitiser plus digital I/O with interface. The front end contains the RF multiplexer and an amplifier for signal conditioning. This unit is located close to the pickup buttons. For easy access the rest of the system is located outside the radiation shielded area. The RF processor performs the conversion of the frequency down to 10.7 MHz, the controlled amplification, the filtering, and the demodulation for stored beam and single turn operation. The analogue signals are either filtered and fed into the AD converter or in the single turn mode, the signals must first be captured by a properly timed sample-and-hold-circuit. The data acquisition is synchronised with the selection of the RF inputs and the normalised beam positions are calculated by the computer from the four individual signals.

10.3.2.2 AM/PM conversion

In this technique the normalised beam positions are obtained from the ratio R/L or Δ/Σ. The AM/PM processing can only be done in the frequency domain because the amplitudes are converted into phase angles. This is accomplished by phase shifting one of the filtered signals by 90 degrees and recombining both channels in phase and 180 degrees out of phase. Thus the phase angle between the two combined signals is proportional to $\tan^{-1}(R/L)$. The phase difference can be detected with a double-balanced mixer and two limiters removing any amplitude dependent effects.[8] The AM/PM circuit is expensive and difficult to implement. The centre frequencies of the filters must be equal to about ±0.1% and phase matching of the cables between pickups and the AM/PM circuit is required. The processing frequency and the dynamic range are restricted by the limiters.

This type of signal processing is foreseen for the APS storage rings and synchrotrons because of its capability to obtain normalised beam positions on each bunch once per revolution.[27] The first stage of this system is a filter-comparator unit to produce the sum and the two difference signals at 352 MHz and a trigger for the timing circuit from the four pickup electrodes. The second stage is the monopulse receiver which uses the AM to PM conversion technique to produce the normalised position and the sum signals. In the third stage these signals are peak detected and digitised with the help of the timing trigger. The results are transferred to the FIFO (first in-first out) memory and can be accessed through the interface by the control system.

10.3.2.3 Log-ratio

The normalised beam position is gained from the ratio of the two amplitudes. The deviation of the ratio from one is proportional to the offset of the beam. Taking the logarithm of the ratio improves the system linearity. In the log-ratio technique the normalised beam position is obtained in a difference amplifier from the logarithms of the two signals.[22,28] This method in its simplest form is only applicable when two pickups are facing each other. The log-ratio technique is not used in any synchrotron radiation facility because it is a fairly new idea and good hybrid logarithmic amplifiers at reasonable price have been available only recently. The APS linac beam position monitor is based on one of those demodulating logarithmic amplifiers but subtraction of the two channels is done digitally.[29] For the application in linacs, transfer lines, and synchrotrons the log-ratio technique has interesting features like normalised real-time response at reasonable costs.

10.3.3 Interfacing of BPMs

The information delivered by any diagnostics monitor is only useful if this information can be processed by the central computer which has access to all other important accelerator components like power supplies, RF master oscillator, and so on.

The control system can access the beam position monitors through an interface. This is easy, if normalised beam positions are available as analogue signals. Easy, because a clear cut can be made between the responsibilities of the instrumentations and the controls group. The definition of the interface is simple: read this signal at a certain speed. The speed varies between ≈1 Hz and the revolution frequency, where timing and synchronisation starts to become a problem. The interface will be more difficult if additional communication links for switching of multiplexers, setting of the gain, and different modes of operation for single- or multi-turn measurements and BPM calibration are desired.

The Brookhaven system is most easily interfaced to the computer. The analogue signals proportional to the normalised beam positions will be digitised at high speed (68040 cpu) and stored in a dual ported memory, so that storage of and asynchronous access to the results by a local computer is possible. The local computer (HP 742rt) is capable to acquire ≈200 closed orbits per second and this information will be used for fast digital global orbit feedback.[30]

In most of the other BPM systems at least the first stage of the interface is an integral part of the monitor. The APS system will offer the control system already digitised beam position and intensity data (see 10.3.2.2). This information is collected by the so-called memory scanner units on VXI platform. The scanner stores the beam history, performs running averages of the positions, and provides high speed digital output for fast global orbit feedback systems.[31] In addition, the unit supports tune and single pass measurements.

The ESRF approach is taken as a more typical example for a BPM system which uses RF multiplexing and a tuned receiver.[17] This system needs a couple of digital inputs for reading the status and outputs for the RF multiplexer, coarse and fine gain adjustment, and the DC multiplexer for the selection of low pass filters and the single turn signals. One AD converter serves 7 units and stores the signals of the 28 button electrodes in the local memory on the same board. Timing is done locally. Thus the selection of the inputs and the AD conversion are synchronised. Sharing the AD converter saves money but slows down or reduces the accuracy of the measurements. The digital I/O and the ADC are interfaced with the low cost G64 standard and they communicate with the central VME crate via a field bus network. All processing of the data, primarily the calculation of the normalised beam positions, is done at this stage because the intelligence is completely centralised in this single crate. Many other systems use distributed intelligence and parallel data processing in order to speed up the data acquisition.

Due to the strong signals picked up by the electrodes at higher beam currents the resolution of the beam position monitor is on the order of a few μm even in the fast closed orbit mode.[18] This extremely valuable information can be used for fast feedback systems only if decentralised computing power (DSP digital signal processing) and a fast direct interface to the power supplies of the corrector magnets are available.[32]

10.3.4 Comparison of BPM Systems

Table 10.1 contains the characteristics of some typical beam position monitor systems. Most of them use parallel processing except the system at Super-ACO where the 64 RF signals are multiplexed to a single peak detector, and the partial parallel interface system of the ESRF. The different approaches for the processing of the signals have been described in the previous sections.

Table 10.1: Comparison of BPM systems.

Light Source (number of BPMs)	**Processing Technique**	**Performance**					**Status**
		closed orbit mode			*single turn mode*		
		resltn. μm	acc. mm	rate Hz	acc. mm	data acquisition speed	
NSLS VUV (24)	parallel processing RF multiplexed tuned receiver analogue Δ/Σ	<20 ≈5 intensity range: ≈25 - 700 mA		2K 300	not available		oper.
Super-ACO (16)	multiplexed peak detector digital Δ/Σ	<20 intensity range: 0.2 - 165 mA (single bunch)	<0.2	0.5	not available		oper.
ALS (96)	parallel processing 4 receivers/BPM digital Δ/Σ	10 dynamic range: 40 dB	<0.15	20	<1 1 orbit display/s	1K turns / shot	design oper.
ELETTRA (96)	parallel processing RF multiplexed tuned receiver digital Δ/Σ	≤10 intensity range: 1-400 mA	≤0.15	2 - 2.4 K	<.25 few orbits/mn	1 turn/4 shots	design oper.
ESRF (224)	partial parallel proc. RF multiplexed tuned receiver digital Δ/Σ	≈1 intensity range: 0.5 - >100 mA	<0.15	0.1	<0.5	1 turn /4 shots	oper.
APS (360)	parallel processing Δ, Σ signal at 352 MHz AM/PM conversion	25 dynamic range: 40 dB	0.2	<0.5 - 2.1 K		16 K turns / 4 shots	design

The BPM systems of the third generation light sources have at least two modes of operation. First, the *closed orbit mode* with acquisition rates ranging from ≈1 to ≈2000 orbits per second. The low speed yields higher precision and is used for the adjustment of the orbit. The higher speeds are foreseen for fast global feedback systems. The second is the *single turn mode*. This mode is necessary for first turn orbit correction.

The resolution and the accuracy of all systems is similar. In the first mode of operation, the resolution is of the order of 10 μm and better, depending on the acquisition speed. The value of ≈1 μm for the ESRF is obtained by averaging many measurements in the central computer. At higher beam currents and longer integration intervals the resolution is limited by the use of only 12 bit-AD converters. The absolute accuracy of typically less than 0.15 mm is similar for all systems because a large part originates from positioning errors of the sensors with respect to the quadrupole magnets. In the single turn mode resolution and accuracy are of the order of 0.5 mm which is sufficient for first turn orbit correction.

The reproducibility of the beam position measurement as other parameters like intensity, time, filling pattern, and mode of operation are changing is difficult to quantify. If reproducibility is approaching the resolution achieved at the ESRF, the BPM systems can be used for orbit feedback.

In many cases the BPM systems implemented in the synchrotrons are identical to systems installed in the storage rings. In synchrotrons the timing of the closed orbit acquisition process is important because the orbit might change during the ramping process. This adds only little complication to the overall systems.[33] All signal processing techniques can be modified in order to be capable to measure the beam positions from single shots. Since nearly all methods use processing in the frequency domain a small bandwidth ringing filter can be included and serves as a stretcher for the short pulses. With proper synchronisation of the data taking process the beam positions can be measured in linacs and transfer lines.[7]

10.3.5 Beam Positions Measured with Fluorescent Screens

The fluorescent screen is an optical monitor. It converts the spatial distribution of the beam flux density by luminescence into visible radiation. The screen can be viewed through a zoom lens with a TV camera.[34] Nowadays charge-coupled devices (CCD) are widely used as detectors. This choice leads to the selection of chromium doped alumina ceramics (Al_2O_3) because the emission of the screen and the sensitivity of the camera both peak at around 700 nm. This leads to a high sensitivity of 10^6 charges/mm^2. In addition[35] the screen material is ultra high vacuum compatible and can be baked to at least 150°C. The spatial resolution of the screens[36] is ≈30 μm. Chromium doped screens are commercially available as Chromox by Morgan Matroc Ltd. in Great Britain. With high intensity, low energy beams charge accumulation is a problem and for discharging metal coatings for example as grid patterns have been evaporated onto the screen.[37]

The fluorescent screens are usually moved into the beam by pressure activators and the beam is observed sideways. If the screen is inserted perpendicular to the particle beam in order to by-pass image distortions, the observation can be done through a mirror.

The CCD camera is a very convenient detector, not only for the observation of fluorescent screens but also as a sensor for images obtained with synchrotron radiation. It has a high geometrical resolution (≈10x10 μm^2 per pixel) and a large dynamic range. Despite the large dynamic range (up to 14 bits if the CCD chip is cooled to 10°C) the camera is forgiving in terms of over exposures.[38] The CCD camera is insensitive to magnetic fields and can be shielded easily against radiation because of its small size. Due to the large commercial market, CCD cameras are cheap and frame grabber cards for many computer systems are available. These cards and the camera can be triggered externally and are able to digitise, to store, and to display a particular frame in real time. Since the decay time of the fluorescent screen is of the order of milliseconds this is important for injectors with low repetition rate in order to freeze the image on a TV monitor until the next shot arrives. The monitor is located in the control room and the various cameras are connected to it through a multiplexer system. In the most advanced systems not only the signal switching and the selection and activation of the fluorescent screens would be done by the central computer but also the image processing and the data analysis. Thus the information would be available for automatic orbit correction and archiving.

10.4 Measurement of the Distribution Functions

In storage rings, the rms width of the transverse distribution of the particles in the beam is related to the emittances, ε, and the energy spread: $\sigma^2 = \varepsilon \cdot \beta + (D \cdot \sigma_p)^2$, where β is the beta function, D the dispersion function, and σ_p the relative rms width of the energy distribution. The formula is valid for both planes. The brilliance of the source is one of the target goals and the transverse beam profiles enter directly into this key figure. The measurement of the profiles is necessary for the estimation of the photon flux density. Usually the profiles are only available on a few locations, for example where the dispersion is large or small in order to distinguish between transverse and longitudinal contributions (emittance and energy spread).

The linear relation between energy spread and bunch length is broken by current dependent effects (instability). Therefore, the measurement of both parameters is important especially if high resolution, time resolved spectroscopy, the peak current for the operation of FELs, or the spectral resolution of higher harmonics from IDs is the target.

Transverse and longitudinal particle distributions can be measured using synchrotron radiation. The limitations of this approach are explained in the next two sections and examples for optical source imaging systems are given. Finally some new ideas to overcome the limitations are presented briefly. This section will be restricted to the non-destructive diagnostics of distributions in storage rings and synchrotrons. The very interesting features of optical transition radiation for time resolved measurements for example of linac beam pulses in connection with free electron lasers can be found in the literature.[39]

10.4.1 Imaging with Synchrotron Radiation

There are two major contributions to the resolution of any optical system imaging the beam with synchrotron radiation: First, diffraction due to the finite acceptance of the system or the limited opening angle of the radiation, θ_λ, is relevant for horizontal and vertical plane. If the wavelength, λ, is large compared to the critical wavelength, λ_{crit}, (see Chapter 1) the natural half opening angle is given by:

$$\theta_\lambda = \left(\frac{3\ \lambda}{4\pi\ \rho}\right)^{1/3}$$

where ρ is the bending radius. Fraunhofer diffraction limits the resolution to about $F \cdot \lambda/\theta$ with $F \approx 0.21$ for a slit and $F \approx 0.26$ for a circular aperture. If the half acceptance angle θ is larger than the radiation angle, F will even be smaller.

The second effect has its origin in the observation of an extended segment of the trajectory. The length of this segment is given by $\rho\theta_h$ with θ_h the effective horizontal acceptance angle given by the half opening angle of the radiation plus the full horizontal acceptance angle. In the horizontal plane this contributes primarily because the trajectory is bend. For the vertical dimension the smear of the image due to the limited depth of focus is more relevant.[40] Rules of thumb formulas for the various contributions are listed below:

Diffraction

$$\sigma_{diff} \approx 0.24 \frac{\lambda}{\theta}$$

Observation of extended segment of trajectory

$$\sigma^{hor} \approx 0.11 \; \rho \; \theta_h^{\,2} \qquad \text{bended path}$$

$$\sigma^{ver} \approx 0.43 \; \rho \; \theta \; \theta_h \qquad \text{depth of focus}$$

As above, θ is given either by the opening angle of the radiation or by the vertical aperture limitation. Very often the beam size of a low emittance ring is as small as the smear of the image. In order to extract the correct size of the beam itself a more careful calculation of the contributions has to be made.[41]

If we choose to image the beam with visible light at around 500 nm and use apertures limiting the acceptance in both planes to the natural opening angle of the radiation at this wavelength the various contributions for typical light sources are listed in Table 10.2. This choice of the apertures is close to the optimum. The total smear, σ_{tot}, is the quadratic combination of the individual contributions. In order to give a feeling for the achievable resolution for the emittance, $\varepsilon_{min}=\sigma_{tot}^{\;2}/\beta_y$, the columns with the beta function, β_y, in the dipoles and the natural emittance, ε_0, were added. The horizontal beta functions are in general ten times smaller leading to a corresponding reduction in the resolution.

Table 10.2: Resolution of imaging systems at 500 nm.

Light Source	ρ m	θ_λ mrad	σ_{diff} μm	σ_{bnd} μm	σ_{dof} μm	ε_0 nm · rad	β_y m	ε_{min} nm · rad
S-ACO	1.7	4.1	28	10	18	39	8	.15
VUV	1.9	4.0	30	10	19	138	15	.08
BESSY1	1.8	4.1	29	10	19	50	15	.08
ALS	4.0	3.1	38	13	25	4.1	23	.09
Elettra	5.5	2.8	42	14	27	7.2	11	.23
BESSY2	4.4	3.0	39	13	25	6.0	22	.10
ESRF	23.4	1.7	68	23	44	7.0	26	.25
APS	39.0	1.5	81	28	52	8.2	18	.52
SPring-8	39.3	1.5	81	28	52	7.0	25	.37

The resolution for horizontal emittance measurements in the visible region of the spectrum is still sufficient for all storage rings. In the vertical plane, however, the resolution is at the limit of what has been achieved in terms of a small horizontal to vertical coupling. The information on the vertical emittance in the high energy X-ray rings is at best qualitative. The only solution is to go to a shorter wavelength. In the following section the limitation for imaging with a pin hole at $\lambda \approx \lambda_{crit}$ will be discussed.

The resolution of 1:1 pin hole imaging systems working at around the critical wavelength will be limited by diffraction, σ_{diff}, and the blur of the image due to the finite size of the pinhole, σ_{ph}. Both effects scale differently with the diameter of the hole, d. The optimum size can be found by making both contributions equal: $\sigma_{diff} \approx 0.5 \cdot \lambda \cdot L/d$ and $\sigma_{ph} \approx d/2.36$. The distance between source and pin hole is L.

In the X-ray region of the spectrum the natural opening angle of the radiation emitted by individual particles might be smaller than the divergence of the ensemble of particles. Under these circumstances not all particles in the distribution will contribute to the image. In this case, however, the opening angle of the emitted radiation is dominated by the divergence of the beam. Thus without any optical components the emittance can be obtained from a measurement of this opening angle if the Twiss parameters are known.[42] On the other hand, in order to get the source size with a pin hole, either the hole has to be moved through the emitted radiation cone or structures with many holes have been used or will be used.[44] If we are only interested in the smaller vertical beam size a single pin hole might be sufficient.

10.4.2 Bunch Length Measurements

The measurement of the longitudinal distribution of particles can best be performed with visible light. The observation of an extended segment of the beam does not cause any problems because the particles move at the speed of light. Photons emitted at the beginning of the segment will arrive at the detector at the same time as those emitted at the end. The bunch length is identical to the duration of the light pulse.[41] Optical single shot picosecond diagnostics is best performed with a streak camera.[12] In this device photo electrons created during the duration of the light pulse are accelerated away from the photocathode. The time resolution is achieved by a fast transverse deflection of the accelerated photoelectrons. They impinge finally on a phosphorous screen. The temporal structure is transferred to a spatial distribution. The screen is observed by an intensified CCD camera. The time resolution is less than 10 ps for single shot measurements.

10.4.3 Examples for Diagnostic Beamlines

In spite of their poor spatial resolution, imaging systems in the visible region of the spectrum are used frequently because they are simple, easy to align, and many different types of detectors are available. At BESSY I imaging with mirrors is preferred,[45] whereas at the NSLS VUV ring focussing is done with a spherical lens and chromatic effects are suppressed with a filter.[46] The alignment is easy, because the light is visible even at extremely low beam currents where an access to the radiation protected area is possible without risk. As detectors, CCD cameras or linear diode arrays can be employed. A frame grabber and commercially available image processing software can be used to extract the beam sizes.

Two examples for imaging at much shorter wavelength will be described briefly. At BESSY we are planning to install a system using Bragg-Fresnel-lenses.[47] The design of the optical system is shown in Fig. 10.3. The Bragg-Fresnel-lens (BFL) in combination with a multilayer (ML) acts as the focussing element for a wavelength around 6 nm. This short wavelength leads to a diffraction limited resolution of a few μm which will be difficult to preserve in the optical components. The image will be observed with a CCD camera.

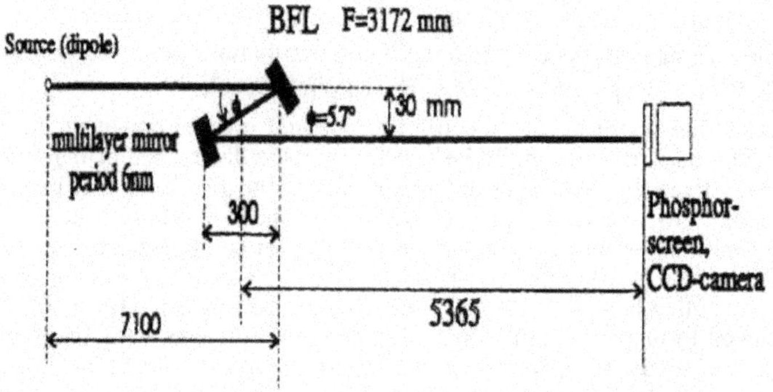

Fig. 10.3: Source imaging with Bragg-Fresnel-lens.

The diagnostics beamline at the ALS[48] consists of an imaging system for a wavelength of 6 nm and a beam port for white light and time resolved measurements. The imaging system employs two crossed spherical mirrors in a Kirkpatrick-Baez configuration. This eliminates astigmatism and the 1:1 imaging gets rid of coma. The resolution and hence the quality of the image depends strongly on thermal loading and residual aberrations. The effects smearing the image mentioned above are on the order of a few μm if the system accepts 1 mrad in the vertical and 2 mrad in the horizontal plane. The expected rms beam sizes are around 50 μm. The intensity distribution is observed with a high resolution CCD camera and a BGO scintillator for the conversion of 6 nm photons to visible light.

10.4.4 Alternatives for Beam Size Measurements

The beam size is one of the primary parameters which must be monitored permanently. If a straight section is available for this purpose an undulator (see Chapter 14 on wiggler and undulator insertion devices) with N periods could be installed and used for diagnostics.[49] The angular divergence of monochromatic radiation emitted on the first harmonic goes down as $N^{-1/2}$. If N is large enough the divergence of the particle beam dominates the emission process and the technique already mentioned can be applied for emittance measurements. In the distance L the distribution of mono chromatized undulator radiation will have a rms width given by:

$$\sigma^2 = \epsilon(\beta - 2L\alpha + L^2\gamma) + L^2 \frac{3}{4\pi\gamma^2 N}$$

where α, β, and γ are the Twiss parameters and the last term is the contribution from the divergence of the radiation.

A technique for the measurement of extremely small beam sizes has been proposed recently.[50] It is based on the Compton scattering with linearly polarised standing electromagnetic waves inside the cavity of a laser, perpendicular to the electron beam orbit. The photon density is spatially modulated at twice the laser wavelength. If the beam profile has a similar size, moving the standing wave pattern would modulate the gamma ray intensity from Compton scattering. The beam size could be inferred from the modulation depth. This technique is non-destructive but only useful for beam diameters of less than 10 μm.

10.5 High Precision Determination of the Energy

The traditional way to determine the energy of particles circulating in a storage ring is the analysis of the measured bending fields. The line integral corresponds to exactly 2π radians of bend. Since usually the particles are moving on a corrected horizontal orbit which is not necessarily the on-momentum orbit, the energy has to be corrected for both effects.[51] The typical total relative error is of the order 10^{-3}.

The resonant spin depolarisation is a very sensitive method to determine the absolute energy in electron/positron storage rings.[52] The method uses the natural polarisation of the electron spins, and a resonant depolarisation of the spins with a frequency which depends on the energy of the particles.

10.5.1 Polarisation and Depolarisation of the Spins

The particles in the beam can become polarised through the emission of synchrotron radiation. The so called Sokolov-Ternov radiative spin-flip effect[53] is rather small and leads at best to a polarisation of 92.4% with a build-up time given by[54] τ_{pol} = $1.63 \cdot R \cdot \rho^2/E^5$ (mn) with R the average radius, ρ the bending radius, both in metres, and E the energy in GeV. In real machines, the spins precess not only in the vertical fields bending the particles in the horizontal plane, but also in all kinds of linear and non-linear fields created by quadrupoles and sextupoles. As a consequence depolarising mechanisms originate, which modify the natural polarisation time and the maximum level of polarisation.[55] Since storage rings for synchrotron light sources are usually small (small average radius R) and in addition to the insertion devices the radiation from bending magnets is a target too (small ρ) the natural polarisation time is comparable to the actual lifetime of the beam. The polarisation time is around 3 h for the VUV light sources of the second and 0.4 h for the X-ray sources of the third generation.

As already mentioned, the spins will precess around the vertical bending fields. After one revolution the precession angle is $2\pi \cdot (a \cdot \gamma)$. The gyromagnetic anomaly of the electron, a, is known to very high precision[56]: a=1159652188 10^{-12}. The spin tune, ν_s, is defined in analogy with the betatron or synchrotron tunes as the number of spin-precessions per revolution: $\nu_s = a \cdot \gamma$. The spin tune can be measured if the beam is polarised, a suitable polarimeter is at hand, and a radial RF magnetic field can be created which resonantly depolarises the spins at a frequency $f_{dr} = |a \cdot \gamma - n| \cdot f_{rev}$ where n is any integer and f_{rev} the revolution frequency. The radial magnetic field can be created most easily with a pair of striplines above and below the midplane approximately 0.5-1 m long, powered with 100 Watts terminated in 50 Ohms.

10.5.2 Monitoring the Depolarisation

The polarimeter as a monitor for the polarisation of the spins is necessary in order to verify the resonant depolarisation. There are a number of spin-dependent phenomena[57] which can be used as a polarimeter and which come effectively for free. As suggested by Baier[57] we propose to use the spin-dependent electron-electron scattering within bunches (Touschek scattering) as a polarimeter, but a careful measurement of the lifetime[58] instead of the more complicated detection of scattered particles as has been done in the past.[59] If the Touschek effect would be the only particle loss mechanism and if the beam originally was completely polarised than the reduction of the lifetime after the resonant depolarisation[60] would be of the order of 10%. Generally the situation is less favourable than that. There are two reasons. Full polarisation is difficult to achieve and the overall lifetime might be also limited by other effects like inelastic scattering of electrons on residual gas molecules. However, at least in the case of single bunch mode operation all conditions for the proposed high precision energy measurement are fulfilled: polarisation time comparable to or shorter than the lifetime, and the lifetime dominated by the Touschek effect.

If, for example, the lifetime reduction due to the depolarisation will be only 1%, the lifetimes have to be known to be much better than that. Following what was said in Sec. 10.2 about the current monitors and their impact on the determination of lifetimes we can estimate a time interval of about 1 minute for the desired 0.1% accuracy.

10.5.3 An Example

The typical experiment will look like this. In step 1 a mode of operation (filling pattern, coupling) has to be chosen which has a comfortable lifetime primarily dominated by Touschek scattering. Step 2 – one has to wait for about 2 natural polarisation times for the spin polarisation to build up. In step 3 the lifetime is measured as a function of the frequency of the depolarising field. Once the beam was resonantly depolarised the lifetime will be reduced by a few percent. Since the polarisation will build up only in times much longer than the measuring interval we will observe a step function.

A typical result from BESSY1 is shown in Fig. 10.4. The lifetime is displayed as a function of time as 21 attempts were made to depolarise the beam with different frequencies. The time when the excitation was turned on is indicated by the bars. The beam was depolarised at a frequency of 3884±1.02 KHz. The uncertainty has its origin in the excitation frequency which is swept over this region. After the beam is depolarised the lifetime is reduced by 5-6%, very close to the expectation.

The high precision determination of the energy is not only of interest for special users but can also be applied for the measurement of the momentum compaction factor, α, which is otherwise very difficult to get. Two energy measurements at different RF frequencies are needed and α is obtained from $\Delta p/p = -1/\alpha \cdot \Delta rf/rf$.

10.6 Beam Loss Monitor

The function of a beam loss monitor (BLM)[61] is to indicate when and where the beam or a fraction of it was lost. This helps in spotting the origin of the loss and in tuning the machines in order to achieve high intensities and long lifetimes. With the BLM system restrictions of the aperture and hot spots can be located and after the problems are cured the level of radiation in and around the accelerator will be minimised.

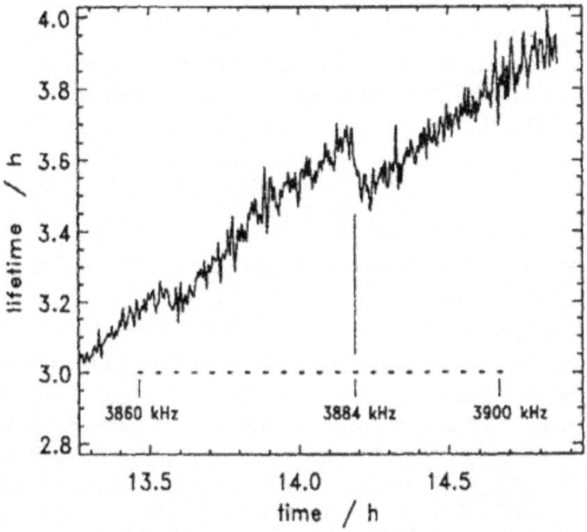

Fig. 10.4: Experimental result of a high precision energy determination at BESSY I with the lifetime as polarimeter.

The principle of the loss monitor is the observation of the showers caused by lost particles interacting with the vacuum enclosure or materials next to it. Therefore, the detectors are mounted outside but close to the vacuum chamber. As a sensor nearly any radiation detector could be employed. The APS is planning to use long ionisation chambers made out of gas-filled coaxial cables installed along the entire extent of beamlines and accelerators.[62] At BESSY we are investigating the use of PIN photodiodes. This is similar to the approach taken at HERA.[61] Obviously the detectors could be mounted in such a way as to be more sensitive to Bremsstrahlung.

In electron storage rings there is always a chance for ion trapping and over the years many observations did indicate even the trapping of clusters of material (usually called trapping of dust particles).[63] A similar situation occurred in the HERA electron storage ring. The current was limited to rather small values and the lifetime was extremely short. The problem was solved using the BLM system. In one of the straight sections a high count rate, inversely proportional to the lifetime was measured. The problem was fixed by replacing a part of the vacuum chamber in this section.[61] In the mean time even the origin of this problem could be found with the support of the BLM system: Under certain operation conditions the ion getter pumps tend to release large clusters of titanium into the vacuum system which were trapped by the beam. A reduction of the high voltage supplied to the pumps cured this problem.[64]

10.7 Measurement of Other Accelerator Parameters

In addition to the parameters already covered in the previous sections which are of direct concern for the synchrotron radiation users' there are other measurements more relevant from an accelerator physics point of view. The results usually will still have an impact for the users community either through the good overall performance or the improvement of the accelerator facility which are based on those measurements. The only additional primary parameters which have to be determined are the tunes of the storage ring and synchrotron. The measurement procedures are explained in the first part of this section.

Important information is related to the tune and the other primary parameters. It can be extracted if these parameters are measured as functions of for example the strength of the magnet elements, the cavity voltage or the frequency of the cavity voltage. This will be explained in more details in what follows. We end with a short description of the diagnostics of instabilities.

10.7.1 Tune Measurements

The transverse and longitudinal tunes, $\nu_{x,y,z}$, belong to the most important accelerator parameters. As a global parameter they give an idea of the correct setting of all the magnet power supplies and the cavity voltage. A poor lifetime of the stored beam or beam losses during the ramping process in the synchrotron will be observed if, for example, the tunes are close to a resonance. Therefore the tune meter must be available even for non-specialists all the time, easy to use, and the result should be accessible by the control system as fast as possible. The last point is mandatory for tune measurements as a function of other parameters in a short time. The process should be completely controlled and analysed by the central computer.

The general procedure for the tune measurement is a dynamic beam response experiment.[65] Either the beam is shock excited, the resulting damped, harmonic oscillation is recorded, and the resonance frequency is obtained from the Fourier transformation of the data or, more often, a harmonic excitation with swept frequency is applied to the beam and simultaneously the response of the beam is observed. This type of experiment will be described in more detail.

A spectrum analyser with tracking generator is the central piece of equipment. The output frequency of the tracking generator is identical to the frequency instantaneously analysed. In the transverse planes striplines are used to excite the beam. Longitudinal excitation is usually achieved by phase modulation of the cavity voltage. The excitation produces an amplitude modulation or a phase modulation of the signals picked up by button electrodes or striplines. Since the excitation frequencies are smaller than the revolution frequency, especially for the longitudinal phase modulation, these HF signals have to be mixed down before they can be fed into the spectrum analyser. In the transverse planes beam motions can be excited at higher frequencies where the sensitivity of the electrodes is large enough to be used directly. The complete measurement takes about 10 seconds. This includes switching of inputs, data acquisition, communication to the spectrum analyser, and analysis of the data. For some important applications like tune feedback during ramping or for the compensation of focussing effects from scanning insertion devices this speed is too small. In this case continuous tune meters based on phase locked loop systems (PLL) have to be employed.[66] In a PLL system the frequency of the excitation is regulated so that the phase shift between excitation and beam motion is locked to 90 degrees.

In synchrotrons tracking errors between bending and focussing and defocussing fields create intolerable tune shifts. If these shifts are too large the beam might be lost completely or partially. Tune measurements during the ramp and special current ramps for additional correcting quadrupoles can cure this problem. The high repetition rates, in most cases larger than one ramp per second, and the large tune shifts make the measurements difficult. Also here the common experimental approach is a dynamic beam response measurement. The beam is excited with bandwidth limited noise and the excitation amplitude varies proportional to the energy of the particles. Simultaneously the response of the beam is observed. This can be done by recording the beam positions turn by turn and performing Fourier transformations of samples in order to extract the tunes.[67] Alternatively, the response of the beam can be analysed with a commercially available digital signal processing system (Tektronix 3052) which is capable of displaying in real time, during the ramp of the synchrotron, the spectra of the beam motion.[68]

All those measurements only give the fractional part of the tune and it is not clear whether the tune is above or below the half integer. This has to be established from the variation of the tunes as quadrupoles are changed. The integer part can be obtained from the analysis of the closed orbit distortion resulting from a variation of corrector magnet settings.

10.7.2 Tune Related Parameters

Many important accelerator parameters can be extracted from tune measurements as other parameters are varied. Tables 10.3a and 10.3b show the additional information obtained from such experiments. I will make only a few comments on some of the measurements and give references in the other cases.

1) The variation of the RF frequency results in a change of the energy of the particles:

$$\frac{\Delta E}{E} = -\frac{1}{\alpha_0}\frac{\Delta \mathrm{rf}}{\mathrm{rf}} - \frac{\alpha_1}{\alpha_0^3}\left(\frac{\Delta \mathrm{rf}}{\mathrm{rf}}\right)^2 + \ldots$$

Usually only the first term is considered, but for some of the following experiments and for quasi isochronous modes of storage ring operation ($\alpha_0 \approx 0$) it is important to include higher order terms of the momentum compaction factor: $\alpha = \alpha_0 + \alpha_1 \cdot \Delta E/E + \ldots$ and similarly the dispersion: $D = D_0 + D_1 \cdot \Delta E/E + \ldots$ In addition to the energy variation the particles will move on a trajectory given by: $\Delta x = D \cdot \Delta E/E$. This is quite different from an energy variation caused by changing the dipole field, where the orbit remains fixed. The observation of the transverse tunes as a function of the energy yields information on the chromatic behaviour and the actual (RF) and natural (dipole variation) chromaticities.[69]

2) The tune variation, $\Delta\nu$, created by a small change of the quadrupole strength, ΔK, is given by[70]: $\Delta\nu \approx 0.080 \cdot \langle\beta\rangle \cdot \Delta K \cdot L$ where L is either the effective length, l, of the individual magnet or $n \cdot l$ with n the number of magnets in the family. The measurements are used to determine the average value of the beta function, $\langle\beta\rangle$.

Another important application of tune measurements as quadrupoles are changed is the investigation of transverse coupling. Close to the coupling resonance, $\nu_x - \nu_y \approx$ integer, the transverse motions are coupled and the strength of the coupling leads to a minimal distance between the frequencies of the two modes of oscillation.[71]

Table 10.3a: Measurement of parameters related to the transverse tunes.

	Varied Parameter	Additional Information
1	RF frequency	tunes as function of energy: chromatic behaviour, chromaticity, higher order chromaticity
	dipole field	in separated function machines as above
2	quadrupole family	average beta functions in the magnets coupling (close to the coupling resonance)
	individual quadrupole	average beta function in individual magnet
3	sextupole family	average horizontal closed orbit distortion in all magnets
	individual sextupole	horizontal orbit offset in individual magnet[72]
4	magnetic gap of ID	focussing properties of ID[73]
5	intensity	transverse impedance of vacuum chamber[74] ion related focussing effects[75]

Table 10.3b: Measurement of parameters related to the longitudinal tune.

	Varied Parameter	Additional Information
6	cavity voltage	momentum compaction factor
7	RF frequency	higher order contribution to the momentum compaction factor
8	intensity	longitudinal impedance of vacuum chamber[74]

10.7.3 Beam Position Related Parameters

Like the linear tunes the beam position is a central parameter because important additional information can be extracted if the positions are measured as a function of other parameters. This can be done by static measurements: the closed orbit is recorded while another parameter is changed slowly. Examples are given in Table 10.4. These measurements can be done pretty fast with state-of-the-art BPM systems, because the closed orbit can be acquired by the control system in about one second. Thus, for example, the measurement of the response matrix becomes feasible within a short time.

1) The horizontal dispersion, and especially its suppression or its correct value in the straight sections, has a strong impact on the emittance and the beam size at the location of the insertion devices (ID). The static measurement of the dispersion is straight forward: the energy is changed (RF frequency variations) and the difference orbit is recorded. If higher order contributions to the dispersion are relevant[76] the measurements have sufficient accuracy not only to extract linear but also the quadratic terms:

$$\Delta x = -\frac{D_0}{\alpha_0}\frac{\Delta rf}{rf} - \left(D_1 + D_0\frac{\alpha_1}{\alpha_0}\right)\frac{1}{\alpha_0^2}\left(\frac{\Delta rf}{rf}\right)^2 + \ldots$$

The quadratic term becomes relevant in the case that the lattice contains harmonic sextupoles. They are placed at locations where $D_0=0$ but there will be chromatic effects due to $D_1 \neq 0$.

The actual values of the vertical dispersion of a storage ring can not be predicted at all, because the distribution of lattice errors like vertical bends, rotated quadrupoles, and vertical closed orbit offsets in sextupole magnets are unknown. More important is the experimental determination of the vertical dispersion. With an accuracy of a closed orbit measurement of 1 μm and a tolerable energy variation of $\pm 1\%$ the error of the resulting dispersion is of the order of 0.1 mm. Values for the spurious vertical dispersion go up to a few cm. In case only some elements contribute to the dispersion these measurements could be used to pinpoint their location. If the transverse coupling approaches the percent level the contributions to the vertical beam profile from the natural energy spread together with a vertical dispersion of 1 cm are comparable in size.

Table 10.4: Measurement of parameters related to the beam position.

	Varied Parameter	**Additional Information**
1	RF frequency	horizontal and vertical dispersion function
2	corrector magnets	response matrix, closed orbit correction algorithms, modelling-Twiss parameters, verification of linear model, calibration of magnets, BPMs coupling-cross correlation hor./ver.
3	(individual) quadrupole	beam position w.r.t. magnet centres absolute zero calibration for BPMs

2) The elements of the response matrix represent the change of the closed orbit at BPM location i as the corrector magnet j is varied. The measured response matrix is extremely useful because even if a satisfactory linear model for the storage ring is missing, the result can be used immediately for the correction of the closed orbit. All correction algorithms are based either directly on the response matrix or the necessary information can be extracted from it.

In addition, the matrix is useful for finding a linear model for the storage ring[77] and its analysis yields the calibration of the strength of corrector magnets and the sensitivity of BPMs. Large deviations of these calibration constants give hints for faulty equipment.

3) One of the central problems in any beam position monitor system is the absolute zero reading because of mechanical or electrical offsets. These offsets can be determined by centering the beam in the quadrupole magnets. In an ideal lattice, the trajectory of the beam would lay in the centres of all magnetic elements. In reality the magnets are displaced by inevitable alignment errors. If the positioning errors are of the order of a tenth of a mm it still makes sense to define the centres of the magnets as the zero position. The determination of the individual offsets is possible by a local quadrupole field variation. If the beam is on axis in this quadrupole the closed orbit will not change. The beam can be steered until this condition is fulfilled.[78] In this context, static orbit measurements can be extended to the measurement of slowly changing orbits: The local field is modulated with a fixed frequency and the motion at this frequency is observed with very high sensitivity by a spectrum analyser or in lock-in technique.[79] This detection technique could be applied to all BPM stations simultaneously. From the observed pattern of the oscillation amplitudes the offset in the individual element can be found with higher accuracy. There are usually many more BPMs than magnets in a family. If the main power supply is modulated, the individual local offsets can be determined from the observed pattern of amplitudes.

There is a second approach using dynamic instead of static orbit measurements. In this case the beam positions have to be recorded turn by turn. This approach is only applicable for some of the BPM systems mentioned in Sec. 10.4. A typical experiment would be the measurement of the phase advance. The beam is excited on resonance and the oscillations are recorded simultaneously on all BPMs. The beta functions and the phases are obtained from the analysis of this motion.[80] Especially the advances of the phase are very accurate because the individual BPM calibrations are not relevant.

Dynamic measurements can be applied to investigate more exotic aspects of the accelerator[69]: Head-tail damping, dependence of head-tail damping on intensity[74] in order to estimate the transverse impedance of the vacuum chamber (see Chapter 12 on instabilities), etc. As for the tune measurements, only one fast monitor for the beam position is required.

10.7.4 Lifetime and Beam Size Related Parameters

Similar to the tune or the beam position the beam sizes and the lifetime are used to extract additional information, though to a much lesser extent. The reason might be that the determination of the latter parameters is usually more time consuming, which need not be the case. The additional information which can be obtained is presented in Tables 10.5a and 10.5b. It is obvious that the insight gained is worth the efforts. Only some of the examples given in the table will be discussed in detail.

4) The scraper is a variable mechanical aperture restriction[83] which is a valuable diagnostic tool. The most direct application is probably the simulation of a small vacuum gap of an insertion device (ID). The observed reduction in lifetime shows whether smaller gaps are tolerable and IDs with shorter magnet periods can be installed. This is an optimistic estimate because only the physical aperture can be simulated that easily.

If the lifetime is observed as a function of the position of the scraper we can find the size of the dynamical or physical aperture, whatever is smaller. The aperture is given by the position where the lifetime just starts to drop.[84] For the scraper closer to the beam we find the expected relation $\tau \propto 1/A^2$ characteristic for elastic Coulomb scattering of electrons on the rest gas molecules,[85] where A is the distance between the centre of the beam and the scraper. The lifetime will be reduced dramatically if the scraper is moved even closer to the beam, into the tails of the Gaussian distribution of the particles. This happens for $A<6\cdot\sigma$ and has been used[86] to measure the beam profile σ. If the experiments are performed as a function of the beam current the pressure dependence of Coulomb scattering and ion trapping effects can be observed. There even is a chance to distinguish between the different lifetime limiting processes.[87]

5) The effective energy acceptance is primarily important for the Touschek effect

since inelastic rest gas scattering only scales logarithmically with this quantity.[88] It can be determined from lifetime measurements as a function of the cavity voltage. In the ideal case, without bunch lengthening and energy widening instabilities, the Touschek lifetime increases linearly with the cavity voltage. The voltage where the lifetime starts to level off is identical to the effective energy acceptance limited either longitudinally or transversely. The relation between energy acceptance and voltage can be found in Chapter 4.

Table 10.5a: Measurement of parameters related to lifetime.

	Varied Parameter	**Additional Information**
1	intensity, fill pattern	lifetime limiting processes
2	quadrupoles, tunes	lifetime limiting resonances
3	magnetic gap of ID	resonances driven by ID[81]
4	scraper position	contributions to the lifetime, dynamical aperture, particles in the halo of the beam, beam profile
5	cavity voltage	longitudinal acceptance, Touschek effect
6	transverse beam size	Touschek contribution
7	bunch length	Touschek contribution
8	local bumps	physical aperture[69]

Table 10.5b: Measurement of parameters related to beam size.

	Varied Parameter	**Additional Information**
9	intensity, fill pattern	longitudinal and transverse instabilities, longitudinal impedance of vacuum chamber[74], ion trapping[75]
10	quadrupoles, tunes	non-linear resonances, coupling
11	time resolved, dynamic measurements	damping times[82]

6), 7) Since the Touschek loss rate depends strongly on the density of particles, the lifetime increases if the vertical beam size is enlarged. This can be done by powering skew quadrupole magnets or with tunes close to the coupling resonance (as done, for example, at Super-ACO[89]) or by a transverse noise excitation (BESSY I). If the vertical blow up is too large, the lifetime will asymptotically approach the limits given by other loss mechanisms. This again allows the discrimination of the different contributions to the lifetime. The increased vertical beam size will reduce the brilliance of the source. In this respect it is better to use the bunchlength to reduce the density. The bunch length can be controlled by a higher harmonic cavity system which has to be phased properly to the main RF system. This technique is applied routinely at the VUV ring in Brookhaven in order to improve the overall lifetime of the beam.[26]

9) The transverse profiles of the beam are sensitive to all kinds of transverse and longitudinal instabilities and to the effects created by ions trapped in the potential of the

beam of electrons. Recording the profiles as a function of the beam current leads to the determination of threshold currents for instabilities and if the fill pattern is modified dramatic changes can occur because ion trapping or instabilities might disappear.

In some cases a short bunchlength plus high intensity are important performance parameters. In this case the bunchlength has to be measured. If the longitudinal impedance is too high bunch lengthening and energy widening will occur spoiling the desired performance.

10) If the beam profiles are observed while the tunes of the storage ring are changed non-linear coupling resonances will show up as an increase in the vertical beam size. The corresponding working points must be avoided in order to produce high brilliant photon beams. The behaviour of the moments of the distribution functions ($<x^2>$, $<y^2>$, and $<xy>$) close to the linear coupling resonance can be used to determine the real part of the complex coupling coefficient.[90]

Table 10.6: Observation of Instabilities.

Instability	Signals and other Observations
Head-Tail (single bunch)	above a clear threshold current particles from individual bunches are lost – at the threshold the smooth fill pattern of the ring will become ragged – threshold current is very sensitive to chromaticity and bunch length (cavity voltage) *cured by slightly positive chromaticities*
Bunch Lengthening, Energy widening (single bunch)	produced by potential well distortion, bunch lengthening from longitudinal instabilities is always accompanied by energy widening - horizontal beam profile blows up ($D \neq 0$) – threshold current but no limit – above threshold strong higher order synchrotron sidebands on either side of revolution frequency harmonics *no cure available*
Transverse Mode coupling (single bunch)	vertical instability leading to beam loss above threshold – below threshold transverse modes approach each other (frequency domain) or beating of the centre of mass motion as the intensity is increased (time domain) *cured by high positive vertical chromaticity*[91]
Coupled Bunch	transverse and longitudinal instability – increased beam sizes, characteristic additional spectral lines at: $\omega/\omega_0 = p \cdot h \pm (n + m \cdot \nu)$, with: $p = 0, 1, 2, \ldots$ integer, h= harmonic number, $n \cdot 2\pi/h$ the phase slip between the motion of neighbouring bunches, n is the mode number, $m = 1, 2, 3, \ldots$ dipole-, quadrupole-, sextupole mode, ν = transverse or longitudinal tune *cure: self stabilising due to non-linearities, damping of high Q resonances, feedback*

10.7.5 Observation of Instabilities

Monitoring and recording the primary parameters (intensity, lifetime, beam profiles) will immediately show the impact of instabilities on the performance of the machine. In addition, the gathered information can be used to discriminate between transverse and longitudinal or single bunch and multi bunch instability. Threshold currents can be found and, if sufficient time resolution is available, we are able to distinguish between permanent and bursting instabilities. The identification of instabilities is only possible if additional equipment is available for more careful diagnostics in time and frequency domain. What we need first are fast, high frequency signals from the beam: position and intensity. The source for these signals could be properly designed circular button pickups[92] and high quality coaxial cables bringing the signals to the analysing equipment: either a spectrum analyser or a fast oscilloscope. The frequency range of the analyser should extent to a few GHz. Time resolution of the scope must be high in order to resolve clearly neighbouring bunches.

The signals produced by the most important instabilities[93] and observed with this kind of equipment are collected in Table 10.6.

References

1. B. Wende, *PTB-Mitteilungen* **103**, 119, 2/1993.
2. R. Talman, *AIP Conference Proceedings No. 212*, 1990, p. 1.
3. A. Hofmann, Lecture Notes in Physics No. **343**, Springer-Verlag, 1989, p. 367.
4. K. B. Unser, *AIP Conference Proceedings No. 252*, 1992, p. 266.
5. K. Scheidt et al., ESRF Annual Report 1992, p. 44.
6. J. L. Revol, ESRF/MAC-19/09, April 1993.
7. J. Hinkson, *AIP Conference Proceedings No. 252*, 1992, p. 21.
8. R. E. Shafer, *AIP Conference Proceedings No. 212, 1990*, p.26.
9. R. Jung, CERN PS/93-35 (DB), p. 54.
10. O. R. Sander, *AIP Conference Proceedings No. 212*, 1990, p. 127.
11. A. Hofmann, IEEE Trans. Nuc.Sci. NS-28, 2132 (1981).
12. E. Rossa, CERN PS/93-35 (DB), p. 34.
13. G. Ulm et al., Rev. Sci. Instrum. 60, 1752 (1989).
14. K. B. Unser, Proc. of the IEEE PAC, March, 1989, Chicago, Vol. 1, p.71.
15. T. Honda et al., Proc. of the 9th Symposium on Accelerator Science and Technology, KEK, Tsukuba, Japan, August, 1993.
16. A. H. Lumpkin et al., *AIP Conference Proceedings No. 281*, 1993, p. 150.
17. F. Loyer and K. Scheidt, CERN PS/93-35 (DB), p. 21.
18. R. Ursic et al., CERN PS/93-35 (DB), p. 110.
19. G.R. Lambertson, LBL, LSAP Note-5, May, 1987.
20. Y. Chung et al., Proceedings of the PAC, Mai, 1993, p. 2304.
21. M.Q. Barton, Nucl. Instrum. Methods **A243** (1986) 278.
22. E. Shafer, *AIP Conference Proceedings No. 281*, 1993, p. 120.
23. J. Darpentigny and L. Cassinari, CERN PS/93-35 (DB), p. 86.
24. T. Ring and R. J. Smith, Proc. of the Workshop on Advanced Beam Instrumentation, April, 1991, KEK, Tsukuba, Japan, p. 269.
25. R. Biscardi and J. W. Bittner, Proc. IEEE of PAC, 1989, p. 1516.
26. S. Krinsky et al., *AIP Conference Proceedings No. 249*, Vol. 1, 1992, p. 762.
27. E. Kahana, *AIP Conference Proceedings No. 252*, 1992, p. 235.
28. G. R. Aiello and M. R. Mills, *AIP Conference Proceedings No. 281*, 1993, p. 301.
29. R. Fuja and Y. Chung, *AIP Conference Proceedings No. 281*, 1993, p. 248.

30. A. Friedman et al., Proceedings of the PAC, Mai, 1993, p. 2284.
31. A. J. Votaw, *AIP Conference Proceedings No. 281*, 1993, p. 242.
32. Y. Chung, "Beam Position Feedback System for the Advanced Photon Source", contribution to the 5th annual Beam Instrumentation Workshop 1993.
33. K. Scheidt and F. Loyer, Proc. of the EPAC, 1992, Vol. 2, p. 1121.
34. C. H. Kim and J. Hinkson, *AIP Conference Proceedings No. 281*, 1993, p. 101.
35. R. Jung, Lecture Notes in Physics No. 343, "Frontiers of Particle Beams; Observation; Diagnosis and Correction", eds. M. Month and S. Turner, Springer-Verlag Berlin Heidelberg, 1989, p. 403.
36. J. Galayda, AIP Conference Proceedings No. 212, 1990, p.59.
37. W. Berg and K. Ko, *AIP Conference Proceedings No. 281*, 1993, p. 279.
38. R.Jung, CERN PS/93-35 (DB), p. 54.
39. A.H. Lumpkin and M.D. Wille, Nucl. Instrum. Methods **A331** (1993) 803.
    Y. Ogawa et al., KEK Preprint 93-37, June 1993.
40. A.P. Sabersky, Part. Accel. 5,199 (1973).
41. R. Littauer, *AIP Conference Proceedings No. 105*, 1983, p. 869.
42. A. Hofmann and F. Méot, Nucl. Instrum. Methods **203** (1982) 483.
43. C. Bovet and E. Rossa, Proc. of the Workshop on Advanced Beam Instrumentation, April 1991, KEK, Tsukuba, Japan, p. 201.
44. A. Ogata et al., Proc. of the PAC 1989, p. 1499.
    E. Gluskin private communication.
45. F.-P. Wolf and W. Peatman, Nucl. Instrum. Methods **A246** (1986) 408.
46. R. J. Nawrocky et al., IEEE Trans. Nucl. Sci. **NS-32**, p. 1893.
47. K. Holldack, W. Peatman, and A. Erko, "Technical Note on Source Imaging at BESSY Using Bragg-Fresnel-Lenses", 12/1993.
48. R. C. C Perera et al., Rev. Sci. Instrum. **63** (1992) 541.
49. A. Hofmann, Proc. of the EPAC, 1988, p. 181.
50. T. Shintake, Proc. of the 8th Symp. on Acc. Sci. and Techn., Nov., 1991, Saitama, Japan, p. 290.
    G. Ya. Kezerashvilli and A. N. Skrinsky, Proceedings of the Workshop on Advanced Beam Instrumentation, KEK, Tsukuba, Japan, Eds. A. Ogata and J. Kishiro, 1991, p. 183.
51. R. Johnson, Lecture Notes in Physics No. 343, Springer-Verlag, 1989, p. 167.
52. A.S. Artamonov et al., Physics Lett. **118B** (1982) 225 and W.W.MacKay et al., Physical Review **D29** (1984) 2483.
53. A.A.Sokolov and I.M.Ternov, Sov. Phys. Dokl. **8** (1964) 1203.
54. B.W. Montague, Physics Reports Vol. 113, No.1 (1984) 1-96.
55. M. Placidi, Lecture Notes in Physics No. 343, Springer-Verlag, 1989, p. 186.
56. S. Van Dyck Jr. et al., Phys. Rev. Lett. **59** (1987) 26.
57. V. N. Baier, Proceedings of the Workshop on Advanced Beam Instrumentation, KEK, Tsukuba, Japan, eds. A. Ogata and J. Kishiro, 1991, p. 365.
58. R. Thornagel et al., contribution to the EPAC 1994.
59. V. N. Baier and V. A. Khoze, Atomnaya Energiya **25** (1968) 440.

60. S. Khan, BESSY TB 177/93.
61. K. Wittenburg, CERN PS/93-35 (DB), p. 11.
62. D. R. Patterson, *AIP Conference Proceedings No. 281*, 1993, p. 194.
63. P. Marin, Rev. Sci. Instrum. **63** (1992) 327.
64. B. Holzer in a talk on the status of HERA given at BESSY in December 1993.
65. M. Serio, Lecture Notes in Physics No. 343, Springer-Verlag, 1989, p. 65.
66. K. Lohmann et al., Proc. of the EPAC, 1990, p. 774.
67. C. Dunnam et al., *AIP conference Proceedings No. 229*, 1991, p. 267.
68. E. B. Blum and R. Nawrocky, Proc. of the PAC, Mai, 1993, p. 2246.
69. A. Hofmann, CERN PS/93-35 (DB), p. 1.
70. M. Sands, SLAC-121, UC-28 (1970).
71. F. Willeke and G. Ripken, "Methods of Beam Optics", DESY 88-114, August 1988
72. P. Kuske et al., Proc. of the EPAC 1988, p. 1214.
73. L. Smith, ESG TECH NOTE-24, Sept. 1986.
74. L. Palumbo and V. G. Vaccaro, Lecture Notes in Physics No. 343, Springer-Verlag, 1989, p. 312.
75. A. Poncet, Lecture Notes in Physics No. 400, Springer-Verlag, 1992, p. 488.
76. J. P. Delahaye and J. Jäger, SLAC-PUB-3585, Feb. 1985.
77. W. J. Corbett, et al., Proceedings of the PAC, Mai, 1993, p. 108.
    J. Safranek and M. Lee, SLAC-PUB-6442, February 1994.
78. D. Rice et al. IEEE Trans. Nucl. Sci.,Vol. NS-30, No. 4, p. 2190.
79. L. Arnaudon et al., CERN PS/93-35 (DB), p. 120.
80. J. Borer et al., Proc. of the EPAC 1992, p. 1082.
81. P. Kuske and J. Bahrdt, Proc. of the EPAC, 1990, p. 1417.
82. M. Minty et al., *AIP Conference Proceedings No. 281*, 1993, p. 158.
83. M. Billing, *AIP Conference Proceedings No. 281*, 1993, p. 55.
84. B. Simon and P. Kuske, CERN 88-04, p. 120.
85. H. Wiedemann, ESRP-IRM-10/83.
86. W. Radloff and W. Kriens, IEEE Trans. Nucl. Sci. **NS-30**, No. 4, p. 2193.
87. P. Kuske et al., Proc. of the International Conference on Insertion Devices for Synchrotron Sources (1985), SPIE Vol. 582, p. 143.
88. J. Le Duff, Nucl. Instr. Methods **A239** (1985) 83.
89. H. Zyngier et al., "Super-ACO Status Report", Proc. of the EPAC 1990, p. 469.
90. A. W. Chao, M. J. Lee, Journal of Applied Physics **47** (1976) 4453.
91. M.-P. Level, KEK Report 90-21, February 1991, p. 101.
92. W. Barry, *AIP Conference Proceedings No. 281*, 1993, p. 175.
93. J. Gareyte, Nucl. Instr. Methods **A239** (1985) 72.

CHAPTER 11: MAGNET SUPPORT AND ALIGNMENT*

ROBERT E. RULAND[◊]
Stanford Linear Accelerator Center, Stanford University, Stanford, CA 94309

11.0 Introduction

From theoretical design of the storage ring and injection system, we move to physical installation. The challenge facing the alignment team is to translate a theoretical storage ring layout designed in Cartesian space into a physical ring in geocentric space—to transform a list of theoretical coordinates into a physical system in which each component lies at its design location to within a specified tight tolerance. How to accomplish this transformation is the subject of this chapter.

As alignment tolerances get ever tighter, the interplay of alignment with mechanical engineering becomes ever more important. In fact, accelerator alignment has advanced so far that mechanical uncertainties now exceed observational uncertainties. Of the mechanical issues bearing upon alignment, one of the most crucial is the magnet supports; these must provide both stability and a fineness of motion substantially exceeding the final alignment tolerances. This chapter therefore includes a section on mechanical support systems and their implications for alignment.

This chapter covers three topics: mechanical schemes to support and align storage ring and injection system components; survey and alignment of those components; and ground motion.

The first section addresses magnet supports (girders and individual magnet stands) and mechanical adjustment systems (shims, struts, and cross slides).

The second section focuses on the alignment of synchrotrons, storage rings and injection lines, and examines the propagation of errors associated with these processes. The relationship of the lattice coordinate system to the selected layout coordinate system, and the subsequent computation of ideal component coordinates are described, followed by a broad overview of the sequence of alignment activities from the initial absolute positioning to the final smoothing. Emphasis is given to the relative alignment of components; in particular, to the importance of incorporating methods to remove residual systematic effects in surveying and alignment operations.

The third section reviews ground motion issues, and describes measures for alleviating disturbances.

11.1 Magnet Supports

Magnet supports are the interface that allows mechanical mounting of components and their subsequent alignment to a nominal position in three-dimensional space. Supports thus provide two functions: that of a spacer to bring the component close to its ideal position, and that of a fine motion system to enable the surveyor to move the component to its ideal location within the required tolerance.

---

*Work supported by Department of Energy contract DE-AC03-76SF00515.
[◊]Author e-mail: ruland@slac.stanford.edu and FAX: (415) 926-4055.

It is essential to understand that Magnets, Supports, and Survey and Alignment are interrelated. Ideally, one person would be responsible for all these functions. In larger projects, beyond the scope of one such manager, the responsible parties must be in regular communication. A magnet designed without supports in mind can be quite impossible to hold onto.[1] A support system that holds the magnets up, but requires a hammer to operate, renders impossible the achievement of tight tolerances. Magnets, Supports, and Survey and Alignment must be designed as a system.

11.1.1 Spacers

Components, with their adjustment systems, are rarely mounted directly to the floor or to an elevated concrete structure. Instead, girders or individual stands are used to hold a component at its approximate position and elevation above the floor. These spacers serve as the backbone on which the more precisely machined adjustment systems can be mounted.

11.1.1.1 Girders

A girder is a strongback or platform onto which a group of components can be mounted at beam height. Girders simplify the installation in cases when many small components need to be supported immediately adjacent to one another, as is often the case in larger size machines (>100 m). The major advantages of a girder support system over individual stands are:

- The girder isolates individual components from ground settlements, since the whole group of components moves up or down together. Any settlement can be corrected by adjusting the position of one girder, rather than many support stands.
- To bring the magnet poles as close as possible to the beam in the latest generation of machines, the clearance between the pole tips and the vacuum chamber is very small, allowing little motion of the magnet with respect to the chamber. A global position adjustment of individual components requires many iterations and much time, unless all the components are mounted together and move as one monolith.
- As vacuum chambers become increasingly complex, it is often impossible to achieve and retain the correct shape in the production process. Whereas magnet supports should generally be kinematic (i.e., provide only the minimal number of constraints), for vacuum chambers, a heavily overconstrained system is often required so that the chamber can be pushed and pulled into shape. Such a system will work satisfactorily only if all constraints connect to the same reference body. This eliminates the use of individual stands.
- Girders can be filled with water to increase their thermal capacity, thereby slowing the rate of response of the girder to temperature variations.
- Girders can be preassembled in a shop before installation. All of the magnets and the vacuum chamber for a girder are installed and aligned to the final relative tolerance in a local girder coordinate system. Water-cooling manifolds and hoses are assembled on the girder at this stage, as are the connections of electrical circuits. All this work can be done in a production line environment rather than the tunnel, making it more efficient

and of higher quality, with a more reliable inspection.[2] Installation of the preassembled girder in the tunnel is also significantly faster.

There are two primary types of girders: steel box and concrete. Concrete girders (Fig. 1) feature two I-beams cast into a rectangular cement block and machined flat. The rail system formed by the I-beams supports the beam line components. This system is widely used at SLAC. Concrete girders have a significant cost advantage, but great care must be taken during the construction and cement curing process, for slow creep and hairline cracking can severely hamper the monolithic quality of the finished girder. The other girder type (Fig. 2) is the stress-relieved structural-steel box girder. During the machining of the top and bottom plates, all the mounting holes can be quickly, cheaply,

Fig. 1. Concrete girder as used in SLAC Final Focus.

Fig. 2. Steel girder as used in LBL ALS. Photo courtesy of Lawrence Berkeley Laboratory, University of California.

and accurately drilled and tapped by NC machines, obviating the need for lengthy prealignment and for manual drilling and tapping of mounting holes.

11.1.1.2 Individual Stands

Individual stands are generally used in situations where components are more spread out; e.g., transport lines. The simplest form of stand is a length of pipe with plates welded to the top and bottom (Fig. 3). The diameter of the pipe is of course a function of stand height and component load. More sophisticated stands are used at SLAC in the FFTB. These stands are made of Anocast, a granite epoxy which gives the stands the appearance of a granite block molded to the specifications of the particular application.[3] In effect, the Anocast stands become a hybrid of stand and girder. In the FFTB some Anocast stands support a group of magnets while still maintaining the typical cross section of an individual magnet stand (Fig. 4). Measurements confirm that these stands have much better damping qualities of vibrations at higher frequencies than steel stands. Furthermore, their thermal mass dampens expansion due to variations in the ambient temperatures. Costs for steel and Anocast stands are comparable.

Fig. 3. Individual steel stand.

11.1.2 Manual Adjustment Systems

All beam components need to be moved and fixed at accurate locations by adjustment mechanisms. These systems should include the following design features:

Fig. 4. Anocast stand in SLAC FFTB.

- Adequate alignment precision: for precise adjustibility, the system's resolution should be ten times the required alignment tolerance.
- Orthogonal motion: there should be no cross coupling between the axes for small adjustment motions. For large motions, any existing coupling must be predictable.
- Kinematic mount: an overconstrained system induces stress into the support and/or component, resulting in a deformation of the component.
- Stability: the support should provide a stiff base when locked down where incidental contact will not cause movement of the magnet. It should also not deform the component during adjustment.
- A small footprint: as real estate is usually at a premium, components must often be placed very close together.
- Vibrational stiffness: typical ground motion frequencies should not be amplified by the support system.

There are two general types of adjustment mechanisms. The most common type separates the horizontal adjustment from the vertical degree of freedom. The second type combines horizontal and vertical adjustments into one system, usually implemented in a six strut layout that holds the component in a kinematic suspension. Other implementations are the CERN Adjuster System and its derivative, the CEBAF 3-D Cartridge, and the SLAC 3-D stage.

Fig. 5. Push-push screw arrangement.

11.1.2.1 One and Two–Dimension Systems

To separate the horizontal from the vertical, a horizontal plane is generated by adjusting the height of three vertical standoffs. In its simplest implementation, the standoffs are either shim stacks or threaded rods. In the case of shim stacks, shim stock is added or removed until the plate is horizontal and the component at its ideal height, a lengthy, iterative process. Where threaded rods are used, the mounting plate rides on three screw nuts that are threaded on vertically mounted rods. Turning the nuts provides vertical translations along the Y–axis and two rotational degrees of freedom, pitch (rotation around the X–axis), and roll (rotation around the Z–axis).

On this horizontal plate slide one or two plates on which the component is mounted. These plates move under the force of adjustment screws to adjust and fix the Z (in beam direction), X (perpendicular to Z), and yaw (rotation around the Y–axis) degrees of freedom. The adjustment screws are often designed in a push-push arrangement (Fig. 5) with two opposing screws pushing on both sides of the component in a colinear arrangement. To achieve a translation, one side is loosened and the other tightened. Tightening both screws locks the position. Often the stand has only one sliding plate; in this case, the X and Z adjustments are not independent, since all adjustment screws must be loosened to permit sliding of the plate. Fine adjustment in the orthogonal direction is usually lost, and must be touched up again. Precise alignment with only a single sliding plate and push-push screw arrangement usually requires many iterations.

This basic design can be refined by replacing the above described horizontal and vertical adjuster with more sophisticated variations. The addition of spherical washers between the horizontal plate and the adjustment nuts makes the system move more smoothly. If the system is designed to carry higher loads, machine screw jacks (Fig. 6) are available that fit almost any application while still providing fine adjustment motion. Less expensive, but more limited in range, are wedge jack adjusters that are made of two wedges with the two sloped planes riding on each other. A horizontal motion pushes the upper wedge higher on the inclined plane, thereby providing a vertical motion. Wedge jack adjusters are available off the shelf in many load travel combinations. The push-push screw arrangement can be improved by a turnbuckle/rail-slide design. The two push screws are replaced by one turnbuckle, which provides both the push and pull force. The fixed end of the turnbuckle can slide on a rail oriented parallel to the other adjustment axis in order to allow two-dimensional adjustments. This design is still relatively simple and inexpensive, while complying with all the above listed requirements. To support the girders in the storage ring of the Argonne Photon Source, a combination of wedge jack adjusters (Fig. 7) and turnbuckle-type horizontal adjustment was used.

Fig. 6. Machine screw jack support.

11.1.2.2 Three-Dimension Systems

Six-strut system A kinematic suspension can be created by arranging six adjustable length links in a 3-D truss. The three vertical struts adjust and hold the vertical translation, and the pitch and roll rotations. The three other struts (Fig. 8) are placed in the horizontal plane, two in one direction, and the third perpendicular. These three adjust and hold the X and Z translations and the yaw

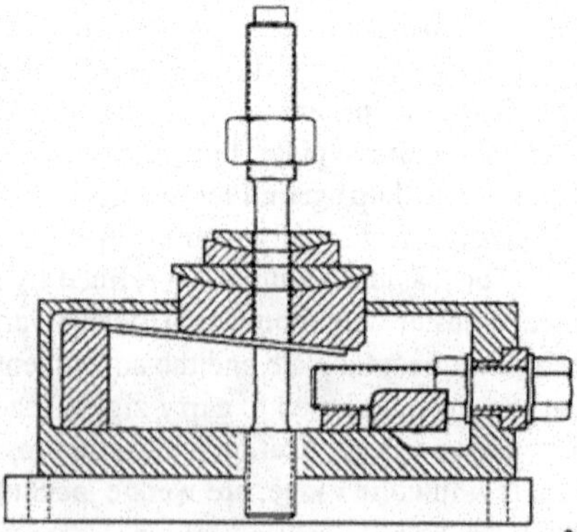

3-94
7633A1

Fig. 7. Wedge jack adjuster as used in APS.

rotation. The orthogonal arrangement of the struts minimizes coupling in motion. Struts are length-adjustable rigid members with spherical joints at each end. A strut will support only an axial load, in axial compression or tension. The spherical joints at either end ensure that a strut never experiences loads in any other direction. Since all struts are in axial compression or tension, they provide very rigid support.

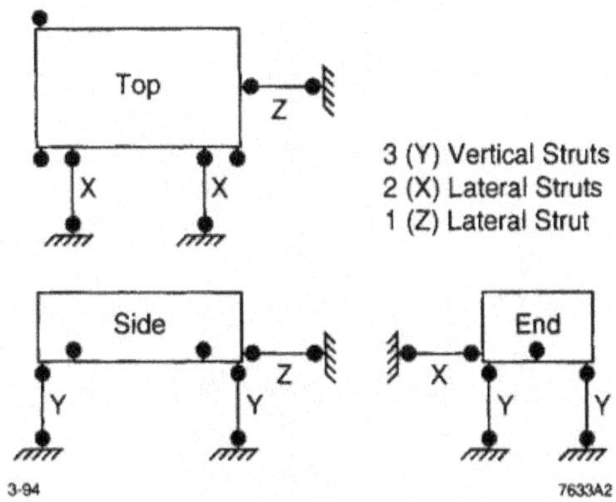

Fig. 8. Kinematic suspension.

11.1.2.3 Typical System Implementations

Advanced Light Source (ALS) strut system. All components and girders at the Advanced Light Source at the Lawrence Berkeley Laboratory are supported by strut systems[4] (Fig. 9), as is the Spherical Grating Monochromator at the SSRL. The struts used for the support systems are not normal stock items. To avoid the backlash present in all regular spherical joints, the spherical rod end bearings have been squeezed in a controlled way to generate friction, which only a specific break-away torque can overcome. A shaft collar has been added at the end of each tube into which the rod end bearings thread. A portion of the tube, at each end, is turned down and slit in two directions so the shaft collar will squeeze

Figure 9. ALS strut supports. Photo courtesy of Lawrence Berkeley Laboratory, University of California.

Fig. 10. ALS 5-ton machine screw jack strut.

Fig. 11. ALS 20-ton machine screw jack strut.

Photos courtesy of Lawrence Berkeley Laboratory, University of California.

the female thread against the male thread of the rod ends to remove any backlash in the threads. The rod end bearings are all right-hand threads with one coarse thread and the other a fine thread, creating a differential threaded device which allows very high resolution adjustments. For the support of heavy loads, the tube and differential threads are replaced by an appropriately rated machine screw jack (Figs. 10, 11).

CERN cartridge. The CERN Adjuster System[5] consists of three cartridges that utilize a combination of the principles in the two styles discussed above. The improvement over the first style mechanism is that the sliding feature is replaced by the three vertically-oriented links of the kinematic suspension. The first or main cartridge works as follows (Fig. 12): the piston-ended link pivots in a socket at the bottom of the base and floats within a hollow cylindrical projection from that base. At the top, the link pivots in and supports a cap whose outer skirt drapes over the cylindrical projection. The device to be positioned is

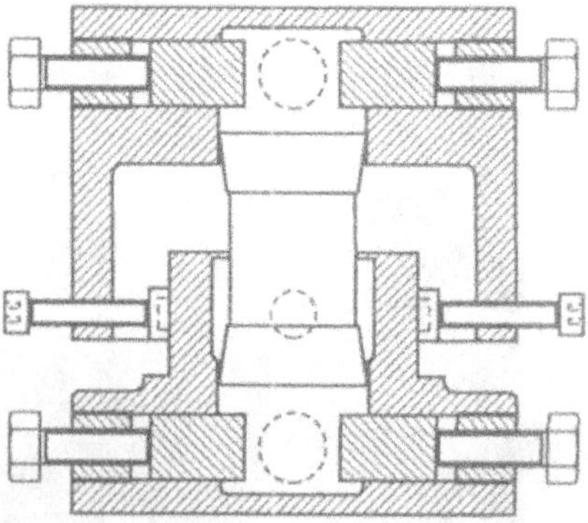

Fig. 12. CERN cartridge adjuster.

placed on this cap. The cap is driven horizontally by four bolts threaded through the skirt of the cap, which press against four flats machined into the cylindrical projection. Lateral and longitudinal adjustment is achieved with one of these pairs of opposing push-push screws. As one bolt is loosened and the opposite bolt tightened, the cap glides easily, rocking on the vertical link. The sockets in which the link is mounted consist of cylinders in the base and cap that are filled with urethane rubber.

Four screws in the base and four screws in the cap drive in and out of this volume, compressing the rubber and driving the link or the cap higher or lower respectively, providing the vertical adjustment. The second cartridge lacks one set of opposing screws and the third cartridge lacks both sets, leaving no restraint on the cap, allowing it to float and provide only vertical adjustment. The three cartridges are placed in a triangular pattern with the set of opposing screws of the second cartridge parallel to one set of screws in the main cartridge. Use of all three cartridges provides pitch, roll, and yaw adjustment. One advantage of the CERN Adjuster System over the kinematic suspension is that there is much less coupling between the adjustments, so that alignment is more easily obtained.

<u>CEBAF cartridge</u> The CEBAF cartridge[6] uses many of the features of the CERN Adjuster System design. Three identical cartridges are attached to a stand through specially bored mounting holes. Each cartridge consists of a vertical cylinder and a cap (Fig. 13). The device to be adjusted is fastened to the caps of the three cartridges. The hollow, vertical cylinder has two opposing flats on its outer wall at the top, and a threaded hole in its bottom, into which is threaded a set screw. Turning this screw raises the cap, via a vertical rod through the cylinder. Lateral adjustment is by a pair of opposing screws through the skirt of the cap, registering against the flats on the cylinder. The cap glides over easily while rocking on the vertical rod. The cartridges are mounted on the stand such that the flats on two cylinders are parallel to each other and the flats on the remaining cylinder are perpendicular to the other two, providing lateral, longitudinal, and yaw adjustment. With this orientation, all degrees of freedom are constrained with no overconstraint. Locking of the movement of all screw threads is provided by locknuts.

<u>SLAC damping ring girder support</u> This design contains the most basic adjustment system construction elements, a push-push screw arrangement combined with a threaded rod[7] (Fig. 14). The girder is supported by three feet. Each foot's baseplate is bolted and grouted to the floor in an approximately horizontal position. Atop this baseplate sits a sliding

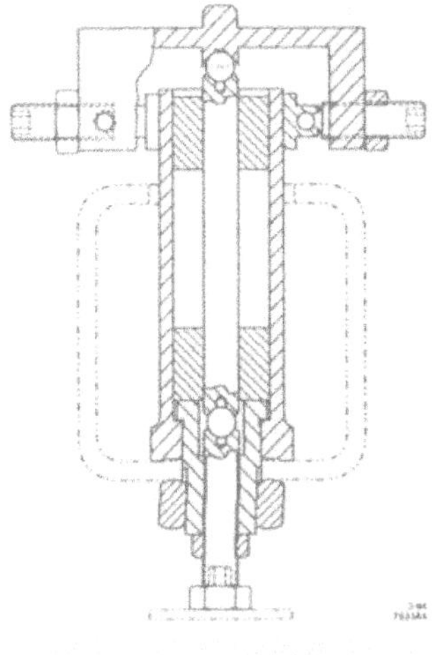

Fig. 13. CEBAF cartridge adjuster.

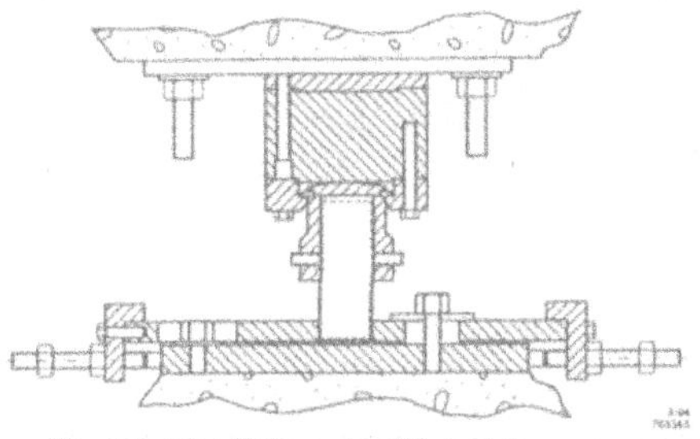

Fig. 14. SLAC Damping Ring girder support.

plate that can be moved relative to the baseplate by the force of a two-dimensional push-push screw arrangement. A short fine-threaded rod of substantial diameter is mounted to the sliding plate at its center. A cap-shaped nut, riding on the threads over the top of the rod, provides the vertical adjustment. The girder is mounted to this nut in a way which prevents any horizontal backlash, while still permitting it to be turned. The system is locked in the horizontal dimension by a bolt holding the sliding plate to the baseplate, and in the vertical dimension by a set screw which prevents the cap nut from turning. While this system allows relatively high resolution adjustment of heavy loads, the total system is significantly overconstrained, and must therefore be operated with great caution.

<u>SLAC Final Focus girder support</u> This design is similar to the Damping Ring supports, but avoids the overconstraints[8] (Fig. 15). The push-push screw arrangement is replaced by one-dimensional stages: two feet have stages oriented for lateral adjustment, while the stage at the third foot provides longitudinal motion. To decouple the cross-motion between stages, the supports are fixed to the girder in only one horizontal dimension, which is accomplished by a rail slide system. The vertical adjustment is functionally the same as on the support discussed above.

<u>CERN LEP dipole support</u> This system[9] can provide kinematic support to a wide variety of applications, from small magnets to heavy girder modules. The general idea and functionality are taken from the CERN cartridge design, but with the vertical adjustment replaced by an adjustable-length link (Fig. 16). To minimize motion correlation, the link is made as long as possible, subject to the restraints of the specific application.

<u>SLAC 3-D stage</u> This is an adjustment system tailored to support a variety of components, from small quadrupoles to long narrow bends that are to be positioned to tight

Fig. 15. SLAC Final Focus girder support.

tolerances[10] (Fig. 17). The horizontal degrees of freedom are provided by a baseplate/sliding plate arrangement. To avoid overconstraint, the adjustment motion is created by three *semi*turnbuckles, in which one end is a conventional rod end bearing, but the other end is a threaded stud (Fig. 18). Two of these *semi*turnbuckles provide the lateral adjustment, and a third gives the longitudinal adjustment. The spherical rod end bearings are threaded into blocks bolted to the base plate. The spherical bearing end is threaded onto a rail that is mounted on the baseplate perpendicular to the rod's adjustment direction. This arrangement allows the sliding plate to be adjusted in one dimension, while maintaining the adjustment in the other horizontal dimension. The vertical adjustment is created in a similar way. Three spherical rod end bearings are bolted vertically into blocks mounted to the sliding plate. Bolts through the spherical rod end bearings support the component.

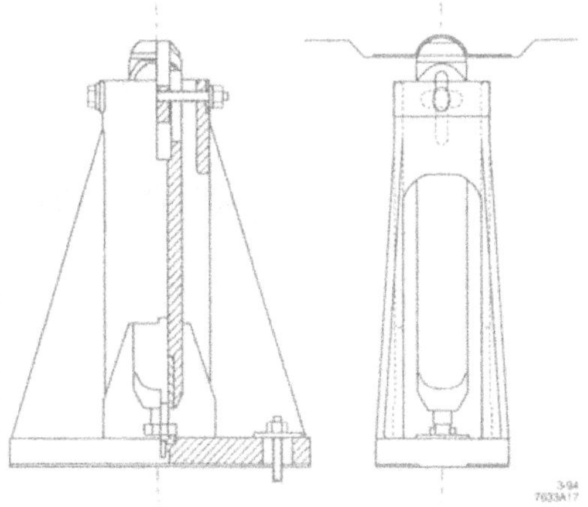

Fig. 16. CERN LEP Dipole support.

Fig. 17. SLAC 3–D stage.

Fig. 18. Lateral adjustment layout.

DESY PETRA single component support system This system[11] has been used to support quadrupoles on single stands and long dipoles on two single stands at either magnet end in the PETRA ring. The underlying scheme is now widely used in other machines at DESY. Shown below in Fig. 19 is a quadrupole sitting with three pads on three vertical screws that provide height, roll, and pitch adjustments. In the horizontal plane, two struts allow motion perpendicular to the beam. No adjustment capability along the beam axis is provided. To create a kinematic mount between the pads and screws, one screw head is resting in a groove, while the other two pads are flat.

11.1.3 Motorized Adjustment Systems

SLAC FFTB magnet positioners The FFTB magnet positioners[12] differ from conventional positioning stages used in instruments and machine tools. The mechanism is designed to support loads exceeding 1 ton, while still providing smooth motion, free of hysteresis, at the micron level. The design is simple and sufficiently reliable for large scale use in the remote positioning of hundreds of magnets. Conventional crossed-slide leadscrew positioning stages are not appropriate for this application. High-resolution piezoelectric positioners[13] cannot meet the load and range requirements. The remote magnet positioning mounts used in the FFTB kinematically support the magnets on roller cams. The magnet rests under gravity in a cradle formed by the cams (Fig. 20). This type of kinematic support is similar to the Kelvin Clamp[14] used in laboratory optics

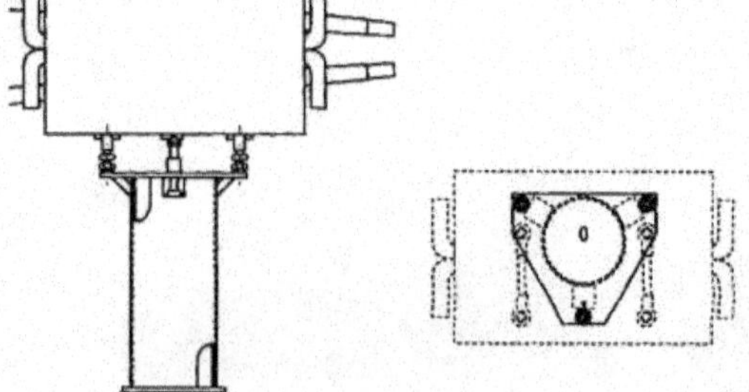

Fig. 19. DESY PETRA support system.

and instrumentation. The V–blocks and flat plates fixed to the magnet make point or line contact with the outer bearing races of the roller cams. Rotation of the eccentric camshafts shifts the magnet position. This type of kinematic support, where the number of contact points balances the number of degrees of spatial freedom, has the advantage of avoiding all free play between the magnet and mount. The magnet always rests in contact with all of the supporting cams, regardless of their position. No precise mechanical dimensions are needed to insure zero play. No clamping forces, other than gravity, can distort the magnet's shape. The magnet can be removed from the mount and replaced without realignment. During operation, only the inner eccentric shaft of a support cam rotates under motor control. The outer cam bearing race remains in contact with the magnet as shaft rotation lifts the magnet. In such a system, failure of the control system will only cause the cam to cycle around again. Magnet motions are strictly bounded by the design geometry. Limit switches are not needed for over-travel protection. All support cams are arranged so that gravity applies a load torque to each cam shaft drive train. This torque removes all backlash, except at the extremes of cam lift. All parts move by pure rolling motion, and are free of the hysteresis

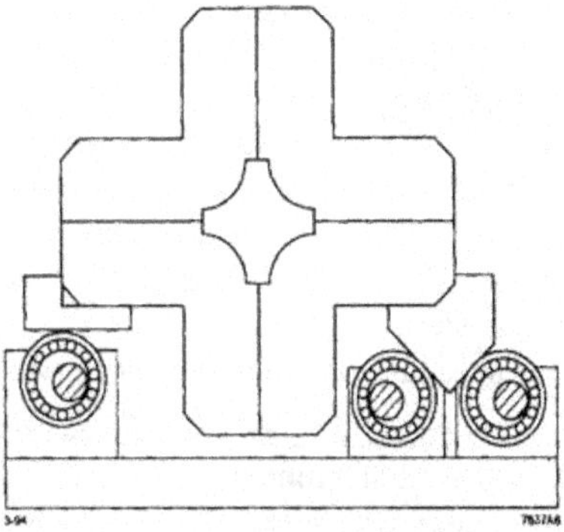

Fig. 20. Magnet positioning mount with roller cams.

Fig. 21. FFTB magnet remote positioner.

typical of intermittent and reversing sliding motion. This mount can adjust the horizontal and vertical position of the magnet, as well as the magnet's roll angle around the beam axis. The magnet's longitudinal position along the beam line, as well as its alignment to the beam direction in this implementation are fixed in the support mount, and not remotely adjustable. Figure 21 shows the three-motor positioning mount used to support FFTB quadrupole magnets. Kinematic roller cam supports can be applied to a variety of geometries. The barrel containing the final triplet of quadrupole lenses for the Stanford Linear Collider is supported on five roller cam supports. This 5-m-long 6-ton assembly is remotely adjustable in pitch and yaw, as well as roll, vertical, and horizontal position.

ESRF servo-controlled jacks Predicted ground motion of more than 1 mm per year led to the development of a remote vertical alignment system. A computer-controlled hydrostatic leveling system was installed in the storage ring with three measurement stations on each girder. These girders are kinematically supported by three vertical motorized screw jacks, which are interfaced to the control system. The horizontal adjustment is provided by a gear-driven X-Z stage mounted on top of the vertical jacks.[15] First results indicate that it takes about two minutes to map the entire ring, and then only two hours to vertically align all girders.[16]

11.2 Alignment

A Survey and Alignment team's charter in building light sources is the physical positioning of all machine components, including magnets, insertion devices, detectors, and diagnostic devices, according to layout specifications. The task of positioning magnets can be broken down into six major subtasks:

- Survey reference frames: the first step is to define and physically establish a survey coordinate system appropriate to the project site and size. Control monuments are established to represent this reference grid.
- Layout description reference frame: the beam line is designed and specified in a lattice coordinate system. Coordinate transformations, including rotations and transformations, need to be defined to relate this to the survey reference frame.
- Fiducialization: the fiducialization of a component relates its effective magnetic or electrical centerline to external reference points that are accessible to subsequent survey measurements.
- Prealignment of girders: after components and vacuum chambers are mounted on a girder, they are aligned relative to a girder coordinate system.
- Absolute positioning: girders are positioned with respect to the global reference grid.
- Relative positioning: local tolerances are achieved by the relative alignment of adjacent components.

11.2.1 Survey Reference Frames

The goal is to define a computational reference frame—a mathematical model of the space in which the surveyor takes his measurements and performs his data analysis. Transformation algorithms and parameters between the surveying space and the machine layout coordinate system must be defined.

11.2.1.1 Surveying Space

Ancient civilizations realized that the earth is round, and geodesy was born when the Greek Eratosthenes (born 276BC) first attempted to determine its size.[17] The earth is actually of a more complex shape, the modeling of which is not easy. Three surfaces are of importance to the geodesist studying the shape of the earth:

i) The *terrain surface* is irregular, departing by up to 8000 m above and 10000 m below the mean sea level.

ii) The *geoid* is the reference surface described by gravity; it is the equipotential surface at mean sea level that is everywhere normal to the gravity vector. Although it is a more regular figure than the earth's surface, it is still irregular due to local mass anomalies that cause departures of up to 150 m from the reference ellipsoid. As a result, the geoid is nonsymmetric and its mathematical description nonparametric, rendering it unsuitable as a reference surface for calculations. It is, however, the surface on which all survey measurements are made as almost all survey instruments are set up with respect to gravity. Even the satellites now used for GPS surveys follow orbits determined by gravity.

iii) The *spheroid* or *ellipsoid* is the regular figure that most closely approximates the shape of the earth, and is therefore widely used in astronomy and geodesy to model the earth (Fig. 22). Being a regular mathematical figure, it is the surface on which calculations can be made. Nevertheless, in performing these calculations, account must be taken of the discrepancy between the ellipsoid and the geoid. The deflection (or deviation) of the vertical is the angle of divergence between the gravity vector (normal to the geoid) and the ellipsoid normal (Fig. 23). Several different ellipsoids have been defined and chosen that minimize geoidal discrepancies on a global scale, but for a survey engineering project, it is sufficient to define a best-fit local spheroid

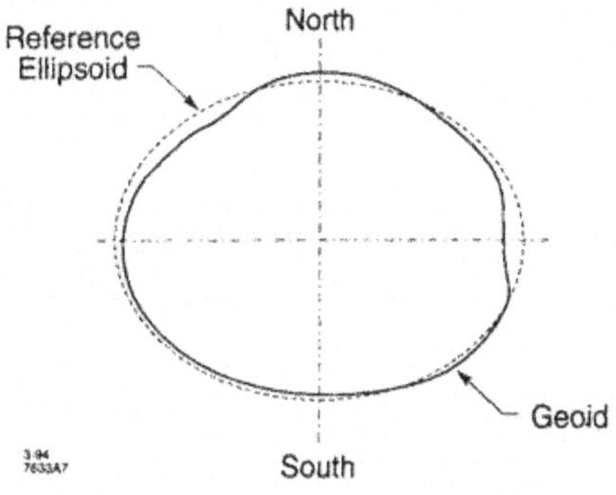

Fig. 22. Spheroid (ellipsoid) and geoid.

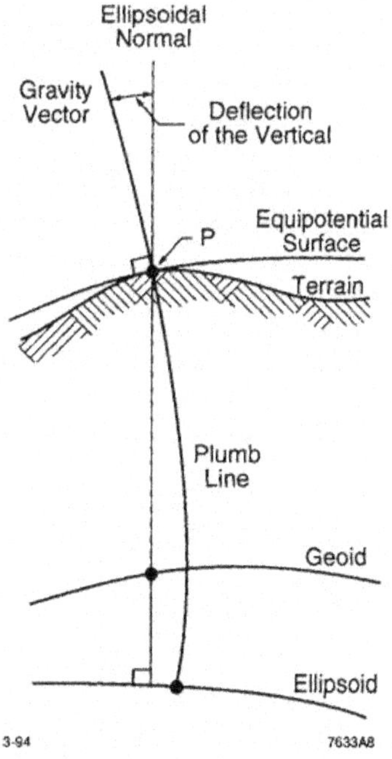

Fig. 23. Spheroid normal and gravity.

that minimizes discrepancies only in the local area. Whatever ellipsoid is chosen, all survey measurements must be reduced to the ellipsoid before computations can proceed. This reduction of observations to the computational surface is an integral part of position determination;[18] the equations can be found in most of the geodetic literature, e.g., in Leick.[19]

11.2.1.2 Surveying Coordinate System

Computations with spheroidal (geographical) coordinates latitude ϕ, longitude λ, and height h are complex. They are also not very intuitive: when using spheroidal heights, it can appear that water is flowing uphill. Especially in survey engineering projects, coordinate differences should directly and easily translate into distances independent of their latitude on the reference spheroid. Therefore, it is desirable to project the spheroidal coordinates into a local Cartesian coordinate system or, going one step further, to project the original observations into the local planar system to arrive directly at planar rectangular coordinates.

A transformation is required to project points from a spheroidal surface to points on a plane surface. Depending on the projection, certain properties of relationship (distance, angle, etc.) between the original points are maintained, while others are distorted. It is simply not possible to project a spherical surface on to a plane without creating distortions[18] (Fig. 24), but since these distortions can be mathematically modeled, it is possible to correct derived relationships, such as distances, angles, or elevations. This situation can be vividly shown on the example of the projection of leveled elevations onto a planar coordinate system (Fig. 25). Table 1 shows the projection errors as a function of the distance from the coordinate system's origin. Notice that the deviation between plane and sphere is already 0.03 mm at 20 m.

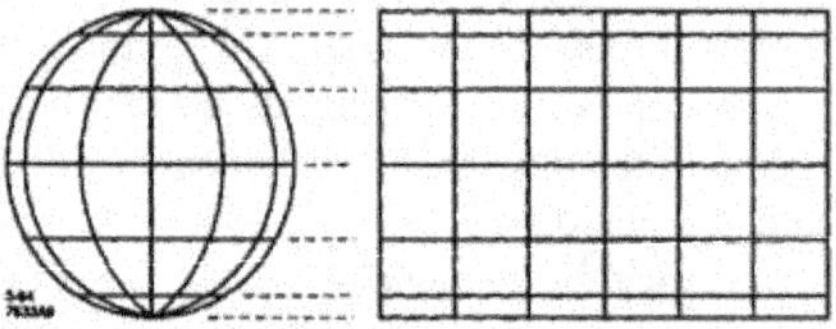

Fig. 24. Projection of sphere onto a plane.

Since further discussion here is focused on small machines, geodetic issues such as the earth's curvature and gravity anomalies can be excluded, thus simplifying the mathematics to planar Cartesian coordinate arithmetic.

Table 1. Curvature correction, plane to sphere or spheroid.

Distance [m]	Sphere H_S [m]	Spheroid H_E [m]
20	0.00003	0.00003
50	0.00020	0.00016
100	0.00078	0.00063
1000	0.07846	0.06257
10000	7.84620	6.25749
25000	49.03878	39.10929

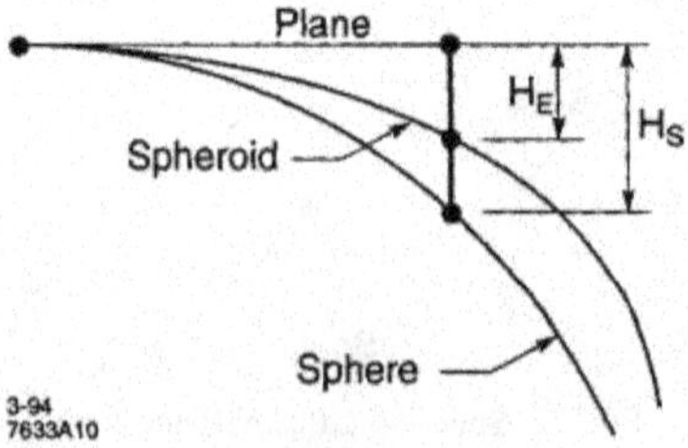

Fig. 25. Curvature correction.

11.2.1.3 Survey Networks

The surveying coordinate system is physically represented by monuments whose coordinates are determined using conventional trilateration or triangulation methods or, for larger size projects, satellite methods like the Global Positioning System.[20]

Surface network In order to achieve the absolute tolerance and the circumference requirements, a surface network with pillar-type monuments (Fig. 26) must usually be established. Traditional triangulation and trilateration methods (Fig. 27) or GPS surveys can be applied to measure the coordinates of the monuments and of tripods over the transfer shafts or sightholes. Differential leveling of redundant loops is the standard method to determine the vertical coordinates. Proper reduction of measured distances also requires accurate elevation difference data.

Using state-of-the-art equipment in a small trilateration network with good intervisibility of monuments can yield standard deviations for the horizontal coordinates in the range of 2 mm + 1 ppm. In medium size applications, it has been shown that GPS, combined with terrestrial observations and careful control of the antenna eccentricities (GPS, too, has its fiducialization problems), can yield positional accuracies of about 2 mm.[21] Trigonometric and differential leveling are the only accurate methods to determine elevations; both methods yield the same accuracies—approximately 1 mm for networks smaller than 2 km.

Fig. 26. SLAC–SLC pillar monument.

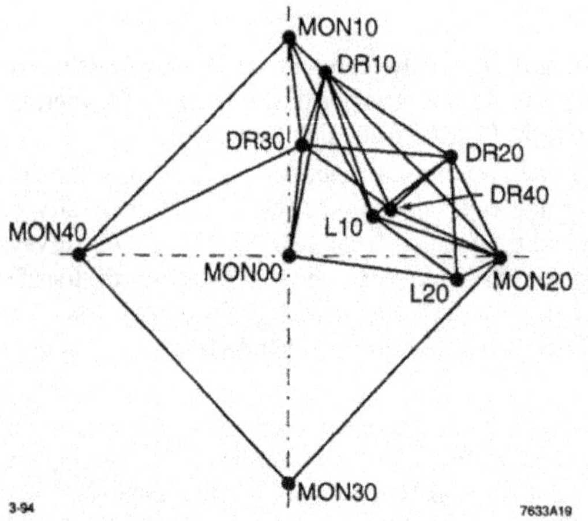

Fig. 27. Example of surface network (Argonne APS).

Tunnel The tunnel horizontal net is usually tied to the surface net by optically or mechanically centering a tripod-mounted translation stage on the surface over a monument in the tunnel through a survey shaft. These tunnel networks are usually long and narrow (Fig. 28), and incorporate points beneath the shafts as connections to the surface net. The floor marks can be 2-D (horizontal only) or 3-D: common designs are the SLAC 2-D marks, the DESY-HERA 3-D reference cups or the standard 1.5 inch floor cups and magnet mounts. Some kind of tripod or column-like monopod is used for the instrument setup. The SLAC setup (Fig. 29) is designed to accommodate slopes of up to 15°; the HERA design is more optimized towards efficiency, virtually eliminating the task of centering instruments and targets over monuments.[22] The elevation of the instrument above the 3-D reference cup is known very accurately, which facilitates 3-D mapping with theodolites.

11.2.2 Layout Description Reference Frame

The layout description of every machine component is given in a document called the design lattice (for details, refer to Chapter 2, *Lattices*) which defines the physical parameters of each machine component, including its ideal position.

For every new machine, various computer programs, e.g., TRANSPORT,[23] are used to simulate the path of the particles. Model components bend, focus, or defocus the particles as they traverse the electromagnetic fields they encounter. Component parameters are manipulated to keep them on the intended trajectory, and to qualify the beam's characteristics. The result of such simulations is a sequential listing of the design components and their parameters. Most commonly, the parameters for the beginning of the magnetic length of a component and of the following drift space are listed, including the six degrees of freedom for the beam following coordinate system. In addition, a magnet's field strength, and, if applicable, its bending angle are given.

Based on experience and the results of lattice simulation runs, position tolerances are determined for each magnetic component and are attached to the lattice specifications

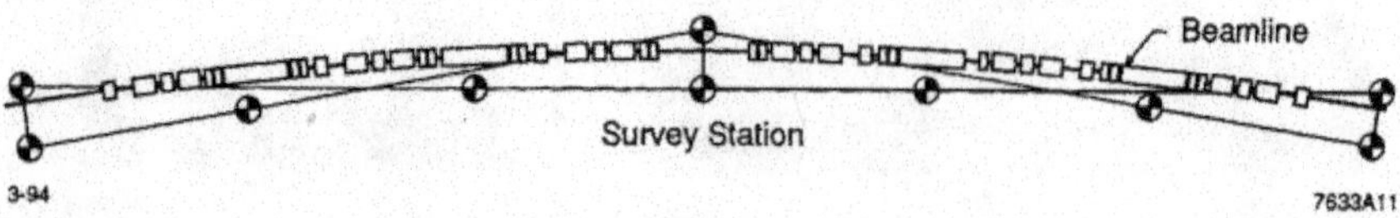

Fig. 28. Tunnel network layout (Argonne APS).

Fig. 29. SLAC tripod setup.

(see Wiedemann[24] for a discussion of the effects of magnet alignment errors). The individually specified parameters are usually the maximum permissible displacements in the direction of the three coordinates and the rotation around the longitudinal axis. The tolerance specifications should distinguish between absolute and relative positioning. The absolute positioning tolerance defines a maximum global shape distortion by specifying how close a component must be to its ideal location, whereas the more important relative tolerance defines the alignment quality of adjacent components. The tolerance definition should also state the required level of confidence, and whether or not the random distribution is truncated.

Surveying measurements, if done carefully with well-calibrated sensors, will show a typical Gaussian distribution, including entries outside the chosen confidence level. Achieving the equivalent of the mathematical truncation requires a means to identify "outliers" and a method to add independent redundant observations. Traditionally, the stochastic computations in surveying are based on a 1σ confidence level. Achieving the same result on a 2σ confidence level requires an exponential increase in survey effort.

The relationship between the surveying and lattice coordinate systems is defined as a transformation matrix.[25]

11.2.3 Fiducialization

Fiducialization is a fancy name for relating the effective internal electromagnetic axes of a component to external marks that can be seen or touched by instruments. It is these reference marks that are then aligned onto their nominal coordinates. It is therefore obvious that the measurement of the magnetic axis to the fiducial marks must be done with at least as much care as the final positioning.

Fig. 30. Fiducialization setup of FFTB magnets at SLAC.

Magnets in storage rings and injection systems have, for the most part, been made with ferromagnetic poles, which are traditionally used as the references for external alignment fiducials.[26] (For more details, refer to Chapter 5, *Magnet Design*.) It is assumed that the magnetic field is well-defined by the poles, but this assumption fails in the presence of saturation, and is invalid for superconducting magnets, which have no tangible poles. Furthermore, since the poles of an iron dipole are never perfectly flat or parallel, where is the magnetic midplane?[27] For quadrupoles, sextupoles, and higher order magnets, there is no unique inscribed circle that is tangent to more than three of these poles; where then is the centerline?

The only way to avoid these problems is to use magnetic field measurement (for details, refer to Chapter 7, *Magnetic Measurements*) to establish fiducials. This has already worked successfully for a number of projects, including the alignment of multiple permanent quadrupoles in drift-tube linac tanks in Los Alamos,[28] the SLC/SLD superconducting triplet quadrupoles, the HERA superconducting proton ring magnets,[29] and the Final Focus Test Beam at SLAC[30] (Fig. 30).

11.2.4 Prealignment of Girders

Girders are commonly used in light sources to support components and the vacuum chamber of one lattice cell of a common plane. These girders are preassembled in a factory before they are transported into the tunnel. After an initial component prealignment, the magnets are split and the vacuum chamber inserted. The chamber can be positioned using gauge blocks held against the magnet pole tips, or optically. If no nonelastic girder deflections are expected during transportation, a fine position alignment is also made. 150 μm is a typical tolerance for the relative positioning of magnets.

Usually, prealignment bays are set up emulating a generic beam line position; i.e., the girder is set up and supported in exactly the same way as it will be in its final beam line position. Traditional optical tooling techniques (Fig. 31) or industrial measurement system measurements (Fig. 32) can provide the required accuracy.

11.2.5 Absolute Positioning

Efficient computer-aided methods and procedures have been developed to increase positioning productivity, accuracy, and reliability. These techniques have been tested and proven in the alignment of many machines, including the ALS and APS light sources and the SLC, HERA, and LEP colliders. The absolute positioning can be subdivided into four steps :

Step 1. "Blue line" survey on the tunnel floor
Step 2. Rough absolute positioning of girders in tunnel
Step 3. Fine absolute positioning of girders
Step 4. Quality control survey

11.2.5.1 Blue Line Survey on the Tunnel Floor

In preparation for the installation of the support systems, a "blue line" survey is performed to lay out the anchor bolt positions. This is done from the tunnel traverse points using intersection methods or, more efficiently, utilizing tachymetry with instruments like the Leica TC2002 or the Chesapeake Lasertracker.[31] A relative accuracy with respect to the monuments of 5 mm can be easily achieved.

Fig. 31. Prealignment with optical tooling (Argonne APS). Photo courtesy of Argonne National Lab.

Fig. 32. Prealignment with industrial measurement system. Photo courtesy of Lawrence Berkeley Laboratory, University of California.

11.2.5.2 Rough Absolute Positioning of Girders in Tunnel

After the blue line survey, the anchors are set and the prealigned monoliths or girders installed, but with the anchor bolt nuts only "hand tight," and the girders' adjustment systems set to midrange to ensure the full adjustment range remains available for fine positioning. This adjustment system should not be used to correct the misalignment of the support system itself; instead, the support system is prealigned by tapping it into position utilizing the slack between anchor bolts and support structure. To determine the actual positions of the supports, direction and distance measurements from monuments are taken, from which actual coordinates are calculated and compared to the ideal coordinates, yielding the adjustment values in the global coordinate system orientation. Before these corrections can be applied in the field, they must be transformed into the local coordinate system of the supports. This process can be greatly accelerated by reducing the data on line in the field, providing immediate in situ coordinate feedback. High accuracy Total Stations, like the Leica TC2002 or the Chesapeake Lasertracker, interfaced to powerful field computers make this possible. The required software has been developed at SLAC and tested with great success in the alignment of the rebuilt SLC Damping Rings and Final Foci.

11.2.5.3 Fine Absolute Positioning of Components

The girders are first aligned vertically: using differential leveling, the girder is set to its ideal elevation with zero pitch and roll. The horizontal positions of the girders are set relative to the tunnel monument system. In principle, the alignment technique here is the same as described above. However, the TC2002 in the on-line feedback loop does not yield the required accuracy; only a laser tracker does. If a laser tracker is not available, traditional time-intensive triangulation techniques will effectively produce the same result.

11.2.5.4 Quality Control Survey

After the absolute positioning of girders is completed in some logically functional section of the machine, a complete resurvey of this section should be conducted to verify the results. Quality control is better achieved by the use of independent procedures, rather than the repetition of the same procedure by different teams. This provides the truly independent observations necessary to check the accuracy of the initial survey. Resurveys with the same methodology do not provide this independent check, and rarely detect the infrequent data gathering errors occurring with today's electronic instruments and field computers.

11.2.6 Relative Positioning (Smoothing)

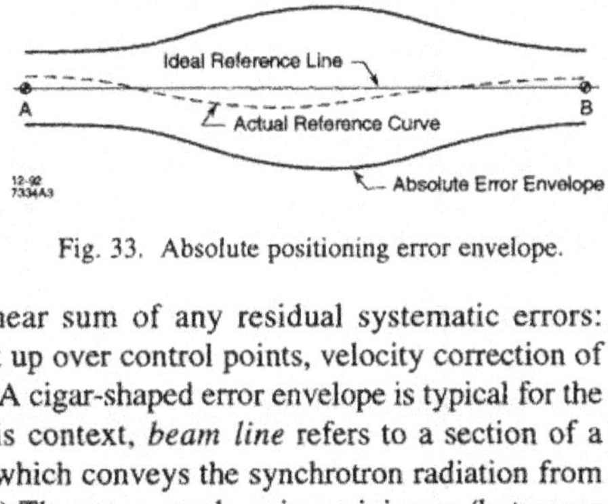

Fig. 33. Absolute positioning error envelope.

The accuracy obtained in the absolute positioning step is the quadratic sum of many random errors (surface network, transfer of control through penetration shafts, tunnel control, magnet fiducialization, magnet layout, etc.) plus the linear sum of any residual systematic errors: instrument calibration, forced centering, set up over control points, velocity correction of light, horizontal and vertical refraction, etc. A cigar-shaped error envelope is typical for the absolute alignment of a beam line. (In this context, *beam line* refers to a section of a storage ring, and not to the tangential port which conveys the synchrotron radiation from the storage ring to the experimental station.) The error envelope is a minimum (but never zero) at the control points, and grows to reach a maximum midway between two successive control points (Fig. 33). The measured reference line oscillates somewhere within this error envelope. Its absolute position cannot be pinned down any more precisely than the size of the error envelope, with deviations within this envelope being statistically insignificant. However, within this absolute error envelope, relative errors between adjacent magnets should be smaller: the major error sources equally affect the positioning of adjacent components, with the result that relative alignment accuracies are significantly higher than absolute alignment accuracies. Consequently, successive surveys will reveal reference lines of different shape whose absolute position floats randomly within the cigar-shaped error envelope. An important implication of this is that the absolute comparison of independent surveys "would be a nonsense"[32] when trying to evaluate differences smaller than the width of the absolute error envelope. If attempts are made to proceed with final absolute alignment, the "nonsense" occurs when successive rounds of survey and alignment do not converge, i.e., do not result in reducing the magnitude of the misalignments. All that is happening in this case is that the components are being moved back and forth within the error envelope.

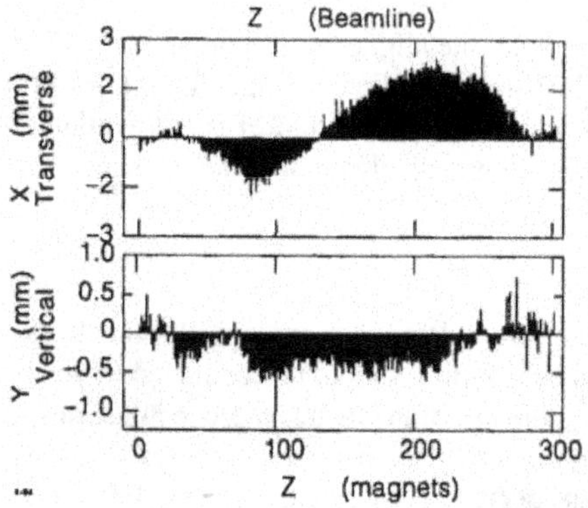

Fig. 34. Residual absolute misplacements.

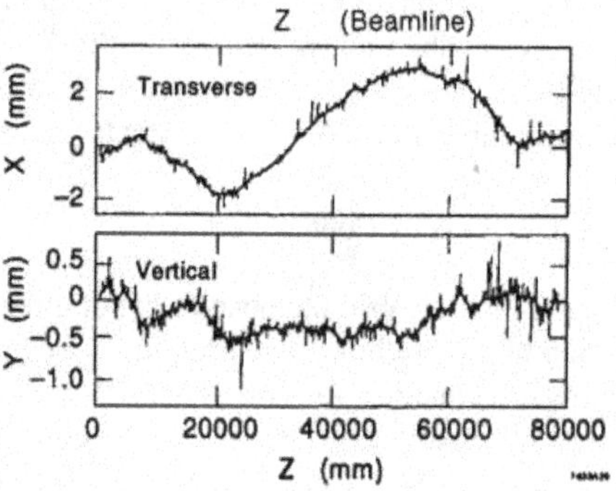

Fig. 35. Trend curve through absolute misplacements.

Because of these problems, the absolute positioning technique is not well suited to achieving the final relative position tolerance. This problem was first recognized when the size of machines increased rapidly, stretching the distance between first-order monuments from 30 m (CERN-ISR) to 1200 m (CERN-SPS), and thereby magnifying and rendering visible this effect. To overcome the problem, techniques were developed to separate relative displacements from the absolute trend curve—techniques which we now refer to as "smoothing." After smoothing, the final distribution of residuals is examined by Fourier decomposition-type analysis to ensure that no significant amplitudes occur at the betatron frequency.

11.2.6.1 SLAC–Style Smoothing

The alignment tolerances set out for the SLC show how smoothness is more important than absolute positioning for beam transport.[33] For this machine, a global positioning envelope is set to ±5 mm for every arc magnet, while the relative alignment of three adjacent magnets should be within ±0.1 mm.

The pitched and rolled sausage-link beam line formed by the arc magnets makes this modeling particularly difficult. The absolute design shape of the path is a series of curves and straight sections in pitched and rolled planes. This form does not readily lend itself to fitting with polynomials or splines. The large coupling of the horizontal and vertical also prevents the separation of smoothing operations into two components.

The complication of an irregularly shaped beam line was eliminated by subtracting out the actual size and shape of the beam line, leaving a series of residual misplacements for a string of magnets (Fig. 34). Correlation between horizontal and vertical misalignments is removed using a spatial fitting routine. Principal Curve Analysis[34] was chosen to simultaneously pass a one-dimensional curve through the horizontal and vertical residual misalignment mapped out along the Z-axis (beam direction). This curve passes through the middle of the data set such that the sum of the squared errors in all variables is minimized (Fig. 35). The curve is nonparametric, with its shape suggested by the data.

The smoothing algorithm provides the options that make it possible to minimize movements of the magnets onto a smooth curve, and to identify outliers. If an outlier (e.g., erroneous measurement) exists, it may artificially bias the fitting routine and draw the curve

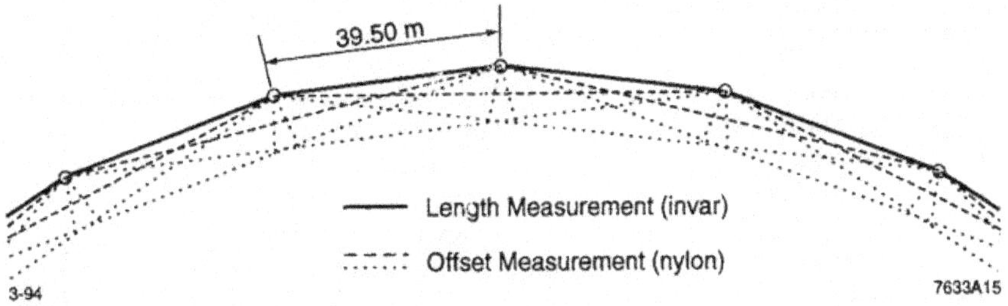

Fig. 36. Radial offset measurements.

away from the general neighborhood trend. For this reason, a robustness estimator is included in the modeling program to weight out these points.

One improvement was suggested through experience. This involved the independent weighting of points, so that a small area of magnets can be "patched in" to existing elements. Other improvements made it possible to deal with irregularly spaced and patterned beam line layouts.

11.2.6.2 CERN–Style Radial Offset Smoothing

The Super Proton Synchrotron (1971–76) presented major new challenges. The SPS was housed in an underground ring of 950 m radius. Six penetration shafts enabled the transfer of survey control from the geodetic network on the surface. The absolute error envelope ranged in size from 1.3 mm at each of the six control points to 2.5 mm midway along each 1152 m-long sextant, far exceeding the 0.15 mm radial alignment tolerance. A procedure of radial smoothing was developed to achieve a relative alignment within this tolerance. Measurements were made directly from each magnet to adjacent magnets with no reference to the control monuments mounted on the tunnel walls[35] (Fig. 36). This gave overlapping measurements of local curvature, which were then entered into a least-squares adjustment, minimizing the sum of squares of both the residuals and the radial offsets.[36] A relative alignment tolerance of about 0.08 mm was achieved using this method. Vertical alignment was undertaken as a separate process, using standard leveling practices for both absolute and relative vertical alignment.[37]

The final alignment of the ESRF ring has been successfully achieved using the CERN method.

11.2.7 Survey and Alignment Toolbox

Table 2 shows the progression through the typical contents of an accelerator surveyor's toolbox. The available space here does not allow a discussion of these tools. However, many surveying textbooks cover these instruments and software tools very competently.[38–46]

11.3 Ground Motion

Only in the last decade with the arrival of high-energy colliders and the third generation of light sources, have ground motion issues become significant and been studied.[47–49]

Table 2. Typical tools in an accelerator surveyor's toolbox.

Hardware tools	Software tools
Geodetic instruments	Integrated database
Theodolite	Data collection routines
Total station	Raw data reduction
Level	Analysis input merging
Plummet	Blunder detection
EDM	Network adjustments:
Distinvar	1-, 2-, or 3-dimension
Distometer	Unconstrained datum
Optical tooling	Overconstrained datum
Jig transit	Bundle adjustments
Spirit level	Graphical output
Alignment telescope	Coordinate database
Interferometry	Data analysis
Photogrammetric equipment	Coordinate transformations
Coordinate measuring machine	Deformation analysis
Stationary	Shape fitting routines
Portable	Special layout programs
Dial gauges	Ideal coordinate calculation
	Alignment movements
Laser tracker	
Industrial measurement system	Smoothing routines
Forced centering system	On-line alignment control program
Targeting systems	Free stationing

Ground motion is conveniently categorized into that due to nature and that due to man. Natural ground motion excites movements with long periods, seconds to years, while manmade ground motion, having far less energy content, is caused locally and generally has frequency components from a few to 50 Hz.[50]

11.3.1 Natural Sources

The main natural sources are:

- Ground settlement Every new construction project experiences some ground settlement. The effects can be minimized by building in areas of competent soil, by minimizing terrain disturbance, and by maintaining ground water levels.
- Tectonic motion Relative motion across a fault can reach several centimeters per year. Since almost all active faults are mapped, it should be possible to avoid these faults. However, there are still many unknown faults (e.g., fault lines discovered at SLAC during the Loma Prieta earthquake[51] in 1989) that could generate ground motion effects at an unknown later time.

- Earth tides The partially elastic body of the earth is deformed by the gravitational attraction of the moon and sun, causing diurnal and semidiurnal tides. The effects are always less than a decimeter and of very long wavelength. Today's light sources are too small for earth tides to be significant.
- Earthquakes Since severe earthquakes happen only relatively seldom, they need not be considered for the daily operation of a light source. However, site selection must evaluate the probability of potential seismic events and its effect on the design of structures.
- Ambient microseismic noise The main source for natural ambient microseismic noise is the coupling of ocean waves to the continents. Since the attenuation inland from the coasts is small, the effect is measurable throughout a continent.[52] Fischer and Morton[53] calculate the time-averaged rms amplitude to reach about 1 mm.

11.3.2 Cultural Noise

Local cultural noise is the dominant signal in the spectral region of a few hertz and above. Figure 37 shows the noise that is measured at DESY over the course of a week.[54] The most common sources are:

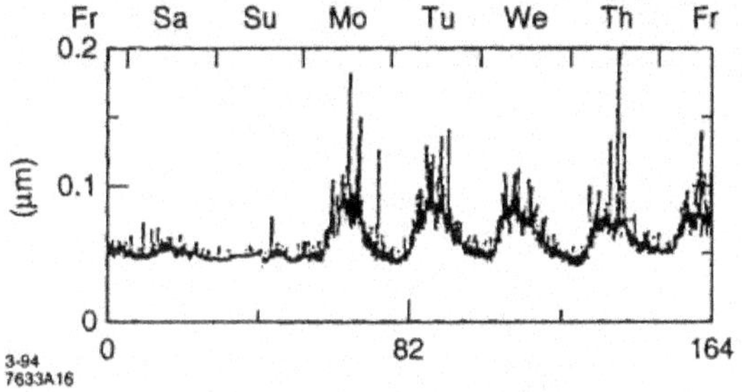

Fig. 37. Cultural noise at DESY.

- Railroad traffic An object as massive as a freight train traveling perhaps at speeds above 100 km/h will couple some energy to the ground.
- Vehicle traffic, on-site and off-site Fischer and Morton[56] report that auto and truck traffic produces disturbances at a tunnel level exceeding 0.5 μm. During site selection, it is therefore important to consider the existence and proximity of public streets, highways, and freeways.
- Continuously operating machinery Compressors, water pumps, fans, and especially all reciprocating machinery, contribute to the noise level of a site. Data given by Fischer and Morton[57] shows that two 75 hp vertical piston compressors operating at 6 Hz can produce a 1 μm peak to peak motion at 30 m. Appropriate isolation from the ground becomes very important.

11.3.3 Countermeasures

11.3.3.1 Prevention

The best countermeasure to ground motion is prevention. First of all, the machine should be designed to make it less sensitive to the positional stability of its components. Secondly, adherence to good engineering and housekeeping principles will prevent, or at least minimize, the effects of traffic noise and of reciprocating machinery. Thirdly, it is important "to prevent politicos from choosing sites that are severely beset by natural and man-made disturbances."[58]

11.3.3.2 Active Countermeasures

- Vibration isolation Fast ground motion can be dampened and significantly reduced with active isolation.[58–60] Ishihara reported that a table was kept stable to 50 nm against a sine wave disturbance with 500 nm amplitude and frequency up to 50 Hz.[61] Such a system was incorporated into the FFTB at SLAC to stabilize the quadrupoles at the interaction point.
- Dynamic alignment system Slow frequency ground motion can be compensated for by dynamic alignment systems. A first step towards a dynamic system was made at the ESRF with the development and deployment of the automated hydrostatic level system and the remotely controlled vertical jacks (see Section 11.1.3 above). A first truly dynamic alignment system for vertical and lateral alignment was implemented in the FFTB at SLAC.[62] First commissioning results indicate that the system can maintain the alignment to a few microns over the course of weeks.
- Feedback on beam derived information Feedback systems have always helped to overcome the dilemma of not meeting tolerances. Hettel reports a successful application at SSRL.[63] However, there are limitations to the application of beam derived intelligence. Fischer warns that "the proliferation of feedback systems will, if not held in check, lead to increasing inoperability since each system adds another layer of complexity."[64]
- Feedforward on beam derived information Since the low-frequency disturbances of an orbit derive primarily from quadrupole magnet vibrations of a certain dominant mode, a scheme was developed to compensate for the quadrupoles' vibration movements. Seismic accelerometers measure the magnets' vibrations and drive the compensation current into the quadrupoles' trim coils accordingly. Yao reports better than 99% canceling of field shaking due to 10 μm magnet vibrations.[65]

Acknowledgments

I would like to thank all the individuals at the universities and accelerator laboratories around the world who shared their ideas and experience with me. The section on magnet supports would not have been possible without the positive response to my request for local support system design examples: thank you to Gordon Bowden, SLAC; Horst Friedsam, APS; Ted Lauritzen, ALS; Michel Mayoud, CERN; Will Oren, CEBAF; Willfried Schwarz, DESY; and Rick Wilkins, SSC. Special thanks to Bernard Bell for painstakingly reading the manuscript.

References

1. Ted Lauritzen, "The ALS Six Strut Support System," presentation at the Pohang Light Source Laboratory (Pohang, September 1992), p. 4.

2. *Conceptual Design Report, 1–2 GeV Synchrotron Radiation Source* (Lawrence Berkeley Laboratory, Berkeley, July 1986), p. 77.
3. Anocast, a Division of Anorad Corp., 110 Oser Ave., Hauppauge, NY 11788.
4. Ted Lauritzen, private communication.
5. Michel Mayoud, private communication.
6. George Biallas, private communication.
7. Charles Perkins, private communication.
8. Bill Davies-White, private communication.
9. Michel Mayoud, private communication.
10. Dieter Walz, private communication.
11. Willfried Schwarz, private communication.
12. Gordon Bowden, private communication.
13. A. Bergamin et al., "Servopositioning with Picometer Resolution," *Rev. Sci. Instrum.* **64** (1993) 168–173.
14. E. Furse, "Kinematic Design of Fine Mechanisms in Instruments," *Phys Sci. Instrum.* **14** (1981) 264–271.
15. Daniel Roux, "Alignment & Geodesy for the ESRF Project," in *Proc. First Int. Workshop on Accel. Alignment* (SLAC, Stanford, 1989), SLAC–375, p. 37.
16. Daniel Roux, "The Hydrostatic Leveling System (HLS)/Servo–Controlled Precision Jacks—A New Generation Altimetric Alignment and Control System," in *Proc. Particle Accel. Conf.* (Washington DC, 1993), pp. 29321f.
17. John E. Jackson, *Sphere, Spheroid and Projections for Surveyors*, (Granada, London 1980), p. xi; Petr Vaníchek and E. Krakiwsky, *Geodesy—The Concepts*, (Elsevier, Amsterdam, 1986), pp. 1–8, 110.
18. Jackson, *op. cit.*, p. 84.
19. Alfred Leick, *GPS Satellite Surveying*, (John Wiley & Sons, New York, 1990), p. 188.
20. Clyde C. Goad, "Precise Positioning with the GPS," in *Applied Geodesy for Particle Accelerators*, (CERN Accelerator School, CERN 87–01, Geneva, 1987), p. 36ff.
21. Robert Ruland and A. Leick, "Application of GPS in a High Precision Engineering Survey Network," in *Proc. First Symp. on Precision Positioning with GPS* (Rockville, MD, 1985), p. 483ff.
22. Franz Löffler and W. Schwarz, "The Geodetic Approach for HERA," in *Proc. First Int. Workshop on Accelerator Alignment* (Stanford, 1989), SLAC–375, p. 117.
23. By "Transport," we mean the generic computer code that performs the ion-optical simulation.
24. Helmut Wiedemann, *Particle Accelerator Physics, Basic Principles and Linear Beam Dynamics* (Springer Verlag, Berlin, 1993), pp. 226–228.

25. Will Oren and R. Ruland, "Survey Computation Problems Associated with Multi-Planar Electron–Positron Colliders," in *Proc. 45th ASP–ACSM Convention*, (Washington, DC), pp. 338–347; and SLAC–PUB–3542.
26. Horst Friedsam et al., "Magnet Fiducialization with Coordinate Measurement Machines," in *Proc. First Int. Workshop on Accelerator Alignment* (SLAC, Stanford, 1989), SLAC–375, p. 206ff.
27. Alex Harvey, "The Magnet Fiducialization Problem," *ibid.*, p. 200.
28. Cliff M. Fortgang et al., "Pulsed Taut–Wire Alignment of Multiple Permanent Magnet Quadrupoles," in *Proc. 1990 Linear Accel. Conf.* (Albuquerque, 1990), p. 48.
29. Franz Löffler, "Referencing the Magnetic Axis for HERA's Superconducting Magnets," in *Proc. First Int. Workshop on Accelerator Alignment* (SLAC, Stanford, 1989), SLAC–375, p. 232.
30. Gerhard E. Fischer et al., "Finding the Magnetic Center of a Quadrupole to High Resolution," *ibid.*, p. 213
31. Robert Ruland, "The Chesapeake Laser Tracker in Industrial Metrology," in *Proc. Third Int. Workshop on Accel. Alignment* (Annecy, 1993), pp. I/101ff.
32. Michel Mayoud, *op. cit.*, p. 138.
33. Horst Friedsam, W. Oren, "The Application of the Principal Curve Analysis Technique to Smooth Beam Lines," in *Proc. First Int. Workshop on Accelerator Alignment* (SLAC, Stanford, 1989), SLAC–375, pp. 152–161.
34. Trevor Hastie, *Principal Curve Analysis*, SLAC–276, 1984.
35. Michel Mayoud, "Geodetic Metrology of Particle Accelerators and Physics Equipment," in *Proc. First Int. Workshop on Accelerator Alignment* (SLAC, Stanford, 1989), SLAC–375, p. 138.
36. Mayoud, *ibid.*, and private communication.
37. Jean Gervaise, "Applied Geodesy for CERN Accelerators," *Seminar on High-Precision Geodetic Measurements* (University of Bologna, 1984).
38. Fritz Deumlich, *Surveying Instruments* (de Gruyter, Berlin, 1982).
39. Peter Richardus, *Project Surveying* (A.A. Balkema, Rotterdam, 1984).
40. Grün/Kahmen, eds., *Optical 3–D Measurement Techniques II*, (Wichmann, Karlsruhe, 1993).
41. J. Uren and W.F. Price, *Calculations for Engineering Surveys* (Van Nostrand Reinhold, Wokingham, 1984).
42. Fritz Hennecke et al., *Handbuch Ingenieurvermessung* (Wichmann, Karlsruhe, 1988).
43. Heribert Kahmen and Wolfgang Feig, *Surveying* (de Gruyter, Berlin, 1988).
44. J.M. Rüeger, *Electronic Distance Measurement* (Springer–Verlag, Berlin, 1990).
45. Ted Busch, *Fundamentals of Dimensional Metrology* (Delmar Publishers, New York, 1966).

46. Philip Kissam, *Optical Tooling for Precise Manufacture and Alignment* (McGraw–Hill, New York, 1962).
47. Helmut Wiedemann, "Tolerances on the Dynamic Stability of Ring Components," ESRP Internal Report ESRP–IRM–81/84 (Oct. 1984).
48. T. Aniel and J.L. Laclare, *Sensitivity of the ESRP Machine to Ground Movement*, Saclay LNS/086, January 1985.
49. Gerhard E. Fischer, "Ground Motion and its Effects in Accelerator Design," 1984 Summer School Lecture at FNAL, SLAC–PUB–3392R (July 1985).
50. Gerhard E. Fischer and P. Morton, "Ground Motion Tolerances for the SSC," SLAC–PUB–3870, SSC–55 (1986).
51. Robert Ruland, "A Summary of Ground Motion Effects at SLAC Resulting from the Oct. 17th Earthquake," in *Proc. Second Int. Workshop on Accel. Alignment* (Hamburg, 1990), pp. 131–156.
52. K. Aki and P. Richards, *Quantitative Seismology* (Freeman and Co., 1980) Vol. 1, Ch. 10.
53. Fischer and Morton, *op.cit.*, p. 9.
54. Jürgen Rossbach, *HERA Errors and Related Experiments during Commissioning*, DESY HERA 91–21 (1991).
55. Fischer and Morton, *op.cit.*, pp. 14–16.
56. Fischer and Morton, *op.cit.*, p. 17.
57. Gerhard E. Fischer, "Ground Motion — An introduction for Accelerator Builders," *Proc. CERN Accelerator School, Magnetic Measurement and Alignment* (Montreux, 1992); and SLAC–PUB–5756.
58. W. Ash, "Final Focus Supports for a TeV Linear Collider," SLAC–PUB–4782 (1988).
59. N. Ishihara et al., "A Test Facility of Active Alignment System at KEK," in *Proc. First Int. Workshop on Accelerator Alignment* (SLAC, Stanford, 1989), SLAC–375, p. 73ff.
60. M. Naganoh et al., "Active Control Microtremor Isolation Systems," *ibid.*, p. 287ff.
61. Ishihara et al., *op.cit.*, p. 76.
62. Robert Ruland, "A Dynamic Alignment System for the Final Focus Test Beam," in *Proc. Third Int. Workshop on Accel. Alignment,* (Annecy, 1993), p. 241ff.
63. R. O. Hettel, "Beam Steering at the Stanford Synchrotron Radiation Laboratory," *IEEE NS–30*, **4**, (1983), p. 2228ff.
64. Gerhard E. Fischer, "Alignment and Vibration Issues in TeV Linear Collider Design," in *Proc. First Int. Workshop on Accelerator Alignment* (SLAC, Stanford, 1989), p. 284; and SLAC–PUB–5024.
65. Cheng Yao, "Compensation of Field Shaking due to the Magnet Vibration," in *Proc. Particle Accel. Conf.* (Washington, 1993), p. 1393ff.

CHAPTER 12: BEAM INSTABILITIES

M. FURMAN, J. BYRD AND S. CHATTOPADHYAY*

Center for Beam Physics
Accelerator and Fusion Research Division
Lawrence Berkeley Laboratory
University of California,
Berkeley, CA 94720, U.S.A.

12.1 Introduction

So far we have considered the motion of the particles in the accelerator in given external electric and magnetic fields. As the particles traverse the ring, however, they interact with their surroundings via the electromagnetic field created by their own charge and current. This field extends for a certain distance behind the particles that created it, and is called the *wake field*. As an example of this interaction, the resistivity of the vacuum chamber causes ohmic losses as the wake field drags along the image currents in the wall of the chamber. In addition, the wake field can act back on the same bunch that created it and/or on the other bunches that come behind. Or, as a bunch traverses an RF cavity, it can excite one or more of its higher-order modes (HOMs); although the electromagnetic fields of these modes are mostly trapped inside the cavity, they typically resonate for a long time, and can therefore influence all the bunches in the beam as they, in turn, traverse the cavity. Thus in general, if certain conditions are met, the wake field can act back on the beam in such a way that an initial disturbance gets amplified and hence an instability is generated. In some cases, the disturbance grows indefinitely, causing beam loss; in others, the disturbance saturates, growing only until a new equilibrium situation is reached.

The key signature for these phenomena is an *intensity dependence*: when the current is low the wake fields are weak, and the beam characteristics are dominated by single-particle dynamics. As the beam current is increased, the wake fields become stronger and can influence the beam dynamics, and hence the machine performance, significantly. Some of these phenomena depend smoothly on current, and some others have a well-defined onset as the current exceeds a threshold value beyond which the wake field forces overcome the damping mechanisms. All of these phenomena arise because the beam, being a collection of charges, acts back on itself via the environment in which it travels; for this reason, these are called *collective phenomena*.

In this chapter we present an outline of the typical instabilities observed or expected in light-source synchrotrons and the techniques used to avoid or mitigate them. Since the physics of instabilities is generic to all rings that store relativistic electrons or positrons, it is

* Work supported by the Director, Office of Energy Research, Office of High Energy and Nuclear Physics, High Energy Physics Division, of the U.S. Department of Energy under Contract no. DE-AC03-76SF00098. Fax: (510) 486-7981. E-mail addresses: miguel@lbl.gov, jbyrd@lbl.gov and chapon@lbl.gov.

important to note that the situation in light sources is very similar to what is found in other circular machines such as damping rings and e^+-e^- colliders.

At the core of any discussion on instabilities is the *impedance of the machine*, which is closely related to the wake field. Once the impedance is known, it is possible to calculate the thresholds and growth rates of the instabilities. The definition of impedance and a discussion of its properties and measurement techniques is presented later.

We also describe briefly ion trapping, intrabeam scattering and Touschek scattering. Although these phenomena do not depend critically on the interaction of the beam with its surroundings, they are nevertheless intensity dependent, and in this sense they can be considered collective effects.

12.1.1 Stability[1]

The actual closed orbit in a real machine deviates from the ideal closed orbit due to inevitable errors in survey and alignment. Typically, the maximum value of this deviation could range from a few mm to a cm, arising from a realistically achievable survey and alignment error of 0.1 mm. This is a time-independent, stationary configuration and can be improved by a closed-orbit measurement and correction scheme, employed in all modern storage rings. An irreducible residue of 0.1–0.5 mm in the maximum closed-orbit deviation is achievable after a convergent series of iterations.

Machine operation would be simple if the orbit, the lattice functions and the RF parameters were independent of time and particle oscillations were linear to large amplitudes. The challenge of control of the photon source stems from the reality of time-dependent perturbations and the essential nonlinearity of the beam dynamics at large amplitudes. There is always long-term ground motion and various vibrations and noise sources at shorter time scales, and particles are subjected to large oscillation amplitudes at injection as well as during the rest of the lifetime of the beam by various scattering processes. In addition, there are other time-dependent processes, such as coherent beam instabilities, oscillations of trapped ions interacting with the beam, etc. The frequencies and time scales of these various processes, their sources, manifestations in the beam and ring properties, monitoring systems and possible cures can be grasped by a look at Fig. 12.1

12.1.2 Overview of Instabilities and Their Effects

Instabilities are usually classified into single-bunch and multi-bunch. Single-bunch instabilities are strongly influenced by short-range wake fields arising from small structures in the vacuum chamber such as bellows, discontinuities, vacuum ports, beam position monitors, etc. Multi-bunch instabilities are strongly influenced by long-range wakefields, or by localized wake fields that last for a long time. As mentioned above, the most important mechanism that gives rise to such wakefields is the excitation of HOMs in resonant structures, especially the RF cavities. The wake fields produced by the finite resistivity of the vacuum chamber are also important in this respect.

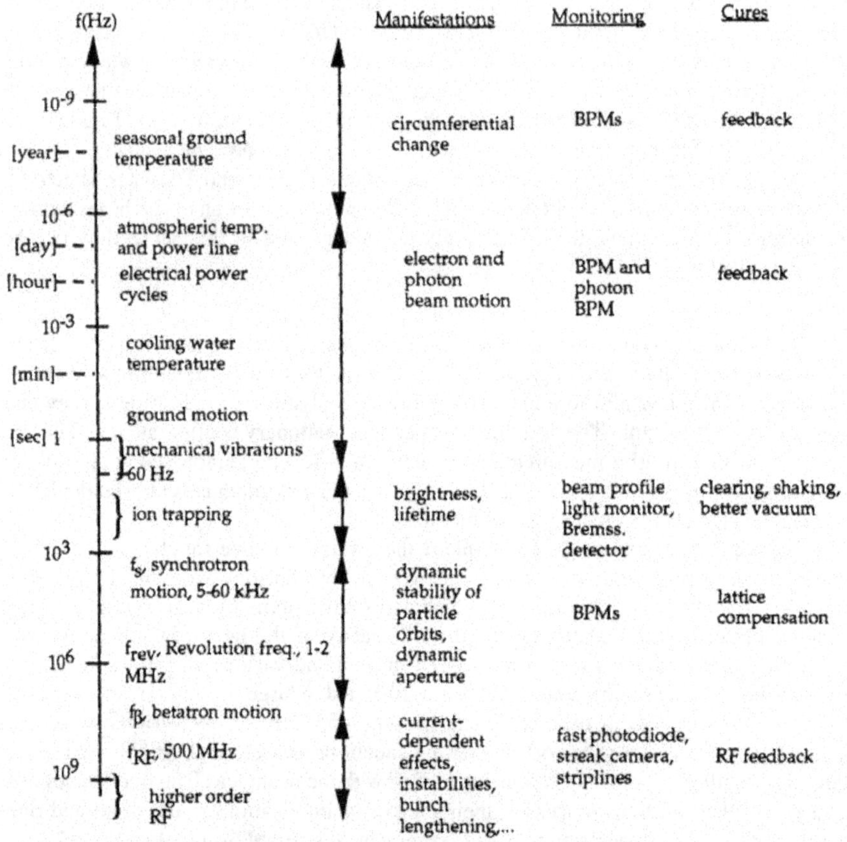

Fig. 12.1. Various frequencies and time scales relevant to storage-ring stability.

Instabilities can have two kinds of unfavorable effects: degradation of the beam lifetime, and degradation of the beam quality. These two effects are not mutually exclusive. Typical design goals of light sources are long beam lifetime, small beam emittance, short bunch length, small energy spread and stable orbits. Instabilities can cause bunch lengthening, increased energy spread, shortened beam lifetime or bunch-to-bunch jitter of the beam orbit or of the bunch arrival time. Examples of mechanisms that can affect the beam lifetime are the transverse mode-coupling instability and multi-bunch coupling. The first phenomenon is single-bunch, the second multi-bunch. In the coupled-bunch instability, all the bunches act together in such a way as to cause a resonance whose typical time scale is quite short. Therefore, unless a feedback system is active, a coupled-bunch instability can lead to sudden beam losses. An example of instability that degrades the beam quality is the longitudinal microwave instability, which increases the energy spread and the bunch length. Longitudinal multibunch oscillations can be stable (finite amplitude oscillations), but they cause a jitter in the arrival time of the bunches at a given point in the machine, thus having a detrimental effect on applications that are sensitive to time resolution. Similarly, stable transverse multibunch oscillations lead to an effective increase in the beam emittance and hence a degradation of the brightness of the emitted synchrotron light.

12.1.3 Damping Mechanisms

The most important way to mitigate single-bunch instabilities is by careful design of the vacuum chamber. Modern designs place a premium on its smoothness, since this leads to smaller impedance and hence a decreased chance for instabilities at a given bunch current. In practice, of course, it is not possible to have a perfectly smooth chamber, and hence certain compromises must be made.

Two mechanisms that help damp instabilities exist naturally in any electron storage ring. The first and most obvious one is the damping provided by the radiation of the synchrotron light. Instabilities whose growth time is longer than the damping time (which is typically on the order of several thousand turns), do not manifest themselves as such and do not lead to a problem. The second mechanism, which is more subtle, is Landau damping. As explained in more detail below, this mechanism requires a spread in the oscillation frequency of the particles within a bunch. Landau damping effectively transforms the coherent motion of the beam into incoherent motion of the particles via phase mixing induced by the oscillation frequency spread. In the case of transverse oscillations, a frequency spread is provided naturally by the unavoidable machine nonlinearities which, in turn, lead to an amplitude dependence of the betatron frequencies. A longitudinal frequency spread also exists naturally due to the sinusoidal RF voltage, leading to nonlinearity of the synchrotron forces at large amplitude. If the natural nonlinearities are not strong enough (in the case of small-emittance beams, for example), there are artificial means of enhancing them, as discussed below.

In any case, these damping mechanisms are typically not enough to eliminate all instabilities in modern light sources, at least not when these machines are operated in high-current, multi-bunch mode. Certain instabilities can be avoided by proper choice of parameters; for example, the Robinson instability is avoided by a slight detuning of the fundamental RF frequency away from $h\omega_0$, where h is the harmonic number (see Chapter 4) and ω_0 is the angular revolution frequency.

Generally speaking, the design of most synchrotron light sources is such that single-bunch instabilities are avoided, or at least are not serious. Coupled-bunch instabilities are alleviated by damping the HOMs of the RF cavities, which can be achieved by clever design of the cavity shape, and by adding damping elements. However, it is typically impossible to avoid all such instabilities by passive methods. An active feedback system (see Chapter 13) is thus required that detects incipient unstable motion and applies appropriate compensating time-dependent forces to counteract it. Although it is in principle possible to design a feedback system that would eliminate single-bunch instabilities, in practice the power and bandwidth requirements on such a system would typically make it prohibitively expensive.

12.2 Wake Fields and Impedances

12.2.1 Definitions[2]

Whenever a relativistic charged particle travels near a material that is not perfectly smooth or not perfectly conducting, an electromagnetic field is created that extends for a certain distance behind it and lasts for a certain characteristic time before it dissipates. This is the *wake field* which, in turn, can act back on the particles traveling behind the one that created it. If the wake field lasts for a sufficiently long time, it will affect the particles in trailing bunches in successive turns. The impedance is essentially the Fourier transform of the Lorentz force caused by the wake field, and is thus a measure of the strength and shape of the frequency spectrum of this time-varying force

In the simplest version, the "beam" consists of a single particle of charge q traveling at the speed of light c down a cylindrically-symmetric pipe. The beam trajectory is a straight line parallel to the axis but is offset transversely from it by $\mathbf{x}$. We consider a "test particle" of charge e also traveling at the speed of light parallel to the beam at a distance z behind it, with some transverse position of its own. The pipe need not be perfectly conducting or smooth; however, we assume that the lack of smoothness is not extreme, and the average pipe radius is b. From Maxwell's equations one can calculate in principle the transverse and longitudinal electromagnetic force on the test particle. If we integrate these forces over a distance $L \gg b$, we obtain, by definition, the *wake functions* $W_\parallel(z)$ and $W_\perp(z)$

$$\int_0^L ds\, F_\parallel \equiv -eqW_\parallel(z)+\cdots, \qquad \int_0^L ds\, \mathbf{F}_\perp \equiv -eq\mathbf{x}W_\perp(z)+\cdots \tag{12.1}$$

where $\cdots$ refers to terms of higher order in $\mathbf{x}$ and in the transverse position of the test particle, and where the integration variable s is the distance along the trajectory of the test particle. The corresponding impedances for the distance L are defined by the Fourier transform of the wake functions, and are given by

$$Z_\parallel(\omega) \equiv \frac{1}{c}\int_{-\infty}^{\infty} dz\, e^{-i\omega z/c}\, W_\parallel(z), \qquad Z_\perp(\omega) \equiv \frac{i}{c}\int_{-\infty}^{\infty} dz\, e^{-i\omega z/c}\, W_\perp(z) \tag{12.2}$$

where the sign convention is that $z<0$ means that the test particle is *behind* the beam.

Obviously, the distinction between the beam and the test particle is a purely mathematical one that allows one to define wake functions and impedances. In a real machine, all particles play both roles, since they produce wake fields and are, in turn, affected by the wake fields of all particles.

Impedances summarize all the electromagnetic effects from the environment traversed by the beam. Thus the vacuum chamber resistivity, RF cavities, bellows, discontinuities, vacuum ports, flanges, curvature of the chamber, synchrotron radiation reaction, etc., all contribute to the impedance. For a circular machine, the distance L in (12.1) is usually taken to be the circumference, so that Eqs. (12.2) represent the whole-ring impedances. In reality, the forces on the test particle fluctuate as the beam and the test particle traverse the different structures along the vacuum chamber. A basic underlying assumption in the usefulness of the wake functions is that the forces on the test particle do not deviate much from their average value. In some cases, however, this averaging does not yield accurate results for certain instabilities such as those caused by coherent synchro-betatron resonances.[3] In these cases, the localized nature of the impedance is important and special methods, such as simulation codes with a time-dependent Maxwell's equations solver, must be used; we will not be concerned with such a possibility here. Therefore, even though wake functions are defined in principle for arbitrary boundary conditions, their usefulness as analytical tools diminishes as the characteristics of the vacuum chamber become more and more complicated. Fortunately most modern accelerators do not fall under this category.

Even for a point charge in a perfectly smooth cylindrical pipe, there is an infinite number of wake functions (and impedances) represented by $\cdots$ in Eq. (12.1). These terms represent a power-series expansion of the transverse or longitudinal force in the transverse displacement of the beam and of the test particle. The leading terms are those shown above, namely the monopole longitudinal wake function, usually labeled $m=0$, and the dipole transverse wake function, usually labeled $m=1$. The corresponding impedances $Z_{\parallel}(\omega)$ and $Z_{\perp}(\omega)$ in Eqs. (12.2) are also labeled $m=0$ and $m=1$, respectively. The words "monopole" and "dipole" refer to the fact that the forces are produced by the monopole and dipole moments of the charge distribution of the beam, respectively (note that the forces on the test particle, Eqs. (12.1), are independent of its transverse position through dipole order). The higher-order impedances become important when the transverse size of the beam is comparable to the vacuum pipe diameter. We shall not be concerned here with any m's higher than 1 for the transverse case or higher than 0 for the longitudinal (the transverse $m=0$ wake function vanishes by symmetry), and we will omit the label m.

On the other hand, the *longitudinal* charge distribution does matter in many cases. Thus, for most purposes, we can view the beam as consisting of needle-like bunches, and the calculation of the electromagnetic force on the test particles requires a superposition over the longitudinal charge distribution, typically assumed Gaussian. The impedance resulting from this superposition is called the *effective impedance.*

12.2.2 Properties and Basic Uses of Impedances[2]

As implied by Eqs. (12.1), the wake functions are *real functions*. Therefore, by taking the complex conjugate of Eqs. (12.2) we conclude that $Z_{\parallel}(\omega)^* = Z_{\parallel}(-\omega)$ and $Z_{\perp}(\omega)^* = -Z_{\perp}(-\omega)$. Therefore the real part of $Z_{\parallel}(\omega)$ is a symmetric function of ω, while the imaginary part is antisymmetric; the transverse impedance $Z_\perp(\omega)$ has the opposite parity properties, on account of the extra factor of i in its definition. These parity properties are generic of all impedances, not just those defined above.

Another generic property of the wake functions is that they are *causal functions*: since the forces ahead of the beam vanish, any wake function satisfies

$$W(z) = 0 \quad \text{for } z > 0. \tag{12.3}$$

A fundamental theorem of the Fourier transform of causal functions then implies that the corresponding impedance is an analytic function in the upper half of the complex-ω plane. One consequence of this is that the impedances satisfy a *dispersion relation* that relates the real and the imaginary parts: if the real part is known for all frequencies, the imaginary part is uniquely determined from the dispersion relation and vice versa. As a by-product of the dispersion relation, one concludes that any impedance must satisfy

$$Z(\omega) \to 0 \ \text{ as } \omega \to \pm\infty. \tag{12.4}$$

Even though there is no generally-valid relation between longitudinal and transverse impedances of different m's, qualitative arguments[2] show that the $m = 0$ longitudinal and the $m = 1$ transverse impedances are related by

$$Z_\perp(\omega) \approx \frac{2c}{b^2\omega} Z_\parallel(\omega) \tag{12.5}$$

This relation is strictly valid for the resistive wall impedance of a smooth, infinitely long, cylindrical pipe of radius b. It is approximately valid for other impedances arising from discontinuities in the chamber wall such as bellows, small cavity-like structures or other objects, provided their characteristic size is small compared to b. For larger objects, such as RF cavities, Eq. (12.5) is valid in an average sense, although it becomes more accurate at frequencies above the cut-off frequency ω_c, defined below.

In calculating beam instabilities, the longitudinal impedance usually appears divided by ω. Therefore, instead of $Z_\parallel(\omega)$, it is customary to deal with the quantity $Z_\parallel(\omega)/n$, where n is defined by $n \equiv \omega/\omega_0$. Thus $Z_\parallel(\omega)/n$ has the opposite parity properties as $Z_\parallel(\omega)$, and the approximate relation (12.5) reads

$$Z_\perp(\omega) \approx \frac{2R}{b^2} \frac{Z_\parallel(\omega)}{n} \tag{12.6}$$

where R is the average radius of the accelerator.

Equations (12.2) imply that $Z_\parallel(\omega)$ (and hence $Z_\parallel(\omega)/n$) is measured in Ω while $Z_\perp(\omega)$ is measured in Ω/m. Modern storage rings, whether colliders or light sources, are typically designed for operation in high-current, multibunch mode. Stable operation requires that the impedance should be kept small, which implies the need for a smooth vacuum chamber. Typically, the *broad-band average* (defined below) of the impedance for these modern rings is $|Z_\parallel/n|_{bb} \approx 0.1 - 2\ \Omega$. Older rings, which were not designed with these requirements in mind, typically have longitudinal impedance values $|Z_\parallel/n|_{bb}$ in the range one to tens of Ω. Figure 12.2 shows a photograph of a joint in the injection region of the ALS vacuum chamber that exemplifies the attention paid to keeping the impedance low.

Fig. 12.2. A transition joint in the vacuum chamber in the injection region of the ALS. The joint must be flexible in order to accommodate a transverse motion of a few cm during injection. The required flexibility and low impedance is achieved by a "wire cage" design (the entire assembly shown is enclosed in a large bellows in order to maintain the vacuum). If the joint had been left open, the discontinuity would have led to a large impedance. Photo courtesy of J. Corlett.

The integral that defines the longitudinal wake function in Eq. (12.1) is also equal to the change of energy of the test particle in the distance L due to the wake fields, so that

$$\Delta E = -eqW_{\parallel}(z) \tag{12.7}$$

(the sign convention is that $\Delta E > 0$ means that the test particle gains energy). For a bunch of particles, the total energy change in the distance L is given by a superposition over its longitudinal charge distribution $\rho(z)$. Assuming that the bunch length and the range of the wake function are both $\ll L$ one obtains[a]

$$\Delta E = -\int_{-\infty}^{\infty} \frac{d\omega}{2\pi} \left|\tilde{\rho}(\omega)\right|^2 \operatorname{Re} Z_{\parallel}(\omega) \tag{12.8}$$

where $\tilde{\rho}(\omega)$ is the Fourier transform of $\rho(z)$. Only the real part of the impedance contributes to the integral because $\left|\tilde{\rho}(\omega)\right|$ is an even function of ω (this follows from the

[a] If the bunch length or the range of the wake function are not small, the integral in this equation must be replaced by a summation over the harmonics of the bunch frequency.

fact that $\rho(z)$ is real). Since the beam as a whole can only lose energy, and this must be true for an arbitrary charge distribution, it follows that

$$\mathrm{Re}\, Z_{\parallel}(\omega) \geq 0 \quad \text{for all } \omega . \tag{12.9}$$

The energy loss of a bunch is often expressed in terms of the *loss parameter* (or *loss factor*) k, which is defined to be

$$k = -\Delta E / q^2 \tag{12.10}$$

where ΔE and q are here the total energy change and the total charge of the bunch, respectively. The loss parameter is always positive, and its typical numerical value for RF cavities is ~a few V/pC. In practice, the bunch-length dependence of the loss parameter can give useful information about the impedance of cavities or cavity-like objects.

Equation (12.1) allows one to define an impedance-induced voltage which is the potential energy change of the test particle in the distance L due to the wake field. For a beam described by a charge distribution $\rho(z)$, this potential energy change is given by

$$V_{\parallel}(z') = -\int_{-\infty}^{\infty} dz\, \rho(z)\, W_{\parallel}(z' - z) \tag{12.11}$$

where we make the same assumptions as in the derivation of Eq. (12.1). In the frequency domain this equation is usually written

$$\tilde{V}_{\parallel}(\omega) = -\tilde{I}(\omega) Z_{\parallel}(\omega) \tag{12.12}$$

where $\tilde{I}(\omega)$ is the Fourier transform of the current, defined by $I(z) = c\rho(z)$.

12.2.3 Resonator Impedance Model

In practice it is impossible to accurately know the impedances for the ring as a whole. However, impedances of individual components can often be calculated or measured at least in a certain frequency range. More typically, one resorts to simple models with a few parameters for a given ring component; the parameters are then determined by fitting the model to the measurements.

A simple and widely-used model for the longitudinal impedance of a resonant structure such as an RF cavity is the superposition

$$Z_{\parallel}(\omega) = \sum_r Z_{\parallel,r}(\omega) \tag{12.13}$$

where $Z_{\parallel,r}(\omega)$ is a *single-resonator impedance*, defined by

$$Z_{\parallel,r}(\omega) = \frac{R_{S,r}}{1 + iQ_r\left(\dfrac{\omega_r}{\omega} - \dfrac{\omega}{\omega_r}\right)} . \tag{12.14}$$

Here Q_r is the *quality factor*, ω_r is the (angular) *resonant frequency*, and $R_{S,r}$ is the strength of the resonator, or *shunt impedance* (measured in Ω). Eq. (12.5) allows one to define a transverse resonator impedance which is of the same form as Eq. (12.14) except for an additional overall ω^{-1} factor.

The real part of the resonator impedance $Z_{\parallel,r}(\omega)$ has peaks at $\omega \approx \pm\omega_r$ with FWHM= ω_r/Q_r. The imaginary part changes sign as the frequency crosses its resonant value. By definition, a *broad-band resonator* has a relatively low Q, typically $Q \approx 1$, and therefore wide peaks. A *narrow-band resonator* has large Q and hence narrow peaks. For example, the typical resonant modes of ordinary RF cavities have Q's of order $10^2 - 10^4$, while those of superconducting cavities have Q's of order $10^6 - 10^9$. By taking the Fourier transform of (12.14) one finds that the decay time of the excitation produced by a resonator is $\tau_r = 2Q_r/\omega_r$. Therefore, narrow resonances last for a long time and thus are a leading cause of coupled-bunch instabilities.

The real part of any impedance is called the *resistive component* while the imaginary part is the *reactive component*. As shown above, only the resistive component can dissipate energy. If the reactive part is positive, it is called *capacitive*; if negative, *inductive*.[b] This terminology arises from the fact that a pure inductor L has an impedance $Z = -i\omega L$, which is negative imaginary, while a pure capacitor C has an impedance $Z = i/\omega C$, i.e., positive imaginary. In fact, a simple model for the single-resonator impedance is an RLC circuit in which all three elements are in parallel, and where the resistance is R_S. In this model the impedance is given by $Z^{-1} = R_S^{-1} + i/\omega L - i\omega C$ which is precisely of the same form as Eq. (12.14). The resonant frequency and the quality factor are given by $\omega_r = 1/\sqrt{LC}$ and $Q_r = R_S\sqrt{C/L}$, respectively.

For a broad-band resonator with $Q = 1$, the reactive part of $Z_{\parallel,r}(\omega)/n$ is inductive and almost independent of frequency in the range $-\omega_r \lesssim \omega \lesssim \omega_r$, while the resistive part has an approximately linear frequency dependence in this range. These properties are also true of $Z_\perp(\omega)$ on account of Eq. (12.6).

The loss factor for a high-Q resonator impedance ($Q \gg \sigma_t\omega_r$) traversed by a Gaussian bunch with rms bunch length σ_t (in time units) follows from Eq. (12.8),

$$k = \frac{R_{S,r}\omega_r}{2Q_r} e^{-(\sigma_t\omega_r)^2} . \tag{12.15}$$

Typically, the fundamental mode of a cavity has the lowest frequency, and is labeled by $r = 0$; this is typically the TM_{010} mode used to accelerate the particles or to replenish their energy that has been lost by the radiation process (see Chapter 4). HOMs, also called parasitic modes, are usually undesirable but unavoidable. Ideally, the Q's should be large for $r = 0$ and small for $r \geq 1$. In practice, one tries to reduce the Q's of the

[b] Many authors define the impedance with i replaced by $-j$ in Eqs. (12.2). This can lead to confusion in the definitions of capacitive and inductive. Whatever convention is used, the defining condition for a capacitive impedance is that the response is ahead of the excitation, while for an inductive impedance the response lags in phase behind the excitation.

HOMs as much as possible by cleverly reshaping the cavity or adding dampers. This is called "de-Qing," or damping, the modes. Typically, it is impossible to de-Q all the HOMs to the point that coupled-bunch instabilities are absent, hence the need for a feedback system. A prototype RF cavity for the PEP-II collider is shown in Fig. 12.3, which exhibits the three wave guides used to damp many HOMs.

Fig. 12.3. Prototype of the PEP-II collider cavity. The three large rectangular wave guides emanating from the body of the cavity are terminated with ferrite, and are used to damp the HOMs. Photo courtesy of R. Rimmer.

12.2.4 *Impedance Beyond Cutoff*[4]

If an RF cavity (or any cavity-like structure) were closed it would have an infinite number of modes. In practice, there are at least two openings needed for the beam traversal. Therefore, those modes whose wavelength is smaller than the pipe radius are not trapped in the cavity and are not resonant. Thus there is a natural *cutoff frequency*, ω_c, above which there are no more resonant cavity modes. It is usually defined by[c]

$$\omega_c \equiv c/b \tag{12.16}$$

In modern storage rings and light sources there is an increasing demand for shorter and shorter bunch lengths. The shorter the bunch, the higher the reach of its frequency spectrum. If the bunch is short compared to the vacuum pipe radius, its frequency spectrum

[c] For a perfect cylindrical pipe the cutoff frequency is $2.405c/b$. Since, in practice, the geometry is much more complicated, the usual convention is to choose the numerical factor to be unity for simplicity.

reaches beyond the cutoff frequency. Therefore the behavior of the impedance at these high frequencies can become important and needs to be examined.

The source of impedance beyond cutoff is the interaction of the beam with the synchrotron radiation that propagates down the vacuum chamber which, in turn, interacts with the different structures in the chamber. In addition, the curvature of the trajectory can make particles resonate with waves having the same angular phase velocity as the particles.

The high-frequency impedance of a curved toroidal vacuum chamber can be understood simply in terms of the far-field radiation in free space of a particle beam on a curved trajectory. Synchrotron radiation along a curved trajectory provides a dissipative mechanism analogous to resistive wall effects. The radiation reaction force follows from a longitudinal self-field that extracts kinetic energy. Neglecting the shielding provided by the vacuum chamber, the resultant *free space impedance* is given by

$$Z_{\parallel}(\omega)/n = \frac{\frac{1}{2}\Gamma\left(\frac{2}{3}\right)}{(3n^2)^{1/3}}\left(\sqrt{3}-i\right)Z_0 \tag{12.17}$$

where $\Gamma(2/3) \approx 1.35$ is the gamma function and $Z_0 = 4\pi/c = 377\ \Omega$ is the so-called *vacuum impedance*. This formula is valid for frequencies below the critical frequency (see Chapter 1), given by $\omega_{\rm crit}/\omega_0 = 3\gamma^3/2$ where γ is the usual relativistic factor of the particle (for $\omega > \omega_{\rm crit}$ the impedance falls off exponentially). The $n^{1/3}$-dependence of $Z_{\parallel}(\omega)$ is a consequence of the well-known $\omega^{1/3}$-dependence of the synchrotron radiation power at large frequency of a particle in circular motion. The real part of the free space impedance has the approximate numerical value

$$\mathrm{Re}\, Z_{\parallel}(\omega) \approx 300\, n^{1/3}\ \Omega . \tag{12.18}$$

The beam pipe, however, provides shielding for low frequencies: radiation is essentially suppressed for harmonics below a cutoff given by $n_c \approx (R/b)^{3/2}$. This expression for n_c is exact for a planar circular trajectory between two infinite conducting planes parallel to the orbit plane. As a result of this shielding, the maximum free space impedance is then given by

$$\left|\frac{Z_{\parallel}(\omega)}{n}\right|_{\max} \approx 300\, n_c^{-2/3}\ \Omega = 300\left(\frac{b}{R}\right)\ \Omega . \tag{12.19}$$

This result happens to be approximately the same for a large class of shielding geometries. Since, typically, $b/R = O(10^{-4})$, this means that the shielded free space value of $|Z_{\parallel}/n|$ is rather small. Nevertheless, prudent ring designers typically assume Eq. (12.19) to provide a *lower* bound for the estimate of the impedance beyond cutoff.

For vacuum chambers with cavity-like structures, several models have led to the generic behavior $Z_{\parallel}(\omega) \propto \omega^{-p}$ for the longitudinal impedance at high frequency. A crucial distinction has been established between a single isolated structure and an infinitely long sequence of cavities: for the first case, the power p has been found to be $p = 1/2$ while for the second, $p = 3/2$.

12.2.5 Impedance Calculations and Measurement Techniques[5]

The modeling and calculation of storage ring impedances has vastly improved over the past 20 years, mostly due to increased computer power and improved algorithms for the solution of Maxwell's equations. For example, electromagnetic modeling codes such as MAFIA[6] can calculate, in principle, the wake function and impedance of any three-dimensional geometry. In practice, if the geometry of the object is sufficiently complicated, the calculations become limited by the power of the computer used to run the codes.

Several bench impedance measurement techniques are used for testing actual or prototype beamline components. For nonresonant components, a wire is passed on axis and the transmission through it is measured as a function of frequency. For resonant structures, the strength of a mode is found by exciting it and measuring the resulting field pattern along the beam axis by introducing a small movable perturbing needle. The frequencies and quality factors of the resonant modes are found by measuring the transmission through the structure. Electromagnetic codes calculate quite accurately the R/Q ratios; by combining these calculated values with the measured Q's one can extract accurate values for the shunt impedances.

The impedance of individual resonant modes with $Q \gtrsim 10$ can be measured by the frequency-perturbation method. This is done by using beads or needles which shift the resonant frequency. From this one can measure the energy density on resonance and extract the R/Q ratios. In practice, since the dimensions of the perturbing object must be much smaller than the wavelength of interest, this technique cannot be used reliably at high frequency.

The pulsed-beam method is conceptually the most attractive since it closely mimics the dynamics in the storage ring, and the definition of the wake field given by Eq. (12.1). This method consists of passing a pulse of electrons through the test object and measuring the energy loss or the deflection of the trajectory to get the real part of $Z_\parallel$ or $Z_\perp$. To obtain the full wake function, one can use a smaller "witness" beam to sense the delayed effects. One can also probe with antennas to study the excitation in the test object. The Wake Field Test Facility at ANL is devoted to this method.[7]

Finally, of course, measurement of instability thresholds, bunch lengthening, etc., in operating storage rings may be the ultimate phenomenological tool to check calculations, predict behavior and evaluate cures. From these measurements one can, in principle, extract the ring impedance if the models used in the calculations are sufficiently complete and the measurements sufficiently accurate.

12.2.6 Broad-Band Impedance Model

In spite of the progress in the calculation and measurement of the impedance of individual components, the determination of the impedance of the storage ring as a whole is a challenging and typically imperfect task. Even more challenging is the calculation of the net effect on the beam given the individual component impedances. Experience has led to the development of the so-called *broad-band impedance model* to account for the entire storage ring impedance. This model provides a simple conceptual and calculational tool and it adequately represents a wide variety of storage rings. It has been particularly successful in describing the beam behavior in storage rings with longer bunches. Machine design reports usually contain an "impedance budget" listing the contributions of the different ring components to the impedance. In its simplest version, the longitudinal broad-band impedance model of the ring has three components:

- A broad-band resonator with ω_r and Q typically chosen to be $\omega_r \approx \omega_c$ and $Q \approx 1$. The shunt impedance R_{bb} is determined empirically from a fit to the data. This broad-band resonator accounts for the impedance contribution of all vacuum chamber components such as bellows, joints, low-Q parasitic cavity modes, vacuum chamber discontinuities, etc.
- A low-frequency contribution from the skin effect of the vacuum chamber known as the *resistive wall impedance*.
- Various narrowband resonator impedances including the fundamental mode of the RF cavity and other parasitic cavity modes.[d]

The contribution from the resistive wall impedance can be easily estimated for a typical vacuum chamber. For all frequencies of practical interest, the longitudinal impedance per unit length for an infinite cylindrical pipe of radius b is given by[e]

$$Z_{\parallel}(\omega)/L = \frac{Z_0}{4\pi b}(1 - i\varepsilon(\omega))\frac{\delta(\omega)|\omega|}{c} \tag{12.20}$$

where $\varepsilon(\omega)$ is the sign function, $\delta(\omega) = c/\sqrt{2\pi\sigma|\omega|}$ is the skin depth of the vacuum chamber whose conductivity is σ.

Determination of R_{bb} involves a wide variety of components. This shunt impedance is conventionally quoted as $|Z_{\parallel}/n|_{bb}$, which is defined to be

$$\left|\frac{Z_{\parallel}}{n}\right|_{bb} \equiv \lim_{\omega\to 0}\left|\frac{Z_{\parallel}(\omega)}{n}\right| = \frac{R_{bb}\,\omega_0}{Q\omega_r} \tag{12.21}$$

whose typical value for modern storage rings is in the range $|Z_{\parallel}/n|_{bb} = 0.1 - 1\,\Omega$. For illustrative purposes we sketch the broad-band impedance for a hypothetical storage ring in Fig. 12.4.

[d] Many authors exclude these narrowband components from the definition of the broadband impedance model.

[e] This formula is not valid for frequencies so low that the skin depth is comparable to or larger than the thickness of the vacuum pipe. It is also not valid at extremely high frequencies.

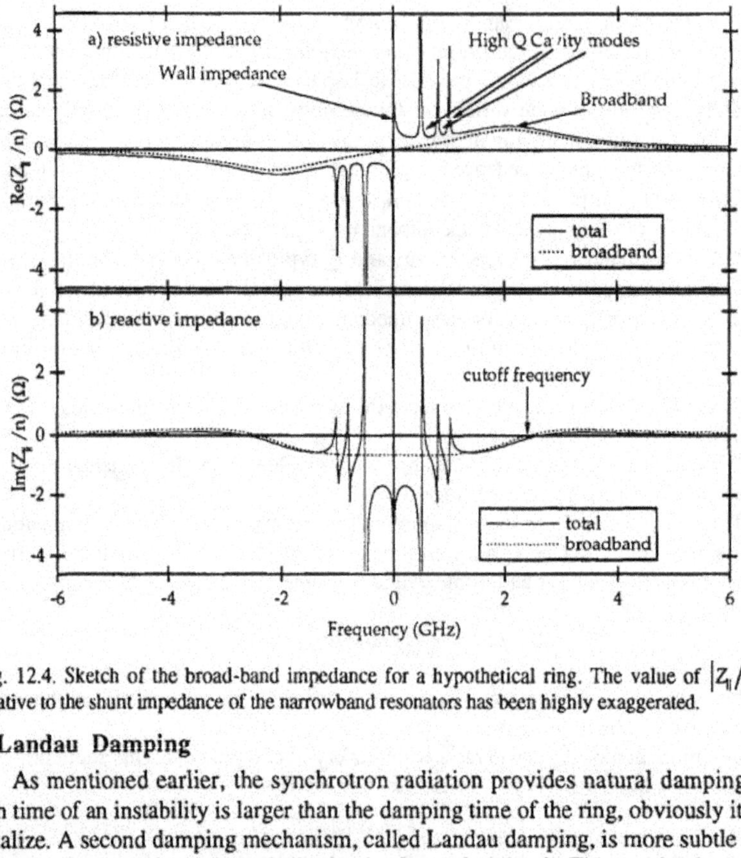

Fig. 12.4. Sketch of the broad-band impedance for a hypothetical ring. The value of $|Z_\parallel/n|_{bb}$ relative to the shunt impedance of the narrowband resonators has been highly exaggerated.

12.3 Landau Damping

As mentioned earlier, the synchrotron radiation provides natural damping: if the growth time of an instability is larger than the damping time of the ring, obviously it fails to materialize. A second damping mechanism, called Landau damping, is more subtle but just as important, and we sketch here the basic physics underlying it. This mechanism requires a *spread* in the oscillation frequency, or tune, of the particles.[8,9,2]

Consider a single particle executing transverse or longitudinal motion at low amplitude. The particle creates a wake field that acts back on itself, driving it on resonance and leading to an instability. This simple picture would lead one to expect essentially all particle motion in an accelerator to be unstable. Landau damping is one of the reasons why, in practice, this expectation is pessimistic.

Concretely, consider a single harmonic oscillator subject to a time-dependent sinusoidal force. The equation of motion is

$$\ddot{x} + \omega^2 x = A\cos\Omega t \qquad (12.22)$$

and we assume that the initial conditions are $x(0)=\dot{x}(0)=0$. If the force drives the particle resonantly, i.e., if $\Omega=\omega$, then the amplitude of the motion grows indefinitely according to

$$x(t)\propto At\sin\Omega t. \tag{12.23}$$

Correspondingly its energy, which is proportional to $\langle x^2\rangle$, grows like $\sim t^2$ for large t. implying instability.

An *ensemble* of particles (such as a bunch), however, can behave in a qualitatively different fashion *even if the particles do not interact among themselves*, provided their natural oscillation frequencies are spread over a certain range. Thus we assume that the oscillators have a narrow frequency spectrum $\rho(\omega)$ of width $\Delta\omega$, and that they are all driven by the same force, $F=A\cos\Omega t$, where Ω lies within the range of $\rho(\omega)$. For times $t\gg 1/\Delta\omega$ one finds that the centroid of the ensemble is given by

$$\langle x\rangle\propto A\left[\cos\Omega t\,\mathrm{P}\int d\omega\frac{\rho(\omega)}{\omega-\Omega}+\pi\rho(\Omega)\sin\Omega t\right] \tag{12.24}$$

where the symbol "P" instructs one to take the principal value of the integral at the singularity and the spectrum is normalized such that $\int d\omega\rho(\omega)=1$. One also finds that the energy of the bunch grows in time like

$$\langle x^2\rangle\propto A^2t\rho(\Omega). \tag{12.25}$$

The ensemble case and the single-particle case behave qualitatively differently in that the power of t with which the amplitude and the energy grow is one less in the former than in the latter. This is the essence of Landau damping: the energy pumped into a bunch of particles goes into increasing its size rather than the amplitude of the motion of the centroid. In most practical cases, this increase in bunch size does not present a problem, and therefore the instability is avoided.

If the oscillators are finite in number and their frequencies take on discrete values over the range $\Delta\omega$, Landau damping works qualitatively in the same way as for the continuum case, except for one difference: the mechanism ceases after a time $\sim 1/\delta\omega$, where $\delta\omega$ is the minimum frequency spacing between the oscillators in the ensemble. The explanation is that, the system being conservative, the driving force and the oscillators exchange energy back and forth in a beating pattern whose period is $\sim 1/\delta\omega$ (we assume that none of the oscillators is exactly on resonance). Therefore, after this time, the ensemble of oscillators comes back to its initial state and the process starts again. For a uniform distribution of N oscillators, $\delta\omega=\Delta\omega/N$. Therefore, for a given finite N, this consideration puts a constraint on how wide $\Delta\omega$ can be for Landau damping to be practical.

For a bunch of N particles, one expects $\delta\omega\approx\Delta\omega/N$ and so the damping ceases after a time $\sim N/\Delta\omega$. In practice, however, since $N\sim 10^{11}$, this constraint is not significant unless the frequency spread $\Delta\omega$ is very large. On the other hand, if $\Delta\omega$ is too small, the long-time limit (12.24) is effectively never reached, and the mechanism does not take effect. The energy is not stored evenly within the bunch: it is selectively stored in particles with a continuously narrowing range of frequencies ω near Ω. The energy stored in these particles grows like t^2, but there are fewer and fewer of them as time progresses. If the driving frequency Ω falls outside the range of $\rho(\omega)$, damping clearly does not take effect and the instability is not avoided.

The analysis for a realistic case is more complicated than what is sketched above because the amplitude A of the driving force in Eq. (12.22) is itself proportional to the bunch centroid $\langle x \rangle$. Furthermore, the force is a superposition of all wake forces from all turns prior to time t. Either one of these two facts imply that the $\cos\Omega t$ and $\sin\Omega t$ terms in Eq. (12.24) get mixed because the force is out of phase with $x(t)$. In the frequency domain, this mixing is a consequence of the complex nature of the impedance. The fact that the driving force is proportional to $\langle x \rangle$ implies that a *consistency condition* must be satisfied by the solution. This condition takes the form of a *dispersion relation.* For transverse motion of a single bunch, we look for a solution of the form $x \propto \langle x \rangle \exp(-i\Omega t)$, in which case the dispersion relation reads

$$1 = i\frac{Ncr_e\overline{Z}_\perp}{2\gamma\omega_\beta T_0^2}\int d\omega \frac{\rho(\omega)}{\omega - \Omega - i\varepsilon} \tag{12.26}$$

where $r_e = e^2/m_e c^2 \approx 2.82\times10^{-15}$ m is the classical electron radius, γ is the usual relativistic factor, T_0 is the revolution period, ε is a small number whose limit $\varepsilon \to 0^+$ is to be taken after the integral is done, and $\overline{Z}_\perp$ is defined by

$$\overline{Z}_\perp = \sum_{p=-\infty}^{\infty} Z_\perp(p\omega_0 + \omega_\beta) \tag{12.27}$$

where the summation is over all integers. In general, the solution for Ω is complex. In practice, Eq. (12.26) is used as follows: one assumes a certain form for $\rho(\omega)$ (say a Gaussian), and one lets Ω vary in the range $(-\infty,\infty)$ through the real numbers. Then $\overline{Z}_\perp$ obtained from (12.26) traces out a line in the complex plane that divides it into two regions. Since Ω is assumed to be real, this line defines a *stability boundary*. On either side of this boundary, Ω has a nonzero imaginary part. If the actual value of $\overline{Z}_\perp$ of the machine lies in the region containing the origin of the complex plane, then $\mathrm{Im}\,\Omega < 0$ and the motion is stable, i.e., it is Landau damped. If $\overline{Z}_\perp$ lies in the other region, then $\mathrm{Im}\,\Omega > 0$ and the motion grows exponentially in time (it is said to be "antidamped"), and an instability materializes if $\mathrm{Im}\,\Omega$ is larger than the damping time of the machine. Therefore, by making several reasonable assumptions about $\rho(\omega)$, one gets an approximate criterion for the allowed values of $\overline{Z}_\perp$ that lead to stability. Note that this method establishes a stability criterion for $\overline{Z}_\perp$ and not for the impedance itself.

For modern storage rings, the main constraint on the practicality of Landau damping is that the spectrum width $\Delta\omega$ is too narrow. A transverse tune spread is provided naturally by the magnet nonlinearities which produce an amplitude dependence of the betatron tune. A longitudinal tune spread is provided by the nonlinearity of the synchrotron forces at large amplitude. However, modern light sources have small emittances and short bunch lengths, and therefore the natural motion of the particles is very linear. As a result, the naturally-existing nonlinearities may not be strong enough to produce an appreciable tune spread. If this is the case, there are means of enhancing the nonlinearities: for transverse motion, one can add octupole magnets; for longitudinal motion, one can add a low-power harmonic RF system that effectively distorts the harmonic-oscillator shape near the center of the RF bucket. Obviously, a delicate compromise is needed in these cases because nonlinearities introduce single-particle resonances or chaotic motion that tend to degrade the beam lifetime.

12.4 Single-Bunch Issues[10]

Single bunch collective phenomena arise from the interaction of a bunch with itself via wake fields whose range is comparable to or shorter than the bunch length. The most ubiquitous single-bunch effect is the so-called *longitudinal microwave instability*, or *turbulent bunch lengthening instability*. This instability does not grow indefinitely: if the beam current is large enough that this instability is excited, the bunch length and energy spread increase until a new equilibrium situation is reached. In the transverse plane, the instability that, typically, has lowest threshold, is the *transverse mode-coupling instability*, or *fast head-tail instability*. This instability leads to fast beam loss; however, the current threshold is typically higher than for the microwave instability, and is easily avoidable.

12.4.1 Calculation of Instabilities

The calculation of thresholds and growth rates of instabilities and other collective effects is codified in codes such as ZAP.[11] A rough sketch of the procedure, applicable to most instabilities (single bunch and multi bunch), is as follows: one first assumes that the low-amplitude particle motion (transverse or longitudinal) corresponds to a simple harmonic oscillator. One then adds the extra force produced by the wake field, and solves for the frequency in lowest-order approximation. For example, the horizontal equation of motion at turn n for a single particle reads

$$x_n'' + (\omega_\beta/c)^2 x_n = \text{const.} \times \sum_{k=-\infty}^{n} W(-kL) x_k \tag{12.28}$$

where the summation over the transverse dipole wake function W represents the superposition of the force from all turns prior to n, ω_β is the betatron frequency, L is the ring circumference, and the primes mean derivatives with respect to the azimuthal position s. By substituting $x_n = A\exp(-is\Omega/c)$, one can solve for the frequency Ω in lowest-order approximation in the impedance.

The real part of Ω implies a frequency shift, which is not in itself detrimental. The imaginary part, however, signals a potential instability whose lifetime τ is given by

$$\tau^{-1} = \text{Im}\,\Omega. \tag{12.29}$$

If $\tau < 0$, the disturbance is damped and does not lead to any problems. But if $\tau > 0$ the disturbance is antidamped and potentially unstable. However, one cannot conclude from this analysis that the disturbance grows indefinitely because other forces may become important at large amplitude that stop it from growing further; this is precisely what happens in the longitudinal microwave instability.

12.4.2 Parasitic Power Loss

As mentioned earlier, the beam image currents dissipate energy into the vacuum chamber components in addition to generating wake fields. This is referred to as parasitic loss, and the dissipated power is proportional to the square of the bunch current. Although this power loss does not inherently affect the beam stability, it can effectively limit the bunch current (and hence the total beam current) because of the excessive heating of the vacuum chamber. This problem usually affects imperfect junctures in the vacuum chamber such as bellows.

For a beam consisting of M identical bunches, one can generally write the power loss for the whole beam in the form

$$P = M I_b^2 Z_{\text{loss}} \tag{12.30}$$

where the bunch current I_b is related to the total beam current I_0 via $I_0 = M I_b$, and where the *loss impedance* Z_{loss} is nothing but the real part of the effective impedance that is causing the energy loss (Z_{loss}, of course, is proportional to the loss factor). We are only concerned here with Gaussian bunches whose rms bunch length is σ_t (in time units). One can then calculate from Eq. (12.8) the loss impedance for various cases.

For the case of a broad-band resonator with shunt impedance R_{bb}, quality factor Q_r and resonance frequency ω_r, the *broad-band loss impedance* Z_{bb} is given by

$$Z_{bb} = R_{bb}\left(\frac{\pi\,\omega_r}{Q_r\,\omega_0}\right)e^{-(\omega_r\sigma_t)^2}. \tag{12.31}$$

This expression is valid provided the bunch is short compared to the bunch spacing, and the resonator bandwidth is large compared to the bunch frequency $\omega_b \equiv M\omega_0$, namely $\omega_r/Q_r \gg \omega_b$. The bunches need not be equally spaced.

There is also power loss due to the HOMs of the RF cavities. In this case the resonators are narrowband, namely $\omega_r/Q_r \ll \omega_b$, and the formula is more complicated.[f] For a beam consisting of M equally-spaced bunches that are short compared to the bunch spacing, the *narrowband loss impedance* Z_{nb} is given by[12]

$$Z_{nb} = 2MR_{S,r}\frac{\Delta^2 e^{-(\omega_r\sigma_t)^2}}{\sin^2(\pi\omega_r/\omega_b)+\Delta^2}. \tag{12.32}$$

[f] In this case one must replace the integral in Eq. (12.8) by a summation over harmonics of the bunch frequency.

where $R_{S,r}$ is the shunt impedance of the resonator, and $\Delta \equiv \pi\omega_r/2Q_r\omega_b$. Since, by definition, $\Delta \ll 1$, the above formula implies that the power loss is substantial only when ω_r/ω_b is very close to (within ~Δ of) an integer, namely when the resonance frequency of the HOM is very close to a harmonic of the bunch frequency. When this undesirable resonance condition is satisfied, the $\sin^2$-term in Eq. (12.32) vanishes, and the power loss is proportional to $M^2 I_b^2 = I_0^2$, which can be intolerably large. Fortunately, the very narrowness of the mode makes it easy to avoid this condition by a slight detuning of the HOM frequency.

The ohmic losses due to the resistivity of the vacuum chamber are, typically, smaller than those from the broad-band resonator described above. The *resistive wall loss impedance* Z_{rw} for a cylindrical pipe is obtained from Eq. (12.20) and is given by

$$Z_{rw} = \Gamma\left(\tfrac{3}{4}\right) Z_0 \left(\frac{\delta(\omega_0)}{2b}\right) \left(\frac{R}{\sigma_z}\right)^{3/2} \tag{12.33}$$

where $\delta(\omega_0)$ is the skin depth at the revolution frequency, σ_z is the rms bunch length ($\sigma_z = c\sigma_t$), and $\Gamma(3/4) \approx 1.23$. The bunches need not be equally spaced.

12.4.3 Longitudinal Effects

It is possible for wake fields to drive coherent oscillations of the bunch shape and density. A general bunch distribution can be analyzed in terms of its radial and azimuthal moments in phase space. Radial modes are characterized by a radial variation of the distribution without an overall variation in shape, while azimuthal modes have the opposite characteristic. In Fig. 12.5 we sketch the first three azimuthal modes of oscillations in phase space along with the projection onto the time axis, which corresponds to the longitudinal charge density. Because each electron in the bunch is oscillating at the synchrotron frequency ω_s, one can see that the m-th mode has a characteristic angular oscillation frequency $m\omega_s$ (this index m should not be confused with the index m that labels impedances). The characteristic signals for radial modes can be found similar to the azimuthal modes.

As explained in Chapter 4, the synchrotron oscillations of individual electrons in a bunch are governed by the voltage in the RF cavity. However, the wake field leads, in effect, to a distortion of the RF voltage seen by the bunch. This is known as *potential well distortion*, and is sketched in Fig. 12.6.

The effective slope of the bunch voltage depends on the characteristic wavelength of the longitudinal wake function and on the length of the bunch. It is typical to assume a $Q = 1$ resonator form for the longitudinal impedance with resonant frequency equal to the cut-off frequency, $\omega_r = \omega_c \equiv c/b$. This implies that,[2] for long bunches ($\sigma_z > b$), the effective voltage is usually such that the bunch is lengthened and the wake function is referred to as inductive. For short bunches, the bunch may be shortened, in which case the wake function is called capacitive. This is qualitatively illustrated in Fig. 12.7. The convolution of the $m = 0$ mode (which represents the distribution itself, rather than any of the moments) of the bunch with the reactive part of the broad-band impedance determines the bunch distortion. If the bunch is long, its frequency spectrum is significant only near zero frequency, where the impedance is inductive and the bunch consequently lengthens. For short bunches, the bunch spectrum extends beyond ω_c, where the impedance is capacitive. If the net effect is positive, there is bunch shortening.

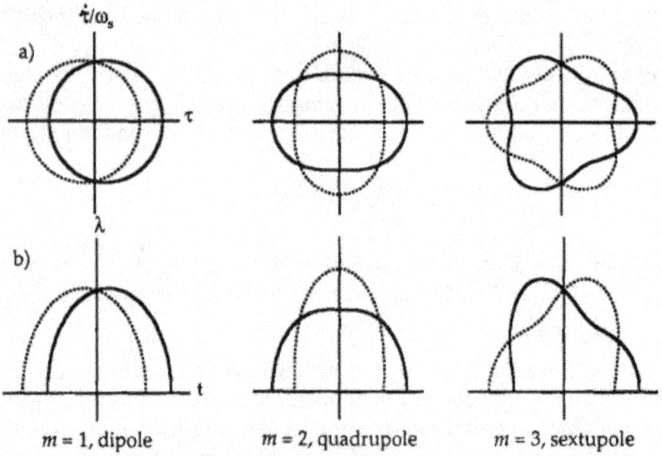

Fig. 12.5. Azimuthal bunch oscillation modes. The m-th mode has a characteristic frequency $m\omega_s$. The solid and dotted lines describe the distributions separated in time by $\pi/m\omega_s$. a) Phase space distribution. b) Line density vs. time

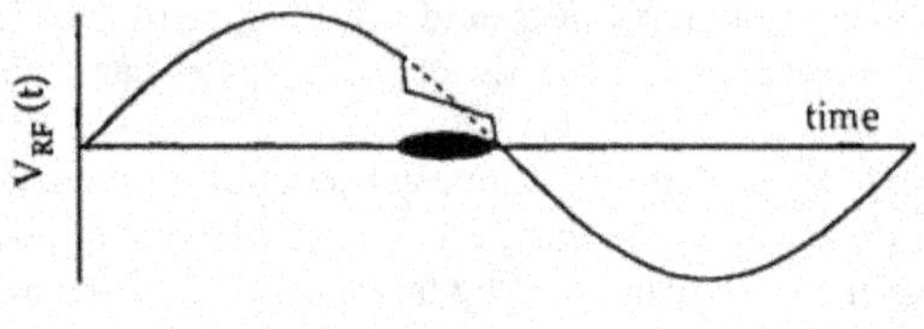

Fig. 12.6. The bunch wake field changes the effective slope of the RF voltage and can lead to bunch lengthening or shortening.

For azimuthal modes with $m \geq 1$, the wake fields also shift the coherent oscillation frequency away from its zero-current value $m\omega_s$. Like bunch lengthening, this frequency shift is determined by the convolution of the spectrum of the mode with the reactive part of the impedance. To first order approximation in the bunch current, this shift is not significant for the centroid of the bunch ($m = 1$ mode) because the wake field moves along with the bunch. However, the effect can be observed by measuring the oscillation frequencies of higher azimuthal modes.

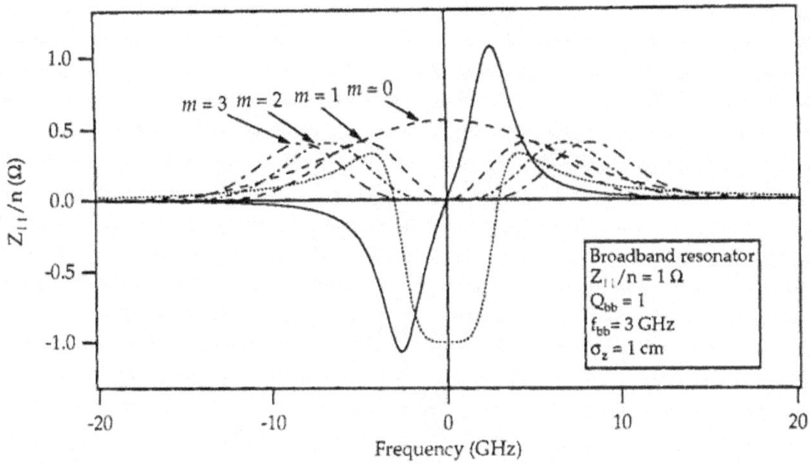

Fig. 12.7. The bunch spectrum for bunch oscillation modes m=0,...,3. The effective impedance for each mode is the sum over the broad-band impedance weighted by the bunch spectrum.

If the current is sufficiently high, the high-frequency components of the wake field can cause ripples in the longitudinal density that can amplify and grow exponentially, leading to an instability. For example, this happens to a coasting (unbunched) proton beam below transition energy (see Chapter 2) if the impedance is capacitive. This instability is called the *negative mass instability*. However, the energy spread in the beam leads to a spread in revolution frequency which, in turn, leads to Landau damping of the perturbation when the growth time is longer than the time necessary for the perturbation to dephase by 180 degrees. The stability limit on the total beam current is given by

$$I_0 < \frac{2\pi\eta(E/e)(\sigma_E/E)^2}{|Z_{\parallel}/n|_{bb}} \tag{12.34}$$

where e is the electronic charge, E is the beam energy, σ_E/E is the relative rms energy spread, and η is the phase-slip factor, related to the momentum compaction factor α by $\eta = \alpha - 1/\gamma^2$ (see Chapter 2). This relation is commonly referred to as the *Keil-Schnell criterion.*[13]

For bunched beams, this instability is called the *microwave instability*, or *turbulent bunch lengthening instability*. The Keil-Schnell criterion can be used in this case if the total beam current I_0 in Eq. (12.34) is replaced by the peak current, $\hat{I} = \sqrt{2\pi}\, I_b/\omega_0\sigma_t$. Thus the threshold for this instability, expressed in terms of the bunch current, is given by

$$I_{b,\text{thr.}} = \sqrt{2\pi}\frac{\eta\,\omega_0\sigma_t(E/e)(\sigma_E/E)^2}{|Z_\parallel/n|_{bb}}. \tag{12.35}$$

If the current exceeds this threshold, both the bunch energy spread and the bunch length grow. This phenomenon is referred to as *turbulent bunch lengthening*. However, this growth stops when the peak current falls below the stability threshold, at which point a new equilibrium situation is reached, and the instability is said to saturate. The bunch length above threshold is given by

$$\sigma_z = \left(K I_b |Z_\parallel/n|_{bb}\right)^{1/3} \tag{12.36}$$

where K is given by

$$K = \frac{\eta R^3}{\sqrt{2\pi}(E/e)v_s^2} \tag{12.37}$$

and the energy spread is given by the usual formula, $\sigma_E/E = v_s\beta^2\sigma_z/\eta R$, where β is here the usual relativistic factor.

Eq. (12.35) assumes that $|Z_\parallel(\omega)/n|$ is independent of frequency, in accordance with the $Q=1$ broad-band resonator model. For the case of a more general broad-band impedance with a power-law frequency dependence of the form $Z_\parallel(\omega) \propto \omega^a$, it can be shown[2] that the bunch length has the dependence

$$\sigma_z \propto K^{1/(2+a)}. \tag{12.38}$$

An example of turbulent bunch lengthening from SPEAR[14] is shown in Fig. 12.8. The power law dependence of the bunch lengthening is clearly exhibited in the results, from which one can extract the value $a=-0.68$. This power-law dependence of the impedance is referred to as "SPEAR scaling," and it is valid only within a limited range of frequencies beyond cutoff. A similar measurement at the Photon Factory[15] yields $a=0.976$, showing that $|Z_\parallel(\omega)/n|$ is essentially independent of frequency in this case.

12.4.4 Transverse Effects

As in the longitudinal case, it is possible for wakefields to drive coherent transverse oscillations within the bunch. However, the situation for transverse oscillations is somewhat complicated by the constant exchange of the head and the tail of the bunch via longitudinal oscillations. Fortunately, this exchange provides a powerful mechanism for Landau damping of transverse oscillations.

Consider an extremely simplified model of a bunch consisting of two electrons, one at the head of the bunch, the other one at the tail. Without longitudinal oscillations, the transverse betatron oscillations of the head would generate a transverse wake field that would drive the tail of the bunch resonantly, as mentioned in Sec. 12.4.1. In the case of linacs, where the synchrotron motion is essentially frozen, this phenomenon leads to the *dipole beam breakup instability.* However, the longitudinal oscillations in circular

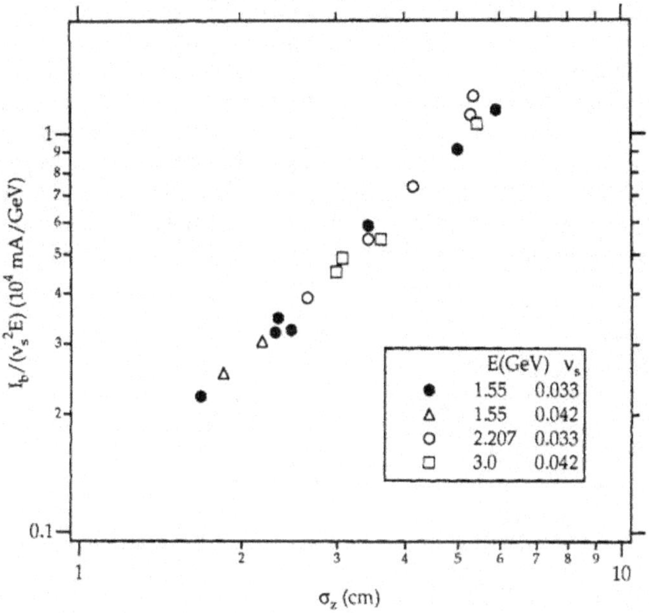

Fig. 12.8. Turbulent bunch lengthening measured at SPEAR.

accelerators cause the head and tail of the bunch to exchange places over a synchrotron period. This continuous exchange does not allow the growth of the oscillation amplitude of the tail to accumulate as quickly, thus extending the stability threshold. Obviously, if the transverse wake fields are so intense that the growth time of the oscillation amplitude of the tail is less than half a synchrotron period, the bunch becomes unstable and is quickly lost. This instability is variously referred to as *transverse mode-coupling instability*, or *fast head-tail instability*, or *transverse turbulent instability.* The threshold for the bunch current is given by

$$I_{b,\text{thr}} = \sqrt{2\pi}\frac{\eta(E/e)(\sigma_E/E)}{\langle\beta_\perp|Z_\perp|_{bb}\rangle} \tag{12.39}$$

where the denominator represents a ring average of the broad-band transverse impedance weighted by the lattice beta function. Typically, the threshold current for this instability is higher than its longitudinal counterpart, given by Eq. (12.35).

As in the case of longitudinal oscillations, it is customary to analyze the transverse oscillations of the bunch in terms of normal modes, referred to as head-tail modes. Each head-tail mode is specified by index $m = 0, \pm 1, \pm 2, \ldots$ (not to be confused with the longitudinal index or with the index of the impedance) which indicates the number of betatron wavelengths per synchrotron period. For mode $m = 0$, all electrons have the same betatron phase (rigid dipole motion), whereas for $m = \pm 1$ the head and tail have opposite phases. The dipole signal for these two modes over several turns is shown in Fig. 12.9 (mode m has m nodes along the length of the bunch). Because of the constant exchange of the head and tail at the synchrotron frequency, each mode has a characteristic angular frequency $\omega_\beta + m\omega_s$. The frequency spectrum of the higher modes peaks at higher frequencies, similar to the longitudinal case described above.

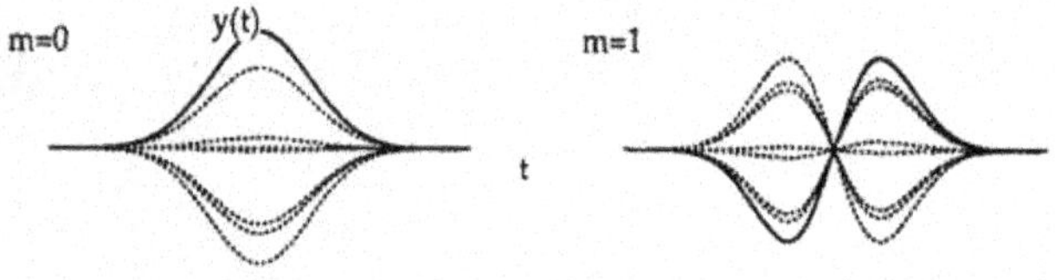

Fig. 12.9. Sketch of the m=0 and m=1 vertical modes of oscillation.

Under certain conditions, coherent bunch oscillations can be excited for currents below the instability threshold. An example of a coherent vertical head-tail oscillation observed at LEP with a streak camera is shown in Figure 12.10.

If the chromaticity is not zero (see Chapter 2), the energy spread of electrons leads to a modulation of the betatron tune at the synchrotron tune. This frequency modulation creates a relative phase shift between the head and the tail of the bunch. Because of this phase shift, the wake field produced by the head of the bunch no longer drives the tail on resonance. The corresponding impedance has a resistive part that can lead to damping or antidamping of the tail oscillation. This effect is referred to as *head-tail damping* and is used quite often to damp coherent transverse motion. Above transition energy, which is the typical situation for electron storage rings, the rigid transverse dipole mode ($m = 0$) is damped by positive chromaticity while the $m = \pm 1$ modes (head and tail out of phase) are antidamped. However, the growth rate must exceed the radiation damping rate for the beam to become unstable. This allows most electron storage rings to operate stably with a slightly positive chromaticity.

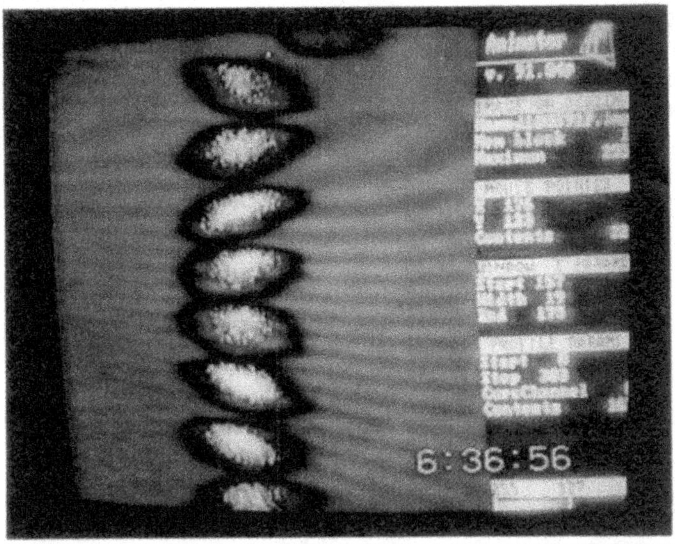

Fig. 12.10. Turn-by-turn pictures of a bunch executing vertical head-tail oscillations in the electron storage ring LEP. The bunch is observed from the side. The synchrotron tune is ~0.1. The horizontal scale is 1000 ps for the total image, not counting the table of numbers at the right. The vertical scale is uncalibrated, but the vertical rms beam size is ~0.2 mm at the observation point. Photo courtesy of E. Rossa.

12.5 Coupled-Bunch Instabilities[10,16]

12.5.1 Basics

Wake fields whose range is long enough to couple the motion of the different bunches in the beam can cause coupled-bunch instabilities. These wake fields are typically produced by narrow resonances in the RF cavities. Even though they remain localized in the cavities, they last for a long enough time that the motion of any given bunch is perturbed by all its predecessors. These long-lasting wake forces can generate a transverse or longitudinal coherent structure in the bunch-to-bunch oscillations. If these coherent oscillations grow indefinitely, they lead to rapid beam loss. If they remain bounded, they degrade the beam quality by inducing a larger effective beam size or oscillations in the arrival time.

Although it is possible for wakefields to couple the bunch shape oscillations from bunch to bunch, the scope of this section is limited to *dipole coupled-bunch oscillations* since these are the dominant concern for the design of a light source (or any other multibunch circular machine). These oscillations are characterized by the motion of the bunches about their nominal centers as if they were rigid "macroparticles." A sketch of a coherent transverse coupled-bunch oscillation is shown in Fig. 12.11.

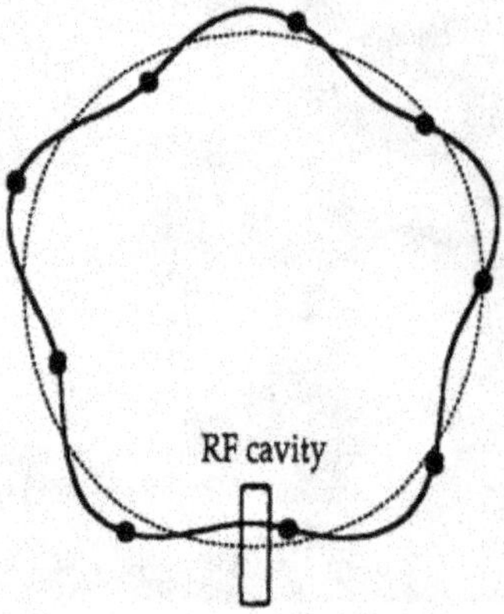

Fig. 12.11. Sketch of a coherent transverse coupled-bunch oscillation.

In analogy with the problem of coupled harmonic oscillators, it is best to analyze the motion of multibunch modes rather than that of individual bunches. The simplest case is that of a beam of M rigid, identical electron bunches spaced equally around the ring. For either transverse or longitudinal oscillations, each multibunch mode is characterized by a bunch-to-bunch phase difference of $\Delta\phi = 2\pi l/M$, where the mode number l can only take the values $l = 0,1,\ldots,M-1$. The net phase advance around the ring is constrained to be a multiple of 2π; when observed from a single point in the ring, each multibunch mode is associated with a characteristic set of frequencies given by

$$\omega_p = (pM \pm (l + \nu))\omega_0 \tag{12.40}$$

where p is some integer and ν is either the synchrotron tune ν_s or the transverse tune ν_β, depending on whether the oscillations are longitudinal or transverse, respectively.

A snapshot view of a multibunch mode in the ring is illustrated in Fig. 12.12 for the case of $M = 3$ bunches. In this case, the longitudinal oscillations have relative phases $\Delta\phi = 2\pi/3$ or $\Delta\phi = 4\pi/3$ (multibunch modes $l = 1$ and $l = 2$, respectively). The corresponding waves are shown in Figs. 12.13a and 12.13b, and the resulting frequency spectrum in Fig. 12.14, where the modes appear as sidebands separated from the revolution harmonics by ν.

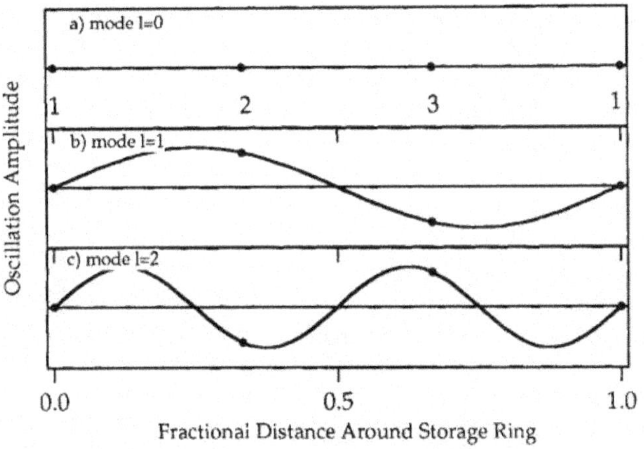

Fig. 12.12. Snapshot of a three-bunch beam executing multibunch oscillations.

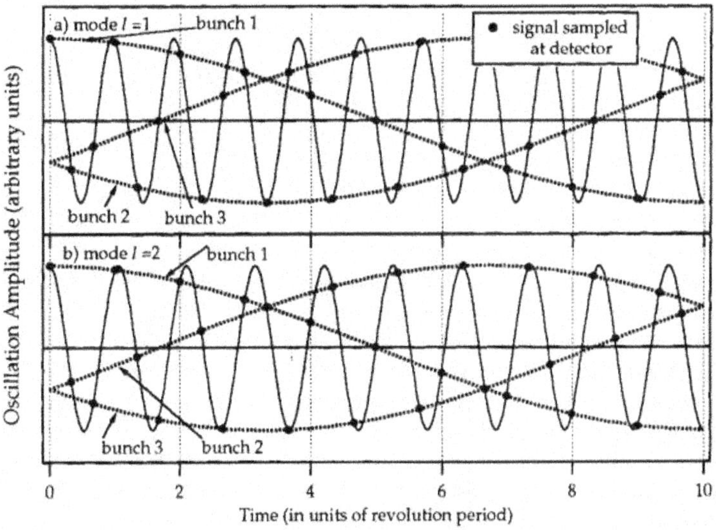

Fig. 12.13. An illustration of the characteristic signal at a fixed location in the ring for each multibunch mode. (a) Mode l=1. (b) Mode l=2. The dots represent the signal sampled at the detector. The thick dashed lines show the motion of individual bunches. The narrow, high frequency line shows the lowest frequency wave that fits the sampled measurement points.

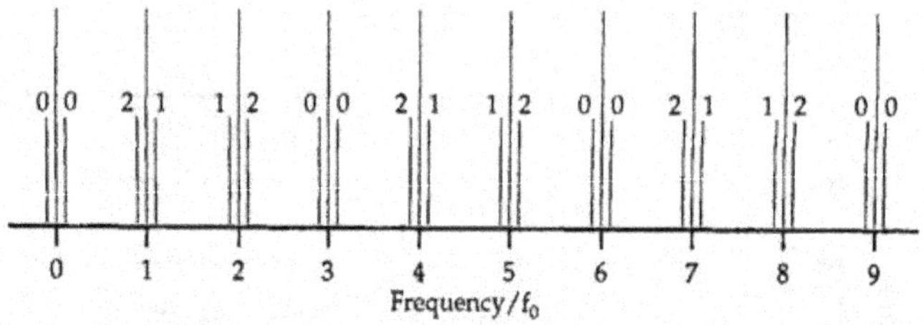

Fig. 12.14. Coupled-bunch mode spectrum for 3 bunches corresponding to Eq. (12.40). The sidebands are labeled with the corresponding coupled-bunch mode index. If the oscillations are longitudinal, the sidebands are split from the harmonics of the revolution frequency by $\pm\nu_s$; if transverse, by $\pm\nu_\beta$.

12.5.2 Longitudinal Coupled-Bunch Instability

It is useful to give a physical picture of a simple longitudinal coupled-bunch instability. Consider a single rigid bunch executing synchrotron oscillations; this bunch is in a storage ring containing an idle RF cavity with a single resonant mode whose frequency can be tuned over any desired range. In addition, the ring is assumed to have at least one other RF cavity that supplies power to the beam but has no other effects. For the sake of illustration, assume that the resonant frequency of the idle cavity is tuned to about twice the revolution frequency. As the beam passes through this element on some arbitrary turn, it induces a voltage that oscillates at the resonant frequency, as shown in Fig. 12.15a (a negative voltage implies deceleration of the bunch; the beam-induced voltage from subsequent passages of the bunch is not shown). During the half of the synchrotron oscillation when the energy of the bunch is smaller than the design energy, the revolution period[g] is shorter than nominal, and the bunch arrives at the idle cavity earlier than an on-energy bunch. The opposite is true during the half of the synchrotron oscillation when the energy of the bunch is greater than the design energy. In either case, the bunch sees the induced voltage from the previous turn as indicated.

In a first example, we assume that the frequency of the resonant mode of the idle cavity is slightly smaller than twice the revolution frequency. When the energy of the bunch is smaller than its design value, the bunch sees less decelerating voltage than when its energy is above the design value. Therefore, over the course of several turns, the energy oscillations of the bunch grows smaller and smaller. Thus the interaction with the resonator damps the oscillations and the motion is inherently stable. This effect is usually referred to as *Robinson damping*[17] when the resonator is the fundamental mode of the RF cavity.

In a second example, let the frequency of the resonant mode be tuned slightly above

[g] We assume that the ring is operated above transition energy since, in practice, this is the typical case.

twice the revolution frequency. The relative arrival times are shifted relative to the previous case, as shown in Fig. 12.15b. In this case, the below-energy bunch loses more energy than the above-energy bunch resulting in an unstable oscillation (if the ring happens to be operated below transition, the two cases are reversed).

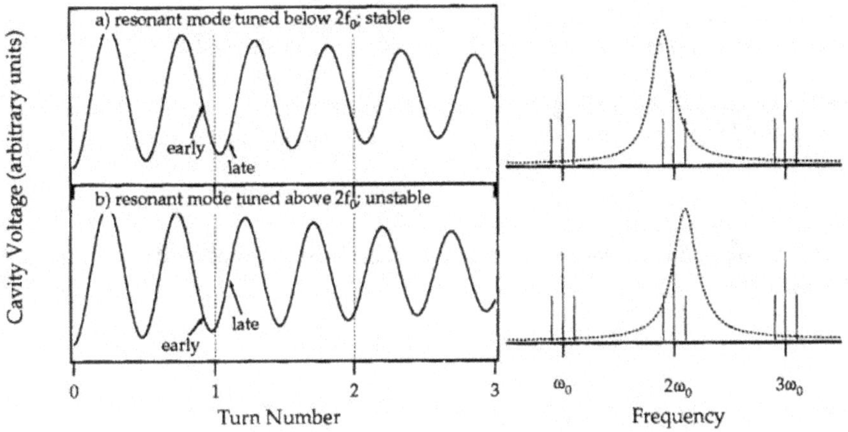

Fig. 12.15. A time-domain view of a resonator voltage driving a longitudinal coupled-bunch instability and the corresponding frequency domain view. The sidebands of the revolution harmonics represent the phase modulation of the beam current resulting from the synchrotron oscillations, and the dashed line is the resistive part of a resonator impedance.

Although this simplified description gives a view of the interaction of the bunch with its own voltage over the course of two revolutions, it is inadequate for the description of the multi-turn cumulative effect. For instance, the net beam-induced voltage in the example above might sum to zero over the course of many turns, or the beam-induced voltage from other bunches in the beam might cancel the voltage from the first bunch. The above treatment also implies a point-like charge. In reality, electron bunches have a distribution in their energy, position, and synchrotron frequency, and beam wake fields can affect the electrons within the bunch.

The obvious difficulties of understanding the summation of beam-induced voltages and the resulting effects on a bunched beam over many turns are greatly simplified by analyzing the problem in the frequency domain. Consider the frequency spectrum of a single bunch and the resistive part of a resonator impedance as shown in Fig. 12.15. The revolution frequency of the bunch increases (decreases) when its energy is greater than (smaller than) the design energy, corresponding, respectively, to the lower and upper sidebands. The energy absorbed by the resonant mode is proportional to the resistive part of the impedance at the frequency of the sideband. When the idle cavity is tuned to the RF frequency, it absorbs the same energy from the upper and lower sidebands. In other words, it absorbs the same energy from the bunch when it is below energy as it does when the bunch is above energy. When the cavity is tuned below a multiple of the revolution frequency, as shown in Fig. 12.15a, it absorbs more energy from the bunch when it is above the design energy than when it is below, and is thus stable. But if the resonant mode is tuned above a multiple of the revolution frequency, as shown in Fig. 12.15b, the situation is reversed and the energy oscillations of the bunch are antidamped.

In general, as mentioned earlier, the interaction of the beam with the wake fields leads to both an amplitude growth and a frequency shift of the longitudinal beam oscillations. For coupled-bunch mode l the *complex coherent frequency shift* is given by

$$\Delta\Omega_{\parallel}^{l} = i\frac{\eta h \omega_0 I_0}{4\pi\nu_s(E/e)}\left[Z_{\parallel}\right]_{\text{eff.}}^{l} \tag{12.41}$$

where the effective impedance is the sum of the impedance weighted by the beam spectrum, and is given by

$$\left[Z_{\parallel}\right]_{\text{eff.}}^{l} = \sum_{p=-\infty}^{\infty} \frac{\omega_p}{\omega_{RF}} Z_{\parallel}(\omega_p) e^{-(\omega_p\sigma_\tau)^2} \tag{12.42}$$

and where $\omega_p \equiv (pM + l + \nu_s)\omega_0$.

The real part of $\Delta\Omega$ yields the shift in the oscillation frequency of the mode, and is driven by the reactive part of the impedance. The imaginary part is the growth rate of the oscillation, and is driven by the resistive part of the impedance. Note that higher frequency resonators have a stronger effect on longitudinal motion because the phase modulation of the beam is larger at higher frequencies. The motion becomes unstable when the growth rate is positive and exceeds the sum of the radiation and Landau damping rates.

For example, in the case of a single high-Q resonator tuned near the frequency $pM\omega_0$, a bunch whose length is short compared to the wavelength of the resonator has a growth rate given by

$$\frac{1}{\tau_{\parallel,l}} = \frac{\eta h \omega_0 I_0 R_{\text{eff.}\parallel,l}}{4\pi\nu_s(E/e)} \tag{12.43}$$

where

$$\begin{aligned} R_{\text{eff.}\parallel,l} = \text{Re}\left[Z_{\parallel}\right]_{\text{eff.}}^{l} \approx (pM + l + \nu_s)\,\text{Re}\, Z_{\parallel}\big((pM + l + \nu_s)\omega_0\big)/h - \\ (pM - l - \nu_s)\,\text{Re}\, Z_{\parallel}\big((pM - l - \nu_s)\omega_0\big)/h. \end{aligned} \tag{12.44}$$

In other words, the growth rate is proportional to the difference in impedance between the upper and lower sidebands of the coupled-bunch mode in question. This agrees with the qualitative argument given above.

In actual storage rings which observe longitudinal instabilities, the oscillations typically grow to an amplitude where some other damping mechanism such as Landau damping limits further growth. Finite-amplitude longitudinal oscillations[18] can affect the average beam size at a point in the lattice with dispersion and thus the average brightness of the photon beam. Furthermore, because of the high magnetic quality of modern insertion devices, the spectral width of higher harmonics of the synchrotron light is sensitive to energy oscillations, even if there is no effective increase in the transverse beam size, as exemplified in initial measurements at the ALS shown in Fig 12.16. The top graph shows a measurement of the electron beam spectrum from a BPM sum signal near several revolution harmonics with all RF buckets filled. The central peaks are revolution harmonics with phase modulation sidebands indicating large amplitude coupled-bunch longitudinal oscillations. The bottom graph shows significant spectral broadening of the synchrotron radiation at the undulator third harmonic in multibunch mode vs. single bunch mode.

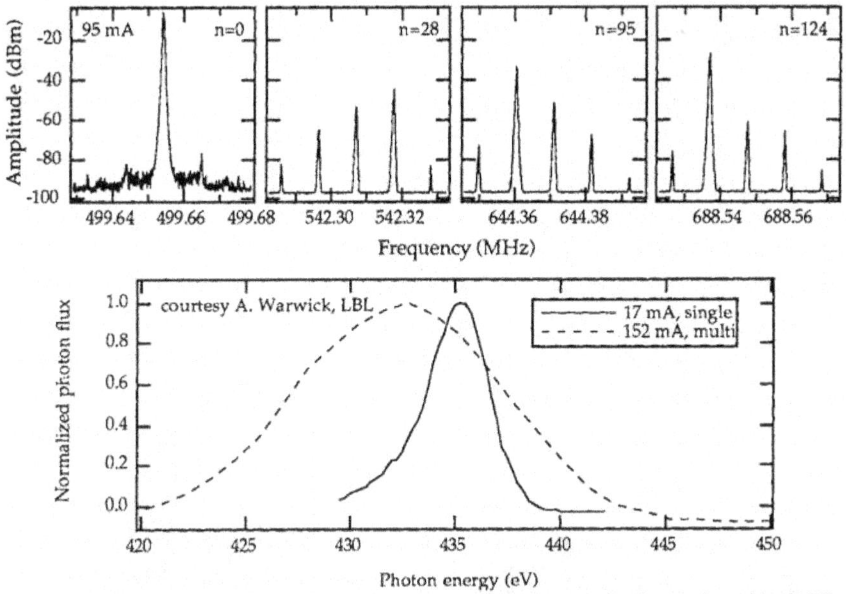

Fig. 12.16. The ALS electron beam spectrum near several revolution harmonics with all RF buckets filled indicates large amplitude longitudinal coupled-bunch oscillations. The central peak in each graph is the signal at the nth revolution harmonic above the RF frequency and the phase modulation sidebands are oscillations of various normal modes of the beam. The bottom graph shows initial measurements of the undulator 3rd harmonic taken under similar conditions during the commissioning of an ALS undulator. A marked increase in the spectral width of higher harmonics of the synchrotron light results from the coupled-bunch energy oscillations.

12.5.3 Transverse Coupled-Bunch Instability

Transverse instabilities are driven by narrow-band dipole HOMs of the RF cavity and also by the resistive wall impedance. For low frequencies, the skin depth is relatively large and hence the wake field can last for a sufficiently long time to couple the motion of different bunches. Since the transverse impedance scales with the chamber radius as b^{-3} (viz. Eqs. (12.6) and (12.20)), it is of particular concern for future light sources which require small chamber sizes to accommodate strong insertion devices.

The physical mechanism for the transverse coupled-bunch instabilities is similar to that for longitudinal instabilities. The transverse complex frequency shift for coupled-bunch mode l and dipole head-tail mode (rigid bunch shape) is given by

$$\Delta\Omega_{\perp}^{l} = -i\frac{\omega_0 I_0 \beta_{\perp}}{4\pi(E/e)}\left[Z_{\perp}\right]_{\text{eff.}}^{l} \tag{12.45}$$

where

$$\left[Z_{\perp}\right]_{\text{eff.}}^{l} = \sum_{p=-\infty}^{\infty} Z_{\perp}(\omega_p)e^{-(\omega_{\xi}\sigma_t)^2}\,. \tag{12.46}$$

Here $\omega_p \equiv (pM + l + \nu_{\perp})\omega_0$, $\omega_{\xi} \equiv (pM + l + \nu_{\perp} - \xi/\eta)\omega_0$, ξ is the chromaticity, $\beta_{\perp}$ is the beta function (x or y) at the location of the impedance, and $\nu_{\perp}$ is the transverse tune. In the case of a single high-Q resonator tuned near the frequency $pM\omega_0$, with zero chromaticity and a bunch length short compared to the resonant wavelength of the resonator, the growth rate is given by

$$\frac{1}{\tau_{\perp,l}} = \frac{\omega_0 I_0 R_{\text{eff.}\perp,l}}{4\pi(E/e)} \tag{12.47}$$

where

$$R_{\text{eff.}\perp,l} = -\beta_{\perp}\,\text{Re}\left[Z_{\perp}((pM + l + \nu_{\perp})\omega_0) - Z_{\perp}((pM - l - \nu_{\perp})\omega_0)\right]. \tag{12.48}$$

12.5.4 Coupled-Bunch Instability Cures[19]

The most obvious remedy for coupled-bunch instabilities is to eliminate or reduce the strength of the HOMs in the design of the RF cavity. However, this reduction usually comes at the expense of the strength of the fundamental mode, thus requiring more total RF power to supply the requisite voltage to the beam. Tuned antennae can be used to couple energy out of the cavity if there is only a single troublesome HOM. Another method is to adjust the frequencies of the HOMs such that they lie in between harmful beam resonant frequencies. However, because the minimum spacing between these frequencies is ω_0, this method is possible only for HOMs with bandwidths much smaller than ω_0. Furthermore, the HOM frequencies shift with changes in the cavity temperature and the position of the tuning rod, thus making them difficult to control. For a storage ring with multiple RF cavities, it is possible to arrange the HOM frequencies of each cavity so that they do not coincide with each other, thus reducing the scale of the problem. Another method is to increase the effect of Landau damping by increasing the effective synchrotron or betatron tune spread. This can be accomplished in the longitudinal plane by either running the RF cavity at lower voltage and thus at longer bunch length, or by adding a higher harmonic RF cavity. The tune spread in the transverse plane can be increased by adding octupole magnets to the storage ring lattice. A variation on this scheme is to create a bunch-to-bunch synchrotron or betatron tune spread by modulating the RF voltage or by using an RF quadrupole.[20] A bunch-to-bunch synchrotron tune spread can also be generated by transient beam loading effects induced by gaps in the beam. Finally, the most powerful method is to add an active feedback system which senses the oscillation of each bunch and provides a corrective kick on the following turn. The cure for coupled-bunch instabilities in a storage ring is usually a combination of all of the above.

12.6 Trapped Ions and Beam Lifetime Issues

12.6.1 Trapped Ions[21]

Positive ions are created when the beam ionizes the gas molecules remaining in the vacuum chamber. Since the beam is negatively charged, the ions remain trapped in the electric potential well of the beam. Possible consequences from this are: reduced beam lifetime due to multiple scattering, tune spread, emittance increase, and electron-ion coherent oscillations. Ion trapping is a poorly understood phenomenon, and has been observed in all synchrotron light sources operating in multibunch mode and in many other electron storage rings. An obvious way to avoid ion trapping altogether is to use positrons rather than electrons in the beam. This solution, however, requires a positron source, which is typically quite expensive. More typical solutions are described below.

Near the beam center, a trapped ion with net charge Ze and atomic number A oscillates vertically in the potential well of an electron beam of total current I_0 with an average angular frequency given by

$$\omega_y^2 = \frac{2Ze}{m_p c A} \cdot \frac{I_0}{\sigma_y\left(\sigma_x + \sigma_y\right)} \tag{12.49}$$

with a corresponding horizontal frequency obtained from the above by the exchange $\sigma_y \leftrightarrow \sigma_x$. In this expression m_p is the proton mass and σ_x and σ_y are the horizontal and vertical rms sizes of the bunch, respectively, at the ring location where the ion is trapped. By applying linear transport theory to the motion of the ions, one can derive a condition for the ions to be stably trapped. For a beam with a uniform bunch population, the condition is

$$\left(\omega_y \Delta t\right)^2 < 4 \tag{12.50}$$

where Δt is the bunch separation in time (it is only necessary to consider the vertical oscillations because, typically, $\sigma_y \ll \sigma_x$ and therefore $\omega_y \gg \omega_x$). This condition implies that all ions with a mass-to-charge ratio larger than a critical value,

$$\frac{A}{Z} > \left(\frac{A}{Z}\right)_c \equiv \frac{\pi R N r_p}{M\sigma_y(\sigma_x + \sigma_y)} \tag{12.51}$$

will be trapped. Here $r_p \equiv e^2/m_p c^2 \approx 1.535\times10^{-18}$ m is the classical proton radius, N is the number of electrons per bunch, and M is the number of bunches in the beam.

If no steps are taken to clear the ions, they progressively accumulate and neutralize the electron beam. As a result, the value of N in Eq. (12.51) effectively decreases and $(A/Z)_c$ becomes smaller, so that ions with lower and lower A/Z ratios can become trapped in turn. This phenomenon is known as the *ion ladder*.

Since $\sigma_y \ll \sigma_x$ the potential well is much deeper in the vertical direction than in the horizontal, and it is deepest where σ_y is smallest, namely near the defocusing quadrupole magnets in the ring. As a result, ions tend to get preferentially trapped at these locations. One way to eliminate these ions is by means of clearing electrodes, typically placed near the defocusing quadrupoles. For low beam currents, a DC voltage of a few kV is usually sufficient to pull the ions away from the potential well. For large beam currents the required voltage might be too large and thus this technique might be impractical or detrimental to the electron beam. In this case, ion clearing is more effective if the electrode voltage has an AC component whose frequency is equal to the ion frequency given by Eq. (12.49), in addition to a DC component. In addition, for beams with many bunches, another method to clear ions is to leave a gap in the bunch train typically equivalent to ~10% of the beam. This *ion-clearing gap* leads to instabilities in the ion motion that are analogous to the betatron motion stopbands arising from resonances in a particle beam. In practice, a combination of all these methods is required.

12.6.2 Intrabeam and Touschek Scattering[22,23]

As time progresses, the particles within a bunch in a stored beam scatter off each other via the Coulomb force, exciting transverse and longitudinal oscillations. As a result of multiple small-angle scattering, particles diffuse in phase space causing a redistribution of all three emittances. This emittance redistribution is usually referred to as the *intrabeam scattering effect*. In addition, on rare occasions, the scattering involves wide angles and, as a result, a particle can fall outside the dynamic aperture or the energy acceptance of the machine, and gets lost. This beam lifetime limitation due to occasional large-angle Coulomb scattering is usually referred to as the *Touschek effect*. Obviously there is no conceptual difference between the two effects. However, since they lead to different manifestations in the beam dynamics, they are traditionally analyzed separately. In the case of the intrabeam scattering effect, the quantities of interest are the damping (or growth) rates for the three emittances; in the case of the Touschek effect, the quantity of interest is the beam lifetime.

Clearly, both effects have a strong dependence on beam energy, becoming more pronounced at lower energies. This can be qualitatively understood by noting that, in the Lab frame of reference, the electric and magnetic forces on any given particle almost cancel each other out at high energies, leaving a net Lorentz force proportional to γ^{-2} (γ is here the usual relativistic factor). As a result, typically, neither the intrabeam effect or the Touschek effect leads to significant problems in the operation of storage rings at energies $\gtrsim$ 1 GeV. However, both effects scale unfavorably with the beam density $N/\sigma_x\sigma_y\sigma_z$, where the σ's are the rms beam sizes. Therefore these effects can become important for modern light sources, which emphasize intense, small bunches.

In all light sources, and indeed in all storage rings built so far, the particle motion is, on average, nonrelativistic in the beam rest frame. In this frame, typically the horizontal and vertical energy spreads are larger than the longitudinal. Therefore small-angle Coulomb scattering predominantly transfers energy from the transverse motion to the longitudinal, leading one to expect damping of the transverse emittances at the expense of growth of the longitudinal. However, a change in the particle energy excites, in turn, horizontal motion due to the dispersion in the ring. Typically, this effect more than compensates the damping of the horizontal emittance, and the net result is a damping of the vertical emittance and growth of both horizontal and longitudinal emittances. During this process, the quantity

$$-\eta(\sigma_E/E)^2+\varepsilon_x/\overline{\beta}_x+\varepsilon_y/\overline{\beta}_y \tag{12.52}$$

remains invariant, where the ε's are the emittances and the $\overline{\beta}$'s are the ring-averaged beta functions.

The Touschek lifetime is given by

$$\frac{1}{\tau}=\frac{r_e^2 cN}{8\pi\gamma^5\sigma_{x'}^3\sigma_x\sigma_y\sigma_z}\cdot\frac{F(\delta)}{\delta} \tag{12.53}$$

where $\sigma_{x'}$ is the rms of the beam divergence, and δ is defined by

$$\delta=\left(\frac{\Delta p/p}{\gamma\sigma_{x'}}\right)^2 \tag{12.54}$$

where $\Delta p/p$ is the momentum acceptance of the machine, which is determined, in turn, by the smallest of the RF bucket height or the energy aperture of the ring. The function $F(\delta)$ depends weakly on δ for small δ. For the range $\delta \lesssim 10^{-2}$, which is typical, it is given by

$$F(\delta)\approx-\ln(\gamma_e\delta)-3/2 \tag{12.55}$$

where $\gamma_e \approx 1.78$ is Euler's constant.

12.7 Acknowledgments

We are grateful to A. Chao, J. Corlett, M. Cornacchia, A. Hofmann, G. Lambertson, H. Winick and B. Zotter for valuable discussions. We are indebted to E. Rossa for providing the photo on Fig. 12.10.

12.8 References

1. S. Chattopadhyay, "Stability of High-Brilliance Synchrotron Radiation Sources," NIMPR **A291**, 455 (1990).
2. A. W. Chao, *Physics of Collective Beam Instabilities in High Energy Accelerators*, J. Wiley & Sons, 1993.
3. *Synchro-Betatron Resonances*, Proc. 6th Advanced ICFS Beam Dynamics Workshop, Madeira, Portugal, October 24–30, 1993, to be published.
4. *Impedance Beyond Cutoff*, Part. Accel. **25** nos. 2–4 (1990), S. Chattopadhyay, guest editor.
5. F. Caspers, "Bench Methods for Beam-Coupling Impedance Measurement," Proc. US-CERN School on Particle Accelerators, Hilton Head Island, S. Carolina, USA, Nov. 7–14, 1990, M. Month, E. Dienes and S. Turner (Eds.), Springer Verlag LNP **400**, p. 80.
6. T. Weiland and R. Wanzenberg, "Wake Fields and Impedances," *ibid.*, p. 39.
7. P. Schoessow, E. Chojnacki, W. Gai, C. Ho, R. Konecny, J. Power, M. Rosing and J. Simpson, "The Argonne Wakefield Accelerator–Overview and Status," Proc. IEEE Part. Accel. Conf., Washington, DC, May 17–20, 1993, p. 2596.
8. L. Landau, "On the Vibration of the Electric Plasma," J. Phys. (USSR) 10, **25** (1946).
9. H. G. Hereward, "The Elementary Theory of Landau Damping," CERN 65-20 (1965).
10. J. Gareyte, "Observation and Correction of Instabilities in Circular Accelerators," Ref. 5, p. 134.
11. M. S. Zisman, S. Chattopadhyay and J. J. Bisognano, "ZAP User's Manual," LBL-21270/UC-28, December 1986.
12. M. Furman, H. Lee and B. Zotter, "Energy Loss of Bunched Beam in SSC RF Cavities," Proc. 1987 Particle Accelerator Conference, Washington, DC, March 16–19, 1987, p. 1049.
13. E. Keil and W. Schnell, "Concerning Longitudinal Stability in the ISR," ISR-TH-RF/69-48 (1969); V. K. Neil and A. M. Sessler, "Longitudinal Resistive Instabilities of Intense Coasting Beams in Particle Accelerators," Rev. Sci. Inst. **36**, 429 (1965).
14. A. W. Chao and J. Gareyte, "Scaling Law for Bunch Lengthening in SPEAR II," Ref. 4, p. 229.
15. N. Nakamura, S. Sakanaka, K. Haga, M. Izawa and T. Katsura, "Collective Effects in Single Bunch Mode at the Photon Factory Storage Ring," Proc. IEEE Part. Accel. Conf., S. Francisco, CA, May 6–9, 1991, p. 440.
16. F. Sacherer, "A Longitudinal Stability Criterion for Bunched Beams," IEEE Trans. Nucl. Sci. **NS-20-3**, 825 (1973); A. Hofmann, "Coherent Beam Instabilities," Ref. 5, p. 110.

17. K. Robinson, "Stability of Beam in Radio-Frequency Systems," CEAL-1010 (1964).
18. J. Byrd, "Measurement of Collective Effects at the ALS," LBL-34954a/CBP Note 054, submitted to the 1994 European Part. Accel. Conf., London, June 27–July 1st, 1994.
19. F. Pedersen, "Multibunch Instabilities," Proc. Joint US-CERN School on Particle Accelerators, Benalmádena, Spain, Oct. 29–Nov. 4, 1992, M. Dienes, M. Month, B. Strasser and S. Turner (Eds.), Springer Verlag LNP **425**, p. 269.
20. S. Sakanaka, T. Mitsuhashi, A. Ueda and M. Izawa, "Construction of a High-Frequency Quadrupole Magnet Used to Cure Transverse Coupled-Bunch Instabilities," NIMPR **A325**, 1 (1993).
21. A. Poncet, "Ion Trapping and Clearing," Ref. 19, p. 202; M. Cornacchia, "Requirements and Limitations on Beam Quality in Synchrotron Radiation Sources," Proc. CAS on Synchrotron Radiation and FELs, Chester College, UK, Apr. 6-13, 1989, S. Turner (Ed.), CERN 90-03, p. 53.
22. A. Piwinski, "Intra-Beam Scattering," Proc. Joint US-CERN School on Particle Accelerators, S. Padre Island, Texas, USA, Oct. 23-29, 1986, M. Month and S. Turner (Eds.), Springer Verlag LNP **296**, p. 297.
23. H. Brück, *Circular Particle Accelerators*, LASL Translation LA-TR-72-10 Rev., 1972.

CHAPTER 13: ORBIT STABILIZING AND MULTIBUNCH FEEDBACK SYSTEMS*

JOHN N. GALAYDA   YOUNGJOO CHUNG

Advanced Photon Source, Accelerator Systems Division
Argonne National Laboratory, 9700 South Cass Avenue
Argonne, Illinois 60439, USA

and

ROBERT O. HETTEL

Stanford Synchrotron Radiation Laboratory
Stanford Linear Accelerator Center, P. O. Box 4349
Stanford, California 94309, USA

13.1 Introduction

The extraordinary brightness of synchrotron radiation sources makes possible measurements of phenomena whose experimental signatures are too weak to be detected with less intense X-ray sources. However, incident intensity, position, angle, and energy of the photon beam must be stable in order to exploit the experimental advantages of synchrotron radiation. Active feedback systems, which detect and counteract undesired beam motions by correcting the particle beam orbit, are essential to achieve the necessary beam stability. In some cases, feedback systems are needed as well to overcome the current-limiting effects of multibunch instabilities (discussed in Chapter 12).

In the first sections of this chapter we will focus on the nature of beam stability problems and identify stability criteria. We will then present design principles and describe components used to implement orbit stabilizing feedback systems. In the last section, we will discuss the basic principles of feedback systems that counteract current-limiting instabilities.

13.2 Nature of Beam Stability Problems

Experimenters can employ the small and highly collimated beams from synchrotron radiation sources to measure the structural, spectroscopic, and dynamic properties of materials with high resolution, sometimes using very small samples, but only if the beams are sufficiently stable with respect to the sample, beamline apertures, and optical elements. In this section we will identify categories and sources of unwanted beam motions, establish stability criteria, and introduce the feedback systems needed to meet those criteria.

13.2.1 Beamline Apertures

Beam motion with respect to small apertures causes intensity variation at the experiment. Intensity changes as small as 0.1% can degrade measurement resolution if they occur during a data collection period. Even in the absence of these limiting apertures, such motion can cause the beam to miss very small (<< 1 mm) samples altogether. Also, variations of a few microradians in beam incident angle on a diffractive grating or monochromator crystal can change the energy of transmitted radiation and diminish spectral resolution.

---

* Work supported by the Office of Basic Energy Sciences, Department of Energy under contracts DE-AC03-76SF00515 and W-31-109-ENG-38.

John N. Galayda   e-mail: galayda@oxygen.aps1.anl.gov; fax: (708) 252-4240
Youngjoo Chung   e-mail: ychung@oxygen.aps1.anl.gov; fax: (708) 252-7187
Robert O. Hettel   e-mail: hettel@ssrl01.slac.stanford.edu; fax: (415) 926-4100

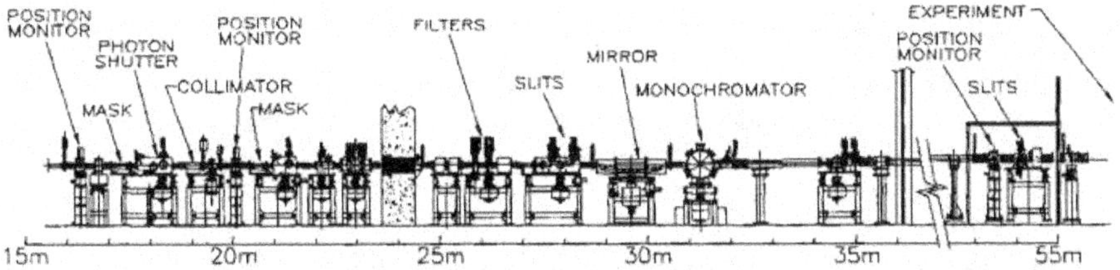

Figure 13.1 Beamline at the Advanced Photon Source. Typical aperture and component locations are shown with their distances from the beamline sourcepoint.

There are many components that restrict the range of positional and angular beam motion in a typical synchrotron radiation beamline (Figure 13.1). They include masks, collimators, mirrors, beryllium windows (that separate beamline vacuum from downstream helium-filled chambers), crystal monochromators, slit assemblies, the experimental sample itself, and detector components. Each of these defines a range of tolerable positional and angular orbit deviations at the sourcepoint, i.e., an area of acceptance in sourcepoint orbit phase space, over which beam flux at the experiment will not vary by more than a small fraction. Each component aperture defines boundaries in orbit phase space; the superposition of all apertures in phase space represents the acceptance of the beamline. It is easy to see that the misalignment of two or more longitudinally separated apertures can create a more restrictive aperture than any one by itself.

13.2.2 Beam Stability Criteria

To see what levels of particle beam orbit stability are needed at the sourcepoint to achieve 0.1% intensity constancy, we consider two typical beamline optical arrangements: a focused beam configuration that will establish orbit position stability needs, and an unfocused configuration that defines orbit angular stability requirements.

Many beamlines are designed to produce a focused image of the beam at a narrow slit; the slit aperture is usually matched to the beam size, although it can be smaller in some cases, and is used to define a fixed source position for the downstream optics. Since the beam image has a Gaussian intensity distribution, the intensity of the beam passing through the aperture will vary as the particle beam undergoes a positional displacement at the sourcepoint. For an aperture whose full opening is three times the rms beam size, i.e., one that transmits 93% of the incident flux of a well-centered beam, displacement of the particle beam must be kept to $\pm$ 10% of its size to limit the intensity change to less than $\pm$0.1%. Sensitivity to small beam motions is much greater if the average beam position is not centered in the aperture, or if the aperture is significantly smaller than the beam size.

If the photon beam propagates a long distance L through an empty beamline to an aperture, the transverse beam size is dominated by the divergence of the beam times the distance L. Intensity fluctuations due to positional displacements of the particle beam become smaller in relation to those caused by orbit angle changes. Maintaining beam position to 10% of its size at a distant aperture amounts to maintaining orbit angle to 10% of the photon beam's divergence.

From the above discussion, one can see that it is natural to express beam stability tolerances and aperture sizes in units of the beam dimensions. Synchrotron light sources have stability goals of about 10% of the photon beam size and divergence. Maintaining beam angle to a few percent of its divergence will usually satisfy energy stability requirements as well for photons transmitted by diffractive components (e.g., monochromators).

To fully exploit low beam emittance, particle beam displacement and angle errors must both be minimized simultaneously. These simultaneous constraints are equivalent to limiting the "effective" emittance of the beam, averaged over all beam motions, to be no more than 10%-20% larger than the ideal photon beam emittance. Third-generation rings are designed for high brilliance; the emittance of undulator radiation in a narrow wavelength (λ) bandwidth ($\Delta\lambda/\lambda \sim 0.001$) sets the scale for stability criteria. For such rings in the 6-8 GeV range, this is equivalent to the particle beam emittance. The wavelength of radiation from undulators in 1–2 GeV third-generation rings can be comparable to the particle beam size. In this case, however, diffraction effects may limit the photon beam emittance to $\lambda/4\pi$.

The photon beam emittance is larger than that of the particle beam, mostly due to an increased divergence, when the bandwidth constraint is relaxed, as is the case for bending magnet and wiggler sources. Under these conditions the synchrotron radiation vertical opening angle is $1/\gamma$, where $\gamma = 0.00051/E$ (GeV) for an electron or positron beam energy E. For rings having energies between 0.7 and 8 GeV, vertical opening angles range from 730 to 64 μrad. In the horizontal plane, the emittance of radiation from wigglers and bending magnets is the convolution of the particle beam emittance with the angular deflection and transverse excursion of the particle orbit. The orbit deflection angle can be several milliradians.

Particle beam sizes and divergences vary from facility to facility. In the horizontal plane, the range is 100-1000 μm and 20-300 μrad. Minimum vertical dimensions of the order of a few microns and microradians are possible in a ring with small linear coupling (see Chapter 2, section 3.6.2.3); typical dimensions are larger than 50 μm and 10 μrad.

To illustrate the stability requirements for some modern synchrotron light sources, consider the goal for the Advanced Photon Source (APS), which is 5% of particle beam dimensions. One may visualize the vertical stability criteria of 4.5 μm and 0.45 μrad as comparable to shooting 9-mm synchrotron radiation "bullet" cleanly through a 10-mm bull's-eye from a distance of one kilometer. Horizontal stability goals, at 17 μm and 1.1 μrad, are just as challenging because horizontal orbit disturbances are expected to be significantly larger than vertical ones.

It is important to keep the beam position fixed during the time it takes to collect enough data to complete an experiment, a period varying from seconds to days or weeks. On the scale of months and years, ground settlement beneath both synchrotron and beamlines produces steering errors that are best compensated by realigning magnets and beamlines rather than by using a steering feedback system.

13.2.3 Categories of Beam Motion

In an ideal storage ring, all the particles would follow exactly the same path and circulate around the ring with exactly the same period. Actual rings are designed to prevent particles from drifting too far away from their ideal positions; the quadrupole magnets and the rf accelerating system apply net corrective forces which oppose the particles' deviations from the ideal. These forces are approximately linearly proportional to the magnitude of the deviation and particle motion can be accurately described in terms of the behavior of simple harmonic oscillators. It is useful to distinguish three categories of resulting beam motion, each having a different characteristic time scale: longitudinal oscillations, transverse oscillations, and closed orbit errors. We will present these basic equations in our discussion of the three types of motion; they will be used in later sections describing feedback system implementation.

The first category of beam motion is synchrotron oscillation, a periodic deviation from the ideal circulation frequency of the beam, which is opposed by corrective forces applied by the rf acceleration system of the ring. As indicated in Eq. (4.4) of Chapter 4, the oscillation of a bunch around the point describing its ideal, constant-period revolution in the storage ring is approximated by damped simple harmonic motion with a frequency on the order of a few kilohertz. Chapter 12 (sections 12.5.1-2) explains that small synchrotron oscillations of the bunches in a sufficiently intense beam can create wake fields which cause the amplitude of those motions to grow exponentially. Such multibunch instabilities are often self-limiting in amplitude and do not result in a catastrophic beam loss. However, since the longitudinal oscillations are accompanied by energy oscillations of a few tenths of a percent, they can cause motions of the order of the beam size in dispersive regions of the ring lattice where, for example, the bending magnet radiation sources are located. Energy oscillations can also degrade the stability of the spectrum from a high performance undulator source. Various countermeasures are applied to prevent this instability, including feedback.

The second two categories of motions, betatron oscillations and closed orbit errors, are related to displacements of the particles perpendicular to the ideal orbit. All the essential features of these two categories can be illustrated by treating the transverse motion (either vertical or horizontal) as simple harmonic oscillation:

$$\frac{d^2}{ds^2}\left(Y(s) + y_\beta(s)\right) + \left(\frac{v}{R}\right)^2\left(Y(s) + y_\beta(s)\right) = \frac{\Delta B(s)}{B\rho} \tag{13.1}$$

where s is the path length along an ideal circular orbit of circumference $2\pi R$; Y(s) is a closed orbit distortion; $y_\beta(s)$ is a betatron oscillation; v is the number of oscillations in the ring revolution (the "betatron tune"); and $\Delta B(s)$ describes the magnetic field error distribution along the ideal orbit. $B\rho$ is called the "rigidity" of the beam and is proportional to the beam momentum. It may be computed using Eq. (2.2a) of Chapter 2.

In a real accelerator the ideal orbit is a polygon and the focusing forces vary dramatically as a function of s. However one can show that a storage ring focusing lattice (discussed more completely in Chapter 2) can be made equivalent to the simple oscillator model using the Courant-Snyder coordinate transformation.[1]

Free betatron oscillations are solutions to Eq. (13.1) for $\Delta B(s)/B\rho = 0$. One may confirm that

$$y_\beta(s) = A\cos\left(v_y\left(s/R\right) + \theta\right) \tag{13.2}$$

is a solution to the equation for any value of A and θ. For the APS storage ring, $v_x = 35.22$ and $v_y = 14.2$. Suppose we observe the position of a vertically oscillating bunch of particles in this ring. The bunch returns to the observation point every 3.68 μsec, the time required to travel once around the 1104-m ring. Successive measurements of y_β would yield, on the Nth return,

$$y_\beta\,(2\pi RN) = A\cos\,(14.2\,(2\pi N) + \theta) = A\cos\,(0.2\,(2\pi N) + \theta)\,. \tag{13.3}$$

The vertical motion repeats itself every 5 turns. Therefore the lowest frequency at which a betatron oscillation can change the position of the photon beam is the rotation frequency multiplied by the fractional part of the betatron tune. In the APS storage ring this is about 0.2×270 kHz, or 54 kHz. With many bunches in the ring, higher frequency motions can appear when each bunch oscillates independently of its neighbors.

The natural emittance of the beam is evidence that the individual particles in each bunch are executing betatron oscillations of the form given in Eq. (13.3). with a Gaussian amplitude distribution and uncorrelated phases. When a transverse "coupled-bunch' instability (see Chapter 12, sections 12.5.1 and 12.5.3) develops, oscillations of all the particles in a bunch become correlated as they respond to the wake field created by all the other bunches; therefore, the centroid of bunch oscillates. Certain correlated oscillation patterns of the bunches (the "coupled-bunch modes"; see Figures 12.11-12.13) begin to grow in amplitude exponentially under the influence of the wake field created by these oscillation patterns, in a runaway process. A feedback system can prevent such instabilities for any pattern ("mode") by effectively adding a damping term of the form

$$|\text{constant}| \times \frac{dy_\beta(s)}{ds} \tag{13.4}$$

to the left-hand side of Eq. (13.1) for each and every bunch.

Closed orbit errors are caused by a stray dipole steering field which is unchanged or at least changes very little in the time it takes the particles to circulate once around the ring. In this limit $\Delta B(s) = \Delta B(s+2\pi R)$, and so $Y(s) = Y(s+2\pi R)$ as well. The effect of the stray field is to distort the orbit away from its ideal shape. If $\Delta B(s)$ is a stray field localized at the point $s = 0$ in the ring, of sufficient strength to deflect the beam by an angle ψ, then

$$\frac{\Delta B(s)}{B\rho} = \psi\ \delta(s) \tag{13.5}$$

where $\delta(s)$ is the Dirac delta function. The resultant distortion of the closed orbit has the form[2]

$$Y(s) = \frac{\psi R}{2\,\nu\,\sin(\pi\nu)} \cos\left(\frac{\nu s}{R} - \pi\nu\right). \tag{13.6}$$

At the APS, where $B\rho \approx 23.4$ Tesla-m, a 0.3 Gauss magnetic field (similar to the earth's field) extending over 1 m will kick the 7-GeV beam by more than 1 µrad, in excess of the horizontal stability goal.

13.2.4 Sources of Beam Motion

There is really only one fundamental cause of unwanted motion of the particle beam in the storage ring: electromagnetic field perturbations. Of course, the photon beam can be mis-steered as a result of unwanted motions of beamline optical components, such as mirrors and monochromators, that are subject to mechanical vibration or thermal distortion; but these components and the measures that are taken to stabilize them will not be discussed here.

Electromagnetic field perturbations in the storage ring arise from a variety of mundane and esoteric sources including unstable or stray magnetic guide fields, electrostatic fields produced by ionized gas trapped in an electron beam, and the aforementioned wake fields. Since their strength increases with beam current, wake fields can also limit the amount of current that can be stored in a ring. Low frequency (< 1 kHz) closed orbit noise can be the most detrimental to a large fraction of beam users; it is caused primarily by stray and guide field perturbations.

A number of mechanical and electrical mechanisms can produce closed orbit disturbances. Thermal expansion of magnets, vacuum chambers, support frames, rigid water and electrical connections, and other accelerator components caused by 1-2 degree temperature variations can cause excessive beam motion.[3] Thermal effects can occur on a larger scale, producing diurnal and seasonal changes in the dimensions of the accelerator housing which are transmitted through the building foundations to the magnets.[4]

Ground vibrations make a significant contribution to beam motion. The spectrum of ground motion is typically such that the mean square amplitude of motion in a bandwidth of 1 Hz is proportional to the inverse fourth power of frequency, and the rms motion at frequencies above 1 Hz is about 1 µm.[5] Vibrating or rotating machinery located near the accelerator can add significantly to the natural ground vibration at frequencies that are usually harmonically related to the local line frequency. These vibrations may be transmitted to the magnets through the ground or through the water cooling connections. Turbulent flow of the cooling water can also cause component vibration.

The electrical equipment surrounding the storage ring can directly produce magnetic fields that deflect the beam. Most storage rings have a cycling booster synchrotron close enough to the storage ring to affect the beam position through both mechanical and electrical interactions. Cables carrying high currents, small current variations in the storage ring magnets, and displacements of heavy equipment such as overhead cranes (which can disturb the pattern of the earth's field, flex the building girders, and induce vibrations) can easily deflect the beam. Magnet power supply ripple, having frequency components at harmonics of 50 or 60 Hz, is another common disturbance source; suppressing this ripple to tolerable levels is a significant factor in the design of high quality magnet power supplies.

Figure 13.2 depicts the vertical photon beam position noise spectrum due to closed orbit disturbance for SPEAR. Other facilities have qualitatively similar noise spectra.

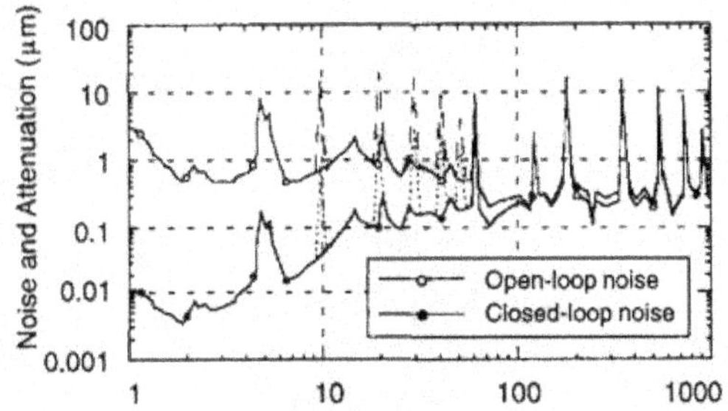

Figure 13.2 Vertical photon beam noise spectrum at SSRL, without and with feedback correction. Frequencies < 60 Hz are due to ground and magnet vibrations; 60 Hz and harmonics come from power supplies. The dashed peaks come from AC power surges when the SLAC linac is pulsed at 10 Hz.

13.2.5 Stabilizing Systems

In the remainder of this chapter, we will describe feedback systems used to minimize beam motion, whether it be caused by wake fields or by more mundane mechanisms, like trucks driving by the laboratory.

Orbit feedback systems are designed to detect beam motions and correct them using steering magnets located around the ring. These systems consist of:

1) one or more beam position monitors (BPMs) that supply a signal proportional to the error in beam position or angle;

2) several steering magnets capable of adjusting the beam position as measured by the monitors;

3) a servo amplifier that connects the position monitors to the correctors, forming a stable negative feedback loop.

Feedback systems designed to suppress longitudinal and transverse multibunch instabilities have similar basic components, although they must work at much higher frequencies. Consequently the beam-kicking components are typically transmission line or cavity structures instead of magnets, and the servo amplifier must operate over a bandwidth of tens or hundreds of megahertz.

In the next sections, we will consider feedback system components separately and describe their functions, capabilities, and flaws.

13.3 Beam Position Monitors

Perhaps the most important components of an orbit stabilizing system are the beam position monitors used to detect small beam motions. The most sensitive monitors for detecting closed orbit instabilities are those that sense photon beam position in a beamline.

Photon position monitors have evolved from simple fluorescent screens that were occasionally inserted in the beamline and viewed by eye to devices that continuously detect beam position with micron resolution. By placing two photon monitors in a single beamline, both position and angle errors can be precisely monitored and corrected.

Particle beam position monitors that employ radio frequency (rf) signal processing techniques (see Chapter 10) originally had accuracies on the order of a few hundred microns but now have been developed to levels of sensitivity that can be useful in orbit control for synchrotron radiation sources. Efforts to improve RFBPM (see section 13.3.2) performance are mainly aimed at reducing systematic errors in the signal pick-up and processing systems. Since photon monitors can be located far back in a beamline and RFBPMs can be located close to the source of radiation, their combined use offers the advantage of the maximum possible separation and hence high sensitivity to angle errors.

13.3.1 Photon Beam Position Monitors

There are two principal categories of commonly used photon beam position monitors: 1) "destructive" monitors that intercept and block the full vertical cross section of the beam over some horizontal range; and 2) "non-destructive" monitors that permit the beam to pass to a downstream destination. Most photon monitors are used to sense vertical position because the beam is highly collimated in that plane. Photon monitors can also detect horizontal position of a horizontally collimated undulator beam, but they are rarely used to do this for a more divergent wiggler or bending magnet beam because of its large horizontal size far from the source. However, they can be used to determine the position of an image of the sourcepoint, produced by a pinhole camera or focusing mirror, for these sources; in this case the monitors are not sensitive to sourcepoint beam angle.

The advantage of a destructive monitor is that, because it senses the collimated central core of the beam, it can provide accurate beam centroid position information even if the beam has an asymmetric profile (for reasons to be discussed). The disadvantage of this type of monitor is that, because it interrupts the beam, unless a dedicated portion of beam is allocated to it, the monitor must be removed to perform a downstream experiment, in which case it cannot be used in a continuous steering feedback system.

Examples of destructive monitors include fluorescent screens or phosphor-coated surfaces, photosensitive diode arrays or area detectors, charge-coupled devices (CCDs), fluorescence scattering monitors, and wire grid photoemission monitors. These monitor types cover a broad photon spectral range, from infrared to hard X-ray. They function by converting primary or secondary photons to position-sensitive currents or voltages either by production of electron-hole pairs in a semiconductor medium, electron-ion pairs in a gas medium, or photoelectrons from in-vacuum metallic electrodes.

Semiconductor detectors are usually used to view visible or near-visible light coming either directly from the beam after a large angle reflection (that prevents hard radiation from striking the detector) or indirectly from a wavelength-shifting phosphor deposited on a screen. The intrinsic spatial resolution using visible synchrotron radiation is limited by diffraction effects to a few tens of microns (see Chapter 10, section 3.5), so the wavelength-shifting method is often used. On the other hand, radiation-hardened versions of these devices are now available and can be used directly in the X-ray beam to sense multi-keV photons.[6] Unsegmented position-sensitive linear and area diodes provide beam centroid information with a few microns resolution by charge division between edge-mounted electrodes. Segmented arrays consist of small individual diodes or CCDs, usually made of silicon, spaced by small distances (25 μm or less is common) covering a linear or square span of a few millimeters. The image detected by a CCD camera can be intensified using a micro-channel plate (MCP) assembly that, in some cases, can also be switched on and off in nanoseconds or less to gate the image for fast time domain measurements.

A device that is particularly sensitive to the highly collimated, high energy photons of the beam (and rejects the more easily corrupted low energy photon pattern) is the fluorescence scattering monitor.[7] Fluorescent photons coming off one or more oblique scattering surfaces, usually made of copper (~ 9 keV) or tantalum (~ 67 keV), are detected through a slit assembly with a pair of photodiodes or equivalent detectors that provide a position-sensitive differential signal having 10-μm resolution.

Windows in the beam chamber are needed to view secondary photons with external detectors. Germanium can be used to transmit infrared light, sapphire or ultrapure fused quartz (ultrapure to avoid darkening from higher energy radiation) is suitable for visible light, and UV-grade sapphire is available for ultraviolet light. In some cases visible light can be channeled through fiber optic cables. X-ray fluorescence photons can be viewed through beryllium or even aluminum windows, and if their energy is high enough, as is the case for the 67-keV photons from tantalum, they can be detected outside the beam chamber with no special windows at all.

Destructive monitors are not suitable for many beamlines that, due to limited horizontal beam size or a particular optical configuration, cannot accommodate a dedicated position monitor port. Non-destructive monitors situated in-line with user beams are needed for these beamlines. The two most commonly used types of non-destructive monitors are split anode ion chamber monitors[8] (see Figure 13.4), which must be located in a gas environment (usually helium so as to minimize absorption of the photon beam) separated from the beamline vacuum by a thin beryllium window, and "gapped" photoemission monitors that operate in the beamline vacuum.

The ion chamber monitor is suitable only for hard X-ray beamlines because of the beryllium window. In contrast, in-vacuum monitors are suitable for VUV and soft X-ray as well as for hard X-ray beamlines and are presently the monitors of choice for most facilities that offer undulator beam sources. These monitors probe only the outer fringes of the beam with metal electrodes, typically made of tungsten, copper, and sometimes high heat-resistant carbon, permitting the central core of the beam to pass through to a downstream experiment. The gap is typically a few times the rms intercepted beam size. Many photoemission electrode shapes and geometries have been developed to sense vertical and horizontal position, ranging from simply a pair of wires stretched across the beam aperture, to horizontally thin blades or vanes that intercept an extremely narrow portion of the beam near its center.[9,10] Electrodes that probe the beam near its center must be capable of withstanding extremely high power densities and heat (of order 1 kW/mm^2 and 2000° C in some cases) if struck directly by the mis-steered beam coming from an undulator; blade electrodes are thin and longitudinally tapered for this reason. Nearby collecting electrodes are often provided to prevent photoelectrons from one blade from crossing to the other blade.

In spite of non-linear off-center responses, in-vacuum photoemission monitors as well as linear ion chamber monitors function very well in steering systems that keep beams precisely centered in the monitors by maintaining nulled difference signals. This operational mode is facilitated if the monitor can be readily translated in the plane of its sensitivity; then it can be moved to match the position of an optimally aligned beam. If the beam is collimated in both transverse directions, translation capability in both planes can be valuable.

Gapped photoemission monitors suffer from the fact that they only sample the lower energy tails of the beam. Any asymmetry in illumination of the electrodes due to occlusion from upstream apertures, low energy scattered or reflected photons, or, in the case of undulator beamlines, from radiation coming from the nearby dipole magnets, leads to an error in determining centroid beam position. It can also cause unwanted sensitivity of a vertical monitor to horizontal beam motion, for example. In particular, corruptive dipole radiation poses the most serious challenge to modern undulator beam monitor designers. The problem becomes evident when the undulator field strength is altered: the relative proportions of undulator and dipole light intercepted by the monitor change, causing the effective null position to shift. The problem is somewhat mitigated by using thin blade electrodes since the most intense portions of the dipole radiation lobes are found on either side of the undulator lobe and are not sensed by the centered blades (Figure 13.3). Several other means have been proposed to desensitize the electrodes to unwanted low energy photons, including polishing electrode surfaces to reflect low energy light,[10] coating the electrodes with graphite,[11] electrically biasing the electrodes to reduce photoemission, and inserting upstream filters.

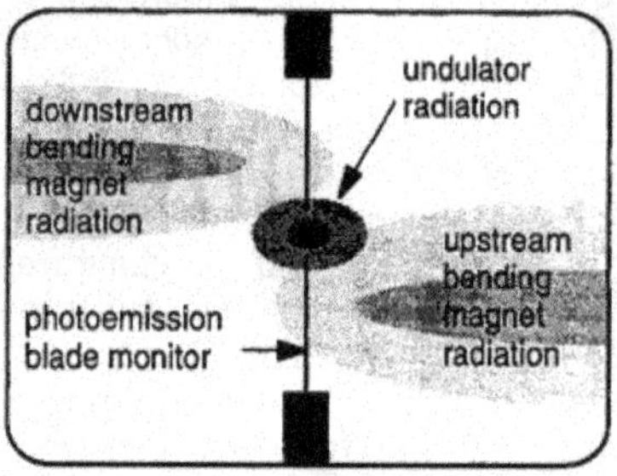

Figure 13.3 Undulator radiation detected with a photoemission blade monitor in the presence of bending magnet radiation.

One can see that achieving 10 μm or better resolution in a photon position monitor is not a simple matter. Besides the need for a high degree of beam profile symmetry in nondestructive monitors, differential signal processing channels for all monitors must be matched to the 0.1% level over short and long periods. This matching can be spoiled by mechanical vibration or thermal distortion of monitor electrodes, by degradation of semiconductor or other conversion media, and by noise and thermal drift in processing electronics.

13.3.2 Radiofrequency Beam Position Monitors (RFBPMs)

Since RFBPMs and their use for monitoring charged particle beam orbits are discussed in Chapter 10, we will add just a few comments on the benefits and limitations of using them for photon beam stabilization. RFBPMs offer one potentially important advantage over photon monitors because the rf fields that excite their electrodes are produced only when the beam is within a few centimeters of the pickup electrode location; consequently, RFBPMs respond to the beam's position at a single point along its trajectory. In particular, provided they have adequate resolution, RFBPMs at either end of an undulator can be used to unambiguously detect the photon beam trajectory since it and that of the particle beam are colinear through the undulator. RFBPMs are used as well for global orbit stabilization which augments the more precise photon beam steering systems.

There are some fundamental properties of rf signal processing that complicate the use of RFBPMs for precise orbit measurement. The signal frequencies of interest from an RFBPM are typically several hundred megahertz. At such frequencies, small changes in capacitance or inductance of the electrodes or cable connections can be misinterpreted as a beam motion of a few microns. Since the effect of such changes is generally proportional to the frequency of the signal, RFBPMs are affected a millionfold more strongly by such changes than are photon monitors, whose electronics are designed for a typical frequency range of 0-500 Hz. Furthermore, the electronic devices used to process RFBPM signals include diode mixers, solid-state switches and limiting amplifiers. These components are used expressly because they respond nonlinearly to input signals. Unless these nonlinearities are precisely eliminated, cancelled, or corrected in post-measurement data processing, changes in beam current or the temperature of the BPM electronics will produce fictitious beam motions of the order of tens of microns or more.

Mechanical motion of the pickup electrode assembly itself will also cause fictitious beam motions. Such motions are caused by temperature changes in the vacuum chamber due to variations in ambient temperature, cooling water temperature, or beam current. Motions of the order of a few tens of microns have been observed at many laboratories.

13.4 Corrector Magnets

The magnets used to correct particle beam orbit are distributed around the storage ring. The orbit-correcting capability of this group of magnets depends on several factors, including the number and locations of correctors and the orbit deflecting resolution, range, linearity, and bandwidth of each corrector.

13.4.1 Corrector Location

The minimum number of corrector magnets needed for orbit correction can be determined from the number of constraints to be placed on the closed orbit: one corrector for each position or angle, vertical or horizontal, to be adjusted at individual points around the ring. Besides the minimum number needed for correcting the orbit at every beamline sourcepoint, more correctors are required to enforce additional orbit constraints, such as orbit position in sextupoles or the localization of corrections around a single beamline so as not to perturb the orbit in the rest of the ring. In general, the optimal locations for correctors used to alter the global orbit are near the quadrupoles, since they are the dominant sources of orbit error. Localized orbit corrections around beamline sourcepoints (section 13.6) are best implemented with correctors nearby and on either side of the source magnets. Optimal corrector placement, together with the efficacy of correction and feedback algorithms, may be checked by simulation of orbit errors in a computer model of the storage ring (Chapter 9).

13.4.2 Dynamic Range and Response

Each corrector magnet must be capable of steering the beam by an amount that is small compared with the beam stability tolerance. This means power supply noise and set point resolution should be about 0.1 μrad or less. The maximum strength of the corrector must be adequate to compensate for ground settlement and magnet misalignments, which may be allowed to evolve and increase for months before magnets are re-aligned. Generally a maximum strength of a milliradian or less is sufficient.

The corrector magnets must operate over a frequency range that matches the important sources of beam motion. Corrector frequency response is not so much limited by power supply bandwidth, which can be a few kilohertz, but by eddy currents in the magnet core iron, and in some cases, by an inductance that requires a high driving voltage for high frequency corrections. For bandwidths greater than a few tens of hertz, laminated cores and low inductance windings are needed; 1-mm laminations are suitable for bandwidths of several hundred Hertz. To reduce losses and to improve response linearity, low-hysteresis silicon steel are often used. The best high frequency performance and linearity is achieved by eliminating the iron core altogether. Air-core magnets produce much weaker fields than do iron-core ones per ampere of driving current; they can be used for vernier orbit correction, with more coarse corrections being made using iron-core correctors.

In some storage rings, orbit corrections are made using dipole trim windings on the cores of main dipole or quadrupole magnets. In this situation, the hysteretic response of the correction field in the presence of the main field is determined largely by the magnetization history of the magnet due to the main coil energization. To obtain correction linearity, it is important to "condition" the core iron by cycling the correction current back and forth over its range of operation a few times to establish a "local" and nearly linear iron hysteresis loop on top of the main energization loop. If this is not done, adding and then removing a correction current increment will not return the field to its original value. This departure from linear response can degrade feedback system performance.

13.4.3 Vacuum Chamber Eddy Currents

In many cases the most important impediment to applying AC correcting fields to the beam are the eddy currents they induce in the storage ring vacuum chamber. Eddy currents attenuate the magnetic fields that actually reach the beam, and the attenuation increases with increasing frequency. If the storage ring vacuum chamber is made of high resistivity stainless steel, like that of ESRF or Super-ACO, eddy current attenuation is minimal up to 100 Hz or so. However, many synchrotron light sources now have vacuum chambers made of rather thick low resistivity aluminum. Eddy currents in these chambers cause severe distortion and phase delay of the correction field and can limit the bandwidth of a feedback system to less than 10 Hz. Fortunately, the feedback electronics can be designed to compensate for eddy current effects, extending the feedback bandwidth to significantly higher frequency (section 13.5.3).[12] A consequence of this compensation is that magnet power supply voltage and current ratings must be increased.

13.5 Feedback Orbit Control

13.5.1 General Feedback Systems

To illustrate feedback system concepts and performance limitations, set mainly by the frequency response of system components,[13,14] we will consider a simple orbit control system consisting of one beam position monitor and one corrector magnet. The feedback system in Figure 13.4 uses the output error signal from the BPM to control a steering magnet capable of changing the beam position at this BPM. If the feedback system is turned off, the beam can move around under the influence of vibrations, stray magnetic fields, etc. The BPM detects this motion which we will call Y. When the corrector magnet is adjusted, the BPM sees both the effect of the random beam motion Y and the motion caused by the corrector, y. The net beam motion is just y+Y. To make the BPM and corrector into a feedback system, we allow the BPM output to change the corrector current by an amount dI such that the measured displacement of the beam by the corrector, y, is proportional to the reading of the BPM, y+Y, and of opposite sign:

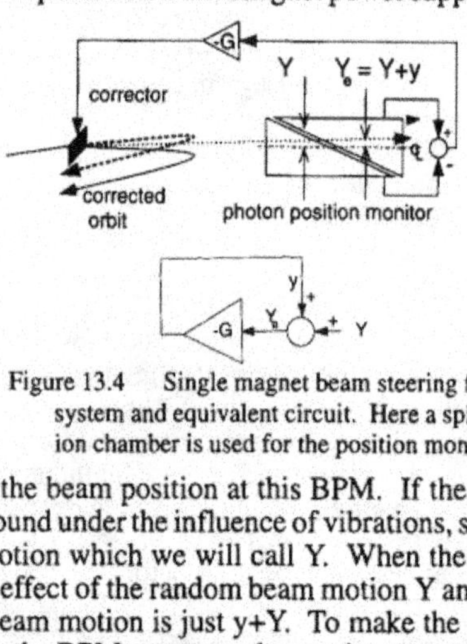

Figure 13.4 Single magnet beam steering feedback system and equivalent circuit. Here a split-anode ion chamber is used for the position monitor.

$$y = \frac{dy}{dI} I = -G\,(Y + y) \tag{13.7}$$

dy/dI is measured or else calculated from a computer model. G is commonly called the open-loop gain of the feedback circuit. We can add and subtract Y in Eq. (13.9) to get

$$-Y + (Y + y) = -G(Y + y) \tag{13.8}$$

$$(Y + y) = Y/(1 + G). \tag{13.9}$$

This expression tells us that if we can make G a large positive number, we can reduce the uncorrected beam position error Y to a much smaller value (Y+y).

13.5.2 Frequency Domain Response of Feedback Systems

We now consider the time dependence of Y and y since the feedback electronics cannot respond instantly to orbit errors. For most feedback systems, the corrective response y, or equivalently the open–loop gain function G, can be described (at least approximately) by a linear differential equation with respect to time. Analysis of such linear, time–invariant systems is simplified if the time domain descriptive functions are transformed into the complex frequency domain using the Laplace transform.[14] The relationship between a time domain function f(t) and its Laplace transform F(s) is given by

$$F(s) = \int_0^{\infty} f(t)e^{-st}\,dt \tag{13.10}$$

where $s = \sigma + j\omega$ (ω (rad/s) $= 2\pi \times$ frequency (Hz)).. This transform is analogous to the Fourier transform (note the similarity if $s = j\omega$ and $f(t) = 0$ for $t < 0$) but is more general in that it permits analysis of transient response and stability in systems that are suddenly "turned on" with a set of initial conditions. By transforming system signals and the open-loop gain function, we can express feedback equations in a simple s-dependent form:

$$Y_e(s) = Y(s) + y(s) = \frac{Y(s)}{1 + G(s)} \tag{13.11}$$

where we have substituted $Y_e(t) = Y(t) + y(t)$ in the time domain equation to denote the corrected "error" orbit. $Y_e(t)$ is then obtained by taking the inverse transform of $Y_e(s)$. $1 / [\,1 + G(s)\,]$ is a complex function that can be expressed in polar coordinates as $e^{j\theta(s)} / |1 + G(s)|$. It can readily be shown that the response $Y_e(t)$ to a complex exponential input $Y(t) = Ye^{st}$ (which includes sinusoidal inputs with exponential growth or decay envelopes) is also an exponential of the same frequency, but modified by an attenuation factor $1 / |1 + G(s)|$ and shifted by in phase by $\theta(s)$.

The transformed gain function G(s), called the "open-loop transfer function," for systems described by linear time-invariant differential equations can be expressed as a ratio of factored polynomials of the variable s:

$$G(s) = K\,\frac{\prod_{m=1}^{M}(s + Z_m)}{\prod_{n=1}^{L}(s + P_n)}. \tag{13.12}$$

G(s) approaches zero as s approaches one of the "zeros" $-Z_m$ of G(s). Similarly, G(s) approaches infinity as s nears one of its "poles" $-P_n$.

The closed-loop behavior of the feedback system can be determined from the location of the poles and zeros of G(s), or equivalently of the function $1 + G(s)$, in the complex s plane. If $1 + G(s)$ becomes zero for some s with $\mathrm{Re}(s) > 0$, then the feedback system will cause y to increase exponentially with time, even if Y is zero. To guarantee stability, $1 + G(s) = 0$ must have no solutions at values of s for which $\mathrm{Re}(s) > 0$. The zeros of $1 + G(s)$ for which $\mathrm{Re}(s) < 0$ are also significant; they indicate the transient response of the system after an abrupt change in Y. The system designer can manipulate the location of complex poles and zeros, and add new ones in series or in parallel, in order to obtain desired response properties. This is equivalent to modifying the differential equation describing the system.

The steady state frequency response of the system can be obtained by letting $s = j\omega$ in the above equations and by letting $Y(t) = Ye^{j\omega t}$ be a sinusoidal input. Then

$$Y_e(t) = \frac{Ye^{j(\omega t+\theta)}}{|1 + G(j\omega)|} \tag{13.13}$$

$$\tan\theta = -\frac{\mathrm{Im}[G(j\omega)]}{\mathrm{Re}[1 + G(j\omega)]} \tag{13.14}$$

If $|1 + G(j\omega)|$ should become less than one for some value of ω, then the feedback system can increase rather than reduce beam motion. As $|1 + G(j\omega)|$ approaches zero, the system becomes unstable. If we express $G(j\omega)$ in polar coordinates as $|G(j\omega)|e^{j\phi}$, instability occurs when $|G(j\omega)| = 1$ and $\phi = 180°$.

Most feedback systems have components whose response magnitudes decrease and whose phase shifts increase with increasing frequency, so the possibility of unstable response is very real. Stability is guaranteed if the system is designed so that $G(j\omega)$ is represented by a single negative real pole in Eq. (13.14):

$$G(j\omega) = \frac{K}{j\omega + P_1} \tag{13.15}$$

since ϕ ranges from 0° (for $\omega << P_1$) to −90° (for $\omega >> P_1$); a system with this open-loop transfer function is unconditionally stable, and $|G(j\omega)|$ decreases at a rate of 20 dB per decade increase of frequency beyond the pole frequency. If $G(j\omega)$ has two or more poles (common for real systems), then ϕ can reach −180° as ω increases; $|G(j\omega)|$ decreases by 40 dB per decade beyond the second pole frequency. The system designer can guarantee system stability by ensuring that $|G(j\omega)| < 1$ when $\phi = 180°$ and that $|\phi| < 180°$ when $|G(j\omega)| = 1$, thereby limiting system bandwidth; this can be done by reducing the DC gain K of $G(s)$ and/or by introducing zeros into the transfer function that cancel all but the lowest frequency pole. Since it is desirable to have a high open-loop gain within the bandwidth, it may seem advantageous to adjust $|G(j\omega)|$ so that it decreases rapidly from a high to low value at the high end of the bandwidth. However, this can only be done by adding one or more poles, together with their subsequent phase shifts, to the transfer function; this in turn endangers closed-loop stability and counteracts any increase in high-end gain.

13.5.3 Eddy Current Compensation — Frequency Domain

The effect of eddy currents induced by oscillating magnetic fields in the ring vacuum chamber, as discussed in section 13.4.3, is to increasingly attenuate and shift the phase of the fields reaching the beam as the oscillation frequency increases. The designer must choose the form of $G(j\omega)$ to compensate for this effect in order to achieve a reasonable feedback system bandwidth. The effect of eddy currents is essentially that of a low pass filter function $H(j\omega)$ in series with the feedback gain function $G(j\omega)$; the overall transfer function is the product of these two functions. The closed-loop transfer function then becomes

$$Y_e(j\omega) = \frac{Y(j\omega)}{1 + G(j\omega)H(j\omega)}. \tag{13.16}$$

In order to increase the bandwidth of the feedback it is necessary to adjust the feedback system amplifier response to compensate for the eddy currents in the chamber by adding a filter function $h(j\omega)$ that effectively cancels $H(j\omega)$:

$$\frac{1}{1 + G(j\omega)h(j\omega)H(j\omega)}; \quad h(j\omega) \approx \frac{1}{H(j\omega)}. \qquad (13.17)\ (13.18)$$

An example of feedback system performance using eddy current compensation is shown in Figure 13.2 for the SSRL local vertical steering system.[15] The combined transfer function of the iron magnet cores plus vacuum chamber resembles a series of single poles at 25, 120, and 200 Hz:

$$H(j\omega) = \frac{K_H}{(j\omega + 2\pi 50)\,(j\omega + 2\pi 120)\,(j\omega + 2\pi 200)} \tag{13.19}$$

where $K_H = (2\pi 50)\,(2\pi 120)\,(2\pi 200)$. The two lower frequency poles are approximately cancelled by adding two zeros:

$$h(j\omega) = K_h\,(j\omega + 2\pi 50)\,(j\omega + 2\pi 120) \tag{13.20}$$

with $K_h = 1/[(2\pi 50)\,(2\pi 120)]$. The resulting bandwidth is limited to ~ 150 Hz since the 200-Hz pole remains uncancelled; without $h(j\omega)$, it would be limited to ~ 25 Hz. The feedback system can suppress 20-Hz noise components by a factor of ten. Steering system performance at other facilities is qualitatively similar.

13.5.4 Digital Signal Processing and Z-Transform

With the advent of fast digital electronics with high resolution capability, it has become practical to filter, process, and distribute signals for beam stabilizing feedback systems using digital signal processing (DSP) techniques.[16] In contrast to the analog systems that use capacitors and inductors to approximate differentiation and integration of the signal waveform, digital systems perform these mathematical operations on a series of measurements taken at fixed time intervals. Since most accelerator signals are analog, they need to be converted to digital format before processing can occur. This is called analog-to-digital conversion (ADC), or "sampling." Mathematical operations such as numerical differentiation and integration may be carried out on the sampled signal, using the repertoire of techniques commonly used in computer programs for data analysis. After processing, the digital signal may be converted back to analog so that it can be used to control analog devices. This inverse process is called digital-to-analog conversion (DAC). The time interval between two samples is called sampling period T and its reciprocal is called the sampling frequency f_s.

Figure 13.5 Processing of an analog signal using a digital filter.

Figure 13.5 shows the schematic of a simple digital signal processing system. The time-varying analog signal x(t) is sampled by an ADC at every interval T. The sampled data is modified by a digital filter and the result is converted back to analog signal y(t) by a DAC.

Digital signal processing in time-domain must be physically realizable (or causal), which means no future samples from either the input or output can be used to calculate the latest output. Such digital filters use a finite number of previous input samples including the latest one and a finite number of previous output samples. Let $\{x_n\}$ be the input sequence, where x_{n-k} is the kth previous input sample. The output sequence $\{y_n\}$ is then given by

$$\begin{aligned} y_n = {} & a_0 x_n + a_1 x_{n-1} + a_2 x_{n-2} + \cdots + a_M x_{n-M} \\ & - b_1 y_{n-1} - b_2 y_{n-2} - b_3 y_{n-3} \cdot \cdots - b_L y_{n-L} \end{aligned} \tag{13.21}$$

or

$$y_n = \sum_{k=0}^{M} a_k x_{n-k} - \sum_{k=1}^{L} b_k y_{n-k} \tag{13.22}$$

The digital filter in Eqs. (13.21) or (13.22) defines the latest output at step n to be a function of the latest and M previous input samples as well as L previous output samples. The real coefficients $\{a_k\}$ and $\{b_k\}$ are called the filter coefficients which, together with the sampling time, determine the behavior of the digital filter.

Just as the Fourier transform and the Laplace transform are important for analysis of analog systems, the Z-transform facilitates the analysis of discrete-time signals and systems. Given a sequence of discrete-time signals $\{x_n\}$, its Z-transform is defined by[16]

$$Z : \{x_n\} \rightarrow X(z) = \sum_{n=-\infty}^{\infty} x_n z^{-n} \tag{13.23}$$

where z is a complex variable and plays a role similar to that of the variable s in the Laplace transform. $\{x_n\}$ is called the inverse transform of X(z). With the substitution $z \rightarrow e^{-j\omega T}$ one may recognize that X(z) is the discrete Fourier transform of $\{x_n\}$, which gives the frequency characteristic of the discrete sequence $\{x_n\}$. The transfer function H(z) of the digital filter is the ratio of Z-transforms of input and output signals:

$$H(z) = Y(z)/X(z), \tag{13.24}$$

and $H(e^{-j\omega T})$ gives the frequency characteristic of the digital filter. Two cascaded digital filter functions $H_1(z)$ and $H_2(z)$ are equivalent to the filter function $H_1(z)H_2(z)$.

Applying the Z-transform to Eq. (13.24) and using the relation

$$Z : \{x_{n-k}\} \rightarrow z^{-k} X(z) \tag{13.25}$$

we obtain

$$H(z) = \frac{\sum_{k=0}^{M} a_k z^{-k}}{1 + \sum_{k=1}^{L} b_k z^{-k}}. \tag{13.26}$$

If b_k is zero for all k, then Eq. (13.26) describes a "finite impulse response" (FIR) filter because the response of the filter function to an impulse input is zero after a finite number M of sample periods. This is not the case if any b_k is non-zero; then the equation describes an "infinite impulse response" (IIR) filter.

The finite sampling frequency, however, limits the frequency range of such analysis to $f_s/2$ (the Nyquist frequency). The response of a digital filter to a sinusoidal input of frequency f is indistinguishable from its response to sinusoids of equal amplitude at frequencies given by $nf_s \pm f$, where n is a positive integer. This is called "aliasing error" and the designer of the digital system must ensure that the input analog signal does not contain frequency components beyond the Nyquist frequency, sometimes by adding a band-limiting analog "anti-aliasing" filter before the digitizer.

13.5.5 Proportional, Integral and Derivative (PID) Control Algorithm

Equation (13.11) can describe the function of a digital feedback control system if Y, Y_e, and $1/(1 + G)$ are all Z-transforms and correspond respectively to X(z), Y(z), and H(z) in Eq. (13.26). If we neglect eddy currents for the moment, we may obtain a reasonable feedback response by defining the open-loop gain G(z) in terms of proportional, integral, and derivative control.

The simplest feedback controller is proportional, in which case the output is a constant multiple K_P of the input (see section 13.5.1). The constant is called the open-loop gain at DC. Another form of controller is integral, the current output of which is the sum of the current input, multiplied by a small positive constant K_I, and the most recent previous output. An integral controller keeps accumulating the input signal times K_I for output. On the other hand, the current output of a derivative controller is a constant K_D multiplied by the difference between the current input and the immediately preceding input.

The digital filter function of the PID controller can be written in the z-domain as

$$G(z) = K_P + \frac{K_I}{1 - z^{-1}} + K_D\left(1 - z^{-1}\right). \tag{13.27}$$

Typically the gains K_P and K_D are set between 5 and 10 and K_I is set less than 0.05 to ensure stability and to obtain optimal feedback performance. The frequency characteristic of G(z) is obtained with z replaced by $e^{-j\omega T}$ as explained in the previous section. At low frequencies, z is close to 1 and therefore the integral control dominates. In particular, the PID gain G(z) at DC is infinitely large with finite K_I, implying that very slow orbit drift is perfectly corrected. As the input signal frequency increases, the action of the integral control becomes less important because the accumulated sum of previous outputs cannot grow large if the input oscillates rapidly. The mid-frequency range, around 10% of the sampling frequency, is covered by the proportional control and the high-frequency range (around 20% of the sampling frequency) is dominated by the derivative control. Design of the PID controller consists of determining the three gain constants K_P, K_I, and K_D according to the desired feedback performance given the sampling frequency f_s and the open loop bandwidth f_b.

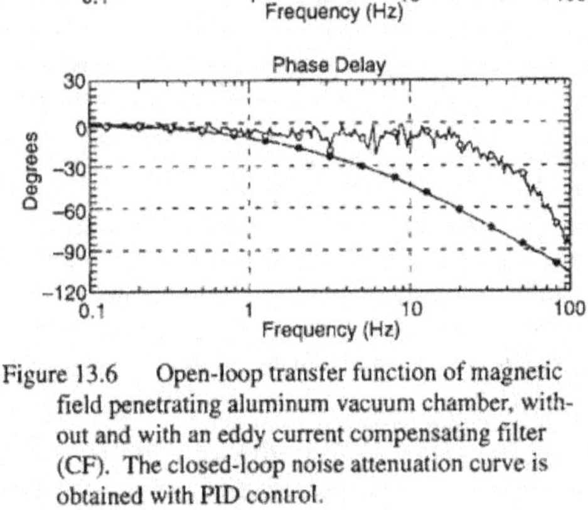

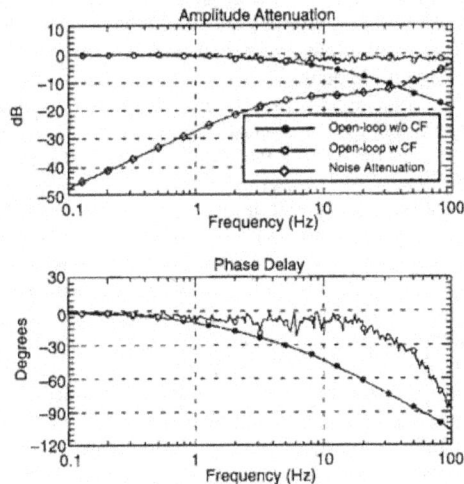

Figure 13.6 Open-loop transfer function of magnetic field penetrating aluminum vacuum chamber, without and with an eddy current compensating filter (CF). The closed-loop noise attenuation curve is obtained with PID control.

13.5.6 Eddy Current Compensation with Digital Signal Processing

Compensation of eddy current effects with DSP can be carried out in in close analogy with section 13.5.3 after replacing G(jω), H(jω), and h(jω) with G(z), H(z), and h(z), respectively. h(z) must take a more complicated form than the PID control described above.

Figure 13.6 shows the attenuation and phase shift of the field of a corrector magnet penetrating the thick aluminum vacuum chamber of the APS storage ring without and with eddy current compensation. The bandwidth of the penetrating field is limited to about 6 Hz at -3 dB. With eddy current compensation, the bandwidth extends beyond

100 Hz; however the phase delay is still significant even after compensation (-60° at 75 Hz and -90° at 100Hz).

13.6 Local Beam Steering Systems

Since many first-generation synchrotron radiation beamlines were operated parasitically during colliding beam experiments, beamline users needed a way to steer the photon beams in one or more beamlines without disrupting the orbit at colliding beam sites. This need was met by using combinations of corrector magnets to create localized orbital "bumps" for the specific steering regions. In dedicated light sources it is still advantageous to use orbital bumps to steer beamlines independently of one another. They can be used in conjunction with BPMs to configure localized feedback steering systems.

13.6.1 3-Magnet, 1-Monitor Feedback

A minimum of three conditions must be satisfied to create a confined orbital bump: 1) the orbit is displaced at the sourcepoint; 2) the orbit position after the last bump magnet must be unchanged; and 3) the orbit angle after the last magnet must be unchanged. Three correctors whose kick strengths are scaled in proportions dictated by the betatron amplitudes and phases at the corrector sites are generally needed to satisfy these criteria.

The 3-magnet bump does not have enough degrees of freedom to independently control sourcepoint orbit position $y(s_{sp})$ and angle $y'(s_{sp})$. Instead, the ratio of position to angle change is a constant. They combine to produce a net displacement of the photon beam centroid at a distance L from the sourcepoint given by $y(L) = y(s_{sp}) + Ly'(s_{sp})$. Clearly the bump shape should be be chosen to avoid the unfavorable situation $y(s_{sp}) \approx -Ly'(s_{sp})$; in this case a small error in the photon beam position at the monitor must be corrected by a large motion of the particle beam.

For large L (> ~ 10m), y(L) is generally dominated by the angular orbit change. Angular bumps (i.e., $y(s_{sp}) / Ly'(s_{sp})$ is small) tend to be more "sensitive" for steering long beamlines in terms of the beam deflection at L per unit bump energization. It is possible to configure angle-changing bumps quite compactly (with phase spans as small as 45°). A compact series of decoupled bumps serving several closely spaced sourcepoints can be arranged by sharing correctors: e.g., the last one or two correctors from one bump can be used as the first one or two correctors of the next as long as the bumps do not overlap each others' sourcepoint. A potentially negative consequence of this close concatenation is that a sourcepoint can end up being situated between the second and third bump correctors where the signs of position and angular orbit corrections are opposite; then the change in photon beam position caused by sourcepoint angle is counteracted by an opposite change due to sourcepoint position (i.e., $y(s_{sp}) \approx -L\, y'(s_{sp})$) resulting in an insensitive steering bump.

A feedback system that uses a local 3-magnet bump can fix the beam position at a single point along the photon beam trajectory (Figure 13.7a). The bump does not fix the angle with which the photon beam passes through the monitor. This means that the position of the photon beam can deviate from the ideal trajectory both upstream and downstream of the monitor. This can be a problem for focused beam experiments, where maintaining a steady beam image is a matter of stabilizing just sourcepoint beam position, and for beamlines having many small apertures that severely limit the range of acceptable sourcepoint orbit parameters required to perform the experiment.

13.6.2 4-Magnet, 2-Monitor Feedback

To achieve stability levels that are small fractions of the beam size and divergence, localized steering systems that correct both position and angle are needed. The extra degree

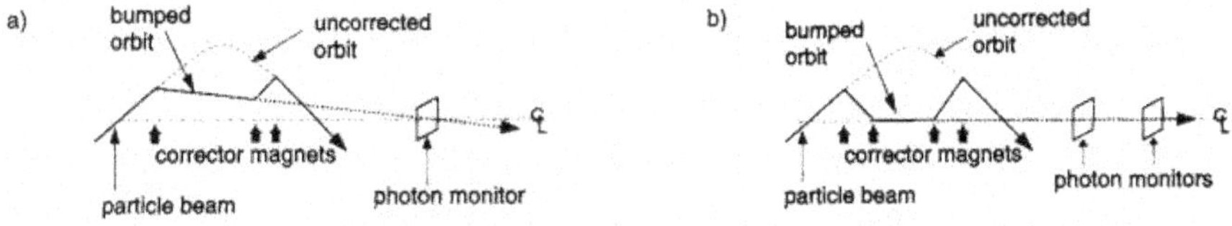

Figure 13.7 a) 3-magnet local bump stabilizes position at single monitor, but not elsewhere.
b) 4-magnet, 2-monitor system stabilizes position and angle in beamline.

of freedom in sourcepoint orbit correction is gained by extending the range of the local bump to include a fourth corrector (Figure 13.7b). The four magnets can be driven with one combination of relative deflections to correct sourcepoint position, and with another combination to correct angle (i.e., the 2-parameter bump is a superposition of two 1-parameter bumps). Two beam position monitors, separated by several meters, are then used to establish a reference beam trajectory and to provide the bump-driving parameters.

An alternative to calculating beam position and angle from the two monitor signals and driving the 2-parameter bump accordingly is to configure the two superposed 1-parameter bumps so that one corrects position only at the first monitor and not at the second, and the other affects beam position at the second monitor and not at the first. The bumps are then driven from their respective monitors. This method has the advantage that each loop using a single monitor can function independently of the other, and that each 1-parameter bump can be empirically tuned using just the position monitor error signals, as opposed to calculated position and angle values, to guarantee minimal interaction between the two feedback loops. Since it is easier to adjust locality for 3-magnet bumps than for 4-magnet bumps (fewer adjustment parameters), a further advantage is gained by implementing each 1-parameter 4-magnet bump as a superposition of two localized 3-magnet bumps.

13.6.3 Feedback System Design

Localized vertical steering feedback systems have been configured using single photon monitors and 3-magnet beam bumps for beamlines in first and second generation synchrotron radiation facilities.[15,17,18] Systems employing 2-parameter, 4-magnet bumps have also been implemented and are presently being planned for new facilities. Localized horizontal systems using photon monitors are uncommon because of the previously mentioned difficulty in monitoring the position of horizontally divergent beams; however, horizontally collimated undulator beams can be stabilized this way, and local systems for undulator or wiggler insertion devices can be configured using a pair of RFBPMs flanking the device as mentioned in section 13.3.2.

The stabilizing capability of a steering feedback system is determined by several factors besides the configuration of its corrective bump as discussed above. They include: 1) the attainable degree of isolation from neighboring local feedback systems; 2) position monitor location; 3) control and monitor resolution, dynamic range, and bandwidth; and 4) the servo processing functions incorporated into the system, as discussed in section 13.5, augmented with a complement of control and interlock functions that are integrated into the storage ring control system.

With sufficient bump isolation, multiple steering loops are decoupled and each behaves as a 1-parameter feedback system. Otherwise the systems become excessively coupled and stabilizing performance is degraded, or an instability can result, because each loop must correct the errors created by all the others in addition to errors from random sources. It is estimated that coupling should be limited to about 10% divided by the number of stabilized parameters over the operational bandwidth. At SSRL, for example, ten 1-parameter loops perform adequately when bump interaction is reduced to the 10%/10 = 1% level;[19] about 10%/64 = 0.15% decoupling will be needed for the 32 vertical 2-parameter loops (i.e., 64 parameters) at the APS. With careful empirical adjustment, bump interaction can be reduced to about 1%, limited by nonlinear or frequency-dependent properties of the magnets (e.g., hysteresis), vacuum chamber (eddy currents), and magnet lattice (e.g., sextupole magnets within the bump). For the APS, adequate isolation can only be achieved by using a global orbit feedback system to cancel residual bump distortion (section 13.7.3).

The vertical angle-stabilizing capability for a 3-magnet, 1-monitor feedback system is defined by the magnitude of possible sourcepoint position excursion divided by the distance to the monitor. The resultant positional stability at the experiment thus depends on the proximity of the monitor to the sourcepoint and to the experiment. In many cases, however, because of the presence of other optical components, and because the monitor must be located upstream of such components that deflect the beam, the separation between monitor and experiment can be tens of meters. If, for example, the monitor is located 10 m from the sourcepoint, and if the orbit position changes by 1 mm, then the angle is only stabilized to 100 μrad; consequently, position at an experiment 10 m from the monitor will also change by 1 mm. On the other hand, the angle-stabilizing capability of a 4-magnet, 2-monitor system is defined by the monitors' position resolution divided by their separation. Two monitors having 10-μm resolution can resolve orbit angle to the 1-μrad level if separated by 10 m. Systems having this level of resolution have been devised using two photon monitors,[20,21] two RFBPMs,[22] and one RFBPM together with a photon monitor.[19] When two photon monitors are used, care must be taken in monitor design to prevent the downstream monitor from being shadowed by the upstream one.

Once sufficiently decoupled corrective bumps are established, more problems can arise if there is coupling between horizontal and vertical correction systems due to beam profile asymmetry in position monitors. Provided coupling problems are resolved, the next gauge of localized feedback system performance is the system's control and monitoring resolution and dynamic range. We have already discussed the fact that it is difficult to achieve 10 μm or better monitoring resolution; on the other hand, it is relatively straightforward to produce orbit corrections on that order. Nonlinear bump response may reduce the attainable dynamic range of decoupled beam deflection by an order of magnitude from the desired value of 1000:1. Then it may be necessary to separate the system into two operating ranges: one that can make millimeter corrections in an open-loop mode where bump isolation is not so critical, and one that handles sub-millimeter disturbances in closed-loop mode.

Preserving 10-μm control resolution and 1000:1 dynamic range throughout the feedback processing system is possible using analog circuitry provided great care is taken to shield, filter, and otherwise reject electrical noise that may enter the system and also by using low drift electronic components. This task becomes more formidable when the various feedback system components are separated by large distances, as they are in the larger facilities such as the ESRF, the APS, or SPring-8; then it is more practical to use digital control links.

13.7 Global Orbit Feedback Systems

Accelerator experts developed computer programs called "global" orbit correction algorithms (see Chapter 10 and references therein) to minimize DC orbit errors everywhere in the ring, not just at the sources of synchrotron radiation. Such algorithms implement various strategies for cancellation of orbit error at or near their source using the corrector magnets. Many of them feature highly sophisticated capabilities such as identification and elimination of erroneous beam position data, automatic adjustment of localized bump parameters in response to changes in quadrupole settings, and accommodation of non-linear ring lattice properties. Global correction systems are generally not configured to bring to zero every BPM reading used as input; consequently, a reduction in rms orbit error by factors of 10-20 is typical. In contrast, a single local feedback loop is limited only by the performance of the BPMs, correctors, and associated hardware, and can achieve a correction factor of over 300 at low frequency.

Global orbit correction algorithms were originally intended to be executed infrequently by an operator to counteract slow drift of the orbit. However, modern accelerator control systems are rapidly erasing the distinction between orbit correction algorithms and feedback orbit control; both the KEK Photon Factory[23] and ESRF[24] have automated their orbit correction algorithms to run repeatedly without human intervention. These algorithms provide minute-by-minute orbit correction to supplement the local feedback systems. Global feedback systems using fast processing techniques to implement orbit correction algorithms have achieved bandwidths of 60 Hz or more.

13.7.1 Global Feedback Basics

Section 13.5.1 gave a description of a one-parameter feedback system, in which the strength of a single corrector is controlled by the reading of a single beam position monitor. Sections 13.6.1 and 13.6.2 described local feedback systems, in which three corrector magnets are controlled by one beam position signal (1-parameter), and four correctors are controlled by two monitors (2-parameter), subject to the constraint that the orbit correction be local. In the 2-parameter system, each parameter is controlled by a linear combination of the two position monitor signals. The next step in generalization is to define an m-parameter system where each parameter is controlled by a linear combination $\mathcal{Y}$ of m (or more) position monitor measurements in a global feedback system. The system will minimize this linear combination $\mathcal{Y}$ in exact analogy to Y in section 13.5.1:

$$\mathcal{Y} + \mathcal{y} = \mathcal{Y}/(1 + G) , \tag{13.28}$$

where $\mathcal{y}$ is a linear combination of corrections applied by a set of n corrector magnets. We must also choose a linear combination $\mathcal{I}$ of corrector magnet currents to generate $\mathcal{y}$.

By intelligently choosing m-parameter sets $\{\mathcal{Y}_m\}$ and $\{\mathcal{I}_m\}$ of such linear combinations, it is possible to design a feedback system that efficiently improves orbit stability everywhere. Using matrix inversion techniques such as singular value decomposition (SVD)[25] (see Chapter 10), one may define these linear combinations to be orthogonal; that is, one may insure that

$$\frac{\Delta \mathcal{Y}_k}{\Delta \mathcal{I}_m} = 0 \quad \text{unless } k = m \tag{13.29}$$

so that the correction of $\mathcal{Y}_m$ is not affected by $\mathcal{Y}_k$.

13.7.2 Global Harmonic Feedback

A simple example of a global feedback algorithm can be illustrated by solving Eq. (13.1) by means of a Fourier transform in the azimuth angle $\phi = \frac{s}{R}$:

$$\Delta B(\phi) = \sum_{m=-\infty}^{\infty} \Delta \mathcal{B}_m e^{im\phi}, \quad Y(\phi) = \sum_{m=-\infty}^{\infty} \mathcal{Y}_m e^{im\phi} \qquad (13.30)\ (13.31)$$

$$(-m^2+\nu^2)\mathcal{Y}_m = \frac{\Delta\mathcal{B}_m}{B\rho}\ ;\quad \mathcal{Y}_m = \frac{1}{B\rho}\,\frac{\Delta\mathcal{B}_m}{\nu^2-m^2}. \qquad (13.32)\ (13.33)$$

In this case the set of orthogonal linear combinations correspond to the Fourier harmonics of the orbit waveform. We see that the mth harmonic amplitude $\mathcal{Y}_m$ of the orbit error is proportional to the mth harmonic of the stray field distribution. We might expect a random, "white noise" distribution of stray field errors around the accelerator. If this is so, the harmonic coefficients $\Delta\mathcal{B}_m$ should be approximately equal in magnitude for any m. Equation (13.33) tells us that we can expect the largest $\mathcal{Y}_m$ to be those for which m is close to ν and $-\nu$. A harmonic orbit correction algorithm forms linear combinations of BPM data that give the best possible approximation of those $\mathcal{Y}_m$'s with m close to ν. The algorithm prescribes adjustments to the correctors which produce the best approximation of pure harmonic field distributions to cancel the orbit error harmonics. Harmonics for which m is far from ν are simply ignored and left uncorrected. In particular, the $m = 0$ harmonic, which corresponds to a change in ring circumference, cannot be corrected using dipole correctors; it must be compensated by changing the ring rf frequency, possibly with a feedback system. If the BPMs and correctors are spaced uniformly in ϕ, fast Fourier transform techniques provide accurate and reliable approximations of the $\mathcal{Y}_m$'s and the $\Delta\mathcal{B}_m$'s. If non-uniformly-spaced BPMs and/or correctors are used, the correction algorithm becomes more complicated in order to avoid problems with aliasing in the harmonic detection process and with producing unwanted higher frequency harmonics.

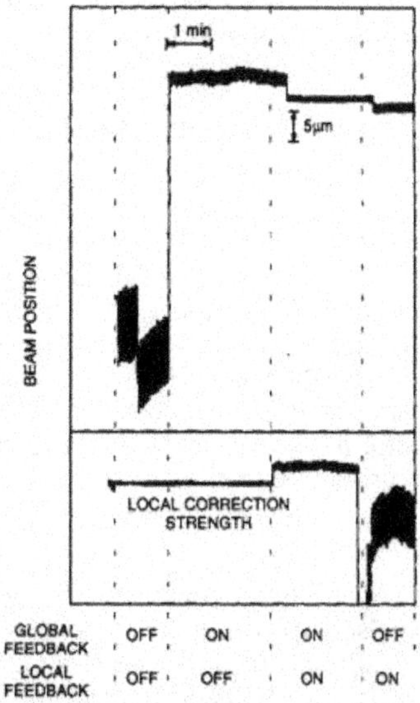

Figure 13.8 Orbit stability using global and local vertical feedback systems at the NSLS. Stability is maximized and the local correction strength minimized when both systems are on.

This kind of feedback system has been implemented at the National Synchrotron Light Source.[26] It is found to improve the performance of local feedback loops operating in the insertion device straight sections while providing a 5x to 10x improvement in beam stability everywhere in the rings (Figure 13.8). The design of the NSLS system mixes analog and digital electronics; the BPM signals are passed to the corrector magnets without having been digitized. However, all the necessary combinations of BPM signals and fanouts of corrector control voltages (in effect the matrix multiplication necessary to obtain the linear combinations of BPM signals and magnet currents proportional to the $\mathcal{Y}_m$'s and the $\mathcal{I}_m$'s, respectively) are implemented by passing the analog signals through analog attenuators under computer control.

13.7.3 Global Feedback Using Digital Signal Processing

The APS feedback orbit control system combines local and global correction strategies in one system.[27] The global correction algorithm is based on measurements of the corrector-BPM "response matrix"; each column of the matrix represents the set of displacements detected by BPMs in response to a single corrector kick; the number of rows is given by the number of BPMs, which in general is not equal to the number of correctors. The algorithm uses SVD[28] to invert the response matrix and to identify the most effective sets of linearly combined orbit readings and correctors (called the monitored and corrected "orbit eigenvectors," which are orthogonal) to be controlled in the global feedback. SVD is described further in Chapter 10.

The APS system uses large numbers of beam position monitors and correctors distributed around the storage ring of 1104 m circumference. Out of the 360 RFBPMs and 320 correctors available for closed orbit correction, 40 BPMs and 38 correctors will be used for real-time global orbit feedback. The system uses digital signal processing; all feedback components between the BPM electronics and corrector magnet power supply interface are digital. Since the correction bandwidth of a digital feedback system is limited by the sampling frequency or the number of corrections per unit time, every effort is made to reduce the time taken for sampling of the BPM data, calculation of the corrector strengths and corrector power supply control. The time budget for beam position feedback system is:

Data acquisition from BPMs $\approx$ 50 μsec
Calculation of corrector strengths by DSP $\approx$ 100 μsec
Corrector power supply control $\approx$ 100 μsec

resulting in 4-kHz sampling frequency for the feedback system. This high sampling frequency is achieved using dedicated digital signal processors and high-speed reflective memory boards (26 Mbytes/sec transfer rate) connected by a dedicated fiber optic link. These are distributed around the ring in 20 VME crates interfaced to BPM processors and power supply controllers through optical fibers.

An important function of the global feedback system is to counteract the local bump closure error in the local feedback systems. However, to take advantage of both the global and local feedback in stabilizing the beamlines, it is necessary to coordinate their operation. The local feedback systems must have access to the global orbit data and subtract the effect of the global system from the correction commanded by the local loops during the next feedback cycle. The feedback system correction bandwidth will be on the order of 100 Hz.

13.8 Feedback Control of Multibunch Instabilities

As mentioned in Chapter 12, multibunch instabilities can be controlled using feedback systems.[29-31] Feedback systems for these types of instabilities are similar in some ways to feedback systems for control of orbit motions. Both systems make use of a beam position monitor to detect the unwanted beam motion, and both systems use a corrector to reduce or cancel the beam motion. The differences between these two types of systems are more numerous than the similarities, however. The sources of orbit disturbance are externally applied forces; multibunch instabilities, on the other hand, are caused by the forces that the bunches apply to themselves and each other via the electromagnetic fields they create within the beam chamber.

Feedback damping is achieved by applying electromagnetic "kicks" to each bunch which cause exponential decay of its longitudinal or transverse oscillation. In the next section we will see how a kicker may be powered to accomplish this.

13.8.1 Basics of Feedback for Instabilities

To achieve damping of transverse oscillations, it is necessary to install a damping kicker of length Δs in the storage ring. The kicker must be powered so as to add a damping term $D(s)$ to the equation of motion of the bunch (Eq. (13.1)) which overpowers any antidamping force from the wake fields:

$$\frac{d^2y_\beta(s)}{ds^2} + D(s)\frac{dy_\beta(s)}{ds} + \left(\frac{v}{R}\right)^2 y_\beta(s) = 0\ . \tag{13.34}$$

$D(s) = D$ for $0 \le s \le \Delta s$, otherwise $D = 0$. We integrate Eq. (13.34) to illustrate the effect of the feedback system on the beam. The limits of the integral extend over the length of the damping kicker:

$$\left.\frac{dy_\beta(s)}{ds}\right|_0^{\Delta s} = -D\int_0^{\Delta s} y'_\beta(s)\, ds - \left(\frac{v}{R}\right)^2 \int_0^{\Delta s} y_\beta\, ds;\quad y'_\beta(s) = \frac{dy_\beta(s)}{ds}\ . \tag{13.35}$$

The integrals on the right side are rewritten in terms of average values of their integrands:

$$\Delta y'_\beta = -D\Delta s\langle y'_\beta\rangle - \left(\frac{v}{R}\right)^2 \Delta s\, \langle y_\beta\rangle. \tag{13.36}$$

The term on the left measures the angular change y'_β of the beam with respect to the ideal orbit. The terms on the right describe the causes of this change in angle. The rightmost term describes the effect of a lens: it represents a kick $\Delta y'_\beta$ proportional to the displacement of the beam, averaged over the length of the kicker $<y_\beta>$. In a storage ring these kicks are administered by the quadrupole magnets which have deflecting field strengths proportional to the displacement of the beam from the ideal orbit. The other term on the right shows that damping is achieved by applying a kick to the beam that changes y'_β by an amount proportional to the average value of y'_β at the time the beam passes through the damping kicker.

The kicker is a discrete element in the ring which applies a magnetic or electromagnetic deflecting field to a beam bunch much as a steering corrector does, but with a much higher bandwidth so that field strength can be changed from one passing bunch to the next. To illustrate the effect of the damping effect of a zero-length kicker, we make use of two equivalent expressions describing a betatron oscillation. Instead of expressing the initial conditions of the oscillation in terms of two constants A and θ, as was done in Eq. (13.2), we may specify these details of the motion directly in terms of the initial values $y_\beta(s)$ and $y'_\beta(s)$:

$$y_\beta(s) = y_\beta(0)\cos\left(\frac{vs}{R}\right) + y'_\beta(0)\left(\frac{R}{v}\right)\sin\left(\frac{vs}{R}\right) \tag{13.37}$$

$$y'_\beta(s) = -y_\beta(0)\left(\frac{v}{R}\right)\cdot\sin\left(\frac{vs}{R}\right) + y'_\beta(0)\cos\left(\frac{vs}{R}\right) \tag{13.38}$$

The connection between A in Eqs. (13.3) and (13.39) is given by:

$$A = \left[\left(y_\beta(0)\right)^2 + \left(\frac{R}{v}y'_\beta(0)\right)^2\right]^{1/2} \tag{13.39}$$

One can confirm that A is unchanged if we substitute $y_\beta(s)$ for $y_\beta(s{=}0)$ and $y'_\beta(s)$ for $y'_\beta(s{=}0)$. This is true because Eq. (13.2) describes betatron oscillations which are neither damped nor antidamped. The betatron oscillations neither grow nor shrink in amplitude after an arbitrary number of turns.

It is generally acceptable to approximate the kicker as having zero length, as is often done for corrector magnets. In this approximation, the damping kicker changes y'_β according to Eq. (13.36) at some point in the storage ring. We are free to set s=0 at this point. Just before the damping kick is applied, A is given by Eq. (13.39). Immediately after the kick is applied, this changes $y'_\beta(0)$ to:

$$y'_\beta(0) \rightarrow y'_\beta(0)\,(1-D\Delta s) \tag{13.40}$$

If we approximate $D\Delta s \ll 1$, then the change in A is

$$\frac{\Delta A}{A} \cong -\frac{\left(\frac{R}{\nu} y'_\beta(0)\right)^2}{A^2} \cdot D\Delta s\,. \tag{13.41}$$

Because the term multiplying $D\Delta s$ is ≤ 1, we see that A is decreased by a small amount each time the beam passes the kicker. The magnitude of the decrease is proportional to $(y'_\beta(0))^2$, which varies from turn to turn. However, its average value is expressible in terms of A, using Eq. (13.2):

$$\left\langle \left(y'_\beta(s)\right)^2 \right\rangle = A^2 \frac{\nu^2}{R^2} \left\langle \sin^2\!\left(\nu \frac{s}{R}\right) \right\rangle = \frac{A^2 \nu^2}{2R^2}\,. \tag{13.42}$$

Substituting this into Eq. (13.41), and assuming that a damping kick is applied once per rotation period T as the bunch circulates in the ring, we find that the zero-length damping kicker reduces the amplitude of betatron oscillations in close analogy with a damped harmonic oscillator. The damping time constant T_y of the feedback is:

$$\frac{1}{T_y} = -\frac{1}{A}\frac{\Delta A}{T} = \frac{D\Delta s}{2T}\,. \tag{13.43}$$

We can apply the same treatment to the longitudinal motion, adding a feedback damping term to supplement the radiation damping term in Eq. (4.4) in Chapter 4:

$$\frac{d^2\tau}{dt^2} + 2\left(\frac{1}{T_\varepsilon} + \frac{1}{T_\tau}\right)\frac{d\tau}{dt} + \Omega^2\tau = 0 \tag{13.44}$$

where τ is the deviation of the bunch from its ideal arrival time at the rf cavity, Ω is the synchrotron frequency, T_ε is the synchrotron radiation damping time constant, and T_τ represents the effect of the longitudinal feedback system. As before, we express the function of a damper as applying a correction to $d\tau/dt$ which is proportional to the average value of $d\tau/dt$ at the instant the correction is applied:

$$\Delta\left(\frac{d\tau}{dt}\right) = -2T\left(\frac{1}{T_\varepsilon} + \frac{1}{T_\tau}\right)\left\langle\frac{d\tau}{dt}\right\rangle - \Omega^2 T \langle\tau\rangle \tag{13.45}$$

where both $\Delta(d\tau/dt)$ and $\langle d\tau/dt\rangle$ are evaluated over a single turn of duration T. $d\tau/dt$ is related to ε, the deviation of a particle from the ideal beam energy, by Eq. (4.3) in Chapter 4:

$$\frac{d\tau}{dt} = \alpha\,\frac{\varepsilon}{E_o} \tag{13.46}$$

so

$$\Delta\left(\frac{\varepsilon}{E_o}\right) = -2T\left(\frac{1}{T_\varepsilon} + \frac{1}{T_\tau}\right)\frac{\varepsilon}{E_o} - \frac{\Omega^2 T}{\alpha}\langle\tau\rangle \tag{13.47}$$

where α is the momentum compaction of the ring. The rightmost term represents the effect of the rf accelerating system, which applies longitudinal "focusing" in the form of an energy correction which is proportional to the longitudinal deviation of the bunch from its ideal position along the particle orbit. The first term on the right side represents the longitudinal feedback damper, which works by applying an energy correction to the bunch proportional to the energy deviation at the time the correction is applied. Since the period of a synchrotron oscillation is usually 100 revolution periods or more, the bunch receives over 100 feedback or rf kicks per synchrotron period; the impulse nature of the kicks can be ignored and the differential equation is a satisfactory approximation. The solution to Eq. (13.44) is especially simple in the limit $\Omega >> 1/T_\varepsilon + 1/T_\tau$, provided we are interested in the evolution of the synchrotron oscillation over a short time interval. Then we may approximate

$$\tau(t) = \tau_0 \, e^{-t(1/T_\varepsilon + 1/T_\tau)} \cos(\Omega t + \psi) \tag{13.48}$$

$$\varepsilon(t) \cong -E_0 \alpha \Omega \tau_0 \, e^{-t(1/T_\varepsilon + 1/T_\tau)} \sin(\Omega t + \psi) \cong -E_0 \alpha \Omega \, \tau(t - t');$$
$$t' = \frac{\pi}{2\Omega} \tag{13.49}$$

13.8.2 Kicker Bandwidth

In a ring filled with bunches an instability can cause each bunch to oscillate. The phase relation between the growing oscillations of the bunches depends on the details of the wake field, and no pattern of phase relations can be ruled out *a priori*. It is therefore sensible to assume that bunches passing the feedback kicker may require an arbitrary pattern of correction kicks in order to damp their oscillations. One extreme instance is that all bunches require exactly the same sign and magnitude kick: i.e., a DC kick. In the opposite extreme, each successive bunch may require kicks of equal magnitude but of alternating sign. Therefore the feedback kicker power supply must have a bandwidth of at least 1/2 the bunch frequency. In practical terms it becomes much easier to insure that the transient response and risetime of the kicker will allow independent correction of each bunch if the kicker bandwidth is greater than this minimum.

13.8.3 Transverse Kick Strength and Kicker Power

With this simple description of a feedback system it is possible to estimate the performance characteristics of the feedback kicker. For the purpose of illustration we will determine kicker specifications needed to achieve a damping time constant of 2 msec for the APS storage ring. Equation (13.43) gives the damping time constant for betatron motion in terms of D, and Eq. (13.36) shows that D is simply the proportionality constant that relates the feedback kicker impulse $\Delta y'_\beta$ to y'_β. The proportionality must hold true when these two quantities achieve their maximum allowable values, $(\Delta y'_\beta)_{max}$ and $(y'_\beta)_{max}$. Combining this fact with Eqs. (13.36) and (13.43), we find

$$\frac{1}{T_y} = \frac{1}{(0.002 \text{ sec})} = \frac{D\Delta s}{2T} = \frac{\left(\Delta y'_\beta\right)_{max}}{2\left(y'_\beta\right)_{max} T} \, . \tag{13.50}$$

The quantity $(y'_\beta)_{max}$ is usually determined by the injection process. Injection bumper magnets move the stored beam close to the injection path for a short time (on the order of one revolution period), after which the incoming and stored beams are returned to the initial stored beam orbit. If this is done perfectly, no betatron oscillation is produced; however, small errors in the settings of the injection magnets will induce a horizontal betatron oscillation which, as a result of linear coupling, will also produce some vertical oscillation in the beam. The injection bumpers for the APS storage ring are designed to deflect the beam about 2 mrad; it is reasonable to assume that, by careful matching of bump amplitude and timing, the uncanceled betatron oscillation of the stored beam after injection will be less than 5% of 2 mrad, or 100μrad. Some of this oscillation will be coupled into the vertical plane, about $\sqrt{0.1} = 32\%$ if the linear coupling is 10%, so the vertical feedback damper should be capable of damping a maximum y_β oscillation amplitude of ~ 30 μrad. To achieve a damping time constant of $T_y = 2$ msec, or about 544 revolution periods, Eq. (13.50) indicates that the maximum required vertical damper kick strength is 30 μrad × 2/544 = 0.12 μrad; that for the horizontal damper is 0.37 μrad.

The most commonly used transverse feedback kicker consists of a pair of 50-ohm transmission lines (often termed "striplines") built into the beampipe and powered in balanced TEM configuration. We can estimate roughly the electrostatic deflection obtained from such an arrangement by assuming that the electric field E applied to the beam is just the potential difference applied to the plates divided by their separation, which is usually about equal to the beampipe aperture d. In terms of the applied voltage V or the current I applied to each of the two plates the stripline:

$$E = \frac{\Delta V}{d} = \frac{2\,(I \cdot 50\Omega)}{d}. \tag{13.51}$$

The transverse momentum p of the beam is changed by an amount eEΔt, where Δt is the duration of application of the field. In terms of the length L of the stripline, the velocity of the beam (approximately the speed of light c),

$$\begin{aligned} \Delta p &= eE\Delta t = e(\Delta V)L/cd \qquad \text{(SI units)} \\ &= \Delta VL/d \qquad\qquad\qquad \text{(units eV/c)}\ . \end{aligned} \tag{13.52}$$

For a TEM transmission line, the direction of the magnetic field is always perpendicular to the electric field and the magnitude of the magnetic field is $|B| = |E|/c$ where c is the speed of light. If the direction of the kicking current is opposite to the beam current, the magnetic and electric fields add to double the kick (and if the stripline signal travels in the same direction as the beam, the electric and magnetic forces cancel!). Substituting E from Eq. (13.51) into Eq. (13.52), and introducing the total beam momentum p, we find the total deflection angle is

$$\Delta y'_\beta = \frac{\Delta p}{p} = \frac{10^{-9}}{p(\text{GeV}/c)} \; \frac{2\,[\,2\,(I \cdot 50)\,]\,L}{d}\ . \tag{13.53}$$

Since the feedback correction signal must travel in the direction opposite the beam, it must have a minimum duration 2L/c in order that the deflection persists without change during the interval L/c that a given bunch passes the deflecting plate. So the beam bunches must be separated by a distance of 2L/c or more, if subsequent bunches are to be unaffected by correction kicks intended for their neighbors. Thus the bandwidth requirement for the feedback system sets the length of the stripline kicker at 1/2 the minimum distance between bunches. For SPEAR, ESRF, and APS this is about 0.43 meters. Using d = 8 cm, Eq. (13.53) tells us the stripline current necessary to produce a 0.37-μrad maximum correction of the 7-GeV APS beam to be about 2.4 amps. This would require two 290 watt feedback amplifiers, one for each plate of the stripline. Of course more striplines can be installed so that more amplifiers of lower power output could be used to achieve the same deflection.

13.8.4 Longitudinal Kick Strength

Longitudinal feedback kick strength requirements can be estimated by reasoning similar to the above. The longitudinal kick "voltage" V corrects the beam energy by an amount $\Delta\varepsilon = eV$. Equation (13.47) shows that the feedback system damping time constant T_τ gives the proportionality between eV and the energy deviation ε. If we arrange for the longitudinal kicker voltage to reach its maximum value V_{max} for some maximum energy deviation ε_{max}, the feedback system damping time constant must be

$$\frac{1}{T_\tau} = \frac{eV_{max}}{2\,T\varepsilon_{max}}\ . \tag{13.54}$$

Energy errors at injection in the form of mismatch of phase or energy of incoming beam, or momentary change in orbit circumference due to the injection bump are the most important sources of error. Let us assume that these errors induce synchrotron oscillations with maximum energy deviation $\varepsilon/E \leq 0.0002$, or 20% of the natural energy spread of the APS positron beam:

$$\frac{eV_{max}}{[\,2\,(0.0002 \cdot 7\text{x}10^9\text{eV})]} = \frac{T}{T_\tau} = \frac{1}{544}\ ;\ V_{max} = 5,147 \text{ volts} . \qquad (13.55)$$

An attractive design for a feedback kicker is a coaxial transmission line, properly terminated at each end, with a cylindrical center conductor through which the particle beam passes. The beam is accelerated or decelerated by the electric fields present at each end of the cylinder as each bunch passes through it. A maximum kick of 2V is administered to a bunch if it passes the upstream end of the transmission line when the voltage at this point is +V, and then passes the downstream end when the voltage there is –V. This can be arranged for a transmission line of length L if the kicking signal on the transmission line has frequency c/4L and travels in the direction opposite to that of the beam with phase velocity c.

A transmission line longitudinal kicker has very good frequency response (3dB bandwidth is $\Delta f = c/8L$), but it does not give a very large kick per unit of power. Byrd, et al.[32] describes an improved longitudinal kicker made up of four 25-ohm transmission lines, each of length 7.4 cm to provide maximum kick at 1 GHz. The four lines are connected in series with 0.5 nsec delay lines so that adjacent ends of the segments have opposite voltage when excited by a 1-GHz signal. The device has a 6-dB bandwidth of about 200 MHz. Thus a 5-ampere peak sinusoidal current applied to the 25-ohm stripline produces a peak longitudinal kick of (4 striplines) × (2 ends) × (5 amps) × (25 ohms) = 1000 V and dissipates 625 W. To obtain 5 kV as specified in Eq. (13.55), it would be reasonable to install an array of five kickers and amplifiers.

13.8.5 Measuring y'_β or ε

As mentioned earlier, the amplifiers and kickers (which correct x'_β, y'_β, or ε) must be excited with a signal proportional to the the value of x'_β, y'_β, or ε that any given bunch has at the instant the kick is applied. This is generally done by direct measurement of x_β, y_β, or τ rather than their derivatives. These measurements are then used to predict the necessary kick strength to be applied some time later.

Standard beam diagnostic techniques generally cannot be used to measure x'_β or y'_β directly, while techniques for measurement of x or y are highly developed. Extrapolating a measurement of $y_\beta(s_{bpm})$ forward to predict the value of $y'_\beta(s_k)$ as the bunch passes the kicker at $s=s_k$ is relatively easy. Suppose we measure the amplitude of the betatron oscillation seen by an RFBPM located at s=0. Equation (13.38) for y'_β tells us that if we choose a BPM for which $\nu\,(s_k\text{-}s_{bpm})\,/\,R = (\text{integer} \pm 1/2)\,\pi$, we find that $y'(s_k)$ is exactly proportional to $y_\beta(s_{bpm})$. If the separation between BPM and kicker does not exactly satisfy this condition, the kick $\Delta y'_\beta(s_k)$ will have a component which is linearly dependent on $y_\beta(s_k)$, and the feedback system will expend some of its power acting like a focusing element, i.e., a quadrupole magnet. In any case it is possible to electronically combine measurements of y_β from two different BPMs to produce a signal proportional to $y'_\beta(s_k)$.

In longitudinal feedback systems it is common practice to measure τ using an RFBPM with its electrode signals added together so that it is insensitive to transverse beam motion. The phase of the oscillator used to excite the rf cavity is measured at the instant the bunch arrives at the BPM. This phase is proportional to $\tau(t)$. Equation and (13.49) tells us that longitudinal damping can be achieved by applying a kick to the bunch with strength proportional to τ measured one quarter of a synchrotron period earlier. This is in close analogy with the conversion of a measurement of y_β to a kick proportional to y'_β, as described in the previous paragraph. However, whereas the delay required for a betatron oscillation is a fraction of a rotation period, the delay for a synchrotron oscillation is tens or hundreds of rotation periods. The separation of the longitudinal kicker from the BPM is therefore not critical.

Feedback damping of a betatron (or synchrotron) instability can now be described as a simplified three-step process: 1) measure the oscillating transverse (or longitudinal) displacement of the bunch from the ideal trajectory of the bunch; 2) wait one quarter of a betatron (or synchrotron) period; 3) apply a transverse (or longitudinal) kick proportional to the measured displacement.

13.8.6 Suppression of "Stable Beam" Signals

The BPM signal used for measuring betatron oscillations is proportional to the sum of the betatron oscillation amplitude and the closed orbit error, $y_\beta(s_{bpm}) + Y(s_{bpm})$. For the APS, the vertical closed orbit error might reach 3 mm under some conditions. We determined in section 13.8.3 that the vertical feedback amplifier will reach maximum power when y'_β reaches 30 μrad. Equation (13.37) tells us that the maximum corresponding value of y_β is (30 μrad) × (R=176 m) / (ν = 14), or about 380 μm. Therefore, the feedback system designer must not allow the closed orbit error $Y(s_{bpm})$ to reach the feedback amplifier. The closed orbit signal power must be suppressed by a factor $(0.38\ \text{mm}/3\ \text{mm})^2$ or about 20 dB in the detection electronics in order that it be comparable to that of the betatron oscillations, and by another 20 or 30 dB so that a reasonable fraction of the feedback system power is available for damping. Even more attenuation is needed if the system is to resolve and respond to smaller amplitude oscillations. For longitudinal feedback, the stable beam signal must be suppressed by as much as 70 dB or more in order that the feedback system respond only to longitudinal oscillations.

By increasing $(y_\beta)_{max}$ in Eq. (13.50) or ε_{max} in Eq. (13.54), or by reducing the stable beam signal in the feedback BPM (by adjusting beam orbit, for example), the stable signal suppression requirement can be relaxed. However, significant suppression (40-70 dB) is almost always necessary to accommodate actual operating conditions. This is perhaps the most challenging aspect of feedback system design since the requirements of having a very high open-loop gain at the instability frequencies and a very low gain at the stable beam signal frequencies are often difficult to satisfy simultaneously in a stable feedback loop.

The primary function of the BPM signal processing electronics in multibunch feedback systems is to artificially set to zero the stable beam signal while minimally affecting the unstable signal. The BPM signal must be modified by the processing electronics so that its waveform resembles, on average, that of a bunch that is centered in the BPM (for transverse feedback), or that has zero stable phase angle (for longitudinal feedback).

Several methods have been used to suppress the signal from an off-center beam in a BPM used for transverse feedback. One approach is to place adjustable attenuators on the individual pickup electrode plates of the BPM. The stable beam signal can be suppressed by adjustment of the attenuators to make all the electrode signals identical in strength.

Another way to suppress the closed orbit error is to measure and electronically subtract the closed orbit signal from the BPM signal as has been done for the LEP transverse damper.[33] This system digitizes position for each bunch with 12-bit (5-micron) resolution; the average of 1000 previous bunch position measurements is subtracted from each bunch signal, leaving the betatron motion signal. The same result can be achieved using an analog "correlator" filter.[34,35] Here the BPM signal is split into two identical signals, one of which is delayed by exactly one rotation period with respect to the other by means of a long transmission line. The two signals are subtracted using a 180° combiner. The result is a signal whose amplitude is proportional to the difference in beam position on subsequent turns. If dispersion in the delay transmission line is carefully controlled, the stable beam signal can be suppressed by 20-45 dB. Delay transmission lines may be made of conventional temperature-compensated coaxial line, superconducting coaxial cable, or optical fiber.[36]

A third way to separate the signal of a betatron or synchrotron oscillation from the stable beam signal is to pass the BPM signal through a narrowband filter. A stable beam follows the same trajectory on every revolution in the storage ring. Therefore BPM signals from such a beam produce signals at the beam rotation frequency and multiples thereof, called "rotation harmonics." Beam oscillations correspond to amplitude modulation (betatron motion) or phase modulation (synchrotron motion) of the stable beam signal. As with any modulated carrier signal from an AM or FM radio station, this modulation produces "sidebands" of the carrier signal, shifted in frequency by the oscillation frequency. A heterodyne receiver can be tuned to the instability signal frequency with sufficient resolution bandwidth to reject the stable beam rotation harmonic. This works like a typical radio receiver, which can selectively amplify just one of a group of radio signals in the range of hundreds of megahertz, separated by just a few kilohertz. This is exactly what is required for detecting synchrotron oscillation signals.

13.8.7 Multibunch Feedback

The feedback system electronics must perform the functions described above correctly for each one of the hundreds or thousands of bunches in the storage ring. The signal from each bunch must contain neither its own stable beam signal nor that of neighboring bunches.

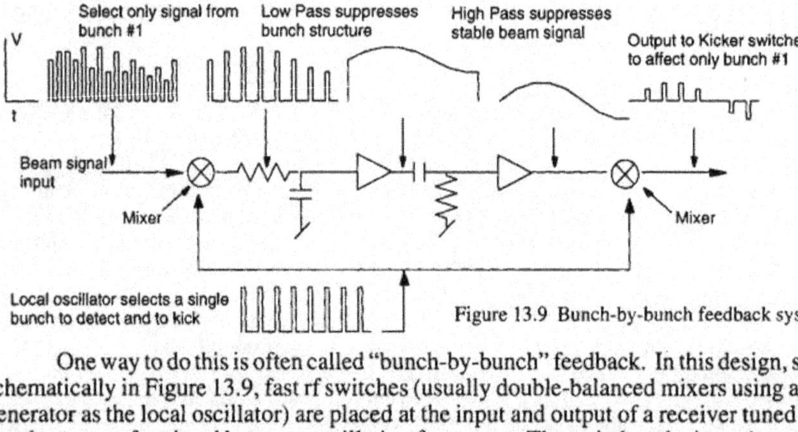

Figure 13.9 Bunch-by-bunch feedback system.

One way to do this is often called "bunch-by-bunch" feedback. In this design, shown schematically in Figure 13.9, fast rf switches (usually double-balanced mixers using a pulse generator as the local oscillator) are placed at the input and output of a receiver tuned to the synchrotron or fractional betatron oscillation frequency. The switch at the input is carefully timed to allow the signal from only one particular bunch to enter the receiver. The receiver must be designed to be unresponsive to the stable beam signals from this bunch, occurring at DC and at multiples of the beam rotation frequency. For betatron feedback, the narrowband receiver simply amplifies the betatron sideband, resulting in a signal proportional to y_β, the betatron amplitude of the bunch. For longitudinal feedback, the receiver input should be proportional to the time jitter of the bunch with respect to the stable phase, τ. This signal is then electronically differentiated to produce a feedback correction signal proportional to the energy error of the bunch. In either case, the switch at the output is timed to power the feedback kicker to correct the very same bunch that was allowed to excite the receiver. With identical electronics timed for each bunch in the ring, all motions of the bunches (i.e., all multibunch modes) may be damped. This processing scheme, has been used at several rings, including UVSOR,[37] CESR, and PEP. This design offers the advantage of design standardization; i.e., each receiver is identical and more bunches or arbitrary bunch patterns may be accommodated by changing the timing of the rf switches. Another advantage is that the feedback may be easily configured to antidamp selected bunches so as to create a desired pattern of bunches.

An alternate strategy is to design a narrowband receiver that detects and damps a single coupled-bunch mode of the bunch motion, rather than the motion of single bunches. Chapter 12, Figure 12.14 is a schematic depiction of the frequency spectrum of a beam with three bunches, affected by longitudinal instability. The signals with labels 0, 1, 2,... below the x axis are the stable beam signals. These signals are located at multiples of the beam rotation frequency, the rotation harmonics. The signals labeled 0, 2, 1, 1, 2, 0, 0, ... above their peaks show the instability sideband signals. The labels are the coupled-bunch "mode numbers" for a beam with three bunches, described in section 12.5.1. The amplitudes of the rotation harmonics depend on the spacing and amounts of charge in each bunch. For equal spacing and identical charges, rotation harmonics 0, 3, 6, 9, ... are proportional to the average stored current and over 50 dB larger than the sideband signals to be detected by the feedback, while rotation harmonics 1, 2, 4, 5, 7, 8, ... are zero. However, if the bunches are not equally spaced or equally filled, none of the rotation harmonics is necessarily zero; they must be rejected by the feedback electronics.

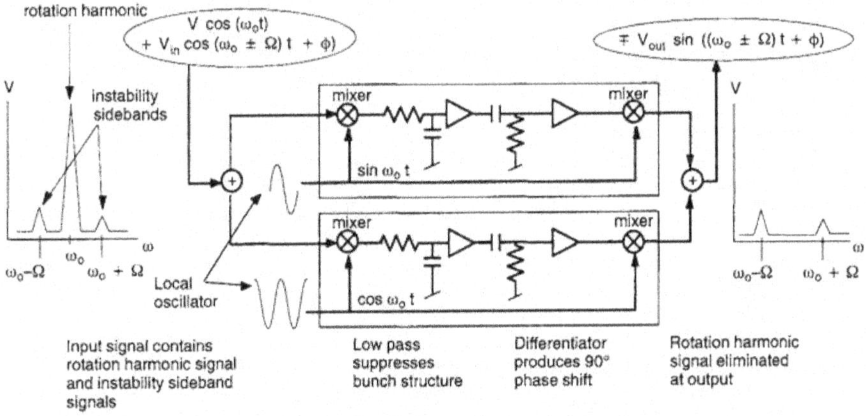

Figure 13.10 Longitudinal mode-by-mode feedback system.

The BPM electronics of a "mode-by-mode" feedback system detects one set of coupled-bunch mode sidebands, (modes m = 0, 1, 2 in Figure 12.14). To do this, the narrowband feedback receivers are grouped into pairs, as shown in Figure 13.10, configured to detect and damp a pair of coupled-bunch modes (upper and lower sidebands of a specific rotation harmonic).[38] The receivers are quite similar to bunch-by-bunch receivers except for the switches at the input and the output. One receiver in the pair has its local oscillator shifted in phase 90 degrees with respect to the other. As with bunch-by-bunch feedback, the stable-beam signal (in this case the rotation harmonic) is reduced to a DC level and eliminated with a filter. For longitudinal feedback, the input is usually the sum of the four pickup electrode signals from one RFBPM. This sum is approximately constant even if the beam moves transversely, insuring that betatron oscillation is ignored by the longitudinal feedback. The instability sidebands of the sum signal provide information on the phase oscillation $\tau(t)$ of each bunch. Figure 13.10 shows that the phase oscillation signal is differentiated in the last stage of amplification to create a signal proportional to the energy oscillation $\varepsilon(\tau)$ of each bunch (see Eq. (13.46)). Modulation of these signals by the output switch results in a kick waveform that applies to each bunch the correction kick appropriate for the detected coupled-bunch mode. Mode-by-mode detection can be applied to betatron feedback as well; to do this the input signal should be the difference signals from top and bottom (or left and right) pickup electrodes of an RFBPM for vertical (or horizontal) feedback. Differentiation of the instability signal is not necessary for betatron feedback because, as mentioned in section 13.8.5, an equivalent effect can be achieved by properly selecting the separation of the kicker and monitor.

In contrast with the critical timing required for the gating pulse for bunch-by-bunch feedback, the phase of the local oscillators with respect to the beam signal is unimportant in a mode feedback receiver pair.

13.8.8 Broadband Longitudinal Feedback

A broadband system is rather loosely defined as one in which a single processing circuit is used to control many bunches. In contrast, a bunch-by-bunch feedback or single-mode feedback requires one circuit per degree of freedom of the beam; the number of heterodyne receivers in such systems must be equal to the number of stored bunches. This becomes impractical for B-factories and synchrotron light sources, which must operate with many hundreds of bunches. While there are many broadband betatron damping systems in use at accelerators, broadband longitudinal dampers are still rather rare. The necessary combination of broad bandwidth and highly selective suppression of stable beam signals has become practical only recently, using modern digital signal processing electronics.

A digital broadband longitudinal feedback system has been designed for the SLAC B-Factory.[39] It has been successfully tested both at SPEAR and at the Advanced Light Source. The system detects and digitizes the phase error of each bunch; the data is routed to and distributed among an array of digital signal processors (DSPs) which implement the digital filtering algorithm that suppresses the stable beam signal from each bunch. Each DSP simultaneously processes the data for 50-100 bunches. The digital outputs of all the DSPs are then routed to a DAC in the correct order, producing a stream of correction kick voltages applied to a stripline kicker array as the corresponding bunch passes through.

The number of digital signal processors needed to control a given number of bunches can be dramatically reduced by "downsampling" the incoming phase information. A downsampling feedback system for the Advanced Light Source (ALS) achieves a factor of ten reduction in the number of processors by measuring the phase error of a given bunch only once per four turns; a kick is applied to the bunch every time it passes the kicker, but the kick amplitude is only recomputed and updated every fourth turn.

The phase error can be digitized with 8-bit resolution using presently available digitizers. This means the ratio of maximum kick strength to minimum is 256. A digital feedback system can be designed to have benign saturation properties by programming it to apply the maximum kick strength in response to a phase oscillation that exceeds the dynamic range of the digitizer. Computer simulations have shown that such a system can damp synchrotron oscillations even when the digitizer saturation threshold is exceeded. This implies that the power and hence the cost of the feedback amplifiers can be reduced without reducing the maximum beam current controllable by the feedback system.

References

1. E. D. Courant and H. S. Snyder, "The Theory of the Alternating Gradient Synchrotron," *Annals of Physics #* (1958), pp. 1-48.
2. M. Sands, "The Physics of Electron Storage Rings – An Introduction," *The Physics of Electron Storage Rings - An Introduction*, SLAC-121, 1970.
3. E. D. Johnson, A. M. Fauchet, and X. Zhang, "Correlation of Photon Beam Motion with Vacuum Chamber Cooling on the NSLS X-Ray Ring," *Review of Scientific Instruments* **63** (1) (1992), pp. 513-518.
4. T. Katsura, Hajime Nakamura, Y. Kamiya, Y. Fujita, in *Proceedings of the 1993 IEEE Particle Accelerator Conference* (IEEE, 1993), pp. 2260-2262.
5. G. E. Fischer, "Ground Motion and its Effects in Accelerator Design," in *AIP Conference Proceedings 153* (1984 and 1985 US Particle Accelerator Schools) (1987), pp. 1047-1119.
6. J. P. Kirkland, T. Jach, R. A. Neiser, C. E. Bouldin, "PIN Diode Detectors for Synchrotron X-Rays," *Nucl. Instrum. Methods in Phys. Research* (1988) NIM0707s.
7. P. Stefan, D. Siddons, J. Hastings, "A New Beam Position Monitor for X-Ray Synchrotron Radiation Facilities," *Nucl. Instrum. Methods* **A255** (1987) pp. 598-602.
8. W. Schildkamp, "Nondestructive Position Monitor for Synchrotron Radiation," *Nucl. Instrum. Methods* **A258** (1987) pp. 275-280.
9. E. D. Johnson, T. Oversluizen, "Compact High Flux Photon Beam Position Monitor," *Rev. Sci. Instrum.* **60** (7), July 1989, pp. 1947-1950.
10. F. Loyer, "X-Ray Beam Position Monitors for the ESRF Front Ends: Design and First Results," *Proc. of the First European Workshop on Beam Diagnostics and Instrumentation for Particle Accelerators*, Montreaux, CERN PS/93-35 (BD).
11. E. D. Johnson, paper presented at the Workshop for Engineering Design of High Power Photon Beam Position Monitor, ESRF, June 1989.
12. Y. Chung, et al., "Compensation for the Effect of Vacuum Chamber Eddy Current by Digital Signal Processing for Closed Orbit Feedback," *Proc. of the 1993 IEEE Particle Accelerator Conference* (IEEE, 1993), pp. 2266-2268.
13. J. J. DiStefano, III, A. R. Stubberud and I. J. Williams, "Feedback and Control Systems," second edition (McGraw-Hill, 1990).
14. G. F. Franklin, J. D. Powell, A. Emami–Naeini, *Feedback Control of Dynamic Systems*, Second Ed., Addison–Wesley, 1991.
15. R. O. Hettel, "Beam Steering at the Stanford Synchrotron Radiation Laboratory," *IEEE Trans. Nucl. Sci.* **NS-30**, No. 4 (1983), p. 2228.
16. A. Peled and B. Liu, *Digital Signal Processing* (John Wiley & Sons, 1976).
17. T. Katsura, et al., "Beam Position Monitoring and Feedback Steering System at the Photon Factory," *Proc. of the 13th International Conference in High Energy Accelerators*, Vol. 2 (1988), pp. 243-246.
18. P. Quinn, et al., "An Advanced Beam Steering System for the SRS at Daresbury," *Proc. of the 3rd European Particle Accelerator Conference*, Berlin (1992).
19. R. Hettel, "Review of Synchrotron Beam Stability and Stabilizing Systems," *Rev. Sci. Instrum.* **60** (7), July 1989, pp. 1501-1506.
20. W. Brefeld, "Stabilization of Synchrotron Radiation Beam at HASYLAB," *Rev. Sci. Instrum.* **60** (7), July 1989, pp. 1513-1516.

21. G. Portmann, J. Bengtsson, "A Closed-Loop Photon Beam Control Study for the Advanced Light Source," *Proc. of the 1993 IEEE Particle Accelerator Conference.* (IEEE, 1993) pp. 2272-2274.
22. R. J. Nawrocky, et al., "Automatic Beam Steering in the NSLS Storage Rings Using Closed Orbit Feedback," *Nucl. Instr. Meth.* **A226** (1988), p. 164.
23. H. Kobayakawa, "Summary of Photon Beam Stabilization at the Photon Factory," *Review of Scientific Instruments* **63** (1) (1992), pp. 509-512.
24. A. Ropert, "Challenging Issues During ESRF Storage Ring Commissioning," in *Proceedings of the 1993 IEEE Particle Accelerator Conference* (IEEE, 1993), pp. 1512-1514.
25. G. H. Golub and C. Reinsch, "Singular Value Decomposition and Least Squares Solutions," *Numer. Math.* **14** (1970), p. 402, and references therein.
26 L. H. Yu, et al., "Real Time Global Orbit Feedback System for NSLS X-Ray Ring," in *Proceedings of the 1991 IEEE Particle Accelerator Conference* (IEEE, 1991), pp. 2542-2544.
27. Y. Chung, "A Unified Approach to Global and Local Beam Position Feedback," in *Proc. of the 1994 European Particle Accelerator Conference*, London 1994.
28. Y. Chung, et al., "Global DC Closed Orbit Correction Experiments on the NSLS X-ray Ring and SPEAR," ibid., pp. 2275-2277.
29. J. Galayda, "Feedback Control of Multibunch Instabilities," in *AIP Conference Proceedings 249* **2** (*publisher,* 1992), pp. 663-692.
30. F. Pedersen, "Feedback Systems," CERN PS/90/49(AR), 1990.
31. J. D. Fox, et al., "Feedback Control of Coupled-Bunch Instabilities," in *Proceedings of the 1993 IEEE Particle Accelerator Conference* (IEEE, 1993), pp. 2076-2080.
32. John M. Byrd, et al., "Progress in PEP II Multibunch Feedback Kickers," SLAC-400 (1992), pp. 220-223.
33. L. Arnaudon, et al., "Commissioning of the LEP Transverse Feedback System," *Proc. of the 2nd European Particle Accelerator Conference*, Nice, June 1-16, 1990, pp. 901-906.
34. D. M. Dykes and J. N. Galayda, "Performance of Correlator Filters in Feedback Systems at Daresbury and NSLS," *Proc. of the 1987 IEEE Particle Accelerator Conference* (IEEE, 1987), pp. 582-584.
35. W. Barry, et al., "Transverse Coupled-Bunch Feedback in the Advanced Light Source (ALS)," *Proc. of the 1994 European Particle Accelerator Conference*, to be published.
36. R. J. Pasquinelli, "Electro-Optical Technology Applied to Accelerator Beam Measurement and Control," *Proc. of the 1993 Particle Accelerator Conference* (IEEE, 1993), pp. 2081-2085.
37. T. Kasuga, et al., "Longitudinal Active Damping System for UVSOR Storage Ring," *Jap. J. Appl. Phys.*, **27** (1) (1988), p. 100.
38. B. Kreigbaum and F. Pedersen, "Electronics for the Longitudinal Active Damping System for the CERN PS Booster," *IEEE Trans. Nucl. Sci.*, **NS-24** (3) (1977), pp. 1695-1697.
39. H. Hindi, et al., "Analysis of DSP-Based Longitudinal Feedback System: Trials at SPEAR and ALS," *Proc. of the 1993 Particle Accelerator Conference* (IEEE, 1993), pp. 2352-2354.

CHAPTER 14: WIGGLER AND UNDULATOR INSERTION DEVICES†

ROSS D. SCHLUETER
Lawrence Berkeley Laboratory
1 Cyclotron Road, MS 46-161
Berkeley, CA 94720, USA‡

14.1 Introduction

The acceleration of a charged particle due to a magnetic field results in photon emission perpendicular to the direction of acceleration. In the case of electrons or positrons with energy γmc^2 traveling at nearly the speed of light in a curved path (thereby accelerated), the spontaneous emission falls within a narrow cone of angular width γ^{-1} centered about the direction of electron travel and is known as synchrotron radiation (Fig. 14.1). The power of the emitted radiation is proportional to the square of the electron energy and to the square of the magnetic field.

14.1.1 Motivation for the Development of Insertion Devices

Synchrotron radiation (SR) was originally considered an undesirable by-product of storage-ring particle colliders which was produced in the bend sections of the storage ring and absorbed in the vacuum chamber wall. When the potential uses of such radiation was realized, sections of the vacuum chamber and ring shielding of these first-generation sources were modified to permit some of the radiation to leave the ring and travel through beamlines to user end stations. Second-generation storage rings were dedicated exclusively to SR generation, primarily from bending magnets. These sources targeted user-specific photon energy ranges and characteristics.

Second-generation synchrotrons led to a large increase in the use of SR for research and, as the users became more sophisticated, the demand for greater intensities and tailored characteristics for the radiation increased dramatically. This demand drove the technological development of special magnetic devices that could greatly enhance the spectral properties of the radiation. These devices consist of arrays of magnets of alternating polarity that repetitively bend electron beams back and forth as shown in Fig. 14.2. The two classes of the special magnet arrays, undulators and wigglers, are collectively known as insertion devices (IDs) because they are placed in the straight sections of storage rings. Today's third-generation synchrotrons are designed with several straight sections to maximize the use of such devices and with lower electron beam emittance to enhance the photon beam brightness, particularly from undulators. Presently, several rings, featuring very small electron beam emittance and multiple

† This work was supported by the Director, Office of Energy Research, Office of Basic Energy Sciences, Materials Science Div., of the U.S. Dept. of Energy under Contract No. DEAC03-76SF00098.
‡ E-mail: ross@lbl.gov, Fax: 510-486-4873.

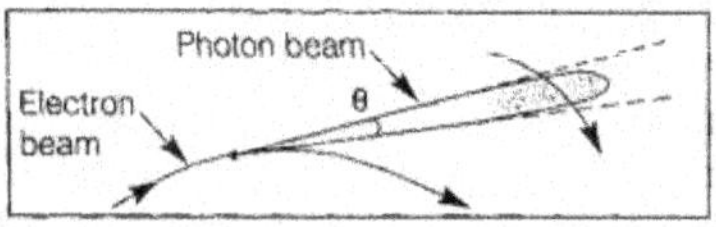

Fig. 14.1 An electron beam traveling in a curved path at nearly the speed of light emits photons into a narrow cone of natural emission angle $\simeq \gamma^{-1}$.

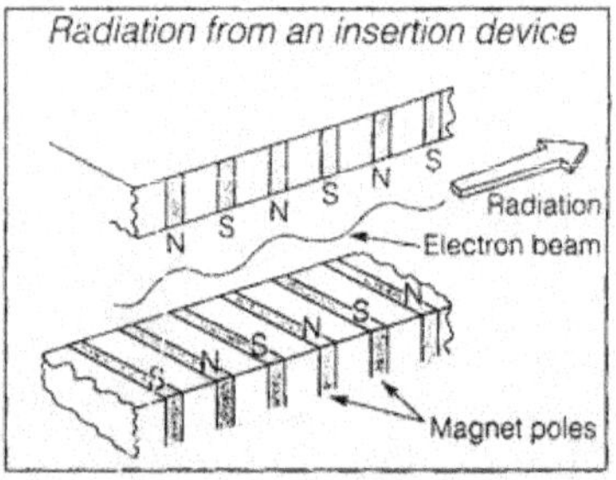

Fig. 14.2 Electrons wiggle or undulate in the midplane along the spatially periodic sinusoidal field in an insertion device.

straight sections for IDs, produce radiation in the soft x-ray regime using electron energies of 1-2 GeV. A few larger machines with electron energies of 6-8 GeV produce hard x-ray radiation. Chapter 1 includes a worldwide survey of SR ring parameters.

An ID magnet array is situated close to the electron beam so as to generate the highest possible field, yet not reduce beam lifetime by compromising the aperture for the circulating electron beam. IDs producing linear polarized radiation are usually designed to have vertical on-axis fields. Beam emittance, closed orbit errors, and injection requirements generally result in an aperture requirement that is smaller in the vertical direction, allowing attainment of smaller magnetic gaps.

The periodic magnetic field of the Fig. 14.2 insertion device is in the + or − vertical direction in the horizontal midplane. The unperturbed electron trajectory is along the longitudinal axis. The Lorentz force on the electron is at right angles to both the direction of electron motion and the magnetic field, causing the electrons to wiggle or undulate in the horizontal midplane. The period and geometrical features are chosen such that along with electron energy and wiggle field strengths, the ID yields desired radiation properties (e.g., wavelength and intensity).

14.1.2 Applications of Insertion Devices

The uses of synchrotron radiation[1,2] are numerous and varied: (a) In the UV/soft x-ray range (10^{-7} m-10^{-9} m), it can be used to explore the electronic structure of solids. (b) In the harder x-ray part of the spectrum ($\leq 10^{-10}$ m) it provides an excellent tool for x-ray crystallography studies to determine the three-dimensional molecular structure of proteins, which can help the pharmaceutical industry develop custom-designed drugs. (c) When absorbed by an atom, SR can cause electrons to detach or other photons to be emitted, thereby giving valuable data about the properties of the sample. (d) Circular polarized SR is extremely useful for imaging magnetic materials, which will allow the computer industry to develop advanced materials for increased data storage. (e) The coherence properties of SR can be exploited in holography to image objects with better resolution than is possible with

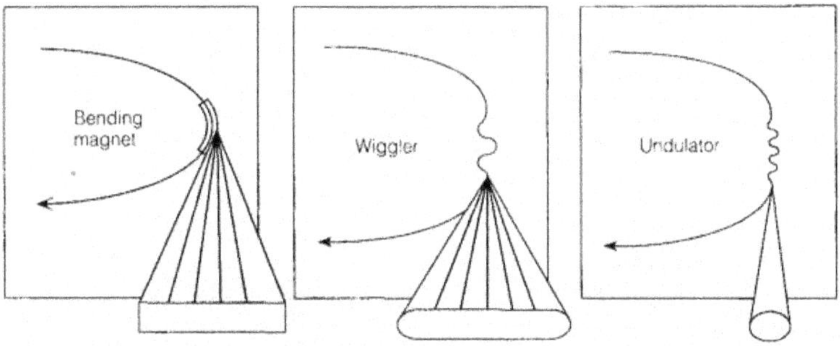

Fig. 14.3 Radiation beam envelope from (a) bending magnet, (b) wiggler, (c) undulator.

visible light. (f) Short SR light bursts allow stop-action spectroscopy of rapid chemical reactions. Chemical reactions can be followed precisely by tuning a free electron laser (FEL) to wavelengths easily absorbed by specific molecules. (g) Extremely short intense pulses from an FEL could make it possible to explore the dynamics of charge carriers in semiconductors.

Specific applications may require very high photon flux, wavelength selectability, spatial resolution, polarization state, time structure, or other radiation characteristics, which can be provided in machines with requisite electron beam properties coupled with specially tailored IDs.

14.1.3 Insertion Device Radiation Overview

Wigglers. Wigglers cause electron deflections that are large compared with the natural emission angle of SR, resulting in broadband emisson of a fan-shaped beam of photons. Wiggler SR is similar to that produced by an individual bend magnet, but $2N$ times as intense due to repetitive electron bending over the length of a $2N$-pole wiggler. The source depth of a wiggler is its length, typically several meters, whereas that of a bend magnet is just a few centimeters of the electrons' curved path.

Wavelength Shifters. A wavelength shifter is a wiggler with just a few poles producing high on-axis fields whose purpose is to shift the radiation spectrum toward smaller wavelengths, e.g., hard x rays, rather than to increase radiation intensity via multiple bends.

Undulators. Undulators cause small electron deflections, comparable in magnitude to the natural emission angle of the SR. The radiation emitted by an individual electron at the various poles in the magnet array interferes coherently, resulting in emisson of a pencil-shaped beam of photons peaked in narrow energy bands at the harmonics of the fundamental energy. Due to the constructive interference, the radiation beam's opening angle at any given wavelength is decreased by $\sqrt{N}$ and thus radiation intensity per solid angle goes as N^2, (assuming that the angular divergence of the electron beam is less than the natural emission angle). Fig. 14.3 contrasts the radiation beam envelope produced by a bend magnet, a wiggler, and an undulator.

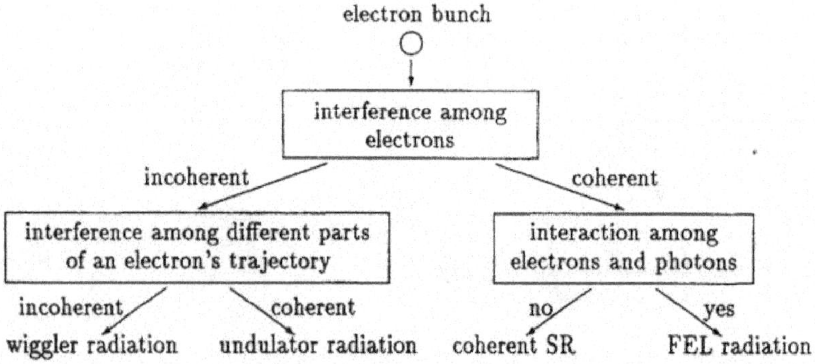

Fig. 14.4 Radiation from relativistic electrons in a periodic electromagnetic field.

Coherent Synchrotron Radiation and Free Electron Lasers. For either the wiggler or the undulator, the SR from individual electrons in the beam add incoherently, and the total radiation power is the sum of the single-particle radiation power from all the electrons in the beam (i.e., total power $\propto n_e$). If instead the electrons at a given pole are all in phase, radiation amplitudes (rather than powers) from the various electrons are additive and the radiation power becomes proportional to the square of the number of electrons, n_e^2. Such coherence, if achieved without the benefit of photon-electron interaction, is known as coherent synchrotron radiation and is bunch-length limited in present SR rings. In a free electron laser (FEL), coupling between the photon and electron beams results in a density modulation of the electron bunch, forming thin pancakes separated by one wavelength of light. The radiation from all electrons in the pancake have the same phase, and radiation amplitudes from different pancakes also add in phase. This coherence of the radiation (i.e., the linear superposition of amplitudes) produces a large gain in radiated power at the FEL frequency.

FELs can be configured as either oscillators or amplifiers. In the oscillator design, the wiggle field is inside an optical cavity. The electron beam traverses the wiggler. Emitted radiation is confined within the cavity and is reflected back through the wiggler between mirrors, thus modulating the electron beam. Coherent radiation builds up and is transmitted through a partially transparent mirror, becoming the usable laser beam. In the amplifier configuration, an electron beam passes through a wiggler coaxial with input laser light of the desired wavelength. The interaction of the laser's electric field and the electron wiggle motion causes the electrons to bunch in pancakes. The emitted photons are in phase with each other, with the radiation emitted from electrons in other bunches, and with the input laser signal. Radiation coherence and intensity increase as the electron beam makes a single pass through the amplifier. Alternatively, radiation can build up from noise in the amplifier configuration without the help of a seed laser, producing self-amplified spontaneous emission (SASE) radiation. FELs can also operate in storage rings, taking advantage of the low emittance electron beam to produce short-wavelength coherent radiation. Figure 14.4 summarizes radiation scenarios from relativistic electrons propagated through IDs.

14.2 Synchrotron Radiation From Wigglers and Undulators

Electron Equation of Motion. Integrating the equation of motion of a relativistic electron moving with average velocity $\langle v_z \rangle$ perpendicular to a sinusoidal on-axis wiggle field of magnitude $B_y = B_0 \cos k_w z$ and period $\lambda_w \equiv 2\pi/k_w$ gives, for the velocity and trajectory in the direction mutually perpendicular to $\langle v_z \rangle$ and $\vec{B}$:

$$\frac{d\vec{p}}{dt} = d\gamma m\vec{v} = e(\vec{E} + \vec{v} \times \vec{B}) \Longrightarrow \frac{v_x}{c} = \frac{K}{\gamma} \sin k_w z \text{ and } x = \frac{K}{\gamma k_w} \cos k_w z, \qquad (14.1)$$

where $\gamma = 1957E[\text{GeV}]$ and deflection parameter $K \equiv eB_0/k_w mc = .934 B_0[\text{T}]\lambda_w[\text{cm}]$.

Peak angular electron deflection and trajectory amplitude are K/γ and $K/\gamma k_w$, respectively. K, therefore, is the ratio between the radiation fan angle and the characteristic radiation cone aperture, $1/\gamma$. Thus $K \gg 1$ for wigglers, and a radiated location downstream sees emitted photons only twice per wiggle period when the sweeping electron velocity points in its direction. For undulators $K \lesssim 1$ and an observer downstream sees photons emitted continuously along the undulator length. The analysis can be generalized to include periodic, non-sinusoidal electron trajectories; essential radiation features remain unchanged.

14.2.1 Wiggler Radiation

The angular distribution of radiation emitted by electrons following a circular trajectory in a horizontal plane, as in a bending magnet, is[3]

$$\frac{d^2F(\omega)}{d\theta d\psi} = \frac{3\alpha\gamma^2}{4\pi^2} \frac{I}{e} \frac{\Delta\omega}{\omega} \left(\frac{\omega}{\omega_c}\right)^2 (1 + \gamma^2\psi^2)^2 \left[K_{2/3}^2(\xi) + \frac{\gamma^2\psi^2}{1+\gamma^2\psi^2} K_{1/3}^2(\xi)\right] \qquad (14.2)$$

where ω is the photon frequency; θ and ψ are the observation angles in the horizontal and vertical directions, respectively; α is the fine-structure constant; I is the beam current; e is the electron charge; the subscripted K's are modified Bessel functions of the second kind; and the argument $\xi \equiv (\omega/2\omega_c)(1 + \gamma^2\psi^2)^{3/2}$.

The two terms in the last bracket correspond to the horizontally and vertically polarized radiation. In the midplane, the second term vanishes and the polarization is purely linear. Off the midplane both terms contribute and polarization is elliptical.

The energy spectrum is smooth and broadband: peaking near, then falling off exponentially above the critical energy given by $\epsilon_c \equiv \hbar\omega_c = 3\hbar c\gamma^3/2\rho$. Half the total power is radiated above and half below the critical energy, which in practical units is

$$\epsilon_c[\text{KeV}] = 0.665 B[\text{T}] E^2[\text{GeV}], \qquad \lambda_c[\text{Å}] = 18.64/B[\text{T}]E^2[\text{GeV}]. \qquad (14.3)$$

In a wiggler, radiation from the various periods interfere incoherently. The spectrum as seen by an on-axis observer, at frequencies of interest $\omega \sim O(\omega_c)$, is for all practical purposes that of $2N$ bending magnets. The harmonic peaks are spaced so closely that they blur together.

Often several experimental end stations share the radiation fan of horizontal angular extent $2K/\gamma$. At nonzero horizontal observation angle θ, the electron's effective radius of curvature at the trajectory locations contributing to the observed radiation decreases and the off-axis critical energy becomes $\epsilon_c(\theta) = \epsilon_c(0)\sqrt{1-(\theta\gamma/K)^2}$.

Off the midplane, circularly polarized components from successive bends cancel and the radiation remains partially linearly polarized and partially unpolarized.

14.2.2 Undulator Radiation

In an undulator, radiation from the various periods interfere coherently. Sharp peaks are produced at harmonics of the resonant frequency, which depends on the electron energy, the undulation period and field strength, and the observation position. The optical wavelength is a Lorentz transformation of the undulation period into the beam frame followed by a relativistic Doppler shift back into the laboratory frame. The velocity used in the Lorentz transformation and the Doppler shift is the longitudinal electron velocity, which is less than the full electron velocity because of the electron's curved path through the undulator. The wavelength of the fundamental radiation peak is

$$\lambda_1(\theta,\psi) = \frac{\lambda_w}{2\gamma^2}\left[1+\frac{K^2}{2}+\gamma^2(\theta^2+\psi^2)\right]. \tag{14.4}$$

On axis, the fundamental energy and wavelength in practical units become

$$\epsilon_1[\mathtt{KeV}] = \frac{0.950E_e^2[\mathtt{GeV}]}{(1+K^2/2)\lambda_w[\mathtt{cm}]}, \quad \lambda_1[\mathring{A}] = \frac{13.06\lambda_w[\mathtt{cm}](1+K^2/2)}{E^2[\mathtt{GeV}]}. \tag{14.5}$$

Relative bandwidth at the nth harmonic is $\Delta\lambda/\lambda = \Delta\omega/\omega \simeq 1/nN$. As K increases, the peak spacing at the energies of interest merge and the insertion device passes from the undulator to the wiggler regime. Additionally, the resonant frequency dependence on observation angle blurs the radiation peaks for any angle-integrated sampling, resulting in a peaked spectrum superposed on top of a broadband continuum. Again, as K increases, the peaks become buried in the continuum.

The angular distribution of the radiation intensity of the nth harmonic on-axis is

$$\left.\frac{d^2F_n}{d\theta d\psi}\right|_0 = \frac{I}{e}\frac{\Delta\omega}{\omega}\frac{\alpha N^2\gamma^2K^2n^2}{(1+K^2/2)^2}\left\{J_{\frac{n-1}{2}}\left[\frac{nK^2}{4(1+K^2/2)}\right]-J_{\frac{n+1}{2}}\left[\frac{nK^2}{4(1+K^2/2)}\right]\right\}^2, \tag{14.6}$$

where the J's are Bessel functions and n is odd. Integrated over the central cone, the flux in practical units [photons/s/0.1% bandwidth] is approximately

$$F_n = 1.451\cdot 10^{14}\frac{NI[\mathtt{A}]K^2n}{1+K^2/2}\left\{J_{\frac{n-1}{2}}\left[\frac{nK^2}{4(1+K^2/2)}\right]-J_{\frac{n+1}{2}}\left[\frac{nK^2}{4(1+K^2/2)}\right]\right\}^2. \tag{14.7}$$

Brightness is a characterization of the effective radiative source size, and is given in units of flux per phase space volume:

$$B_n(0,0) = F_n/4\pi^2\sigma_{Tx}\sigma_{Ty}\sigma_{Tx'}\sigma_{Ty'} \tag{14.8}$$

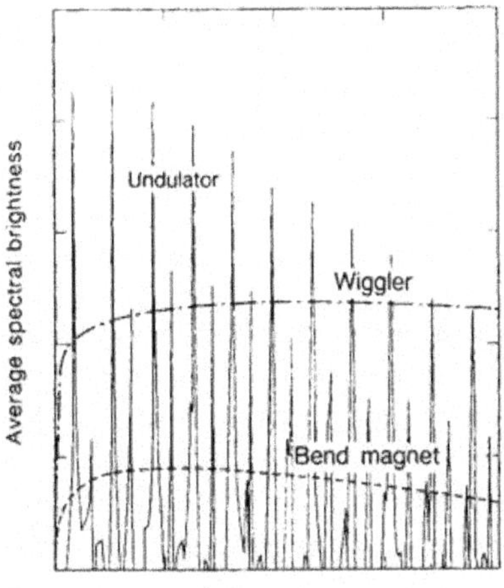

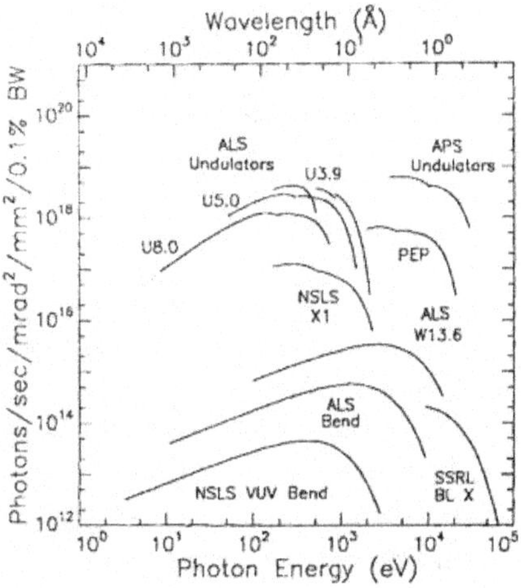

Fig. 14.5 (a) Typical radiation spectra from a bend magnet, a wiggler, and an undulator; and (b) spectral brightness from IDs in various facilities.

where $\sigma_{Tu}^2 \equiv \sigma_u^2 + \sigma_r^2$, $\sigma_{Tu'}^2 \equiv \sigma_{u'}^2 + \sigma_{r'}^2$, and σ_u and $\sigma_{u'}$ are the beam size and divergence in the u-direction, u being either x or y. The σ_r and $\sigma_{r'}$ are extended-source size and divergence limits for single-electron radiation. Thus, small emittance is essential to achieving a high brightness undulator radiation source. Temporally, the radiation maintains phase over a distance $l_c = \lambda(\lambda/\Delta\lambda) \simeq nN\lambda$, the coherence length.

The theory and characteristics of synchrotron radiation have been developed in detail from first principles and are reviewed elsewhere.[4,5] Figure 14.5 shows typical on-axis photon spectra from a bend magnet, a wiggler, and an undulator along with spectral brightness from IDs in various facilities.

14.2.3 Free Electron Lasers

The interaction between electron beams and light beams underlies the physical mechanism that causes electrons to bunch into pancakes, resulting in coherent emission of radiation from the various electrons. Light amplified in the FEL propagates in the same direction as the beam, and is polarized such that its electric field lies in the horizontal plane as does the wiggle motion of the electrons. The electric field of the light interacts with the electrons, causing those electrons lagging in phase with the light to gain energy and those leading in phase to slow down, resulting in a spontaneous electron bunching in thin pancake-shaped packets separated by an optical wavelength and a concurrent shift of the electric field's phase. If, in a frame moving with average electron velocity $\langle v_z \rangle$, $\vec{v} \cdot \vec{E} > 0$, then energy is transferred from the electron beam to the optical beam, and electron motion is retarded, i.e., $d\gamma/dt < 0$. The FEL resonance condition for which this coupling occurs is

$$\frac{v_z}{c} = \frac{\omega/c}{k+k_w} \simeq 1 - \frac{\lambda}{\lambda_w} \quad \Longrightarrow \quad \gamma^2\Big|_{res} = \frac{\lambda_w}{2\lambda}\left[1 + \frac{K^2}{2} + \left(\frac{\gamma v_{\perp,\epsilon}}{c}\right)^2\right], \tag{14.9}$$

where ω and k are the electric field frequency and wavenumber, respectively; and $v_{\perp,\epsilon}$ is an individual electron's incremental transverse velocity due to emittance. Energy spread, emittance, and wiggle field errors must be small for the bulk of the electrons to be resonant. Thus, high brightness beams and high accuracy IDs are required for FELs.

An electron's axial velocity determines the phase relationship between its wiggle motion and the light's electric field. As electrons radiate photons they lose energy, and if B_0 is constant along the device, their average axial speed decreases and the resonant frequency for bunching changes. Axially tapering the magnetic field, so as to keep electron bunches moving at a constant average axial velocity, makes the electric field that electrons see invariant, and resonance at the original frequency is maintained along the entire device.

The lasing medium in FELs is the electron bunch in vacuum and there are no power-density limitations associated with going to high power. In contrast, conventional lasers utilize atomic transitions and, as power density increases, the heated medium interacts with the laser light, distorting the phase fronts and destroying beam quality. Thus, FELs can achieve higher power than conventional lasers at greater efficiency and are tunable over a range of wavelengths, providing unprecedented power and flexibility.

14.3 Pure Permanent Magnet Insertion Devices

The wiggle field in pure permanent magnet (PM) IDs is generated by the direct flux emanating from the PM material, some of which crosses the device's midplane. These novel devices enabled light generation in new regimes upon their introduction[6] in the early 1980's because they are capable of higher fields than electromagnet (EM) counterparts as device dimensions (e.g., period, gap, and pole height) shrink.

14.3.1 Advantages

In addition to superior performance for short-period devices, pure PM devices are compact and require no peripheral equipment such as power supplies or cooling systems. Operating energy costs are not incurred other than the negligible amount associated with varying the gap to adjust peak on-axis field. Near unity permeability of the PM material makes these IDs immersible in other fields (e.g., a focusing quadrupole), with linear superpositioning of fields. This same characterisitic, $dB/dH|_{PM} \simeq 1$, also facilitates analytical calculation of the ID field magnitude and distribution.

14.3.2 Design Methodology

Presently, the PM material of choice for IDs is neodymium-iron-boron (NdFeB), which features a remanent field $B_r \simeq 1.0 - 1.3$ T, and a coercive force $H_c \simeq$

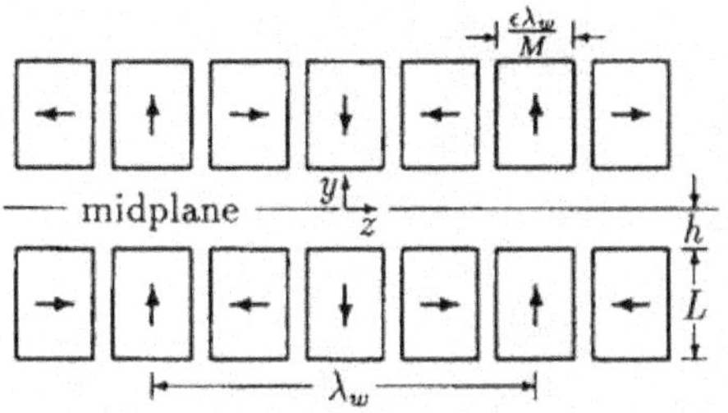

Fig. 14.6 Permanent magnet IDs: (a) PM ID showing block magnetization directions, (b) KEK's pure PM, 179-pole, 4-cm-period undulator is situated "in-vacuum" so as to achieve 0.82 T operating at its small 1 cm magnetic gap (reprinted with permission[7]).

0.95–1.25 T, with differential permeability $dB/dH|_{PM} \simeq 1.03$–1.05. An alternative material samarium-cobolt (SmCo), with a strength 10% less than NdFeB, features superior field-strength temperature stability (0.03% per °C versus 0.11% per °C) and superior radiation resistance.

A PM block with $H_c = 1$ T is equivalent to current sheets of strength ±8 kA-per-cm block length, separated by the block width. As ID dimensions shrink, heat could not be dissipated in analogous high-current-density EM devices.

An ideal PM ID would consist of two rows of PM material, separated by the gap, with the easy-axis direction of the material continuously varying along the axis, rotating once per wiggle period length. In practice a pure PM ID is sectioned into several homogeneous blocks per period, as shown in Fig. 14.6. The field strength in the gap region of a two-dimensional pure PM ID of period λ_w and half-gap h, with M blocks per period, having vertical dimension L and axial dimension $\epsilon\lambda_w/M$ is[8]

$$B_z - iB_y = i2B_r\Sigma_{\mu=0}\cos(nk_w[z+iy])e^{-nk_w h}\frac{\sin(n\epsilon\pi/M)}{n\pi/M}(1-e^{-nk_w L}) \qquad (14.10)$$

where harmonic $n \equiv 1 + \mu M$. The last term in parentheses quantifies the diminishing return of increasing block height L. Using a smaller number of homogeneously magnetized blocks introduces field harmonics. Still, with respect to attainable field strength, using just four blocks per period gives a peak on-axis B_0 over 90% of that for infinite segmentation. Finally, note that since on-axis ID fields arise directly from flux emanating from the PM blocks, field errors are sensitive to imperfections in PM material (e.g., dipole strength variations and easy-axis orientation errors) and to block mispositioning. A range of field strengths is usually attained by varying the gap, where peak on-axis field B_0 is proportional to $e^{-2\pi h/\lambda_w}$.

Fig. 14.7 Hybrid IDs: (a) Assembly of the lower half of the 30-pole, 12.85-cm-period, 1.4 T Beam Line X hybrid wiggler for SSRL[10] (1987), (b) An early hybrid device is the 55-pole, 7-cm-period LBL/EXXON/SSRL Beamline VI wiggler[11] at SSRL (1982).

14.3.3 Design Variations

Alternatively, fixed-gap adjustable-phase IDs vary peak on-axis field strength by sliding one row of magnets axially along a rail.[9] The sinusoidal variation of peak strength as the longitudinal phase between upper and lower magnet rows is shifted from 0 to $\lambda_w/2$ enables precise control of field magnitude. Electron beam vertical focusing, and thus electron beam tune, are phase-independent. Likewise, the integrated field along the beam axis, and thus beam steering, do not change with phase adjustment. The on-axis longitudinal field component in these devices gives rise to small higher order terms that alter the radiation spectra slightly from their adjustable gap counterparts.

14.4 Hybrid (Permanent Magnet Plus Iron) Insertion Devices

Hybrid insertion devices (Fig. 14.7) utilize PM material to generate flux, which by geometrical design enters directly into high-permeability, soft-iron pole pieces. An energized pole then assumes a magnetic scalar potential with respect to the midplane of the device. This potential difference drives an indirect flux out of the pole, some of which crosses the midplane, creating the desired wiggle field. The magnitude of the pole-tip scalar potential, and thus that of the on-axis field is such that direct fields into the pole from the PM material are balanced by indirect fields out of the pole arising from the potential gradient (Fig. 14.8). This equality is a physical requirement following directly from $\vec{\nabla} \cdot \vec{B} = 0$.

14.4.1 Advantages

Hybrid designs, like their pure PM counterparts, outperform EM devices as spa-

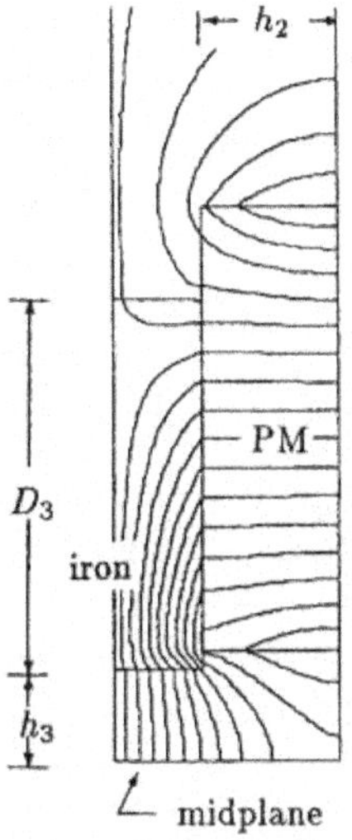

Fig. 14.8 Upper-half, quarter-period section of a typical hybrid insertion device showing net magnetic flux.

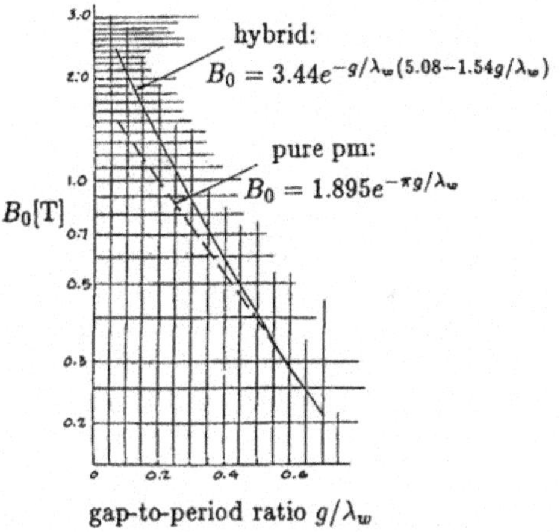

Fig. 14.9 Attainable on-axis field in pure PM and hybrid insertion devices ($B_r = 1.1$T, $H_{pm} = -0.8H_c$).

tial dimensions shrink, and do not require electrical energy and associated cooling paraphernalia to produce field.

Additionally, for devices with small gap-to-period ratios g/λ_w, hybrid IDs outperform pure PM designs because of their ability to channel flux through the high-permeability pole pieces to the pole tip and then across the small gap. A comparsion of attainable on-axis field versus gap-to-period ratio for pure PM and hybrid devices is reproduced in Fig. 14.9, assuming $B_r = 1.1$ T and hybrid PM material operates at $-0.8H_c$ (above which point increased PM volume results in only marginally higher on-axis field).[12] For the hybrid design, attainable peak on-axis field is characterized emperically by $B_{0_{max}}[\mathrm{T}] = 3.44e^{-g/\lambda[5.08-1.54g/\lambda]}$ over the range $0.07 \leq g/\lambda \leq 0.70$.

In hybrid devices, the field shape and quality in the region of interest is governed by the shape of the soft iron pole tip rather than the PM positioning and quality. Deviations in strengths and easy-axis orientation errors of PM blocks, problematical with pure PM devices, are more readily accommodated in hybrid designs where the PM blocks ideally do not contribute directly to the field in the gap region. Low on-axis field errors are attainable by sectioning the PM material in contact with a given pole and sorting PM blocks so that (1) average strength entering all poles is highly uniform and (2) easy-axis orientation errors of the PM material near the gap region are minimized. More subtly, the effect of perturbing sources producing direct error fields is largely negated by indirect fields arising from the presence of the iron itself.

14.4.2 Design Methodology

A quarter-period of a typical hybrid design is given in Fig. 14.8. The period is chosen such that along with electron energy and wiggler field strengths, the ID yields desired radiation properties. Again, a range of field strengths is usually attained by varying the gap.

Given ID period λ_w, a high-permeability soft-iron pole tip surface is shaped so as to obtain the desired field distribution. Subsequently, the pole scalar potential necessary to obtain the desired field strength is determined. Finally the remainder of the iron and PM is designed so as to produce this pole potential.

High-field hybrid IDs tend to saturate magnetically at the pole tip (near the gap). Analytical designs using closed-form expressions, assuming infinite permeability in the soft-iron poles, accurately predict on-axis field to the extent that the pole tip does not begin to saturate.[13] Incorporation of results from a single 2-D magnetostatic code run in the analytical formulation can account for pole-tip saturation, while still allowing for perturbations in the design parameter space. Full three-dimensional IDs can be designed analytically to within an accuracy of $\simeq 1-2\%$. Of course any 3-D design can be modeled using a 3-D magnetostatics code. However, analytical 3-D modeling is accurate, efficient, and provides greater insight as to the effect of perturbations in the design parameter space, thereby aiding design optimization.

Low-carbon iron or vanadium permendur are the materials of choice for pole pieces. These high-permeability materials have saturation fields $B_s \simeq 2.1$ T and 2.3 T, respectively. The vanadium permendur with higher B_s, though more expensive and size-limited in availability, is magnetically preferable. Saturation for these materials effectively limits on-axis fields in hybrid IDs to $B_0 \simeq 2.0$ T$-$2.2 T; beyond this only marginal increments of on-axis field are attainable at an exorbitant premium in PM volume and cost, and control over the field distribution is also sacrificed.

14.4.3 Design Variations

For most ID designs, it is desirable to attain as high an on-axis field as possible, given the period λ_w, pole-tip shape, etc. Several features can be incorporated in a hybrid ID design to increase the magnetic scalar potential of the pole tip V_0, and thereby on-axis field B_0, within the given parameter envelope.

Permanent Magnet Overhanging and Packing. One approach to increase V_0 is to increase the direct flux Φ_d into the pole. Indirect flux Φ_i out of the pole likewise increases, and if pole size is kept invariant, the pole-pole and pole-midplane magnetic capacitances c_1 and c_0 remain unchanged. Since $\Phi_i = V_0 \Sigma_j c_j$, then V_0, and thus B_0 increase. Overhanging PM material above the top of the pole and in the 3rd dimension beyond the lateral pole edges (as shown in Fig. 14.8) has the desired effect.[13] Fig. 14.10(a) illustrates maximum packing of poles with PMs on top and on the sides to increase B_0 by increasing direct flux. The 2 T, 27-cm-period SPEAR Beamline 9 wiggler at SSRL employs this technique.

Wedged Poles. Another method of increasing V_0 is to decrease the pole-pole capacitance. This can be accomplished by tapering the pole as shown in Fig. 14.10(b),

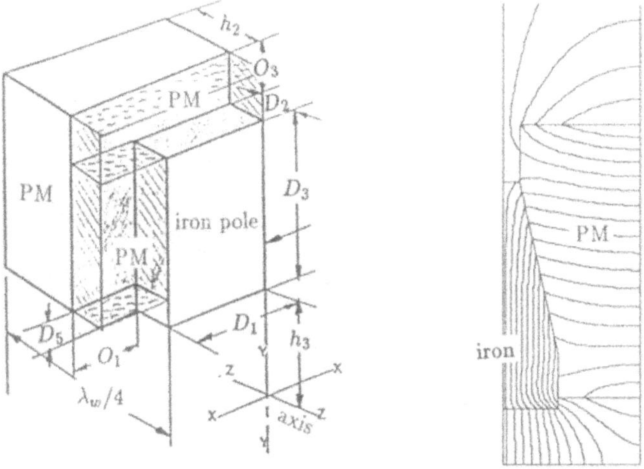

(a) pole packing with PM material, (b) pole tapering.

while keeping the pole tip shape and thus the on-axis field distribution unchanged.[14] Untapered, flux density in the pole decreases as one moves away from the gap. Away from the gap the pole thickness can be decreased, but not so much as to cause magnetic saturation and choke off the flux being channeled down the pole to the tip. Since direct flux Φ_d from the PM is unchanged from that in the untapered version, indirect flux Φ_i out of the pole also is unchanged, but as pole-pole capacitance decreases, pole scalar potential V_0 and on-axis flux density B_0 increase. The 0.7 T, 3.3-cm-period APS Undulator A devices at Argonne employ this technique.

14.5 Warm Electromagnet Insertion Devices

The first IDs were electromagnets arranged in an alternating array of +/− dipoles. These devices generate magnetic field via $\vec{H} \cdot d\vec{s}$ around a loop enclosing the energizing current, in contrast to permanent magnet designs which generate a given amount of flux at block faces.

14.5.1 Advantages

For general SR sources, any large-period ($\lambda_w \geq 20$ cm) ID has ample real estate to house EM coils with sufficiently low current density to make energy operating costs reasonable. Capital costs for such an EM ID are significantly lower than those of a PM or hybrid counterpart. A range of field strengths is attainable by simply varying the coil current, rendering unnecessary gap changes and associated support and drive systems. Furthermore, adjustment and fine tuning of end section energization is straightforward in electromagnet IDs. Wavelength shifters, being few-pole, long-period devices, are typically electromagnets.

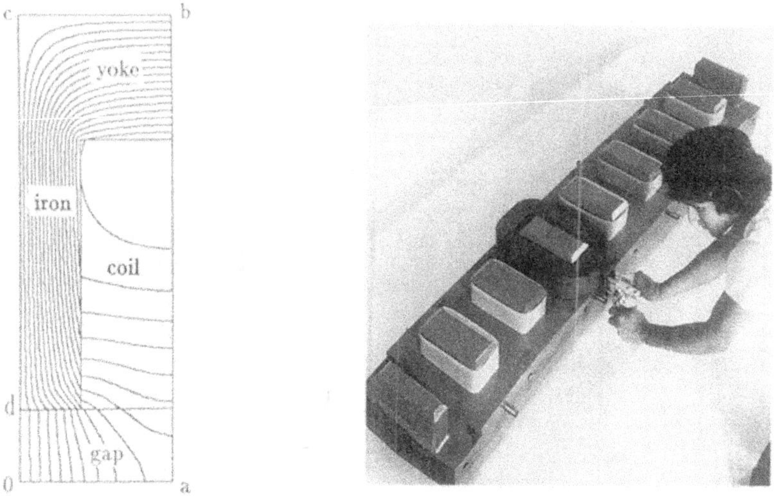

Fig. 14.11 Electromagnet IDs: (a) Magnetic flux lines in the upper-half quarter-period of an EM ID, (b) assembly of the lower half of the 7-full-strength plus two-half-strength pole, 45-cm-period, 2.8-cm-gap, 2-m-long, 1.9 T SSRL wiggler.[15]

If an FEL is to convert a significant fraction of the electrons' energy to light in a single pass, the ID's magnetic field must be tapered axially to maintain resonance (see section 2.3 of this chapter). An EM device can feature a variable, readily-tunable taper along the axis.

14.5.2 Design Methodology

In principle, EM ID design procedure parallels that of hybrid devices: (1) the pole tip surface produces the desired field distribution, (2) the pole tip scalar potential produces the desired on-axis field strength, and (3) the remainder of the coil and pole design provides the required pole tip scalar potential.

Poles of EM ID designs are connected via a high-permeability yoke to shunt the flux to the zero scalar potential plane midway between adjacent poles. (Refer to Fig. 14.11.) Assuming low $\vec{H} \cdot d\vec{s}$ losses in the iron, Ampere's law gives $\mu_0 I_{coils} = \int_{0abcd0} \vec{H} \cdot d\vec{s} \simeq \langle B_{gap} \rangle h_3$, and coil location along the pole is immaterial as far as the on-axis field B_0 is concerned. However, when coils are situated further from the pole tip, flux between adjacent poles increases and poles could saturate at the base (near the yoke), invalidating the low-pole-losses assumption. In EM designs coil current density J_c is usually set as high as heat transfer/cooling allows (typically 1000 A/cm^2), thereby minimizing required coil cross-sectional area and pole height. For a given coil current, and thus on-axis field, pole flux is likewise minimized, delaying saturation of the pole base. Required pole height D_3 is given approximately by $D_3 h_2 J_c \mu_0 = \langle B_{gap} \rangle h_3$, assuming negligible pole saturation.

In an FEL, adjustment of on-axis field at any axial location along the ID should

pole number:	01	02	03	04	05	06	07	08	09	10	11	12	13	14	15	16	17	18
series coil winding circuits	1	3	3	1	1	3	3	1	1	3	3	1	1	3	3	1		
			1	3	3	1	1	3	3	1	1	3	3	1	1	3	3	1

Table 14.1 Steering-free, overlapping, 2nd-order binomial expansion coil winding pattern.

be steering-free so as not to cause displacement of the electron beam from the axis of the laser. Winding coils in series around several adjacent poles with turns ratios in a binomial expansion pattern (e.g., $+1, -2, +1$ or $+1, -3, +3, -1$), and overlapping patterns, enables steering-free local field tapering (see Table 14.1).[16]

14.5.3 Design Variations

Maximum attainable on-axis field in electromagnet IDs is limited by magnetic saturation in the iron pole at the base and/or by current density/heat-transfer limitations in coils. Several design techniques delay the onset of saturation until a higher pole tip scalar potential V_0 is reached.

Pole Tapering. Widening poles towards the base allows more flux to be carried where the base attaches to the common yoke, and thus delays saturation onset. Of course real estate for the coil must be sacrificed and flux between adjacent poles increases, though minimally, since the driving pole scalar potential also tapers to zero at the pole base. The tradeoff can be advantageous in certain operating regimes.

PM Reverse Biasing. PM material laced between coils or placed on lateral pole sides as in Fig. 14.12 results in a direct flux input into the pole.[17,18] This flux does not find its way across the gap, but rather short circuits through the pole to the yoke and back through adjacent poles, thereby providing a reverse-bias flux in the pole which opposes the EM-induced pole flux. As a result, current in the EM coils, and therefore V_0, and thus $B_{0_{max}}$, can be increased before the pole begins to saturate. Of course the minimum on-axis field possible $B_{0_{min}}$ also increases in magnitude, and in fact at a much quicker rate than $B_{0_{max}}$ because laced PM displaces the coil away from the gap. Tunability range thereby decreases. Generally, reversed-biased EM IDs cannot be turned off because, in the absence of coil current, the pole saturates in the reverse direction and part of the PM induced pole flux is redirected across the gap.

Tunability Enhancement. Tunability range can be re-expanded by independently controlling the various coils around a given pole.[19] B_0 depends on coil current, while the field in the iron pole B_{pole} depends on both the coil's current and its vertical location along the pole. To exploit these different functional dependencies: (1) B_0 can be maximized by locating ampturns as close to the gap as possible thereby minimizing EM-induced B_{pole} and increasing the total coil current at which the onset of saturation occurs, and (2) B_0 can be minimized by locating ampturns as far from the gap as possible thereby maximizing EM-induced B_{pole} and decreasing the total coil current

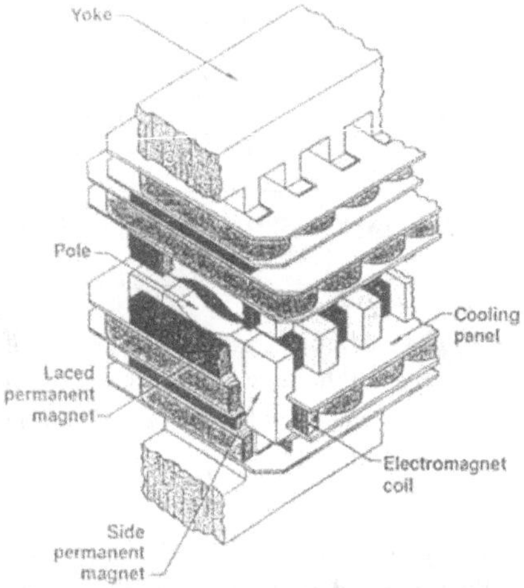

Fig. 14.12 Electromagnet ID with laced and side reverse-bias PMs enables attainment of higher B_0 (reprinted with permission[19]).

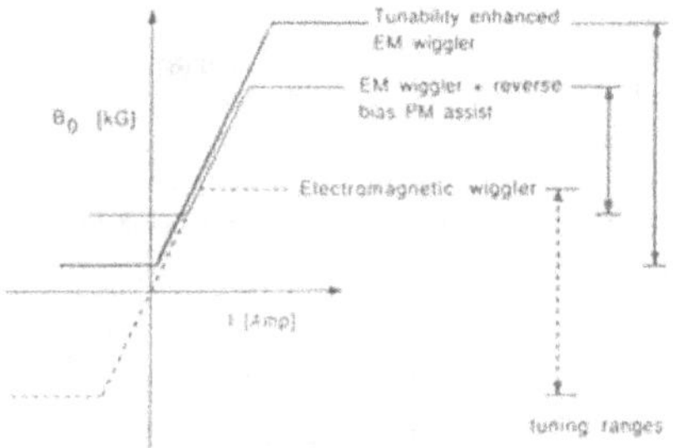

Fig. 14.13 On-axis magnetic flux density B_0 versus coil current I, showing EM wiggler tuning ranges (reprinted with permission[19]).

at which the onset of reverse-direction saturation occurs (Fig. 14.13). Figure 14.12 depicts a reversed-biased, PM-laced, tunability-enhanced EM wiggler used in an FEL for electron-cyclotron heating of a Tokamak plasma at Lawrence Livermore National Laboratory (LLNL).

Pulsed Devices. Pulsed, air-core EM devices offer the possibility of high fields for pulsed radiation applications. The ELF wiggler at LLNL employs air-core solenoids to generate a pulsed, 10-ms-period, 0.5-T on-axis field in a 10-cm-period device.[20]

14.6 Superconducting Electromagnet Insertion Devices

14.6.1 Advantages

Long-period, few-pole superconducting wigglers (e.g., Fig. 14.14(a)) are capable of producing ultra-high fields of ~ 5 T and are ideal devices for wavelength shifters (see section 1.3 of this chapter).

At the other extreme, microundulators with $\lambda_w \leq 2$ cm offer the possibility of ultra-compact FELs for laboratory, industrial, and medical uses. Dramatically reduced-cost systems are feasible by coupling low-emittance, low-energy electron beams and high-field, short-period undulators. The entire infrared spectral range could be covered using a short-period (< 1 cm) superconducting microundulator capable of deflection values of $K \simeq 1$, with accelerator energies and undulator lengths as low as ~ 15 MeV and ~ 1 m, respectively. These superconducting DC devices are capable of the high repetition rates necessary for superconducting-linac and high-voltage electrostatic

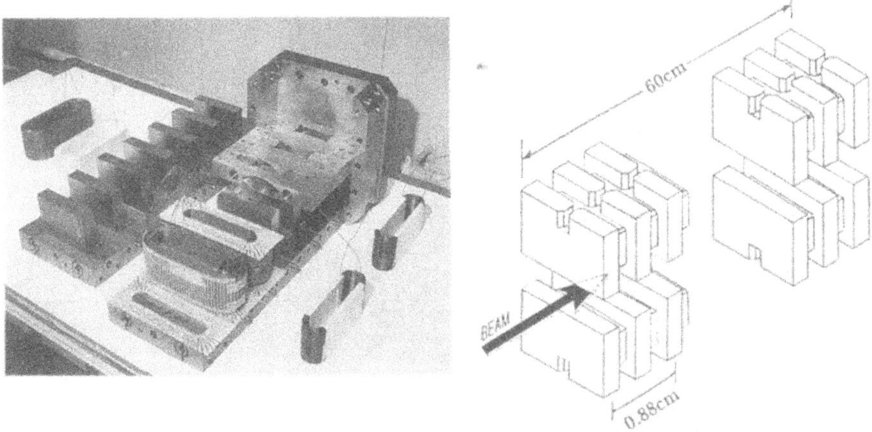

Fig. 14.14 Superconducting IDs: (a) NSLS's 5-full-strength plus 2-half-strength pole, 17.42-cm-period, 3.2-cm-gap,1.4-m-long, 6 T superconducting wiggler[21], (b) BNL's continuously wound superconducting microundulator (reprinted with permission[22]).

accelerator based FELs. The superconducting coils operate in relatively benign, low field environments. An inherent drawback of these devices is lack of accessibility. Alignment and magnetic characterization are difficult because the undulator is housed in a cryostat.

14.6.2 Design

A 60-cm-long superconducting microundulator with $\lambda_w = 8.8$ mm, $g = 4.4$ mm, and current density $J_c \simeq 1400$ A/mm^2 capable of a deflection parameter $K = 0.4$ has been tested in a visible-region FEL oscillator experiment at Brookhaven National Laboratory (BNL).[22,23] This elegantly simple design features a continuously wound multifilament superconductor on a slotted soft iron bar that simultaneously serves as the poles, yoke, and coil mandrel (Fig. 14.14(b)). In Russia, an 8.0-mm-period device with $g = 2.5$ mm has achieved $K = 0.57$ using the same concept, while tapering the poles to delay pole saturation (see section 5.3 of this chapter).[24] Materials of choice for windings are Nb_3Sn tape superconductor or NbTi multifilament wire in a copper matrix.

14.7 Circularly Polarized Radiation From Insertion Devices

The amount of circularly polarized light absorbed by materials that exhibit dichroism, such as DNA and amino acids, depends on the handededness of the radiation; their presence can thus be identified, and molecular structures determined. Magnetic materials, with their inherent directional spin, also exhibit dichroism– allowing the study of magnetic characteristics of microscopic regions with important applications to magnetic storage capacity for computers.

Off-midplane bending-magnet radiation has a circular polarization component and

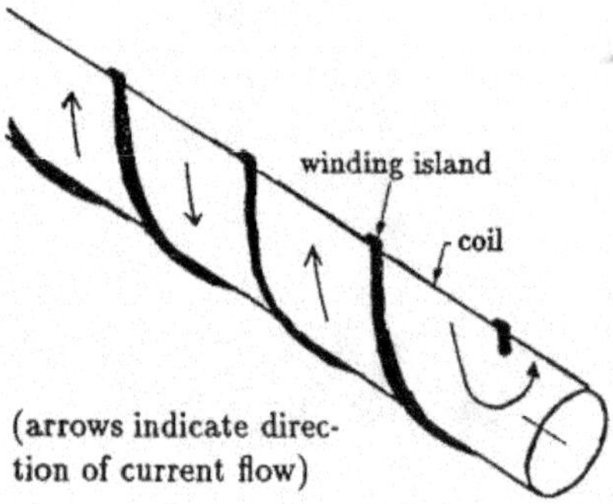

Fig. 14.15 A bifilar helical undulator produces circularly polarized light.

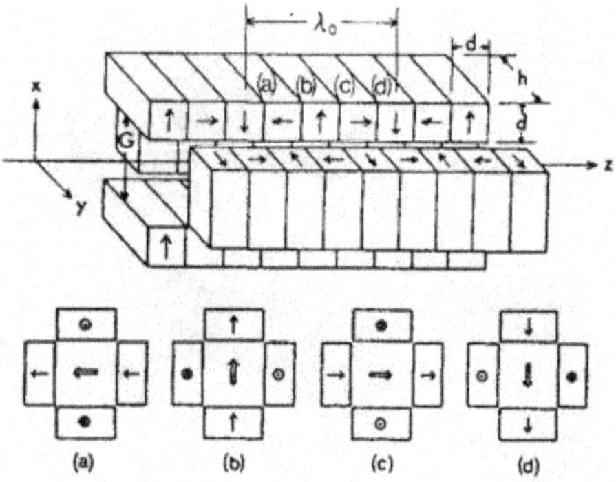

Fig. 14.16 Two planar IDs rotated 90° produce arbitrary elliptical or arbitrarily oriented linear polarization (reprinted with permission[27]).

is often used in dichroism experiments, though high polarization degree comes by sacrificing the intensity considerably. IDs that can produce circularly polarized light (often with enhanced characteristics such as high flux intensity and/or degree of polarization at certain wavelengths) include: (a) helical devices (bifilar solenoids, planar helical undulators, and elliptical wigglers), (b) asymmetric devices, which have positive and negative poles of unequal magnitude, and (c) interference devices, in which radiation from several devices is mixed.

14.7.1 Helical Insertion Devices

The electric field component of the SR of the devices discusssed thus far oscillates in only one direction in a plane perpendicular to the direction in which the light is moving. When electrons traverse a helical magnetic field, the electric field rotates azimuthally, generating elliptically polarized light with a right- or left-handedness that makes it an ideal tool to investigate materials exhibiting dichroism.

Bifilar Solenoidal Undulators. Two identical coaxial solenoids separated by half the coil-pitch λ_w, carrying current in opposite directions give rise to a circularly polarized transverse field with no longitudinal field component (Fig. 14.15). The on-axis transverse field magnitude is $B_0[\mathrm{T}] = 8\pi 10^{-7}(I[\mathrm{A}]/\lambda_w[\mathrm{m}])[\epsilon K_0(\epsilon) + K_1(\epsilon)]$, where $\epsilon = \pi g/\lambda_w$ and K_0 and K_1 are modified Bessel functions. A very early FEL experiment at Stanford employing a superconducting bifilar solenoid design of 3.2-cm-period and 1.1-cm-gap with a current density of 700 A/mm^2 achieved a peak on-axis field of 1.3 T.[25] Analogous devices employing pure PM or PM plus iron, and consisting of an array of adjacent dipole magnets incrementally rotated azimuthally by an angle $2\pi/M$, are also possible.[8,26] These devices produce relatively weak fields.

Elliptical Undulator/Wiggler. An elliptically polarized SR source consisting of two superposed planar undulators oriented 90° about the beam axis with respect to one another produces crossed magnetic fields (Fig. 14.16).[27] The undulators phase-shifted longitudinally one-quarter period produce arbitrary elliptical polarization by independently varying vertical and horizontal gaps. Likewise, in-phase the undulators can produce linear polarization along an arbitrary x-y plane. Handedness could be switched slowly by mechanically varying the relative phase shift between undulators through zero, or rapidly if one of the undulators were an AC electromagnet. Horizontal field strength is limited by the vacuum chamber, which bounds the minimum horizontal magnetic gap dimension. A prototype of this design using PMs with $\lambda_w = 8.0$ cm, $g = 5.7$ cm and $B_0 = 0.15$ T in the helical mode and $B_0 = 0.21$ T in the linear mode was installed in Japan's 600 MeV Teras storage ring producing $\lambda = 63$ nm circular and linear polarized radiation, respectively.[28]

Analogous longer-period higher-field helical devices, i.e. elliptical wigglers,[29] have been built for the 6 GeV TRISTAN accumulator ring and the 2.5 GeV Photon Factory at KEK.[30] A strong vertical field produces broadband wiggler radiation, as a relatively weak horizontal field phase shifted a quarter-period alternately bends the electron trajectory up, then down. Locally "off-axis" elliptically polarized light in the device midplane adds to that of like-handedness polarized light from subseqent half-periods also directed in the device midplane. Intense radiation with a high degree of circular polarization $P_c \simeq 0.5-0.9$ from the VUV to the hard x-ray regime has been produced in these devices with $\lambda_w = 16$ cm, $g = 3$ cm, and $B_0 = 1$ T ($K_y = 15$).

Pure PM Planar Helical Insertion Devices. Pure PM planar helical devices, simple to build, can generate purely circularly polarized light ($P_c = 100\%$). An advantage of these helical devices is that no part of the magnetic structure lies in the device midplane, simplifying both insertion device and vacuum chamber construction and making possible operation at smaller magnetic gap.

The European Synchrotron Radiation Facility (ESRF) pioneered the development of a planar device with (a) a split-lower jaw consisting of two PM arrays with blocks whose magnetization directions rotate longitudinally in the x-z plane, producing a vertical field component in the vertical midplane, and (b) a split-upper jaw consisting of two PM arrays with blocks whose magnetization directions rotate longitudinally in the y-z plane and differ in polarity by a half-period, producing a purely horizontal field in the vertical midplane, phase-shifted 90° (Fig. 14.17).[31] Generation of a circular helix necessitates vertically adjusting upper and lower jaws independently so that maximum horizontal and vertical field strengths are equal "on-axis" in the desired horizontal midplane. The upper jaw assembly could be moved longitudinally over a period with respect to the lower jaw to vary helicity and/or change handedness.

Sincrotrone Trieste proposed a planar device capable of higher fields consisting of two adjacent magnet arrays each in split-upper and split-lower jaws whose blocks have magnetization directions that rotate longitudinally, alternately in the x-z and y-z planes (Fig. 14.18).[32] Each jaw assembly individually produces maximum horizontal and vertical fields equal in strength at any y location in the vertical midplane,

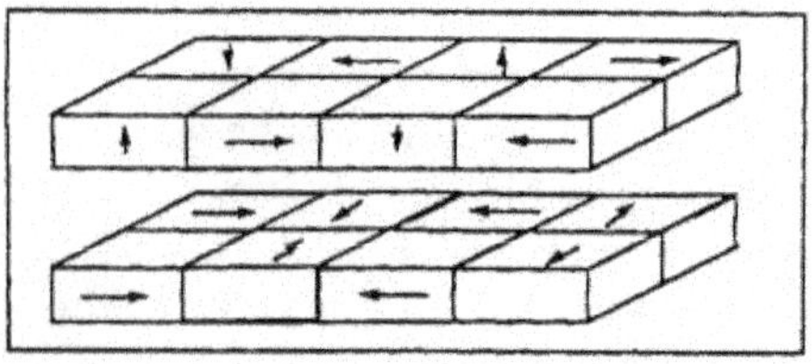

Fig. 14.17 ESRF's planar device yields elliptically polarized light, yet does not impinge on the vacuum chamber.

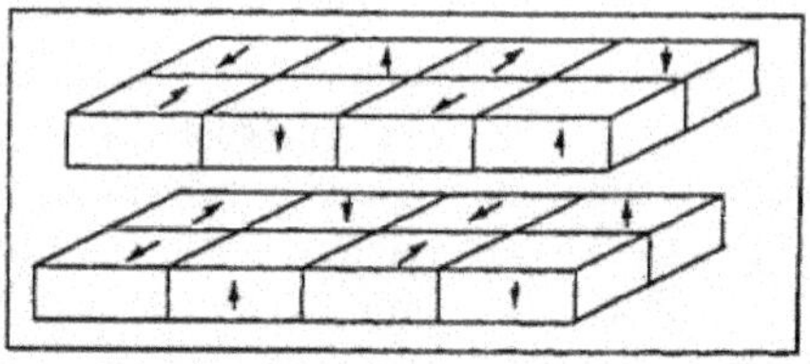

Fig. 14.18 Trieste's proposed planar device produces gap-independent, circularly polarized light (handedness is invariant).

obviating independent control of jaw assemblies. Magnetization direction in PMs differs by 180° in otherwise identical upper and lower jaw assemblies, yielding maximum on-axis field strengths, variable via gap adjustment. Helicity remains perfectly circular, independent of gap, and handedness is invariant in this device.

At the Japan Atomic Energy Research Institute (JAERI), planar devices capable of even higher fields have been built[33] and proposed.[34] Their proposed APPLE-II planar device consists of two adjacent Fig. 14.6-type magnet arrays each in split-upper and split-lower jaws (Fig. 14.19).[34] Lower-front and upper-back arrays are fixed in space with block magnetization directions in the orientation of the Fig. 14.6 design. Lower-back and upper-front arrays have blocks with this same magnetization direction pattern, but shifted together longitudinally a fraction of a period to maintain a circular helix. If the longitudinal shift is fixed, varying the gap changes ellipticity somewhat. Ellipticity and handedness can be changed by longitudinally shifting together diagonal arrays.

SSRL recently commissioned a planar undulator source of circularly polarized soft x-rays in the 500-1000 eV range.[35] This elliptically polarizing undulator (EPU) is capable of producing linearly polarized soft x-rays in the vertical or horizontal planes, and right- and left- circularly polarized x-rays. Additionally, the four rows of NdFeB magnets are independently positionable, enabling changes in photon wavelength at a single gap. Magnet block orientations are as shown in Fig. 14.19. Axial movement of magnet rows have not measurably interfered with SPEAR electron beam parameters or other SSRL users.

Single vs. Chicane Configuration for Planar Helical IDs. Dichroism experiments usually require that the handedness of radiation can be changed. Fast cycling of the polarization's handedness can be accomplished using chicane configurations, in which the radiation from two nearly colinear devices of opposite handedness is mechanically chopped downstream of the ID, alternately giving right- and then left-circularly polarized light. Chicane configurations sacrifice monochromatic flux because radiation from only half the total periods is used at any one time; the peak width is inversely porportional to the number of periods.

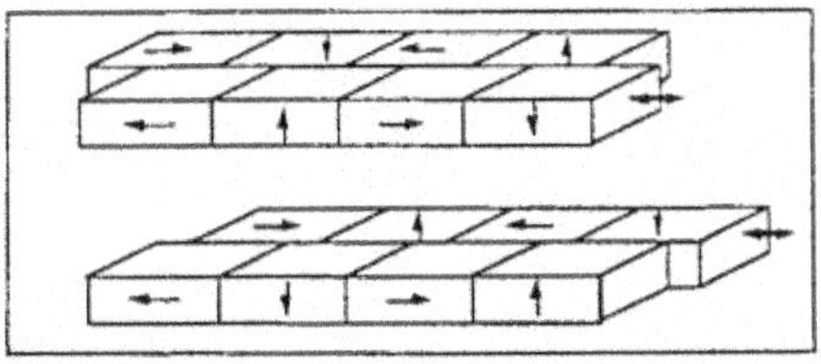

Fig. 14.19 Block orientations for JAERI's proposed APPLE-II planar device and SSRL's elliptically polarizing undulator.

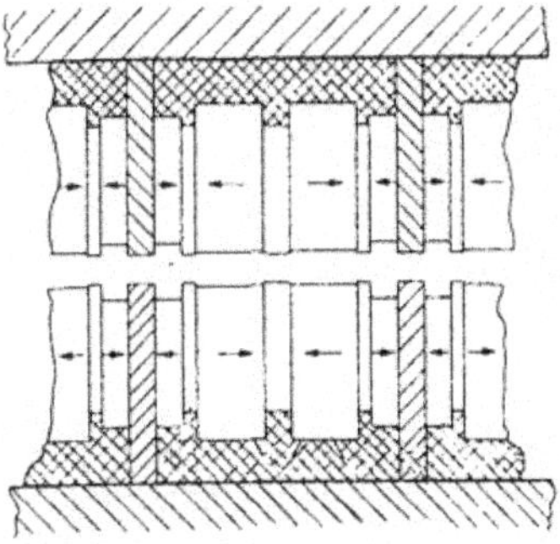

Fig. 14.20 Asymmetric hybrid structure generates broadband elliptically polarized light (reprinted with permission[37]).

Relaxed handedness cycling on a time scale of the order of several seconds or minutes can be accomplished through longitudinal shifting of selected magnet arrays. This technique has the additional disadvantages of (1) possibly affecting the users on other beamlines due to slight changes in electron beam dynamics and (2) having to contend with massive, rapidly moving parts. Advantages include producing more than twice the flux of a corresponding chicane design, and use of the center of the monochrometer optics, rather than the edges as with the chopped beams of the chicane design, thereby minimizing elliptical distortion.

14.7.2 Asymmetric Insertion Devices

Asymmetric Wiggler. An asymmetric wiggler produces a large positive field over a longitudinally short distance, followed by smaller negative field extended over a longer distance, so as to have a net zero integrated field over the wiggler length (Fig. 14.20). The highly asymmetric on-axis field leads to a high degree of circular polarization at photon energies from the VUV domain through the soft x-ray regime in the Super-ACO SR ring at Orsay[36] and in Doris III at Hasylab.[37] Adjustable correctors are used to keep integrated field through the ID nulled over the range of gaps.

14.7.3 Interference Devices Producing Circularly Polarized Radiation

Crossed Undulators. Crossed undulators separated by a variable phase shifter produce any desired polarization (Fig. 14.21).[38,39] The polarization vectors of the radiation beams from a pair of in-line planar undulators oriented at right angles azimuthally are orthogonal. On-axis, the linearly polarized radiation from the two undulators can be mixed in a monochrometer to produce the desired polarization state. Modulating the electron path length between the two IDs with an EM phase shifter varies the phase delay between the two radiation beams, enabling rapid, variable-waveform selection of polarization states.[40] The polarization state is very sensitive to the electron-

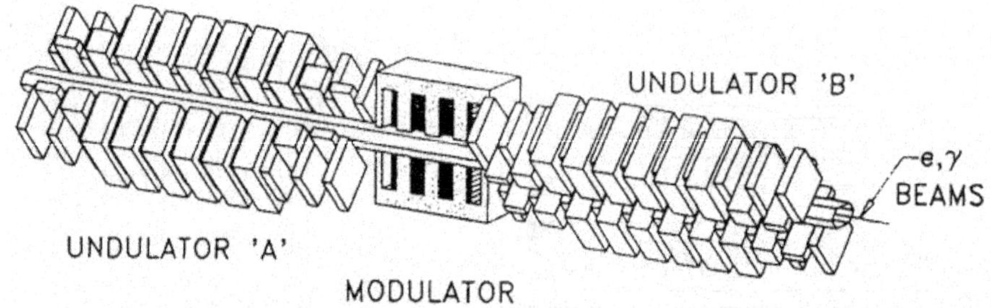

Fig. 14.21 Pair of crossed undulators separated by a tunable EM phase shifter (reprinted with permission[40]).

beam divergence in this device, which must be restricted to $\sigma' \leq (\gamma N)^{-1}$. Thus, use is limited to low-emittance storage rings and lower energy wavelengths. A crossed undulator of this type is operating at the BESSY storage ring in Berlin, producing circularly polarized radiation in the VUV–soft x-ray range.[41]

14.8 Other Novel Insertion Device Design Ideas

Harmonic Generation. One approach to reaching very short wavelengths ≤ 1 μm is FEL amplification of higher harmonics. BNL's high gain harmonic generation experiment uses two superferric undulators (see section 6.2 of this chapter) separated by a dispersive section to amplify the third harmonic of a CO_2 laser.[42] Modulation of the electron-beam energy occurs in the first (modulating) undulator, followed by electron bunching through an achromat in the dispersion section, and finally coherent radiation from the bunched electrons in a second (radiating) undulator tuned to resonate at the third harmonic, $\lambda = 3.47$ μm. For the radiating undulator, the wiggle period $\lambda_w = 1.8$ cm, length $L = 1.5$ m, deflection parameter $K \simeq 1$, and full gap $g = 0.8$ cm.

High Electron-Beam-Energy, Short-Wavelength FELs. Alternatively, using an available high-energy, low-emittance electron beam, laser action could be extended to wavelengths as low as 0.1 nm by SASE (see section 1.3 of this chapter) or harmonic generation from a longer wavelength seed laser. Recent developments in producing low emittance beams in RF photocathode guns and in transporting and compressing beams open the possibility of such a linac-based FEL[43] on SLAC's 50 GeV linac in a 60-m-long, 8-cm-period ID.

Pulsed Electromagnet Microundulators. Pulsed, air-core microundulators also offer the possibility of generating ultra-high fields for compact FEL systems, even surpassing the ~ 1–2 T fields of the small-period superferric designs. FEL gain increases as K^2, saturating when $K_{rms} \simeq 1$. Obtaining maximum gain while decreasing wiggle period below 1 cm requires on-axis fields $B_0 > 2$ T. Circularly polarized bifilar designs (see section 7.1 of this chapter) in the pulsed mode are capable of producing these high

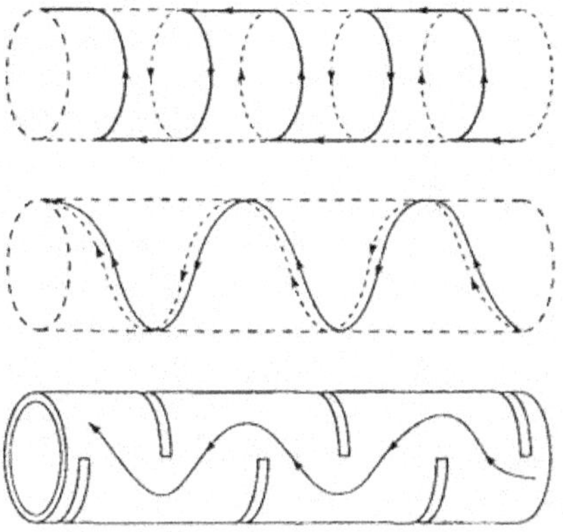

Fig. 14.22 Conductor configurations producing planar fields in pulsed microundulators (reprinted with permission[44]).

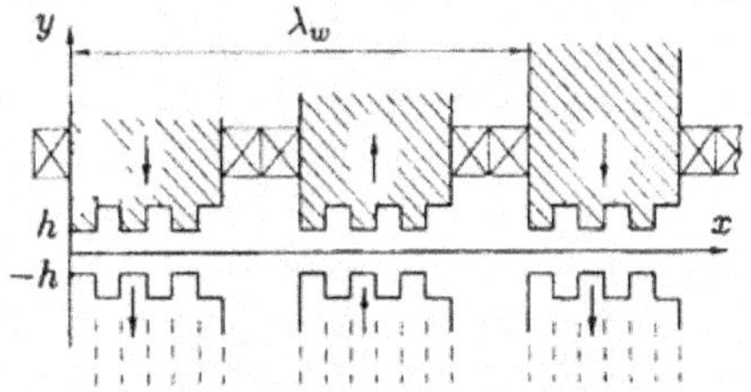

Fig. 14.23 Millimeter-period dipole microundulator immersed in chain of alternating-polarity background fields (reprinted with permission[46]).

fields. Analogous planar field designs capable of harmonic generation which feature conductors distributed over a cylindrical surface have been proposed,[44] with on-axis fields as high as ~ 5 T and $\lambda_w = 3$ mm yielding $K \simeq 1.4$ (Fig. 14.22). However, FEL repetition rate is limited in these devices so as to avoid thermal stresses and potential conductor melting associated with heat generation in small spaces. Field stability in these pulsed devices is also problematical, requiring expensive high-stability pulsed-power supplies. Finally, the absence of field-shaping iron poles makes coil-positioning fabrication tolerances exceptionally tight.

Submillimeter-Period Undulators. Submillimeter devices[45] have an inherently low deflection parameter $K \leq 0.1$, producing on-axis radiation consisting almost exclusively of the fundamental wavelength. If coherence length (see section 2.2 of this chapter) is sufficiently long, a photon-eating monochrometer could be rendered unnecessary, thereby yielding reasonably high net coherent power at beamline end stations in low-energy rings ($\leq 0.1 - 1.0$ GeV). A potential submillimeter-period ID design consists of upper and lower jaws of PM material in which all poles are magnetized in the same direction, immersed in an external reverse-biasing field that zeros the average field on-axis. Field strength is inherently limited to half the value of remanant field B_r.

An alternative design employs an electromagnetically excited series of grooved ferromagnetic poles producing a wiggle field superposed on a dipole background.[46] A chain of such dipole-undulators with alternating polarity background fields zeros the net steering through the ID (Fig. 14.23).

At submillimeter-period dimensions these IDs are inside of the vacuum chamber. Since the aperture requirement for the circulating beam varies along the electron beam path and is proportional to the square root of the local beta function (see Chapter 2),

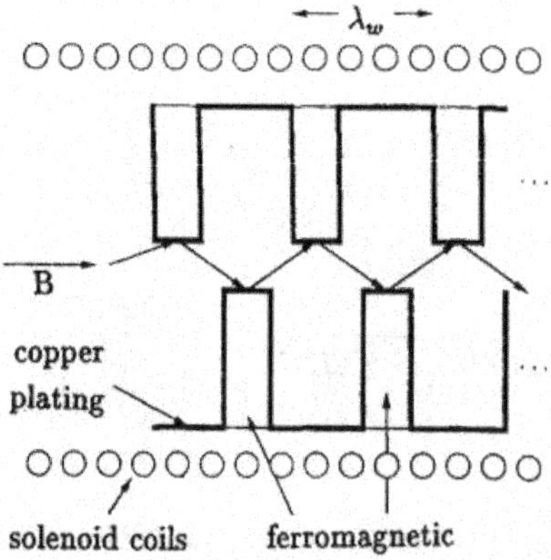

Fig. 14.24 Solenoid-derived wiggler: copper-plated staggered poles deflect the B-field and simultaneously support microwaves (reprinted with permission[47]).

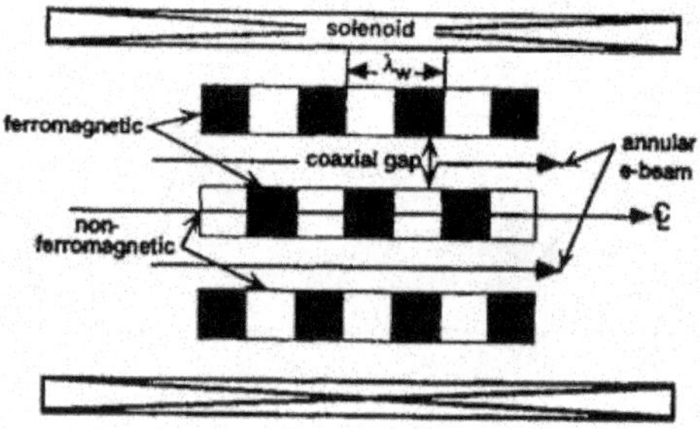

Fig. 14.25 Coaxial hybrid iron microundulator for annular beams (reprinted with permission[48]).

such short-period, small-gap devices are best located in so-called low-beta sections to minimize beam scraping.

FEL IDs with Microwave Acceleration of Electrons. Axially tapering the wiggle field in an ID maintains FEL synchronism as electron beam energy is extracted (see section 2.3 of this chapter). Alternatively, microwave acceleration of electrons traversing an untapered wiggler doubling as a waveguide can keep electron energy invariant and likewise maintain electron-photon resonance. This latter technique is not susceptible to the reduction of small signal gain or to the electron detrapping phenomena associated with large-field-tapering of the former. For this microwave application, a hybrid wiggler design (see section 4.2 of this chapter) would require greatly recessed PMs to support the microwave field, and ensuing pole tip saturation would drastically limit attainable on-axis field B_0. Alternatively, a Stanford design employs a solenoid derived wiggler structure comprising a periodic array of staggered, copper-plated ferromagnetic pole pieces which also support microwaves at the klystron frequency (Fig. 14.24).[47] Pole-tip geometry is optimized balancing relative desires for high microwave and high magnetic-field strengths. An advantage of this design is the simplicity of solenoidal field generation giving a substantial transverse wiggle field. Drawbacks include a large periodic variation in the axial B-field component and large off-axis transverse and axial gradients, leading to focusing complexities.

Annular Wiggle Devices. The coaxial hybrid iron wiggler generates large wiggle fields by imbedding a periodic array of coaxial ferromagnetic rings in a solenoidal magnetic field.[48] The central portion of the coax is shifted relative to the outer by a half period, giving rise to a periodic axisymmetric radial field (Fig. 14.25). This design, which requires a non-standard annular beam, can produce ~ 0.4 T wiggle fields in millimeter-period devices. Its relatively small field gradients give superior beam focusing and transport properties. Simulations predict that a wiggler based amplifier in the K_u-band at $\sim$ 15 GHz and 200 kV could yield high gain ($>$ 1dB/λ_w), efficiency ($>$ 10%), and power (several megawatts).

Inverse Free-Electron Lasers. Wigglers can also be used as an inverse free electron lasers, where the energy from a high power laser beam transfers energy to a coaxial electron beam. A fast-excitation pulsed iron-core EM wiggler for this purpose has been developed at Brookhaven.[49,50] Eddy currents induced in copper field reflectors laced between wiggler poles effectively uncouple the axially alternating polarity wiggle field, significantly enhancing the maximum achievable on-axis field. An on-axis field of B_0 = 1.8 T with a 200-μm pulse length has been achieved in a 3.7-cm-period, 0.4-cm-gap device. Poles consist of thin laminations which are removable so as to allow axial tapering of the wiggle period.

Multi-Undulators. Several facilities, including SSRL, HASYLAB, and the Photon Factory, have incorporated in their rings multiple interchangeable undulators mounted on a readily movable carriage. These devices allow a single beamline to use a range of different undulator periods, or polarizations in the case of the SSRL device, and hence extend the energy and polarization available to the user.

14.9 Insertion Device Fields and Errors

14.9.1 Field Requirements

Undulator Radiation Driven Field Requirements. The intensity of light in a spectral peak is limited by both electron beam characteristics (see Chapter 2) and ID field errors. An undulator should produce high brightness in several harmonics so as to span a broad energy range. Attaining the appropriate phase of the emitted light throughout the device necessitates maintaining the appropriate electron path trajectory. This in turn requires a high-quality on-axis field, establishing error tolerances on the undulator structure.

Random field errors vertically perpendicular to the beam axis cause light production over a broad range of frequencies, thereby reducing brightness in spectral peaks. Kincaid derived a relationship from a stastistical ensemble average between allowable random field error σ_{rms} and reduction in peak intensity.[51] Results for a 30% intensity reduction of the nth harmonic peak in an N-period device are reproduced in Fig. 14.26. The effect of magnetic field errors on emitted radiation is actually quite complicated; it is the change in trajectory length downstream of a perturbation

$$\Delta s(z) = \int \left\{ \left[1 + (x_0' + \Delta x_0')^2\right]^{1/2} - \left[1 + x_0'^2\right]^{1/2} \right\} dz \qquad (14.11)$$

that affects the light. Here $x_0(z)$ is the electron trajectory in an ideal field $B(z) = B_0 \cos k_w z$, and $\Delta x_0(z)$ is the trajectory perturbation caused by a field perturbation $\Delta B(z)$. The nature of the effect on the light, (e.g., the production of harmonics, the reduction of the fundamental's peak, and broadening and shifting of spectral lines)

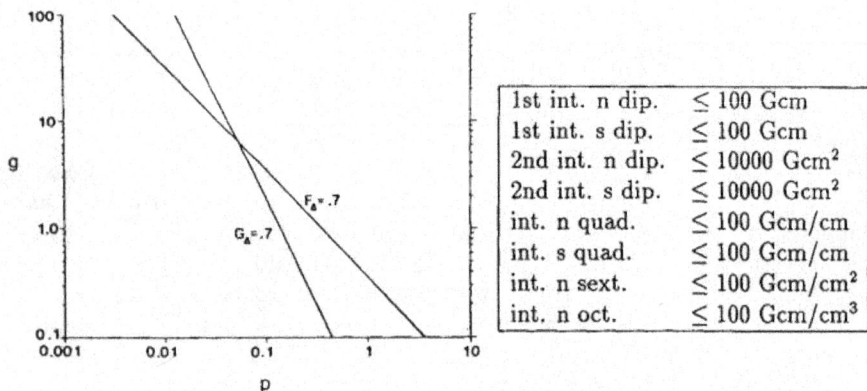

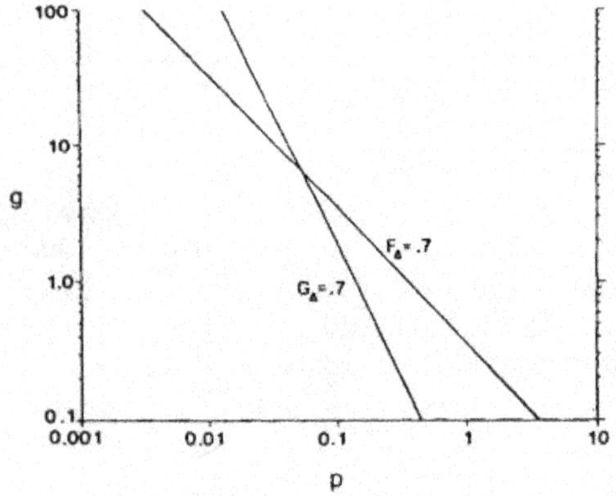

Fig. 14.26 Peak spectral intensity reduction, random field errors of magnitude σ: G_Δ, F_Δ characterize intensity loss and peak position shift for small, large orbit walks [$g \equiv \sigma^2 N^3, p \equiv n/N(1+2/K^2)$] (reprinted with permission[51]).

1st int. n dip.	$\lesssim$ 100 Gcm
1st int. s dip.	$\lesssim$ 100 Gcm
2nd int. n dip.	$\lesssim$ 10000 Gcm^2
2nd int. s dip.	$\lesssim$ 10000 Gcm^2
int. n quad.	$\lesssim$ 100 Gcm/cm
int. s quad.	$\lesssim$ 100 Gcm/cm
int. n sext.	$\lesssim$ 100 Gcm/cm^2
int. n oct.	$\lesssim$ 100 Gcm/cm^3

Table 14.2 Typical storage ring driven integrated field multipole requirements (int $\equiv$ integrated, n $\equiv$ normal, s $\equiv$ skew, dip $\equiv$ dipole, quad $\equiv$ quadrupole, etc.)

is dependent on the nature of the field perturbation (e.g., steering or steering-free, symmetry with respect to unperturbed field, and perturbation magnitude), and can be determined explicitly by evaluating the expression given above.

Systematic field errors such as axial variation of B_0 or λ_w produce a shift in radiation wavelength, thus broadening spectral peaks, whereas total power in peaks remains virtually unchanged. The characteristic linewidth for a perfect ID is determined by electron beam characteristics, the number of periods N, and harmonic number n. For a single electron, $\Delta\lambda/\lambda \simeq \Delta\omega/\omega \simeq 1/nN$. Machine dependent restrictions on variations in K and λ_w can limit broadening of the spectral peak FWHM to a specified percentage beyond that of a perfect ID with the given electron beam and accelerator operating in its normal mode (e.g., multibunch).

FEL Driven Requirements. The field kick from each pole of an ID changes the angle of the electron's trajectory with respect to the beam axis. Each individual bend must be of precisely the same size to keep the beam from receiving a net deflection. Random errors in the wiggle field can result in an undesired "random walk" of the electron beam which, in the case of an FEL, would eventually result in the separation of the electron beam and the laser light, degrading FEL performance.

Additionally, in any ID the shape of the electron beam must be preserved. Even in the absence of ID field errors, the nonzero beam emittance may require focusing in long devices.

Storage Ring Driven Field Requirements. Interestingly, the first wiggler in a storage ring was used to control the beam.[52] Specifically, this "Robinson wiggler" redistributed damping rates to simultaneously damp synchrotron and horizontal and vertical betatron oscillations in an alternating-gradient ring for which the horizontal betatron oscillation is normally anti-damped. Linear and nonlinear beam dynamics can be studied quantitatively through the use of SR integrals;[53] high field wigglers can selectively modifiy these integrals and are still used today to control emittance, bunch length, energy spread, and spin polarization in non-SR applications. First use of a wiggler as a light source was on the SPEAR storage ring at SSRL.[54]

Ideally, IDs for SR would be decoupled from the accelerator electron optics, and thus could be inserted or removed at will. Yet, since ev_xB_z always points towards the midplane in the periodic magnet array, where z is the axial direction and x is the transverse (wiggle) direction, IDs naturally provide vertical electron beam focusing. Adjustment of storage ring quadrupoles can compensate for this natural ID focusing, however this breaks the periodicity of the storage ring, thereby reducing the dynamic aperture. Additionally, spatial field harmonics in the periodic ID structure contribute to perturbing forces that limit the acceptance of the storage ring. Both effects can lead to a reduction in beam lifetime. A good lattice design will cause the ID-induced dynamic aperture reduction to be dwarfed by that produced by field errors in other realizable ring magnets (see Chapter 2).

ID error fields cause a variety of storage ring problems. Nonzero first and second field integrals through the ID missteer and displace the beam, respectively. The horizontal distortion of the electron orbit, so produced, can be amplified at other ring locations, adversely impacting the performance of other beamlines which require a highly stable photon flux. Nonzero integrated quadrupole, sextupole, and octupole integrals focus and missteer off-axis particles. Furthermore, as the field level in a given ID changes, these integrated fields could change, thereby requiring active magnet tuning control around the ring. Therefore, it is desirable to design IDs with integrated field errors small and independent of gap so as to minimize disruption of the rest of the ring during operation.

Typical specifications for integrated multipoles in third generation rings are listed in Table 14.2. Orbit stability is discussed fully in Chapter 13.

14.9.2 Insertion Device Design Implications

Insertion Device Wiggle Plane Focusing Schemes. Beam focusing in the short IDs of ≤ 5 m in storage rings is not usually an issue. Longer undulators and very low emittance beams in FELs require focusing to preserve the small size of the electron beam. Vertical structure devices naturally provide vertical focusing. Several design techniques can provide horizontal (wiggle plane) focusing. Pure PM IDs can be imbedded in a conventional external quadrupole and the superpositioning of fields is nearly linear. In hybrid devices, PM material may be positioned inside the pole tips, oriented so as to give an on-axis gradient.[55] Alternatively, poles can be canted or wedged in the transverse direction to produce x-direction focusing. Finally, focusing can be introduced by employing parabolic-shaped pole tips.[56] For FEL applications, this latter technique has the advantage of preventing modulation of the electrons' longitudinal velocity, which could result in a change of the phase of the electron with respect to the electromagnetic wave, thereby causing electron debunching and loss of gain.

Insertion Device End Effects. The entrance and exit of an ID should be steering-free so emitted photons are directed accurately to the beamline and so the electron orbit stays on track, respectively. This can be accomplished in iron dominated IDs by energizing end poles of a uniform array using a binomial expansion of pole tip scalar potentials.[16] Using an odd number of poles coupled with a second order binomial expansion end excitation pattern ($\hat{V}_i = 0, +1/2, -1, +1, -1, ... + 1..., -1, +1, -1, +1/2, 0$) has the advantage of yielding zero net electron displacement at the ID exit, in addition to being steering-free. Conversely, an even number of poles coupled with a second order binomial expansion end excitation pattern ($\hat{V}_i = 0, +1/2, -1, +1, -1, ... + 1, -1, +1, -1/2, 0$) results in a net electron displacement $2\langle\Delta x\rangle$ at the ID exit, where $\langle\Delta x\rangle$ is the average electron displacement through the ID. However, it has the advantage of being steering-free even if the ID ends have steering errors, since their opposite polarities negate one another. Regardless of the number of poles, third order binomial expansion end excitation patterns ($\hat{V}_i = 0, +1/4, -3/4, +1 - 1, +1, ... \pm 1, \mp 1, \pm 3/4, \mp 1/4, 0$) have the additional desirable feature of zero average electron displacement $\langle\Delta x\rangle$ throughout the ID, thus eliminating the odd poles advantage seen in the second order pattern. EM devices achieve these end pole potentials without difficulty by scaling end pole energization currents. Hybrid designs may achieve this by scaling PM material size at the last few poles[57] or by recessing tips of end poles;[58] either method attempts to approximate the designed ideal excitation over a range of gaps. Errors in placing a pole on its correct scalar potential propagate through an ID, decaying exponentially. Iron-free PM designs can be made steering-free and displacement-free independent of gap by appropriate orientation, sizing, and spacing of the PM material at device ends.[59]

ID Field Errors. ID error fields arising from (1) iron pole, EM coil, and/or PM positioning/orientation errors, (2) PM strength and global/local easy-axis misorientation errors, and (3) iron pole nonuniformities, including saturation effects, collectively must cause neither excessive spectral performance degradation nor storage ring electron beam perturbations.

Any field perturbing source can be described by equivalent magnetic charges, which can be decomposed into charge distributions that are symmetric and antisymmetric with respect to the midplane, as illustrated in Fig. 14.27. Symmetric field errors $\Delta B_y(z)$ are perpendicular to the midplane in the midplane and lead to both a horizontal displacement and steering of the beam. Antisymmetric field errors are parallel to the midplane in the midplane. Those errors $\Delta B_z(z)$ that are also parallel to the electron trajectory have no detrimental effects. Those errors $\Delta B_x(z)$ parallel to the midplane, but perpendicular to the electron trajectory, cause vertical beam steering and displacement.

In pure PM or ironless EM IDs, the induced on-axis error field strengths and distributions are calculable in closed form. For systems with iron, hybrid theory[13] enables

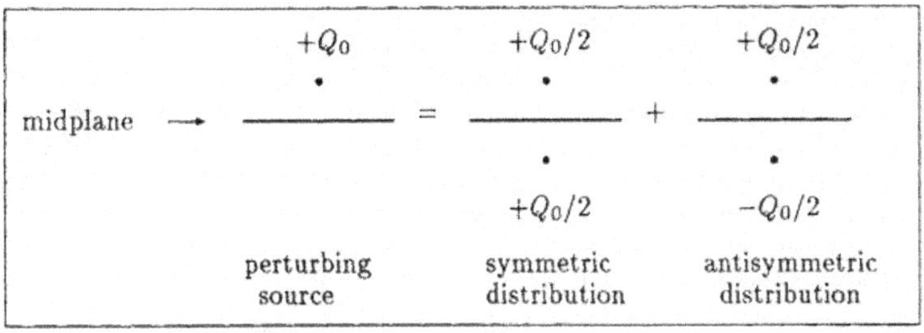

Fig. 14.27 Field perturbing sources can be decomposed into symmetric and antisymmetric components for field error analyses.

semi-analytical calculation of error fields. Mechanical and magnetic specifications can be made in conformance with given spectral and accelerator requirements. PM block sorting and dimensional tolerancing minimize field errors.

Additionally, contoured pole surfaces can provide local multipole field correction. After ID assembly, transverse arrays of vertically adjustable trim magnets[60] positioned at ID ends can also correct for any residual integrated multipole (i.e. normal and skew dipole and quadrupole) errors over a range of gaps.

Permanent Magnet Characteristics and Quality. A PM easy-axis orientation error at axial position z_q of a block with magnetization direction parallel to the beam axis gives rise to an even on-axis field error, $\Delta B_y(0,0,z-z_q)$, causing both a beam displacement and (only by virtue of 3-D effects) a net steering. Such orientation errors are important only on PM block surfaces near the midplane and in principal could be corrected before assembly by grinding block surfaces; in practice one sorts blocks and positions only the best (i.e. those with low misorientation) near the gap. Typically the easy-axis of good quality PM blocks deviates from the nominal direction by less than 1%−3%.

An error in the magnetization strength of a PM block with the magnetization direction parallel to the beam axis gives rise to an odd on-axis field error, $\Delta B_y(0,0,z-z_q)$, causing a beam displacement, but no net steering. In a hybrid design, sorting PM blocks so as to average the strength of the various blocks in contact with a given pole reduces this effect to a neglibile level. Typically the strength of good quality PM blocks deviates from the nominal magnitude by less than 2%.

In a pure PM device, strength and magnetization-direction orientation errors in a block with the magnetization direction perpendicular to the beam axis give rise to even and odd on-axis field errors, respectively.

Quality PM blocks must maintain linearity well into the third quadrant of the B-H curve and must exhibit good temperature stability.

Tolerances. Both gap and pole thickness errors result in an even on-axis field error distribution. A space between PM and a pole also gives rise to a non-zero $\int_z B_y(0,0,z)dz$. Typical tolerances on these dimensions are 50 μm−100 μm.

Iron Characteristics and Quality. Magnetic saturation of soft iron components limits attainable fields. Variations of the saturation behavior of different parts of the system's iron (e.g., due to voids, or nonuniform heat treatment or composition) can give rise to random field errors, which are difficult to predict. Additionally, saturation can give rise to a gap-dependent integrated sextupole. When the iron permeability $\mu \gg 1$, this potential error source vanishes.

14.10 Power and Power Density Considerations

Thermal load on beamline and optical components intercepting the ID's photon beam must be accommodated. Lower energy photons impinging on high-atomic-number materials deposit their energy on the materials' surface. Conversely, the energy of higher energy photons can be absorbed and scattered throughout the depth of a lower-atomic-number material such as carbon. The angular distribution of power in an ID is[61]

$$\frac{d^2P}{d\theta d\psi} = P_T \frac{21\gamma^2}{16\pi K} G(K) f_K(\gamma\theta, \gamma\psi), \tag{14.12}$$

where $G(K)$ and $f_K(\gamma\theta, \gamma\psi)$ are given by

$$G(K) = K \frac{(K^6 + 24K^4/7 + 4K^2 + 16/7)}{(1+K^2)^{7/2}} \quad \text{and} \tag{14.13}$$

$$f_K(\gamma\theta, \gamma\psi) = \frac{16K}{7\pi G(K)} \int_{-\pi}^{\pi} d\alpha \left(D^{-3} - 4D^{-5}(\gamma\theta - K\cos\alpha)^2\right) \sin^2\alpha, \tag{14.14}$$

and where $D = 1 + \gamma^2\psi^2 + (\gamma\theta - K\cos\alpha)^2$ and f_K is normalized so that $f_K(0,0) = 1$. P_T is the total ID power. In practical units, power and power density are, respectively,

$$P[kW] = 1.267E^2[\texttt{Gev}]\langle B^2\rangle[\texttt{T}]I[\texttt{A}]L[\texttt{m}] \quad \text{and} \tag{14.15}$$

$$\frac{d^2P}{d\theta d\psi}[\texttt{W}\cdot\texttt{mr}^{-2}] = 10.84B_0[\texttt{T}]E^4[\texttt{Gev}]I[\texttt{A}]NGf_K. \tag{14.16}$$

14.11 References

1. *Handbook on Synchrotron Radiation*, ed. G.S.Brown and D.E.Moncton, (North-Holland, 1991).
2. *Synchrotron Radiation Research: Advances in Surface and Interface Science*, ed. R.Bachrach, (Plenum, 1992).
3. J.D.Jackson, *Classical Electrodynamics* (Wiley, 1975) 674.
4. S.Krinsky, *IEEE Trans. Nucl. Sci.* **NS-30** (1983) 3078.
5. K-J.Kim, *Nucl. Instr. Meth.* **A261** (1987) 44.
6. K.Halbach, et al., *IEEE Trans. Nucl. Sci.* **NS-28 3** (1981) 3136.
7. S.Yamamoto, *Rev. Sci. Instr.* **63** (1992) 400.
8. K.Halbach, *Nucl. Instr. Meth.* **187** (1981) 109.
9. R.Carr, *Nucl. Instr. Meth.* **A306** (1991) 391.

10. E.Hoyer, et al., *1987 IEEE Part. Accel. Conf.* **3** (1987) 1508.
11. E.Hoyer, et al., *Nucl. Instr. Meth.* **208** (1983) 117.
12. K.Halbach, *J. Physique C1* **44-2** (Feb. 1983) C1-211.
13. K.Halbach, *Law. Berk. Lab. Rep. V-8811-1.1-16* (1989).
14. N.A.Vinokurov, *Nucl. Instr. Meth.* **A246** (1986) 105.
15. G.Brown, et al., *Nucl. Instr. Meth.* **208** (1983) 65.
16. K.Halbach, *Nucl. Instr. Meth.* **A250** (1986) 95.
17. K.Halbach, *Nucl. Instr. Meth.* **A250** (1986) 115.
18. G.Deis, et al., *IEEE Trans. on Magetics* **24-2** (1988) 1090.
19. R.Schlueter and G.Deis, *Nucl. Instr. Meth.* **A331** (1993) 711.
20. T.J.Orzechowski, *Phys. Rev. Let.* **57-17** (1986) 2172.
21. H.Hsieh, et al., *IEEE Trans. Nucl. Sci.* **NS-28 3** (1981) 3292.
22. K.Batchelor, et al., *Nucl. Instr. Meth.* **A296** (1990) 239.
23. I.Ben-Zvi, et al., *Nucl. Instr. Meth.* **A318** (1992) 781.
24. V.A.Papadichev, et al., *Nucl. Instr. Meth.* **A331** (1993) 748.
25. L.R.Elias and J.M.Madey, *Rev. Sci. Instr.* **50** (1979) 1335.
26. G.Bekefi and J.Ashkenazy, *Appl. Phys. Lett.* **51-9** (1987) 700.
27. H.Onuki, *Nucl. Instr. Meth.* **A246** (1986) 94.
28. H.Onuki, et al., *Appl. Phys. Lett.* **52** 173.
29. S.Yamamoto, et al., *Jap. J. Appl. Phys.* **26-10** (1987) L1613.
30. S.Yamamoto, et al., *Rev. Sci. Instr.* **60-7** (1989) 1834.
31. P.Elleaume, *Nucl. Instr. Meth.* **A291** (1990) 371.
32. B.DiViacco and R.P. Walker, *Nucl. Instr. Meth.* **A292** (1990) 517.
33. S.Sasaki, *Nucl. Instr. Meth.* **A331** (1993) 763.
34. S.Sasaki, *Proc. 8th Nat. Conf. on Synch. Rad. Instr.* Gaithersburg (1993).
35. R.Carr and S.Lidia, *SPIE* **2013** E-Beam Sources of Hi-Brit. Rad. (1994).
36. J.Goulon, et al., *Nucl. Instr. Meth.* **A254** (1987) 192.
37. J.Pfluger, *Rev. Sci. Instr.* **63** (1992) 295.
38. M.Moissev, et al., *Sov. Phys. J.* **21** (1978) 332.
39. K-J.Kim, *Nucl. Instr. Meth.* **A219** (1984) 425.
40. M.A.Green, et al., *1991 IEEE Part. Accel. Conf.* **2** (1991) 1088.
41. J.Bahrdt, *Rev. Sci. Instr.* **63** (1992) 339.
42. X.Zhang, et al., *Nucl. Instr. Meth.* **A331** (1993) 689.
43. C.Pellegrini, et al., *Nucl. Instr. Meth.* **A341** (1994) 326.
44. R.W.Warren, *Nucl. Instr. Meth.* **A304** (1991) 765.
45. R.Tatchyn, P.Csonka, in *Und. Mags. for SR & FELs*, World Sci. (1988) 71.
46. N.V.Smolyakov, *Nucl. Instr. Meth.* **A308** (1991) 80.
47. A.H.Ho, et al., *Nucl. Instr. Meth.* **A296** (1990) 631.
48. R.H.Jackson, et al., *SPIE* **2013** E-Beam Sources of Hi-Brit. Radiation (1994).
49. A.vanSteenbergen, US Pat. 368618 (Aug. 1990).
50. J.Gallardo, et al., *IEEE Trans. on Mag.* **30-4** (1994).
51. B.Kincaid, *J. Opt. Soc. Am. B* **2** (1985) 1294.

52. K.W.Robinson, *Phys. Rev.* **111-2** (1958) 373.
53. M.Sands in *Proceedings of the International School of Physics "Enrico Fermi"*, Course XLVI, ed. B.Touschek, (Academic Press, 1971).
54. M.Berndt, et al., *IEEE Trans. Nucl. Sci.* **NS-26-3** (1979) 3812.
55. R.Tatchyn, *SLAC-PUB-6186* (1993).
56. E.T.Scharlemann, *J. Appl. Phys.* **58-6** (1985) 2154.
57. U5.0 Conceptual Design Report, *Lawrence Berk. Lab. PUB-5256* (1989) 30.
58. I.Vasserman and E.R.Moog, internal APS note, unpublished (1993).
59. K.Halbach, unpublished.
60. E.Hoyer, *Law. Berk. Lab. Rep. LSME-570.* (1993).
61. K.-J.Kim, *Nucl. Instr. Meth.* **A246** (1986) 67.

CHAPTER 15: CONVENTIONAL FACILITIES

V. Saile and J. D. Scott
Center for Advanced Microstructures and Devices
Louisiana State University
3990 West Lakeshore Drive
Baton Rouge, LA 70803, USA

15.1. Introduction

The overall success of a research facility is measured in its scientific output, and this output certainly depends to a large extent on the availability of state-of-the-art research equipment such as modern accelerators. However, other factors such as the intellectual environment and, in general, the well-being of the individuals working at a research facility contribute significantly to its productivity. The *conventional facility* for a synchrotron radiation (**SR**) source is not simply a building; it expresses an attitude. Its primary function is to house an accelerator system and the experimental beamlines and to provide space for additional functions such as support laboratories; it is also the environment for the staff and scientists involved in operating the hardware and performing research. The overall success of the facility requires an integrated approach for accelerators, beamlines, staff and users needs such as offices, meeting and conference rooms, ancillary laboratories and shops.

Figure 15.1. The BESSY facility in Berlin, Germany. The building in the center houses an 800 MeV electron storage ring, beamlines and experimental areas. The structure on the right is a water cooling tower. (courtesy A. Gaupp, BESSY GmbH)

Conventional facilities for synchrotron sources have changed dramatically over the past 25 years. Synchrotron-radiation pioneers used available particle physics accelerators for their research and the "laboratories" attached to these machines illustrate very well the mode of operation, dubbed "parasitic use"; structures ranging from "do-it-yourself" garages to small "bunkers" were attached to these large accelerators and it became, in fact, the time of the pioneers. With the advent of more-general users and with the increased complexity and precision of experiments, requirements and demands on the conventional facilities changed significantly. More space, temperature control, vibration control and better working conditions became mandatory. Consequently, the buildings for the second generation of light sources, the dedicated storage rings, are bigger and some are actually attractive. **BESSY** in Berlin (see Figure 15.1.), the National Synchrotron Light Source (**NSLS**) in Brookhaven and the Photon Factory (**PF**) in Tsukuba are vivid examples from this period. However, mistakes were made as well, in particular, the needs for tight vibration control, floor stability and climate control were sometimes underestimated. For today's third generation sources based on low emittance storage rings, these stability issues have gained overwhelming importance to preserve the high brightness capability of these rings.

Writing a chapter on conventional facilities for synchrotron light sources presents a formidable task because the broad ranges of differences in sources and applications are reflected in the buildings. The sizes of the storage rings range from 3-m circumference to well over 1 km; there are rings applied to industrial production, others preferentially to applied research and others to cutting edge basic research. The site can be on a university campus, part of a larger laboratory, at an industrial company or it can be a "green field" site without adjacent activities. The specifications for the conventional facilities vary accordingly from light source to light source and from site to site; furthermore, the laws and codes differ from facility to facility. There are, however, also many commonalties and it is the intent of this chapter to give the reader an in-depth look into matters which play an important role in the design and building of a conventional facility for a synchrotron-radiation source. The authors are scientists and synchrotron radiation users rather than architects or engineers. Their practical-experience base comes from the work done to bring **CAMD** in Baton Rouge, LA, USA from an idea to a successfully working facility.

15.2. Planning the Conventional Facility

Planning the conventional facility should not proceed without a close involvement of the users in all phases to satisfy their technical requirements as well as to meet other needs for their future workplace. In practical cases, however, compromises are unavoidable such as compromises on the size of the buildings and aesthetics caused by constraints in budgets and the availability of land. Without integrated planning, such constraints can result in super-modern accelerator systems with inadequate accommodations for the experiments or in gorgeous buildings housing second-rate equipment. This is why staff, scientists, users and all other personnel working at or using the synchrotron-radiation facility need to be involved in the design of the conventional facility rather than leaving the decisions to administrators, the staff or a sub-group of the facility staff or the architects and engineers only.

The conventional facility encompasses a number of systems. Some of them are:

Physical (structural) support of the accelerators
Physical support of beamlines and experimental equipment
Enclosure and climate control
Utility distribution to accelerators (electricity, water, air, gases)
Utility distribution to beamlines and experiments

Cranes (lifting and moving equipment)
Communications
Radiation safety shielding, equipment, and installation
Infrastructure for specialized applications such as clean rooms
Central-control infrastructure
Equipment and building safety
Security and access control
Experiment preparation
Mechanical, vacuum, electrical and other shops
Storage space
General purpose laboratories (Chemistry, Test labs, Darkroom, etc.)
Specialized laboratories (molecular biology, medical suites, clean rooms)
Offices
Meeting rooms
Conference rooms
Computer rooms
Library
Cafeteria
Accommodations for users and guests

This list may give an impression of the complexity of buildings for a synchrotron radiation source. Depending on the applications of the source, some of the systems may be unnecessary or less important, but for the architects and engineers, such buildings are in any case highly *unconventional* and they will succeed with an acceptable solution only if supported and advised by staff and users.

In general, realization of the conventional facility will commence with site selection, design and engineering, leading finally to construction, acceptance by the owner and occupation of the facility. The sole purpose of the light source in a synchrotron radiation facility is the production of x-rays, i.e., ionizing radiation. Gamma radiation and neutrons are emitted as by-products by the accelerators and components of the system may become radioactive after some period of operation. The experiments often employ chemically or biologically hazardous materials and sometimes radioactive samples. Furthermore, governments and the public become increasingly concerned about environmental issues to such an extent that irrational and even hysterical reactions must be anticipated. Convincing the public and government that a planned facility is safe in every respect requires proper information and documentation to avoid serious delays in the construction schedule. The importance of producing accurate and quality environmental and safety reports cannot be over-emphasized. Laws and codes are certainly different for different countries and states, but the authorization for capital outlay is in most countries contingent on an environmental statement. Commissioning and operation of the accelerators require further permits.

All successful synchrotron radiation facilities expand their utilization with time. Therefore, additional beamlines, accelerators and buildings need to be anticipated early on. A lack of vision will adversely affect the future of any synchrotron facility.

Realization and integration of the various systems for the conventional facility require testing of the proposed building site and careful electrical and mechanical engineering of the highly specialized facility. For a state-of-the-art facility, vibration testing of the site followed by specific planning for vibration control, evaluation of the impact of the site's geology on the buildings and calculations of energy consumption and cost are mandatory. Finally, aesthetics, traffic control and public access to the facility are important, in particular, for facilities close to populated areas.

15.3. Site Selection

Selection of a site may be governed by a number of criteria, such as

Geological suitability of the site
Vibrational suitability
Future expansion potential of the site
Access, proximity of major roads, train stations and airports
Support by local government
Laws and codes
Availability of utilities such as electricity
Availability of technical infrastructure, shops, competent local industries, etc.
Proximity to research facilities, universities and potential industrial users
Construction and operational cost for the facility in this area
Cost of living in this area
Proximity of hotels
Attractiveness of area
Proximity and quality of schools and universities
Proximity of shops and stores
Language spoken in this area (a European issue)
Availability of international schools (a European issue)
Political issues

Decisions regarding factors such as site region (e.g., city, university or industry campus) may have been made previously and as this is not a political discussion, the many forces driving such decisions are not considered. The selection of the exact site may be within the power of the group which can make decisions based on physical criteria; therefore, attention is given to these important items only.

If the **SR** facility is to be part of the research, analytical and/or manufacturing infrastructure of a university, industry or existing laboratory, there may be a number of features which make building on the campus an attractive choice. Convenient access by local users is not the sole benefit from such location of the facility. Major university, laboratory and industry complexes usually have many desirable features which can be shared by an **SR** facility. Workshops and available technical specialists from other departments are on-site, police and fire security systems are generally part of the local community and can handle the addition of the facility. Universities and national laboratories can provide rooms for seminars, meetings and conferences; dormitories, accommodations for guests, libraries and cafeterias are already part of their infrastructure and frequently can handle some additional load. Also, proximity of the physical plant or facility services is most beneficial. Equipment such as large fork lifts, etc. does not require duplication as is required at a remote, a so-called "green field" site. Furthermore, and most importantly, availability of (already existing) utility services will save greatly on cost, both in capital investment and in operational budget. Universities and industries have contracts with utility companies or may even own their own utility supply. Addition of a modest-sized **SR** facility may add very little to the overall cost of the entire complex's bill, whereas supply to an isolated facility requires payment of premium utility cost. Cooling of environment and equipment can be accommodated at an on-campus site; the requirements include supply and return chilled-water pipes and distance to the supply will, of course, determine the cost savings when compared with remote-site situations. However, the age and dependability of the infrastructure should be taken into consideration as well. If cost is not a primary concern, then, of course, having unique control of the systems is certainly ideal. Repair, modification or other impacting activity involving the infrastructure can cause severe scheduling problems when primary control is not in the hands of the facility staff.

Geological features may be among the most important criteria in site selection. An example of the complexity of this issue involves the complete relocation of the **CAMD** facility from on the campus of Louisiana State University (**LSU**) to a location 8 km from the campus.[1] A site on the campus had been selected after extensive vibration and soil testing. The accelerator design group suggested placing the 200 MeV-linac injector in a tunnel beneath the experimental hall and storage ring area. The desirability of this relocation was two-fold; first, it allows a completely clear area around the storage ring for installation of beamlines and second, it provides excellent neutron and gamma-radiation shielding. The available on-campus site was in the ancient bed of the Mississippi River. This area is hydrostatically coupled to the river which is currently approximately 2 km away. This geologic feature, while offering no problem to an on-grade installation, would have caused stability problems to a basement structure. This feature was considered serious enough to cause relocation away from the benefits of existing infrastructure on campus.

Geologic features which are of more general concern involve ambient vibrations and soil stability. Vibration testing should be performed by an engineering firm capable of making and interpreting such tests. Such a firm can also recommend possible solutions to problems and can estimate the final vibrational motion a facility might experience after following the recommendations. After abandoning the on-campus site, **CAMD** was allowed to conditionally accept the donation of a 6 hectare (15 acres) tract of land located approximately in the center of a larger (ca. 200 hectares) vacant area located in the heart of Baton Rouge. The conditions for acceptance were the results of geologic testing. Figure 15.2. provides locations of points where vibrations were measured and gives plots of the results. Soil testing of the new site revealed excellent over-compacted-clay structure[2].

Other criteria need to be considered in site selection. Availability of adequate electrical power is necessary unless an on-site power plant is planned. *Adequate* implies more than simply available power. It also refers to the quality of the supply. This will be covered more thoroughly in the section on utilities. The surrounding neighborhood may be a factor in selecting a site. For example, if the potential site is in or near a residential area, public forums to educate the public about the nature of the facility are important to begin a long-term good relationship. Remember, terms such as *high-energy accelerator*, *x-ray source*, *radiation*, etc. may be alarming to the general public. Because the **SR** facility is moving into the neighborhood, it is the responsibility of the facility management to be open to questions and concerns of the public. "Town meetings" can provide an excellent forum for the dialogue with concerned neighbors.

15.4. Choosing the Architectural and Engineering (AE) Firm

The choice of a suitable **AE** firm is one of the earliest keys to success. As for any major building project, it is advisable to have at least part of this team be local. **CAMD** was designed by a local firm of architects who employed a large non-local engineering firm having experience in the construction of microcircuit and synchrotron facilities. Furthermore, they worked closely with the vibration-analysis specialists and coordinated efforts among the mechanical, electrical, vibrational and soil engineering teams. Of course, the owner of the facility does not passively sit-by during the design and construction period. The **AE** firm expects input to guide it during design and construction of this "unconventional" conventional facility. It is beneficial to travel with representatives of the

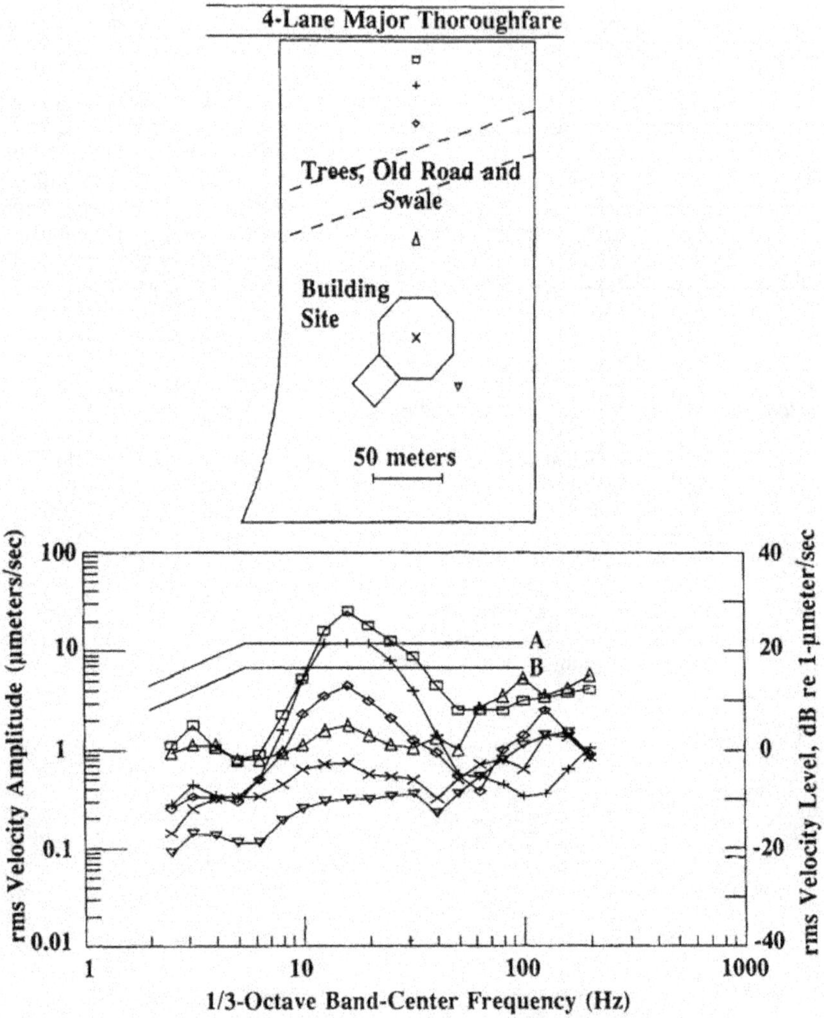

Figure 15.2. CAMD site map indicating pre-construction vibration measurement points and plots of corresponding measurements (reference 2). Vibration-standard curves, **A** and **B**, were established by vibration specialists based on beamline length and tolerable beam motion, also on motion criteria for fabrication and metrology of integrated circuits. Curve **A**, 1μm peak-peak displacement at low frequency and 12.5 μm/sec rms velocity at higher frequencies. Curve **B**; 0.5**A**.

firm and others involved in the design process to **SR** facilities in operation (or under construction, if possible). This not only provides insight to the designers but it also gives members of the firm and owners of the facility a chance to develop better communication skills between one another. Also, personnel at the visited facilities can be identified for future reference when consulting is needed.

While on the subject of *consultants*, it should be mentioned that the hiring, as consulting engineer(s), of one or several excellent facility engineers that are at **SR** facilities around the world or have retired from such a position can be a very important step. This is especially true if the staff of the facility under design does not include such an experienced engineer. Consultants can be used directly with the **AE** firm or can be used during design reviews to help in acceptance by the owner of the various components of design.

In summary, the most successful projects are a testimony to successful interactions with good **AE** firms. Experience of the team is important; however, this does not mean that the entire team must be experienced in designing **SR** facilities. There needs to be experience in the areas of engineering for vibration control and isolation, utility distribution, mechanical stability and temperature control of the environment.

Finally, close interactions among the **AE** firm, the contractor and the builder of the storage ring are imperative during all phases of the project. If the storage-ring construction is done by the facility staff, then this interaction should be natural. However, if the ring is built by an outside source, such as a commercial vendor, then the responsibility lies with the **SR**-facility staff to ensure effective communication and integration. Scheduling periodic meetings involving representatives of all parties is highly recommended.

15.5. Design Parameters

Design parameters of major importance for the conventional facility of a synchrotron radiation center are:

Function of the buildings
Future modifications and expansion
Time for construction
Date of occupancy
Appearance
Cost for construction
Lifetime of the buildings
Cost for operation and maintenance

The primary functions of the buildings are (1) to house the accelerators, the beamlines with experimental stations and specialized facilities such as x-ray-lithography exposure tools and (2) to provide space for laboratories, shops, preparation areas, staff and user offices and conference and meeting rooms. The results for an optimized facility can be rather different as demonstrated in Figures 15.3.-15.5.

Figure 15.3. shows IBM's Advanced Lithography Facility (**ALF**), East Fishkill, NY.[3] Its centerpiece is a very compact, only 5 m long super conducting storage ring. The light source and entire facility are optimized for one specific application, namely, x-ray lithography. The lower left part of the approximately 5000 square-meter building is used for x-ray lithography related research activities, the upper right sector is occupied by large clean room areas housing steppers and processing equipment. **ALF** was constructed as an addition to the Advanced Semiconductor Technology Center (**ASTC**), which is IBM's new pilot line development center for semiconductor devices. Climate and vibration control for **ALF** are equivalent to those of the **ASTC**. In particular, **ALF**'s floor slab is isolated from the rest of the building structure to comply with the specification of less than 0.13 μm vibrational displacement. Another special feature of **ALF** is the massive accelerator-radiation shielding that is the result of IBM's specification to operate **ALF** as a non-radiation facility.

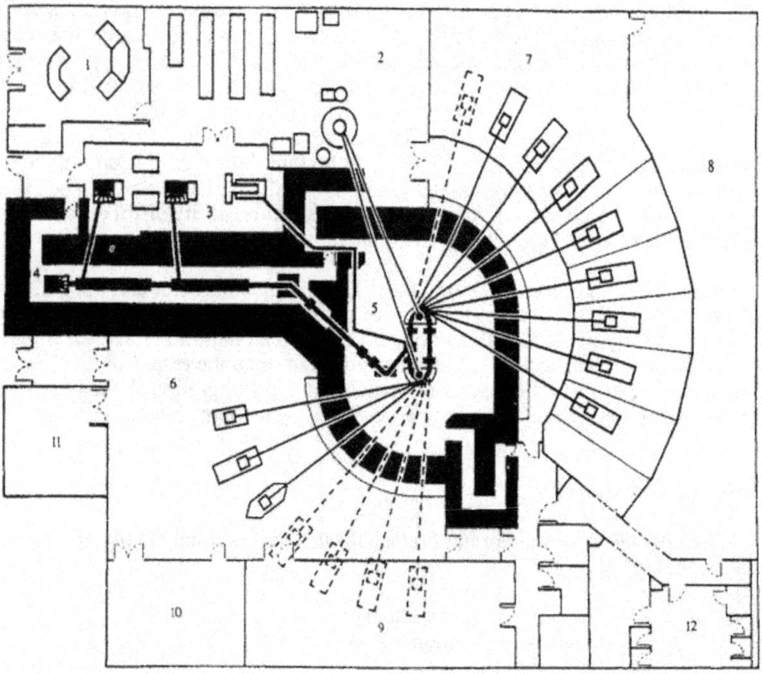

Figure 15.3. IBM's Advanced Lithography Facility (ALF) optimized for x-ray lithography applications in East Fishkill, NY, USA. The 700 MeV super conducting storage ring in the center is approximately 5 m long. The layout of the ALF building is as follows: 1, control room; 2, plant room housing power supplies and cryogenic equipment; 3, rf equipment room housing klystrons for linac and TV transmitter; 4, linac room; 5, storage ring room; 6, research area; 7, process area (class 1000); 8, tool core (class 10000); 9, laboratory 1; 10, laboratory 2; 11, laboratory 3; 12, change area and lockers. (Copyright 1993 by IBM, reprinted with permission; courtesy J. Leavey, IBM Corp.)

The number of third-generation light sources for the production of hard x-rays has grown to a respectable size. The Advanced Photon Source (**APS**) is under construction at Argonne National Laboratory and the building layout is shown in Figure 15.4. The electron storage ring has a circumference of approximately 1104 m; the principal building is an annular metal structure with an outer circumference of 1244 m and a width of 27 m. It houses the electron storage ring which is enclosed in a concrete tunnel and the beamlines with experimental end-stations. Several laboratory/office modules are distributed around the outer perimeter of the experimental hall. They provide offices, laboratories, a conference area, and service support space within walking distance of the respective experimental stations at this very large synchrotron radiation facility. Inside the large annular building are additional metal buildings for the linac injector and booster synchrotron. These accelerators are shielded by concrete as well. A central laboratory and office building can house 300 staff and includes laboratories, library, meeting facilities, remote control room, assembly areas for experiments, stock room, machine shop, truck airlock, clean rooms, and a mechanical room for air conditioning and service utility equipment. There are additional services buildings for power supplies, radio-frequency equipment and electrical substations. A utility building contains the mechanical and electrical support equipment for the **APS**. Housing and transportation, including roads and parking for many hundreds of staff and users require an adequate infrastructure.

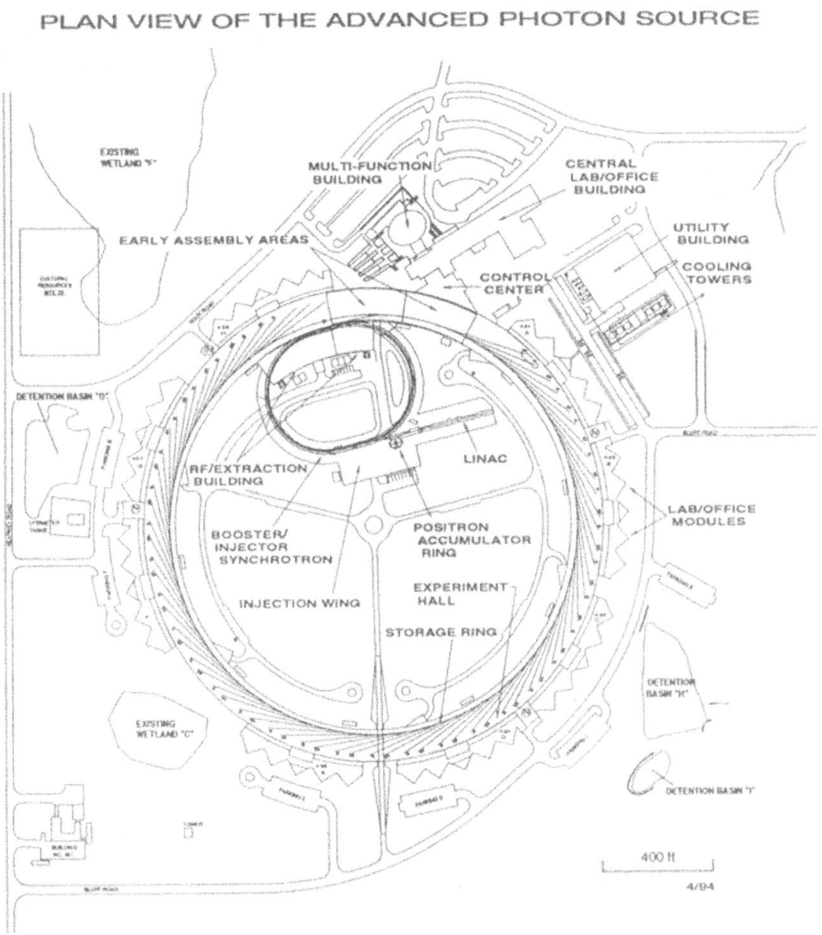

Figure 15.4. The Advanced Photon Source (APS) under construction at Argonne National Laboratories, Argonne, IL., USA. The doughnut-shaped building houses the 7 GeV storage ring and experimental hall. External buildings are the project center and laboratory-office buildings for visiting experimenters. The linear accelerator and booster synchrotron are in the infield. (courtesy R.Fenner, APS)

Some sites present special geological and other challenges to the planners of a conventional facility for a synchrotron light source. The Advanced Light Source (**ALS**) now in operation at Lawrence Berkeley Laboratory in Berkeley, CA[5] is such a project. The task was to incorporate a national landmark, the famous dome built for the E. O. Lawrence 184-Inch Cyclotron, into the new building. The result is shown in Figure 15.5. In some areas, such as California and Japan, the potential for earthquakes requires special construction and installation procedures including securing of equipment. Severe weather conditions including humidity, dust, temperature extremes, rain, ice and snow can best be overcome by separating sensitive experimental areas from the outside world by airlocks.

Figure 15.5. The Advanced Light Source (ALS) at Lawrence Berkeley Laboratory, Berkeley, CA, USA operates a third generation, 1.5 GeV electron storage ring. The inner section under the historical dome (1900 m^2 area) houses a linear accelerator and a booster synchrotron.; the annular section (5700 m^2 area) shelters the storage ring and the experimental areas. (courtesy B. Kincaid, ALS)

15.6. Structural and Vibrational Considerations

The floor for the accelerators and connected beamlines needs to be very stable. In this context, both long- and short-term variations as well as higher- frequency motions, i.e. vibrations, must be considered. The performance of the storage ring depends critically on the alignment and position stability of its components (see Chapter 11), in particular, the position of the quadrupole magnets. Some of the third-generation light sources employ gradient dipole magnets with stringent requirements on alignment and stability. The source of light in the accelerator and in any specific beamline must be treated as one unit: The source must be stable with respect to the beamline and, furthermore, significant displacement of essential beamline components such as slits, mirrors, gratings or crystals can be detrimental. At many facilities beam position control and feedback systems are employed (see Chapter 13) to keep the source stable in space and angle. However, their bandwidths in amplitudes and frequencies are limited. A stable floor and rigid support structures are essential for reliable operation.

If the conventional facility entails building a structure for housing beamlines which are to be attached to an existing large, particle-physics accelerator for parasitic operation or for converting such a machine to dedicated service as a light source, the primary concern is the coupling of the new structure to the existing one. High-energy machines are usually surrounded by thick concrete shield walls; therefore, the support floor for beamline installation will, most likely, begin just outside of this wall. The two most critical factors that need be considered are long-term stability of the floor and isolation of vibration sources such as on-site mechanical equipment, roads and other, off-site, sources. Long-term stability involves differential settling of the new facility with respect to that of the original facility. **SSRL** at **SLAC** (Stanford, CA), **CESR** at **CHESS** (Ithaca, NY), **HASYLAB** at **DESY** (Hamburg) or **VEPP-2m** and **VEPP-3** (Novosibirsk, Russia) serve as models for adding synchrotron radiation facilities to existing large electron/positron accelerators.

The dimensions of the light sources for synchrotron facilities vary from several meters to greater than one-kilometer circumference. This has, certainly, consequences on the methods by which the floors are constructed. For small machines, a monolithic floor slab is technically feasible and desirable, for larger rings, such a slab would crack during curing. Therefore, other strategies have to be employed such as pouring the slab sequentially in segments with locking mechanisms between the segments, e.g., steel dowels and key and channel structures at the joints. The performance of such a floor can be very close to that of a monolithic slab if properly engineered and constructed.

After construction, the floor will settle over a long time.[6] Furthermore, the installation of heavy equipment such as accelerator magnets and shielding will cause settling. Alignment and re-alignment of the accelerator components and beamlines require a grid of survey monuments on or in the floor. These marks must be well protected from transport activities on this floor but should remain easily accessible. Survey relies on optical equipment such as theodolites; therefore, lines of sight between the monuments and equipment to be surveyed must not be obstructed. Surveys are performed to determine the positions of equipment with respect to the monuments. Re-alignment occurs when the survey shows movement of the equipment that has to be positioned within certain tolerances. Surveys of storage rings are mandatory on a regular basis, typically every 6 months to several years, depending on the particular geological features and construction techniques. Such re-alignments correct for long-term movements of the floor.

There are, however, instability problems with shorter time scales, shorter than a couple of months. They can affect seriously the performance of the storage ring and the beamlines. An example of such a floor instability with a characteristic time scale of hours is shown in Figure 15.6. The roof of the Photon Factory (**PF**) in Tsukuba, Japan expands and contracts with changing temperature of the roof structure.[7] These changes in dimensions tilt the side walls and couple into the floor which is rigidly connected to these walls. The floor buckles and displaces the locus of the stored positron beam with respect to the beamlines. Consequently, the beam position depends on meteorological changes, the outside temperature, the time of day and the position of the sun. The **PF** management decided to alleviate this problem by thermally insulating the outside of the roof rather than the inside as it is done usually. However, one important lesson can be learned from the **PF** experience; *Separate the floor or that part of the floor that supports accelerators and beamlines from any structure that potentially can move or vibrate.*

Meteorological changes distort the PF building.

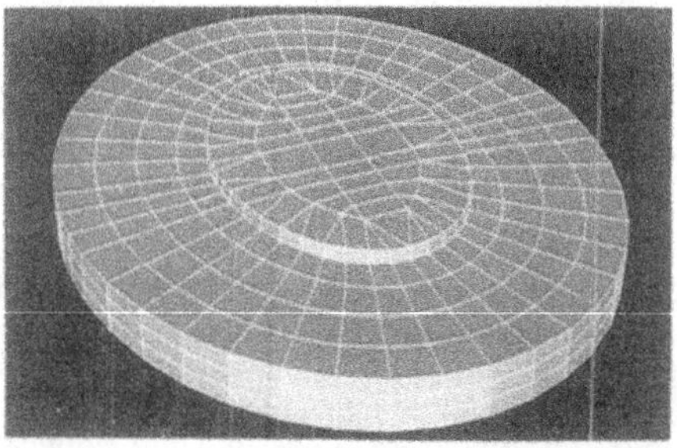

Undistorted

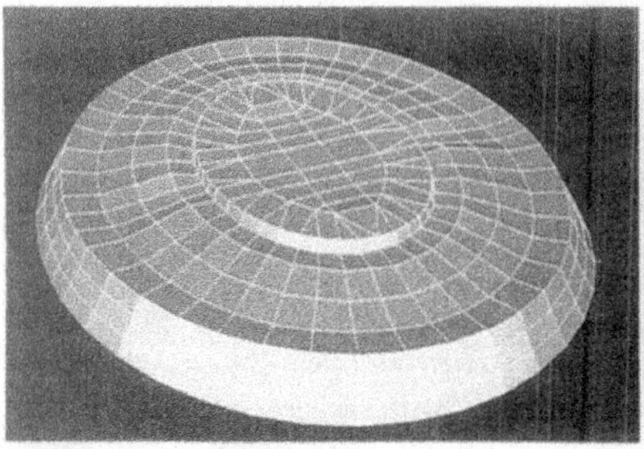

6:00 A.M. Contracted

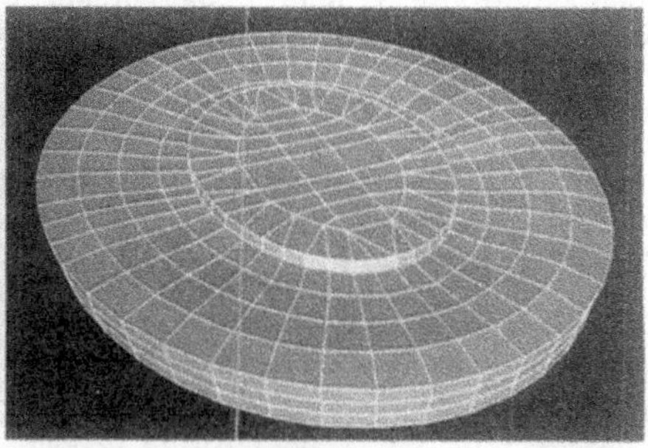

3:00 P.M. Expanded

Figure 15.6. Meteorological changes distort the Photon Factory (PF), Tsukuba, Japan, building. When the temperature changes, the PF buildings expands and contracts. This distortion changes the locus of the orbiting particles with respect to the beamlines. In the three computer simulations the distortions are magnified by approximately 1000 times. (with permission of H. Kobayakawa, PF/KEK)

Vibrations and their control are major concerns at any synchrotron facility. Vibrating equipment, such as compressors, chillers, water pumps and others, must be specified properly with their eigenfrequencies given. *In general, rotating equipment such as, e.g., compressors are better than reciprocating devices. The mechanical equipment should be located on a stable floor separated from the sensitive experimental-hall slab. The mounting of vibrating equipment on support structures such as concrete blocks of sufficient mass with vibration attenuation material such as rubber mats, springs with the appropriate spring constant and other approaches require careful engineering. All pipes connecting the storage ring or the beamlines with vibrating equipment should be de-coupled by flexible bellows and be supported with flexible hangers, e.g., springs. Air flow in the air ducts and water flow in the water pipes should be regulated to avoid resonant vibrations.*

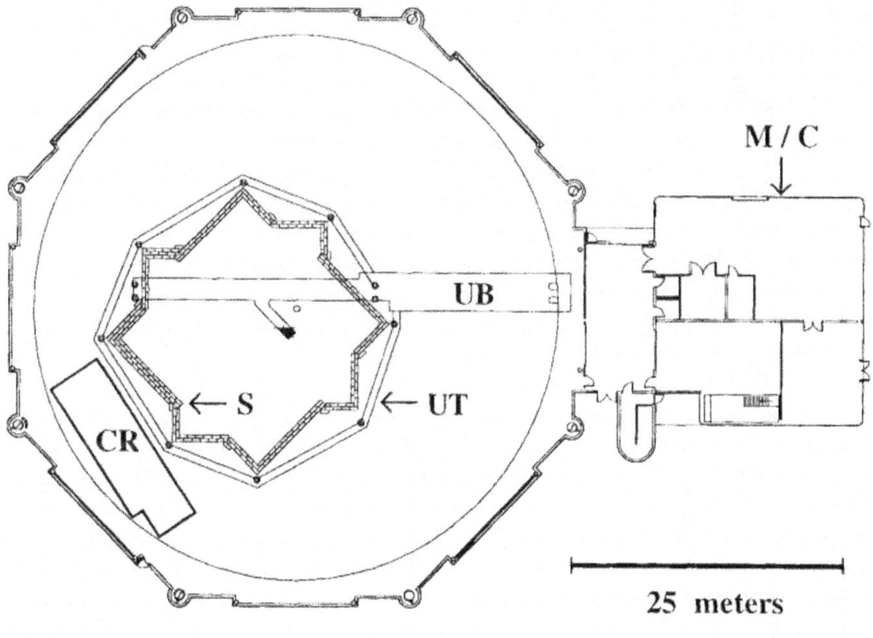

Figure 15.7 Layout of CAMD at Louisiana State University, LA, USA. The 1.5 GeV electron storage ring is housed inside the ratchet-shaped shield wall. An off-center installation was chosen to accommodate a few very long (28m) beamlines. **UB** and **UT** denote utility bridge and truss, respectively. Storage ring, shield wall (**S**) and beamlines are supported by the circular vibration isolated floor which has a diameter of 50 m. The rectangular building (**M/C**) houses the principal mechanical and control systems. **CR** denotes the 100-square-meter clean room necessary for x-ray lithography procedures.

CAMD might serve as an example for the implementation of these concepts, as shown in Figure 15.7. The 50-m-diameter floor of the experiment hall supports, on its central part. the very heavy storage-ring magnets and radiation shielding; at the perimeter only relatively light-weight structures are installed. These requirements were met by pouring a central, 25-m-square by 46-cm-thick slab to support the storage ring (eight bending magnets have mass of 11 metric tons apiece) and concrete shield wall (total mass of over 150 tons). The four slabs poured around this thick slab are tapered from 46-cm thick next to the "central" slab to 30-cm thick at the perimeter. Because of the nature of the soil, over-compacted clay, and terrain, very flat and fifteen feet above the 100-year flood plane, this large slab was poured on grade. This large disk is actually an annulus with an ca. 2-m-diameter hole in the center to accommodate a central roof support. The roof and wall support are independent of the monolithic slab. The supports consist of sixteen piling foundations and caps around the perimeter of the experiment hall and a single piling foundation and cap in the center. An integral part of the perimeter support is a concrete apron that extends to the edge of the monolithic disk. A vibration damping, elastic sealing material forms a vibration isolating and water-tight joint between the monolithic slab and the perimeter and central roof/wall supports. Similar seals form joints between the monolithic slab and piers for utility-bridge and truss support and duct banks transporting power through the slab. Sewage pipe penetrations into the hall all are through non-critical areas. The two-story mechanical/receiving/control part of the facility is built along more conventional lines, having a piling-supported foundation.

The goals of such a construction are separation of the critical foundation of the storage ring and beamlines (i.e., the experimental-hall floor) and all other structures to as great a degree as possible. This is done to isolate the ring and beamlines from mechanical instabilities such as those caused by thermal fluctuations and mechanical stresses commonly experienced by a large building. Furthermore, such construction isolates the critical area from vibrations produced by machinery in the mechanical area and by wind against the outside walls of the building. To enhance this isolation, all machinery is mounted on spring supports and pipes and air ducts are installed with vibration isolating joints between the experiment hall and the outside world. Connection of these pipes and ducts to holders is done with special vibration-isolating connectors. The vibrational spectra measured in the hall with all mechanical devices in operation[8] are given in Figure 15.8.

Even the most sophisticated anti-vibration measures will result only in attenuation of vibrational amplitudes and not in their complete elimination. The question is, consequently, what is the tolerable range for amplitudes and frequencies? The answer depends on the light source under consideration, in particular, its emittance and its most critical planned application. For lateral motions one would recommend maximum amplitudes not to exceed 1/10th of critical dimensions; these could be the dimensions of the synchrotron light source or of a slit in a monochromator. Beam dimensions in a third-generation storage ring have small standard deviation of approximately $\sigma = 50$ µm; slits and apertures in beamlines can be as narrow as 5 µm. This leads to specifications of less than 0.5 to 1.0 µm for vibration amplitudes. In specific applications, such as x-ray microscopy, very small foci of the order of 10 nm are produced and a sample is scanned through this focus. However, such a unit consisting of the imaging elements and the sample is operated on a common base plate and, consequently requires extreme stability only on and within this plate and not for the floor supporting it. Other parameters may yield somewhat different specifications. In the cases of **ALF** and **CAMD,** both operating less demanding, medium-emittance storage rings, the vibration criteria come from industrial standards for steppers printing microcircuits with critical feature size of 0.25 µm.

Not only lateral but also angular stability have to be considered. As an example, a typical Bragg-crystal has a Darwin-width of several arc seconds. Consequently, the required angular stability between light source and crystal is of the order of 1 µrad. Undulators in third generation rings emit very narrow beams with a typical divergence of tens of µrad. Again, this requires an angular stability in the µrad range. To what extent floor vibrations can cause angular instabilities must be investigated case by case.

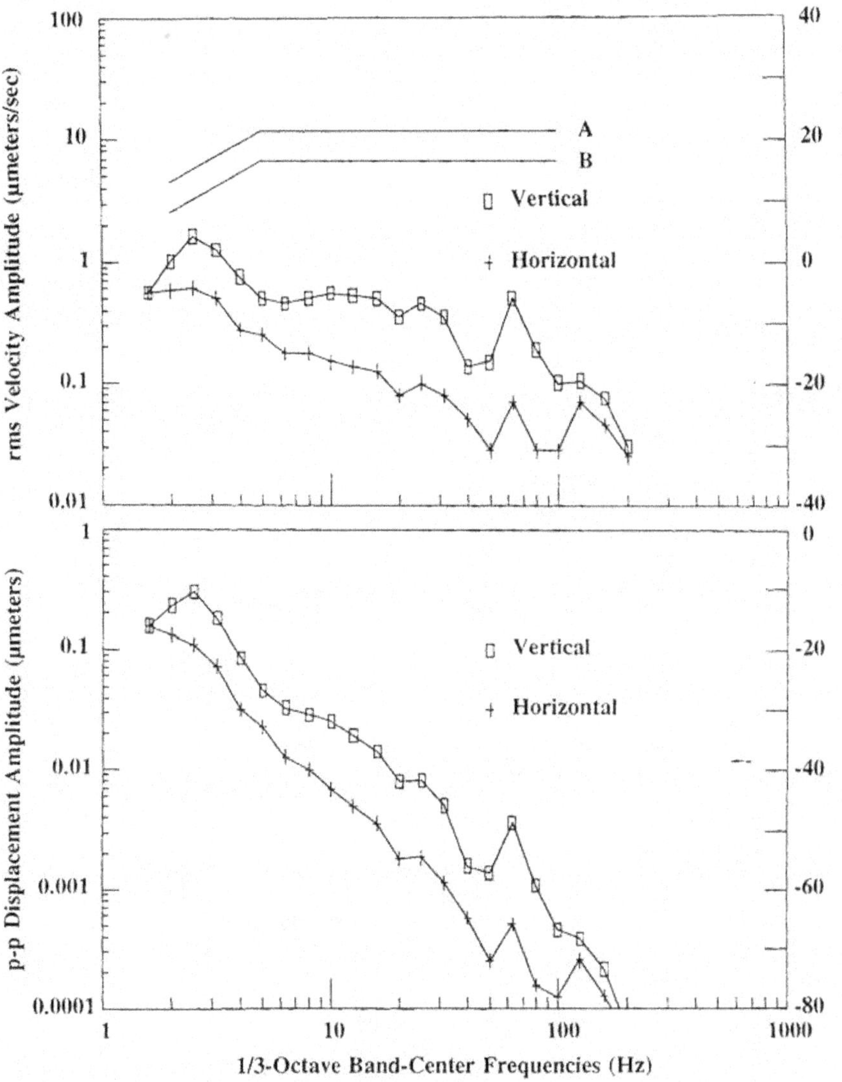

Figure 15.8. Plots of CAMD post-construction vibration measurements. See Figure 15.2. caption for explanation of standard curves **A** and **B**. Values given are averages over experimental hall and are separated into vertical and horizontal components. Data are taken from reference 8. The peak-peak displacement d is related to the rms velocity v by: $d=\sqrt{2}\cdot v(\mathrm{rms})/\pi\cdot f$ at frequency f.

Several synchrotron-radiation facilities have experienced serious problems with floor design, construction, and performance. This is not a basic problem and the required floor stability is easily achievable; however, the site must be suitable and engineering and construction must be performed with necessary care. In any case, the users of a future facility should pay close attention to floor stability and vibration issues. In summary, many factors affect the mechanical stability of a facility. The soil or substructure is, of course, of primary importance. Solid rock geology may provide excellent long term stability but attention should be given to potential vibration coupling through the rock. On the other hand, clay may provide good vibration damping from external sources but attention must be given to long-term stability such as settling. The importance of having, on the team, knowledgeable engineers and architects to design the correct structure for the existing geology is imperative. Once the floor and support is planned, attention must be given to isolation of sources of medium and short-period motions. Again, good engineering is required.

15.6.1.Hutches, clean rooms and other enclosures built on the experimental hall floor

Often it is necessary to build small buildings on the floor inside the experimental hall. Two important concerns merit attention:

- If structure requires new utility runs, make certain vibrations are isolated.
- If air conditioning is required, make certain that vibration sources are isolated properly and that the structure itself does not introduce mechanical stress on the floor, on equipment or on beamlines.

The largest installation of this kind at **CAMD** is a 100 m^2, class-100 clean room for x-ray lithography. The structure is actually a building, 6 m high. The beamlines penetrate the wall through special interface structures. The walls and roof are panels designed for clean-room construction. The air conditioning required consists of high-velocity passage of air through HEPA filters (*high-efficiency particulate air* filters) and temperature and humidity control. Chilled water from the existing refrigeration units linked with small electrical heaters accomplishes the temperature and humidity control. This control is made simpler because of the fact that the hall itself has excellent control of temperature and moderate control of humidity. The filters, fans, fan-coil units and heaters are contained in seven-foot tall rooms atop the clean room. The equipment is built and mounted in such a way to minimize vibration. Also, de-coupling joints between the structure itself and the experimental hall floor and the beamlines are utilized. The mechanical equipment of the clean room was in full operation when the vibration spectra were measured (Figure 15.8.).

15.6.2.Movement of heavy equipment within the facility

Constant change is the mode of operation at an **SR** facility. Beamlines are built or modified, new experiment stations must be installed and the accelerators themselves may undergo changes, such as addition of insertion devices to the storage ring. A modest-sized (2-3 ton lifting capacity) electric-powered forklift will find very frequent use. Certainly, there will be items exceeding the load limit; however, when this occurs, renting or borrowing the necessary forklift is superior to having the inconvenience associated with using the larger machine in limited space. Once equipment is installed in the experimental hall, there will be places where the forklift access is restricted. Many facilities have installed a bridge crane that runs overhead and accesses the entire area of the experiment-hall floor.

The crane reflects the radial symmetry of the experiment hall at **CAMD**. It pivots 360° around the central roof-support column and is attached to a track around the perimeter. The block and tackle move the full length of the ca. 27 m long radial arm. The crane is rigged for lifting two tons; however, the structure is capable handling loads of five tons. A second convenience of the crane is that the top of the radial arm has been fitted as a walkway giving workers access to the space-frame-structured roof support (i.e., ceiling) for purposes of inspection, cleaning and lighting-fixture replacement and repair. The entire motion in the polar grid can be controlled from a portable radio transmitter. The crane is capable of moving all but the heaviest equipment (dipole bending magnets and power supply) and can be used to transport items over the seven-foot-tall shield wall as well as to function as a "sky hook" to hold heavy beamline assemblies in place as they are being installed. Movement of the very heavy magnets, if necessary, will require disassembly of part of the shield wall. This involves use of the crane as the wall is modular, consisting of 2.2 m tall by 30 cm thick by one of three widths (30, 60, 90 cm) concrete blocks. These blocks have eye bolt hooks on the top to enable lifting with the crane.

A program of scheduled maintenance and inspection should be instituted by the facility. There probably exist laws regarding periodic inspection and certification of lifting equipment, particularly overhead cranes. Regardless of law which may or may not exist, safety considerations of lifting equipment should be taken very seriously. Using outside, expert inspection agencies is advised.

15.7. Utilities; Quality, Distribution and Access

Utilities are taken to mean both the externally supplied and refined and internally self-contained utilities. For clarity of presentation, this subject is divided into two sections; *storage-ring utility supply* and *beamline utility supply.*

Storage-Ring Utility Supply. One of the most important reasons for maintaining close interactions among the facility-design/construction teams and the storage-ring builder is to ensure that adequate and easily accessible utilities are made available for operation of the storage ring. For a super-conducting ring, electrical supply of sufficient power to operate the cryogenic refrigerators and the *rf* system is the primary concern. In the case of a warm-magnet storage ring, electrical power in the range of 0.5 to 1 MVA is required for a moderately sized ring such as the **CAMD**, **Aladdin** or **NSLS-VUV** rings to 20 to 50 MVA for large facilities with many-GeV rings such as the **ESRF**, **APS** or **SPring-8** rings. Of critical concern should be the power supplied to the bending-magnet voltage source. The loss of a few cycles in the AC utility supply to the facility can cause a loss of stored beam. It is , therefore, important to consult the electrical utility company to determine if their normal service meets the standards required by the magnet power supply for successful, uninterrupted service. Of course, power failures often cannot be helped; however, there may be service nearby that is more dependable than that which is available at an existing sub-station.

For example, the local utility company offered three potential supplies for **CAMD**; a standard 13.8 kV residential substation which could supply the facility as the sole customer on a loop, the 69 kV supply to the substation and the primary 235 kV lines of the intra-city grid. Each of these sources was increasingly more dependable in terms of both major outages and short (few cycle) interrupts. They are, however, each increasingly more expensive to connect because of the need for more transformers and/or longer distances to connection points. In practice, the 13.8 kV supply turned out to be of sufficient reliability with, by far, less than one-interrupt-per-month caused by power fluctuations or outages.

A second concern is isolation of large switching-type DC power supplies from the outside source and, possibly, back into other supplies to the facility. **CAMD**'s SCR power supply operates daily at approximately 750 kW. The only filter between the supply and the outside world is the 750 kVA step-down transformer that drops the 13.8 kV line voltage to 440 V 3-phase. No deleterious effects have been observed either by the power company or into the other (4 MVA) service for the facility. The DC power supply (located inside the storage ring area) is connected to the transformer and associated breakers through duct banks buried beneath the experiment-hall floor. Once inside the ring, power wires as well as signal cables are distributed with cable trays mounted ca. 2-m above the floor. This leaves the floor free of obstructions and provides safety of electrical insulation from water leaks.

A second utility required of all storage rings is cooling water for the *rf* system and magnets. The requirements for the magnet cooling are de-ionized water with a quality of typically 1-MΩcm pumped in such quantity that the temperature rise is a few tenths of a degree with a heat load of 500 kW or greater. It is best if the cooling water temperature is maintained at near ambient room temperature to obviate the need for insulation and also to remain above the dew point in the experiment hall. The requirements are met at **CAMD** through the use of a heat-exchanger, the primary side of which is supplied by chilled water from a 1000-ton capacity refrigeration system and the secondary side is an ultra-pure (continuously filtered and de-ionized) water closed loop which directly supplies the magnets. Similar, but smaller, systems are used to cool the *rf* system (ca. 40°C ethylene-glycol coolant) and the linac injector.

Beamline Utility Supply. Often, the major nuisances (and sometimes safety hazard as well) at an **SR** facility are the lack of sufficient convenient electrical power, the presence of wires and pipes on the floor of the working area and the necessity of installing extensive systems (long runs) to handle new utilities, if unavailable. Ideally, an **SR** facility would have every conceivable utility available at any beamline when needed. The most practicable realization of such an ideal situation is to have all "common" utilities available near each beamline. To determine what are "common," a poll of a sizable sampling of potential users and/or consultation with facility engineers of other **SR** centers is strongly advised. A table of those available at the **CAMD** beamlines is given below:

electric power	three 110 v (AC) legs, 40 kW total, evenly distributed each phase
	100 kW service for bakeout, 1 service per approximately 4 beamlines, portable distribution cabinet to transport service near beamline to be baked
room-temperature supply and return water loops	nominal supply of 40 l/minute, maintained at 4 bar pressure differential between loops
chilled-water loops	nominal supply of 20 l/minute, parameters maintained at 5°C and 4 bar pressure differential between supply and return
air	dry, oil-free air supplied at ca. 8 bar for operating pneumatic devices and other uses
nitrogen gas	dry, oil-free nitrogen supplied at ca. 8 bar for purging vacuum systems and for diluting reactive gases

helium gas	pure helium supplied at ca. 8 bar for protecting Be windows and for minimizing absorption of x-rays in experiments
exhaust manifolds	as required by fire code; one for flammable gases and one for oxidizers and nonflammable gases
communications	telephone, computer network and closed-circuit television networks and other communication from the control room (radiation/safety- interlock) are available to each beamline

The actual distribution of all beamline utilities at **CAMD** involves a large steel truss and bridge construction. The pipes and wires enter the experiment hall from the mechanical area and are supported on the underside of the bridge structure. They extend out to the annular truss which has a diameter slightly larger than that of the 2.2-m-tall concrete shield that surrounds the storage ring. The underside of the truss has 3-m clearance from the floor and its cross-section is a 1.2-m square. The beamline power services originate at transformers located on the top-side of the bridge (easily serviceable). The communication services run in rf-shielded conduits. The chilled-water supplies and returns are thermally insulated. The gas and water supplies are equipped with large valves above the area of each of the eight bending magnets and will supply approximately four beamlines apiece. The valves obviate the need for plumbing work on the primary supply pipes when connecting individual beamline services. Cable trays are erected to run above each beamline (at approximately 2.2-m above the floor) as standard part of beamline installation. The tray installation allows power, gases and water to be transported to a convenient location (usually the terminal end of the beamline) and there connected to distribution/control centers mounted in cabinets. A second level of distribution in the tray consists of power, water and gas service to individual elements (valves, pumps, shutters, filters, monochromator and mirror drives, detectors, etc.) as well as control signals along the beamline. Finally, the remaining communication lines are included; closed-circuit television for dissemination of facility information, computer network and telephone.

15.8. Safety and Security

Safety hazards at a synchrotron radiation facility result from the presence of ionizing radiation, electrical and mechanical installations, chemical and, sometimes, biological materials. Radiation-safety issues are discussed in this book in Chapter 16. A large number of electrical power supplies is required for the operation of the accelerators and beamlines. Many are operated at high voltage and/or high current. Electrocution accidents and electrical fires must be primary concerns at any synchrotron- radiation facility. All equipment must be installed and operated in compliance with the applicable codes; however, these codes can be very complicated and the extent of their applicability to specific situations is sometimes questioned, especially for short-time experimental set-ups. Such attitudes can result in very dangerous situations such as "installing" high-voltage power supplies for ion pumps on wooden chairs, a real example. Some very simple installation requirements, policies and rules should be mandatory at any facility:

Crash buttons to switch off the electrical power of the entire facility or part of it should be installed at strategic locations (inside the accelerators, at the end of each beamline, in the control room). Additional electrical crash buttons to interrupt beamline power should be located at the end of each beamline

Proper grounding of all electrical equipment and all conducting hardware such as electronic racks, beamlines, support structures, metal radiation safety hutches, etc.

Interlock switches and systems for potentially dangerous power supplies.

Proper cable runs with high-voltage and high-power cables well separated from others and from water lines and pipes. Penetrations through walls should be sealed to isolate and reduce fire hazard.

Clear administrative procedures must be established and enforced for maintenance and repair, including use of warning tags, padlocks at the breakers, etc.

Access to potentially dangerous mechanical equipment such as cranes and forklifts. must be limited to well trained personnel.

Handling of chemical, biological and radioactive materials must comply with codes and administrative policies. As an important part of the approval process of any project, the user or staff member should declare what materials he/she intends to bring to the facility. The safety officer will evaluate admissibility of the material and, in case it is approved, decide how it will be handled. Storage of hazardous materials requires designated fire-resistant storage cabinets with controlled access. The duct pressure for fume hoods and for exhausts of hazardous-material pumps must be monitored to alarm the operators in the event of over-pressure or fan-failure. In general, the amount of hazardous materials within the facility should be minimized.

For very hazardous materials, such as some of the gases used in semiconductor processing, very severe safety precautions are mandatory, including requiring special containers for the entire equipment operated at under-pressure and continuously monitored by sensors. Proper installation and use of scrubbers on exhaust systems may be required. The experimentalists should be aware of the fact that it may take several months to prepare a facility for use of a particular hazardous material. In any case, the safety officer should have the authority to reject any experiment that he/she deems to be "too dangerous". Such an authority must be based on an adequate organizational structure of the facility; the safety officer reports to the director or a unit outside the **SR** facility. The specific organizational structure usually has to comply with government regulations. Reviews, inspections and site visits by independent experts are highly recommended and are probably mandatory at most facilities.

There may be scientific reasons to study very dangerous materials, e.g., radioactive samples that are not enclosed in a safe container or extremely poisonous gases. Such materials can be handled in highly specialized laboratories but generally should not be used at a synchrotron-radiation center which is a public facility having many people working in the same laboratory. If there is a necessity to perform experiments on high-safety-risk samples, a dedicated beamline is mandatory and needs to be spatially separated from the common experimental hall,[9] if risks cannot be reduced to an acceptable level by other means.

Pressurized gas cylinders that are in continuous use should be mounted outside the building, whenever this is possible and feasible, and connected with appropriate pipes to the point of use. Cylinders used only for a short time can be connected directly to the experimental station but must be secured to prevent falling, e.g., in an earthquake. The connection of the cylinders should be approved by the safety officer and may not be modified without his/her approval. Similar procedures apply to the use of exhaust lines for hazardous materials. Pressurized gas cylinders must be transported with safety caps and adequate equipment and finally tied rigidly to a wall or other structure to prevent falling.

Fires ignited by a burning power supply, by overheated cables, or by chemicals are a serious danger. A sufficient number of fire extinguishers for electrical and chemical fires must be distributed throughout the facility. Emergency exits are required as well. Adequate training of staff and users in safety systems and rules is imperative. Smoke detectors at strategic locations in the facility and in the air-ducts have proven to be very useful. The alarm triggered by these detectors should be connected to a fire station or another safety office that can react properly on such an alarm at any time, even when the facility is vacant. Water sprinkler systems are installed in some facilities. They are undoubtedly useful in regular offices and some laboratory areas; however, their applicability in the accelerator tunnel or the experimental hall can be questionable. There, the most likely fire is an electrical fire and to spray water onto such a fire seems to be not an optimum solution. Furthermore, the ceiling height in these facilities can be as high as 10 m and, consequently, the sprinklers are very far from the source of fire. Codes and regulations may require the installation of a water sprinkler system; however, care should be taken to protect personnel from the combination of water and high voltage. Most of the material in a **SR** facility is concrete and steel. Fire hazards can be significantly reduced by preventive measures such as prohibiting staff and users from bringing inflammable materials into the facility including wooden furniture, wooden and cardboard boxes and by installing cables with fire resistant insulation where applicable.

Safety is governed by laws, codes and administrative procedures, however, the attitude of the people working at a facility, their willingness to improve safety, and their common sense are equally important to ensure safety. It is one of the most important tasks for the management to make people safety conscious, to encourage them to participate in safety issues rather than to literally bury safety in bulky documents only.

Security systems prohibit unauthorized individuals from access to the facility. Besides having standard intrusion alarm systems to prevent burglary or vandalism, special access systems allow only researchers, staff and authorized visitors to enter the experimental areas. The most elegant solution is to issue magnetic access cards. This allows not only controlled access but also permits documentation of all entries.

15.9. Construction Oversight by Owner

In certain cases, particularly when the owner of the facility is a state or federal government or one of its agencies, the scientists and engineers actually involved in the building process are restricted in their access to the construction site, the **AE** firm and/or the building contractor. Remember, there may be good reason for this restriction; often the final occupant, although knowledgeable about the facility, can cause undue conflict and confusion around a construction site. One may be certain that there is an owner representative with direct access to the process. It is best to determine who this is and to establish good relations with her or him. The first step in establishing these relations is to admit ignorance of the process (which is usually true) and to ask for instructions in how best to behave. This will be a great help in all phases of the building project; from **AE** selection through design and development, construction and final acceptance. It is far better to explain the needs of the facility in non-architectural terms to an owner representative in private than to attempt to do such in front of a number of people who might include the contractor and several of his sub-contractors. It is in the best interest of everyone involved in the project that the building be "perfect" in the eyes of the final user and at the same time meet all building codes, design and structure codes, safety codes, etc., and be constructed within the time of the contract established between owner and builder. While, if permitted, it is very useful to talk to individual workers and sub-contractors, one must, however, know the "chain-of-command" if a mistake or needed change is perceived. Frequent meetings, usually held at the construction site after building commences, are common. If the final occupant group has shown responsibility and not become preoccupied with unreasonable or "nit-picking" demands, a representative of this group probably will be invited to attend these meetings. If such meetings are not held, it is best to suggest they be. There are complex features in the design and structure which must be given proper attention.

15.10.Occupation and Operation

Before final acceptance of the building housing the storage ring, it is imperative that all systems be operating properly. This, of course, is true for any construction project; however, in the case of an **SR** facility, the systems include low-vibration operation. As the limits of vibration should have been explicitly stated in the building contract, acceptance should require that they be met. This may require hiring the engineering firm doing the original measurements to perform post-construction measurements. The measurements should be made under "normal-operating" conditions and under "extreme-operating" conditions. Also, having vibration spectra made when the building is as quiet as possible will allow one to determine if the structure has damped or amplified ambient conditions.

Cost for utilities for lighting, cooling, heating and dehumidifying the facility can be estimated by the **AE** firm; however, to obtain realistic estimates, utility requirements of equipment must be known. Furthermore, one must realize that a large percentage of the electrical power consumed by the equipment will be converted to heat which must be considered in calculating the cooling requirements of the facility. Failure of important components is of concern not only for accelerator systems but also for the conventional facility. Back-up systems for chillers are highly desirable. Emergency generators to provide electrical power are a necessity to run critical systems such as over-pressure fans for clean rooms and exhaust systems, for sump pumps and for emergency lighting. A diesel powered generator providing approximately 150 kW is an adequate solution for a smaller **SR** facility.

Most countries require a review before operation of the **SR** facility commences. For the DOE facilities in the US, this is the *Operational Readiness Review (ORR)*. The emphasis of such a review is on environment, safety and health. Proper documentation during design and construction facilitates this review process. Reliable and safe operation of a **SR** facility depends to a large extent on the attitude of staff and users. A dedicated facility manager is a key person to ensure proper operation of the facility with regard to personnel and equipment safety, the overall integrity and appearance of the facility including cleanliness, and support of the experiments. He or she will also be responsible for regular inspection and maintenance of potentially dangerous systems such as lifting equipment (e.g. cranes and forklifts), for electrical installations and distributions, and other systems.

References

1. B.C. Craft, V. Saile, J.D. Scott and E. Morikawa, Nucl. Instr. and Meth. B56/57(1991)379.

2. H. Amick and S.K. Bui, "Site Vibration Evaluation, Center for Advanced Microstructures and Devices, Jefferson Highway Site, Louisiana State University", Report No.32, February 1990, Acentech Inc., 21120 Vanowen Street, Canoga Park, CA 91303

3. J. A. Leavey and L. G. Lesoine, *IBM J. Research and Development* **37** (1993) 385.

4. Technical information on the APS conventional facilities is summarized in: "Environmental Assessment of the proposed 7-GeV Advanced Photon Source", prepared by Oak Ridge National Laboratory, Oak Ridge, TN 37832-6285, for the US Department of Energy, DOE/EA-0389, October 1989; see also: D.M. Mills, Nucl. Instr. and Meth. A319(1992)33.

5. A.L. Robinson and A.S. Schlachter, Nucl. Instr. and Meth. A319(1992)40.

6. To give a typical number: For BESSY II in Berlin, Germany, a ground settlement of approximately 0.3 mm/100 m/year; from "The BESSY II Parameter List", No. 1, 6 December 1993.

7. T. Katsura and Y. Fujita, Rev Sci Instrum. **62**(1991)2550; T. Katsura et al., Rev Sci. Instrum. **63**(1992)530; Photon Factory Activity Report 1988, No.6, National Laboratory for High Energy Physics, Tsukuba, Japan, 1989

8. H. Amick and S.K. Bui, "Post-Construction Evaluation of Vibration and Noise, Center for Advanced Microstructures and Devices, Louisiana State University", Report No.64, July 1991, Acentech Inc., 21120 Vanowen Street, Canoga Park, CA 91303

9. Photon Factory Activity Report 1992, No.10, National Laboratory for High Energy Physics, Tsukuba, Japan, 1993, I-2.

CHAPTER 16: SAFETY

THOMAS DICKINSON
National Synchrotron Light Source
Brookhaven National Laboratory
Building 725D
Upton, NY 11973

16.1 Introduction

At a research facility, or any other enterprise for that matter, safety is part of the way of doing things, and not a separate activity. Like research integrity, scientific discipline, and fiscal responsibility, safety is a product of culture and sound management.

Most of the elements of the safety program at a synchrotron radiation source will be the same as at other industrial and research facilities, and many of the safety practices and programs developed elsewhere will serve well. There are differences, however, and much of the discussion in this chapter will be on these differences and their effects on safety programs.

16.1.1 Safety Management

Facilities and equipment are important to safety: fire protection, proper exits, guards on machinery, and suitable tools should be provided so people can work without risking injury. However, the essential part of any safety program is in getting people to act safely. This involves communication and training in the recognition of hazards and in how to deal with them, in the assignment of responsibility for the safety of activities and processes, and in the development of the attitudes which lead to a safe operation.

Effective safety programs act through the line of management and supervision, and safety should be defined as a line management responsibility. Employees usually get most of their information, motivation, and inspiration about their work from their supervisor, and safety should be part of this stream. But a large part of the population of a synchrotron radiation source is not employees, it is visiting users. Not only are these visitors not subject to local employer supervision, they are accustomed, and even required, to behave with independence of thought and action. In addition, many beamlines are not "owned" by the synchrotron radiation source but are funded and managed by outside teams of researchers. In the absence of a conventional line of management which covers the players in the research activities, a structure must be created to carry out these functions. There will be roles in this structure for beamline management and resident staff, for the light source management and operations staff, and for visiting researchers. The objective is to provide a coherent way of accomplishing the assignment of responsibility for safety and proper operation, and to insure that training, information and resources are available to the users.

16.1.2 Built In Safety Facilities

One of the strengths of a synchrotron radiation source is its versatility, the wide variety of experiments which can be carried out there. This means that the experimental area will probably not have some of the specialized facilities often available for research: single pass ventilation and the specialized plumbing of the chemistry laboratory, for example. The means to handle hazards must be available for those experiments that need them, but in many cases these can be ad hoc rather than built in. However, there are cases where an effective research program can be carried out only if there are custom-built facilities. Examples include work involving transuranic isotopes, high level biohazards, and semiconductor fabrication. Feasibility experiments can be carried out on a limited basis with these materials using ad hoc precautions, but the restrictions and dangers become too great for continuing programs. The design and construction of specialized research facilities is also discussed in chapter 15.

16.1.3 Research Operations

Research laboratories have accommodated visiting scientists for many years, but synchrotron radiation sources bring a new scale to this arrangement. The number of visiting users at a light source is usually very large - much larger than the number of resident staff. There are many experiments, most of short duration, and many experiments going on simultaneously. Together these factors make research activities at synchrotron light sources qualitatively different from those at other facilities such as high energy physics accelerators. The challenge is to accommodate the transient users and their varied research into a safe and efficient operations program. Much of the effort will be in providing a suitable infrastructure so that visiting users receive safety orientation and training and the technical and physical support for effective research. Day to day operations must include comprehensive safety review of experiments so that potential hazards are identified, and continuous monitoring of research activities to insure that shielding and other safety provisions remain effective. Technical resources must be available so that experimental hazards can be addressed and problems with equipment can be solved. Procedures and controls must be in place so that experimental activity can be tracked and safety requirements enforced.

Research operations comprise this collection of activities and procedures, and should be distinguished from core or machine operations, which is the business of running the storage rings and injector. Machine operations at particle accelerators is a well defined discipline, using techniques developed during many years of experience at accelerator facilities and reactors. An example is the U.S. Department of Energy (DOE) order 5480.19 "Conduct of Operations Requirements for DOE Facilities". [1] This lore does not adequately define research operations, however, and new techniques and strategies are needed. These are illustrated by example in the following pages.

16.1.4 Laws, Codes, and Government Regulation

Laws and regulations vary from one jurisdiction to another, but they are always a consideration when building and operating a research facility. During the planning stages it is important to have the services of staff or consultants who are experienced in preparing for reviews of safety features and environmental impacts. These reviews can extend over months or years, and involve their own formats and cadences. A missed step along the way can lead to frustrating delays.

The body of safety and environmental regulations which apply to the facility can provide the framework for the safety program and can be quite helpful in that role. In the ideal case, these regulations are the result of national experience and wisdom in these subjects. Often however, the regulations are generalized from experience with rather different facilities, and are overly burdensome or miss the mark in other ways. In addition, while the regulations may dwell on issues which are irrelevant to the light source, they may also fail to address real safety concerns. A creative way to respond is to analyze the regulations for the underlying safety concerns and to develop an effective response to each of those concerns. The objective is not for relief from burdensome requirements but rather to create a safety program which has credibility and responds effectively to real hazards. A program developed in this way will almost certainly be more effective than one which merely "goes by the book." There is no choice but to comply with all legal requirements, and facility practice must be complete in this respect. A responsible program will focus on all the safety issues and respond to all of them.

16.1.5 Ordinary Safety Issues

Fire safety provisions, electrical installations, guards on machinery and similar matters are common to industrial and research establishments, and apply to a synchrotron radiation source as well. These need to be fitted to the layout and intended use of the laboratory, but are not extraordinary. There is further discussion of fire protection and electrical installations in chapter 15. The operational aspects of these matters do require careful tailoring to the facility because of the unique mix of population and activities. Fire drills and other training, oversight of electrical safety, enforcement of safe housekeeping, and the like all must be adapted to be effective.

Hazards from natural phenomena include earthquakes, violent storms, and floods. In areas of high seismic activity, continuous attention to earthquake safety is required, with standing committees to review design and to monitor housekeeping and other practices necessary to minimize earthquake dangers. Facilities subject to tornadoes and hurricanes should have suitable building design and have emergency response plans in place. Floods are inevitable, whether from storms, failed drainage systems, or cooling water spills. Electrical connections and outlets should be above floor level, and storage and handling of materials hazardous to the environment should be done in a way to avoid environmental contamination in the event of a flood.

16.1.6 Radiation Protection

Compared to other radiation producing facilities, a synchrotron radiation source is relatively low hazard because high energy electrons are not very efficient in producing radioactive materials. Some components of the injector will become active enough to require restrictions on handling and disposal, and the positron converter, if one is used, will become quite radioactive. These problems are localized, and in general, operation and maintenance of a synchrotron light source is not complicated by residual activity. For all practical purposes the radiation stops when the machines are shut off. For this reason, light sources operated under DOE regulations qualify as low hazard, non-nuclear facilities.

There are three kinds of ionizing radiation which must be dealt with at a synchrotron radiation source: bremsstrahlung radiation, consisting of high energy photons or gamma rays produced when the electron beam encounters matter; neutrons which are produced in turn when the bremsstrahlung interacts with shielding; and synchrotron x-rays in the experimental stations and elsewhere along the beamlines and electron orbit. The ultraviolet synchrotron radiation is normally confined within the vacuum enclosure and does not present a personnel hazard.

The first line of protection against ionizing radiation is shielding: stacks of lead bricks and the concrete walls of the storage ring and injector enclosures confine the bremsstrahlung and neutron radiation, and beam pipes and hutches shield the x-rays. Access control interlocks play an important role, and radiation monitoring and configuration control round out the protection scheme.

16.1.7 Interlocks

Interlocks are suitable where a danger is not obvious in itself, and access to the hazard area is frequent and routine. Access to particle accelerators is usually controlled by interlocks, and for the x-ray experimental hutches at a synchrotron radiation source, interlock protection is even more critical. Hutches are often entered several times an hour to deal with the experiment, then secured and returned to beam-on status. The only practical arrangement is for the experimenter to perform these access and secure cycles and it is particularly important that the interlocks be reliable and not subject to user error. The use of an interlock system for protection carries with it the responsibility for competent design, proper maintenance, and periodic testing, and this can be a considerable administrative and technical burden. One of the problems is that people become dependent on interlocks and may become lax in awareness of danger or in attention to proper procedure. In such cases an unreliable interlock will actually increase the danger.

16.2 Safety Management

16.2.1 Safety Analysis

Management of the safety program for a synchrotron radiation source should begin while the facility is still in the planning stage. A safety analysis document can be the vehicle for shielding design and other safety matters which have a bearing on the overall technical design of the facility. Environmental impacts are also addressed at this stage and the safety document can provide a major part of the environmental assessment. The projected operating parameters should include a "safety envelope" which places limits on critical factors used in the safety review such as beam current and energy. Since the design of shielding and other features depends on these factors, the safety envelope defines operational boundaries which should not be crossed without additional safety review. In practice, the main machine parameters are determined by technical and economic considerations, and it is important to have independent criteria which trigger a safety review in the event these requirements or limitations change in the future. The most useful resource in developing the safety analysis will be similar documents generated during the design and construction of other synchrotron radiation sources.

16.2.2 Organizational Structure

The organizational structure of the safety effort should also be addressed in the safety analysis. The backbone of this organization is line management: communication, direction and responsibility for safety are carried along with all of the management functions of the organization. There will be specialized parts of the safety effort as well: safety reviews, training, inspections, and response to emergencies for example. These will be performed by standing committees, by the safety staff, and by assignment as part of the jobs of other staff members. The safety staff and committees will have strong interactions with the design-engineering and operations groups of the facility, but reporting should be to management to avoid a conflict of interest with productivity goals. A common arrangement is for the safety committees and staff to serve as advisors to the facility director, who must ultimately make the decisions on acceptance of the risks associated with systems and operations.

The safety effort should evolve along with the facility as a whole: design of the core facility with its interlock systems and other safety features will be replaced with design and reviews of beamlines and facility modifications; the efforts to organize and insure safe construction will be turned to operations of the machines and the research program.

16.2.3 Safety and Design Reviews

The components and systems of the facility should be reviewed for technical soundness as well as safety, and this review process is usually the purview of standing committees. Safety should be considered as part of the overall design process rather than as a subject to be approached separately. Otherwise, safety provisions will be "tacked on" and are likely to be more complicated and restrictive than if they were included from the beginning. Much of the safety and efficiency of operating systems depends on how people interact with them, and this must be considered early in the design. A safety review will be required of the facility as a whole before it goes into routine operation. The United States DOE calls this an Operational Readiness Review (ORR) and it involves careful scrutiny of design and construction records, test results, and operations and safety organizations as well as physical inspection of the facility. Formal reviews will probably also be required during the commissioning process.

There should be formal requirements for review and documentation of safety systems, such as interlocks, where reliability is a matter of great importance. Much of this activity falls under the discipline of Quality Assurance (QA) and a formal QA program can handle much of the organization and documentation required for the technical and safety reviews.

16.3 Built in Safety Facilities

Often the first thoughts of the user, or the traditional industrial hygienist is of massive fixed apparatus for handling hazardous materials: fume hoods and exhaust systems, isolation rooms, and the like. In most cases, it is worth the effort to try to arrive at smaller, more flexible solutions for containing and managing the hazards. This approach keeps all of the research opportunities at the facility open, including the beamlines with specialized instruments and optics, not just those with the safety apparatus. The exercise of developing the safeguards and the procedures for dealing with difficult materials leads to an awareness of how to avoid danger that probably is more important to safety than the engineered aspects of the protection system.

However, there are circumstances where specialized safety facilities must be provided. Materials which require absolute containment such as transuranics or high level biohazards are one example. Medical research brings with it a whole array of specialized facilities and regulations. Most experiments use small quantities of material, but molecular beam studies may use gases in quantities which require ventilated gas cabinets and robust exhaust systems. Any of these facilities can be put in place after the light source is built, but often it will be more convenient and economical to provide for them during construction. In any case a substantial amount of planning ahead is needed. Simple toxic exhaust systems are needed frequently for experiments and there should be some provision for these made during construction. Mounting platforms for blowers and roof penetrations for ducts are much less expensive to install during building construction than to add later, and if these are in place then blowers and duct work can be easily added when and where needed.

16.3.1 Experimental Support Labs

For most experiments the chemical work is done during sample preparation rather than during the actual experiment, and laboratories must be provided for this activity. Fume hoods, sinks, benches, reagent storage and the like are needed. The problems here are much like those which afflict student and user machine shops: failure to clean up, abandoned materials, and conflicting or incompatible activities. Effective supervision is needed, a responsible person who provides scheduling, oversight, and enforcement. Often this role is best filled by a research support person, a technician or associate, rather than by a scientist. In addition to general preparation labs, other support facilities may be needed including biological, photographic, optical, crystal preparation, and clean room facilities. Location and access are not just important for time critical operations, convenience is also a factor in encouraging compliance with good safety practice.

16.3.2 Setup and Storage Space

Setup and storage space are also very important to a safe and efficient research operation. These are needed because of the nature of the experimental cycle at most beam lines, which requires that the experiment and the research team change every week or two. Often major changes of equipment occur as well: vacuum chambers, detectors, and refrigerators are moved into the experimental station. The problem is that much of this equipment requires set-up time and space. In the absence of dedicated staging areas this set-up activity will occur in the vicinity of the beamline, where another experiment is in progress. There is apt to be overcrowding and incompatible activity, causing inconvenience, reduced efficiency, frayed tempers, and compromised safety. Related requirements include storage space for equipment not currently in use, and for packing and shipping materials. If these spaces are not available the problems are particularly frustrating, since compliance with good safety practice means significant compromise of the research program. To avoid unsafe overcrowding, setup cannot start until the previous experiment is done. To prevent the accumulation of combustible packing material in the experimental hall, expensive crates must be discarded, then built anew when the equipment is shipped again. Since these are not rational options, there is pressure to compromise safety so the research can proceed.

Figure 1 shows the storage rings and experimental floors of the National Synchrotron Light Source. Most of the space at the top and at the lower right of the figure is dedicated to preparation and support labs, setup space, and storage for experimental equipment.

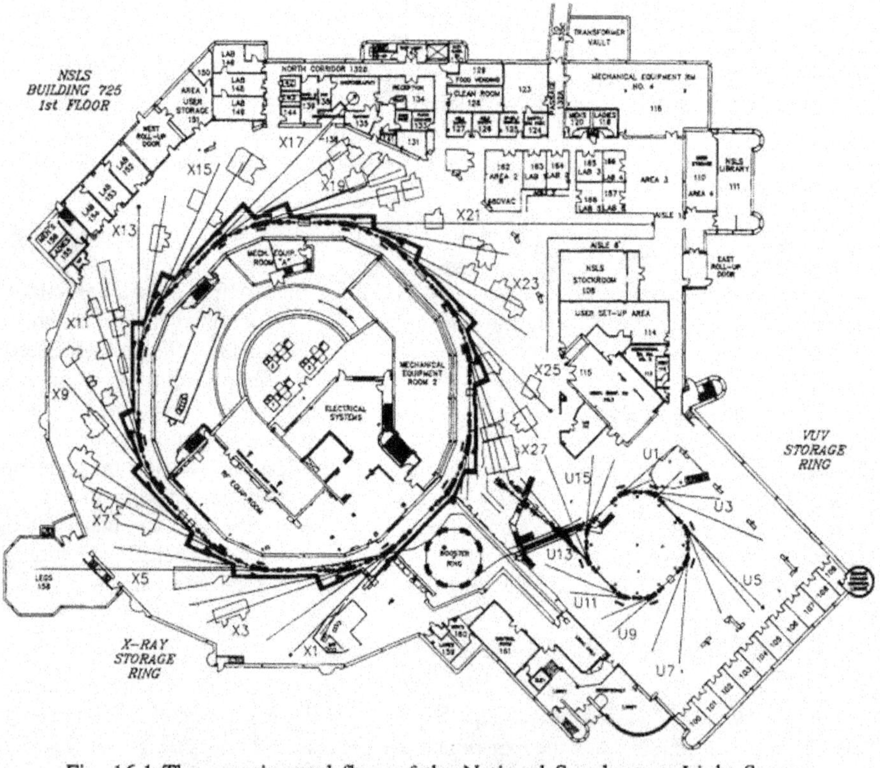

Fig. 16.1 The experimental floor of the National Synchrotron Light Source.

16.4 Research Operations

The elements of research operations at a synchrotron radiation source include the safety review of experiments, safety orientation and training of visiting users, oversight of experimental activities, configuration control of radiation protection and other safety features, and technical support of research activities. Staff must be available to perform these functions, and problems of scale influence the way in which many of these matters are approached. With many users and experiments to keep track of at once, records and information retrieval take on more importance than they do where a small number are involved. The large scale user facility is a recent development, and in the following discussion the research operation at the National Synchrotron Light Source (NSLS) at Brookhaven National Laboratory is used to illustrate the issues involved and some ways of dealing with them. The objective is to present an effective description of what goes on at an active user facility.

16.4.1 Parameters of the NSLS Affecting Research Operations

The NSLS has two storage rings, at 750 MeV and 2.5 GeV, providing radiation in the vacuum ultraviolet and x-ray regions. At this writing in early 1994 there are 91 experiment stations of which about 80 are active nearly continuously. There are about 3000 registered users performing 1200 experiments per year, making it one of the most active user facilities in the world. The experimental program started in 1983, and research operations has been evolving since that time.

In figure 1, the lines tangent to the storage rings represent the center lines of the beam ports on those rings. Each of these ports may be divided into two or more beamlines which provide radiation to the 91 experiment stations.

16.4.2 Research Operations Staff: Safety Professionals

The staff requirements for research operations include safety professionals, "research operators", and supervisors. Other facility personnel serve important safety roles on standing committees or in technical support. Professional safety training is important for the safety staff. This is worth saying because it is common at research facilities for people to move into jobs from other professions and to make do with on-the-job training. Legal issues, fluency with regulations and the literature, and technical methods are areas where professional safety training is important. Industrial Hygiene is the "general practice" of safety, and is a good background. Expertise in health physics is important, and experience with accelerators and research facilities is helpful. Technical help on some of these issues can be obtained from outside the facility. The NSLS is part of Brookhaven National Laboratory which has an extensive Safety and Environmental Protection (S&EP) Division. This serves as a technical resource in many safety disciplines, and much of the routine health physics service for the NSLS is provided by the S&EP division. On the other hand, experience has shown that the safety review of experiments must be done in-house. The number and pace of experiments and the unique situations at the experimental stations make it very difficult to have these reviews done by an outside agency. In addition to safety of research operations, the professional safety staff is also responsible for the general safety program including fire and life safety, periodic inspections, and related matters.

16.4.3 Research Operations Staff: Research Operators

The staff members who exercise on-the-spot oversight and safety control of experimental activities are known at some facilities as Safety Operators and at the NSLS as Operations Coordinators (OPCOs). For lack of a universal term, we refer to them here as research operators. The required qualifications are difficult to define since there are very few comparable jobs, so examples from the NSLS are included in this discussion. At the NSLS, there are nine OPCOs, providing two around-the-clock covers. Their training and experience includes mechanical, electronic, safety, and machine operation; some were hired directly from school while others came with extensive experience as technicians. This variety of skills and backgrounds has proven to be an advantage since the operations group functions as a resource as well as overseer. The mission of the research operators is different from that of the machine operators; it is important that this distinction be clear to all involved. The machine operator's job has production goals (get more and better beam) and the required technical skills are more defined. On the other hand, the OPCO controls the start-up of each experiment after initial set up at the beamline and oversees the continuing activity, as detailed in section 16.4.7. The procedures and safety requirements are clearly stated, but each experiment is different, and interpretation and judgement is often needed in a given situation. The scientists are highly motivated to get on with their work, and it is important that the OPCOs be self confident and skilled at conflict resolution. The OPCOs have the responsibility and authority to stop any activity where safety or compliance is in question, but having done that they are expected to start the process of resolving the problem. Often this can be done on the spot, once the situation is understood, but sometimes members of the safety staff or others must be called for assistance. To help provide definition of mission and value in their role, a research scientist is included in the line of supervision of the OPCOs. A weekly staff meeting of the OPCOs, their supervisors, and the safety staff has been very valuable, and most of the procedures used in research operations at the NSLS have originated in these meetings, often at the initiative of the OPCOs.

16.4.4 Experiment Safety Review

The safety review of experiments at a synchrotron radiation source is most effectively done separately from the scientific review. This is because safety considerations are primarily an operational concern as opposed to a programmatic one: how an experiment is done rather than why. The information necessary for this review includes the materials to be used (including those used in sample preparation and any other aspect of the experiment), chemical reactions or processes to be undertaken, and any special equipment to be used. It is important that information be provided on all these things rather than just those which the user believes are hazardous. The judgement of hazard must be made by safety personnel at the light source rather than by the user since standards and risks may well be different at the light source than they are at the user's home laboratory.

16.4.5 Control of Experimental Hazards

Experiments which present significant hazards can be approached at three levels. The first is to insure that the experiment is designed to avoid mishap. This may involve sophisticated safety provisions in the apparatus, but often mundane precautions like dressing away dangling cables to avoid snags and falling objects and planning ahead to avoid spills are more important. The second consideration is to minimize the effects of an accident, usually by providing secondary containment. Finally, if a worst case accident could have serious consequences, then there should be contingency plans to deal with those consequences and avoid injury to personnel. If there is a lack of confidence in any of these stages then the experiment should be redesigned or deferred until better facilities are available.

One powerful technique for reducing the risk from a hazardous material is to minimize the quantity which is brought to the experimental floor. For example, a commercially available lecture bottle size cylinder of nitric oxide contains about 30 liters of the gas. This could cause a very dangerous, even life threatening accident if the contents were released on the experimental floor through a broken valve or fitting. The same gas in a 100 milliliter sample cylinder at one atmosphere absolute pressure would present relatively little hazard, and the experiment could be carried out with only simple precautions. Most synchrotron radiation experiments require only very small quantities of sample material, and this feature can be exploited for safety and convenience.

The same example can be used to illustrate another aspect of experimental floor safety. A lecture bottle of nitric oxide may present an acceptably low risk in the experimenter's home chemistry lab because in that situation the airflow is single pass, the experimenter has good control over access to the area, and people nearby are experienced in identifying and responding to chemical odors and hazards. None of these conditions are likely on a light source experimental floor. Similar access control and bystander knowledge considerations also govern the handling of laser beams, high voltage, and other hazards.

Figure 2 shows about one-third of the experimental floor of the NSLS VUV ring and provides an idea of the close proximity of experimental stations. In the foreground, from left to right are the stations for U5, U4B, U4A, and at top center on the platform, U4-IR. In the background and not clearly visible are beamlines U15, U16A, U16B, U1, U2A, U2B, U3A, and U3C. In general, the research programs are completely independent, with one to four workers at each experimental station. Each of these people may be an "innocent bystander" with respect to uncontained hazards at nearby experiments.

16.4.6 The Experiment Safety Approval Form

The Experiment Safety Approval Form shown in figure 3 has been in use at the NSLS for over ten years. This represents the minimum which will serve for an effective review. Experience has shown that in 10 to 20 percent of the cases more information is needed, but this is best obtained by direct contact with the user. An advantage to keeping the form simple and on a single page is that it also can serve as a status and tracking document. About three quarters of the experiments done at the NSLS are arranged

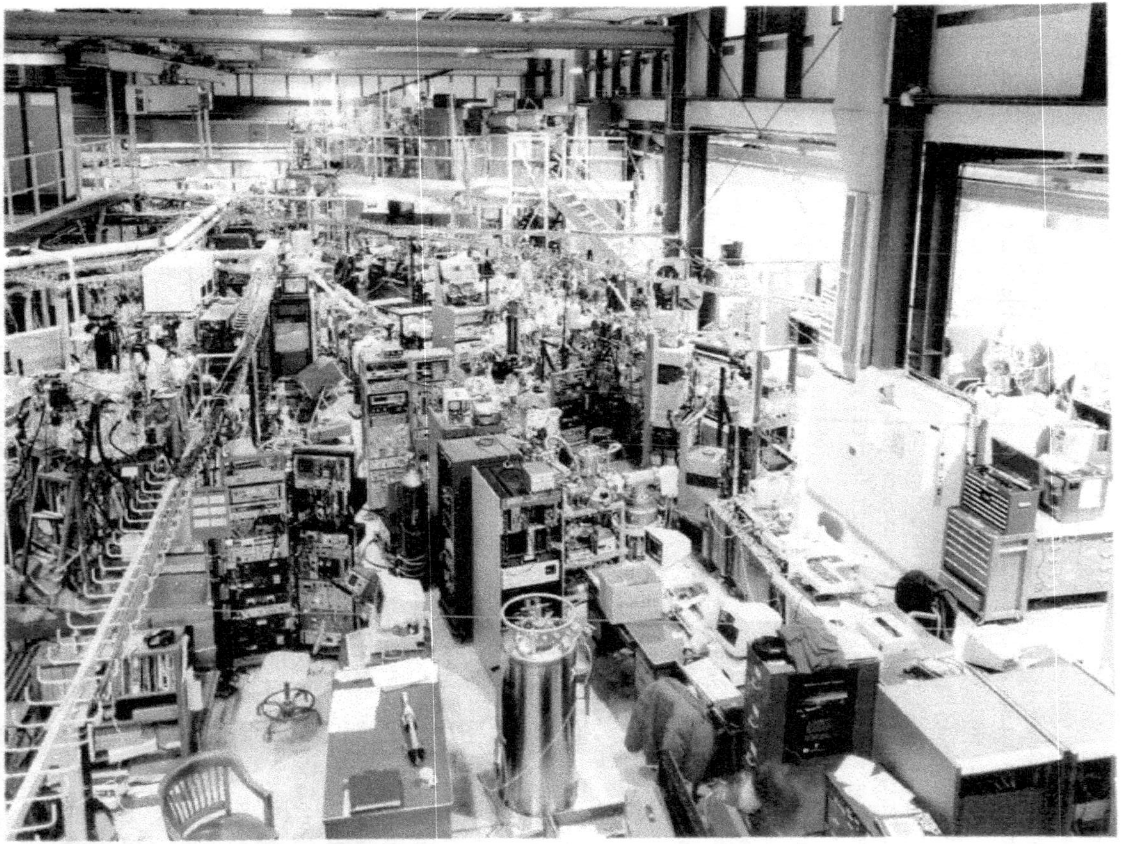

16.2 A portion of the experimental floor of the NSLS Vacuum Ultraviolet Ring

NSLS EXPERIMENTAL SAFETY APPROVAL

NOTE: No experiment will be allowed to run without a properly completed and approved experimental safety approval form. This form must be submitted at least one week prior to commencement of experiment or experiment will not run.

Date Submitted: ______________________

Safety Officer Approval: ______________________ Date

1. Experiment ID# (General Users leave blank)

   ______ beam line ______ year ______ sequence

2. Expected Start Date: ______________________

2A. Actual Start Date: ______________________ (Users leave blank)

3. Expected Duration: ______________________ (must be reviewed every six months)

4. List Experimenters who will be working at the NSLS. Include Beamline personnel.

First and Last Name	Affiliation	Phone Number

5. Experimental Title: ______________________

6. Materials to be used:* ______________________

7. Equipment which will be used in the experiment or is not a permanent part of the beam line:* ______________________

8. User Comments: ______________________

9. Safety Officer's Comments: ______________________

*Note instructions on reverse side pertaining to hazardous materials or equipment which present safety concerns.

BNL F 2671D Rev. 10/91

Fig. 16.3 The form used at the NSLS for experiment status and safety review.

through the Participating Research Teams, and the NSLS does not participate in the scientific review or assignment of beam time. The Experiment Safety Approval form is the only document which identifies these experiments to the NSLS. It serves as a notice of intent and as a starting point for the safety review, it is then posted at the beamline while the experiment is on line, and finally it is the basis for annual activity reports. At the end of an experimental session, the form is retrieved and replaced in a file in the control room, and the beamline is locked out. A safety approval automatically expires after six months, and often an experiment will have several on-line sessions over that time. After expiration the form is placed in an archive file and becomes the record of that experiment.

16.4.7 Oversight of Experimental Activities

Oversight of experimental activities is one of the most important aspects of research operations. This is done by the Operations Coordinators who check that safety approvals and checklists are in order and that shielding and other safety features are properly in place. The OPCO is the first contact for users when they have questions and problems, and is the person who has "real time" knowledge of what is happening on the experimental floor. Many of the procedures which are in use on the experimental floor have been initiated by the OPCOs who feel strongly about their responsibility to stay on top of the activities there. A good example of this is information about unattended operation. Most experiments can be safely left unattended, and at many beamlines controls and instruments are set up so that data can be taken automatically over long periods of time. When this mode of operation first began, the OPCOs found that they were in the dark about the status of unattended experiments, and were uncomfortable about this lack of knowledge and the inability to respond effectively to problems at these beamlines. The users were required to notify the OPCO when they left the beamline, but there was poor compliance, and handling this information was awkward, particularly at OPCO shift change. The answer was to provide a printed notification card which the users fill out with the time of expected return, the place where they can be reached, and any other pertinent information. The card is posted next to the beamline shutter controls, and if an experiment is left with the shutter open for more than an hour without a card posted the OPCO will lock out the beamline. These cards are pink, and a beamline or experiment which is approved for unattended operation is said to have "pink card status".

The Experiment Safety Approval Form is the tracking document for individual experiments at the NSLS, and the details about how the form is handled illustrate some of the dynamics of an ongoing operation. After the safety review, the approval is signed in red, to distinguish the original from copies. The form then goes into a file in the control room, ready for the start of the experiment. The sequence number near the top left of the form is there only to assist in retrieving the form from this file. When the experimenter is ready to start, the OPCO brings the form to the beamline, and if the safety checklists (described in section 16.4.8) confirm that the line is ready for beam, the form is posted and the beamline is unlocked or "put on line."

The posted form defines the experiment in progress from an operations point of view: the personnel, the materials, and the activity. A person must be listed on the form in order to run the experiment unsupervised, and if there is need to add someone to the list this may be done by the OPCO only if the request comes from someone who is already on the list. The NSLS does not attempt to qualify users at particular experiments, but the fact that a name is listed on a form is considered to be an assertion by the users involved that that person is competent to conduct the experiment. Other changes that commonly are requested are the addition of materials and changes in equipment. Both of these require approval from the NSLS Safety Officer or Safety Engineer, and this often can be given by telephone. Over all, the approved safety form can be viewed as an agreement between the users and the NSLS which defines those parameters of the experiment which are important to safety and operations. The way in which these changes to the form are made illustrates a subtle point about where responsibility resides. Judgments about who is qualified to work on a particular experiment must come from the research group, and requests for changes in the roster must come from them. The assessment of hazards on the experimental floor is the responsibility of the Light Source, and any changes in materials, equipment, or processes must be approved by the safety staff. The front line member of the operations staff, the OPCO, is responsible for keeping track of what is happening on the floor, and once an experiment is in progress any changes on the safety form may be made only with the knowledge and approval of the OPCO on duty.

16.4.8 Configuration Control

Much of the OPCOs activity comes under the heading of configuration control, and is vital to safe operation. The safety features of each beamline are determined in reviews conducted before the beamline is constructed and periodically thereafter. Many of these safety features are not evident from casual inspection and a formal configuration control program must be in place. Each beamline has a safety check list which accounts for shielding and other radiation protection features, with photographs posted on location to assist the check. The beam pipes themselves provide protection against synchrotron radiation, and flanges are padlocked to prevent uncontrolled opening. The padlocks on a given beamline are keyed alike, and the padlock key is attached to another key in a lock switch which disables the beamline when it is removed. This provides protection against confusion about which beamline is to be opened, a credible mistake on a crowded experimental floor. These procedures have developed with experience over several years of research operations. In the early days, when few beamlines were in operation, configuration control was mostly informal after the initial beamline safety review. At that time there was a requirement that in-line vacuum valves be locked open under operations control to prevent uncontrolled reconfiguration of the beamline. With the implementation of formal configuration control, there is a high level of confidence in the continuous oversight of beamline safety features, and the locked valve requirement has been dropped.

16.4.9 User Safety Orientation

Safety orientation and training for the visiting users is very important but also difficult to arrange. The visits are often short and at odd times, and the conventional scheduled training sessions are not practical. In many cases, the usual content of training also is not suitable, and in fact the term "safety orientation" is more appropriate. Responsibilities of the visitor, hazard awareness, emergency procedures, access to resources, and a few defined rules provide the main content. Ideally, this orientation should be available on demand so it can be provided immediately on arrival. An effective vehicle is the video presentation, backed up by written material. These must be revised frequently to reflect changes in personnel, conditions, and training requirements; modular organization of the video will greatly facilitate these revisions. The safety orientation will also be more effective if the presentation is concise and to the point: the material in use at the NSLS in early 1994 consists of a 15 minute safety video accompanied by a 20 page "Guide to the Experimental Floor." About a third of the guide is about safety, reinforcing and providing more detail on the material in the video. The remainder covers resources and services available to the user with the idea that the booklet will continue to be useful and will be kept close at hand. After the orientation the user must sign the following "NSLS Safety Orientation Certification":

> Each person who works at the NSLS is responsible for his/her own safety. This includes acting in a prudent and responsible way when dealing with hazards and seeking help when unsure of proper procedures. In addition each person is responsible for insuring that his/her actions do not endanger others, and for reporting unsafe conditions and activities.
>
> Users are responsible for the safe conduct of their experiments, and for providing the necessary knowledge and planning for dealing with hazards or potential accidents connected with their experiments. This requirement is in addition to the general safety responsibilities of all workers at the NSLS.
>
> I certify that I have viewed the Safety Video and have read, understand, and will abide by the safety rules and procedures of the NSLS outlined in the video and the Guide to the NSLS Experimental Floor.
>
> I also certify that I have been assigned an NSLS encoded ID card _____ and that it will not be used by someone other than myself and that it will be returned to the User Administration Office on ________.
>
> Signature_________________________
> Date
>
> Issued By_________________________

In addition to the signature, the user is required to provide name, address, and affiliation. This orientation and certification meets the minimum requirements for unescorted access to the experimental floor of the NSLS, and is available around the clock. Emphasis is on the acceptance of responsibility for adequate knowledge of the safety requirements and agreement to abide by them. Additional orientation is given at the beamline on operations and safety specific to that experimental station. This instruction is the responsibility of the beamline personnel, and completes the minimum required training for a user. Detailed information is provided on only a few subjects, and the emphasis is on awareness, recognition of problems, and access to further information and resources.

At the NSLS, the operations staff can provide safety orientation and radiation badges outside of regular working hours for users who arrive then. About 10% of the visitors arrive for the first time on evenings or weekends, and need this service in order to be able to commence work without delay. Most visitors receive this service through the User Administration Office, and all users must complete their registration there. This includes guest appointments, financial arrangements, personnel records, and identification cards. This information is entered into a data base and is widely used during facility operation.

16.5 Radiation Protection

The design of shielding at a synchrotron radiation source is a major undertaking. This design must be tailored to the particular facility, not only because it depends on the parameters of the storage ring and injector but also because the shielding will have a significant impact on conventional construction and the use of space in the facility. Other aspects of radiation protection such as interlock systems and radiation monitoring also require significant effort. Documentation and records form a larger part of this effort than they do of most other aspects of building a light source. Shielding and safety system design will constitute a major part of the safety analysis document, and up to date documentation and records must be maintained on interlock systems and radiation monitoring. This is a matter of good practice and, in some cases, of regulatory requirement. The discussion that follows is a general overview of radiation safety issues, with emphasis on issues peculiar to synchrotron radiation sources. It should not be considered to be a design guide for any particular facility.

16.5.1 Bremsstrahlung Radiation

Gamma radiation is produced when high energy electrons encounter matter, either molecules of residual gas in the vacuum chamber or the walls of the chamber itself. For electrons at the energies used in synchrotron radiation sources, the dominant loss mechanism is bremsstrahlung radiation, the production of photons. This process involves the same mechanism as the production of synchrotron radiation: force on a moving electron. In this case, instead of the magnetic fields in a storage ring, the force comes from the Coulomb fields of the nuclei of the matter through which the electron is passing. In a close atomic encounter the electron experiences a force enormously greater than that from the storage ring magnets, and in fact can give up much of its energy in a single event, producing a high energy photon. If this occurs in a thick target, this gamma photon goes on to produce an electron-positron pair, which in turn produce more gammas, forming what is known as an electromagnetic shower. This shower builds up, and "shower maximum", the point of greatest intensity of ionizing radiation, is some distance into the material being traversed by the bremsstrahlung radiation. This has practical consequences where bremsstrahlung shielding is not contiguous, and can be a factor in the shielding of beamlines, as discussed below.

The bremsstrahlung radiation is strongly peaked in the forward direction and if the shielding is placed to block tangents to the electron orbit in the storage rings and injector then most of the radiation will be captured. However, radiation at large angles cannot be ignored, and shielding transverse to the beam direction must be provided as well. Synchrotron light also emerges from the storage ring on a tangent to the electron orbit, so that each beamline represents a hole in the bremsstrahlung shielding. Shielding collimators should be placed along the beamline, and the in-line portion terminated with a backstop to confine any bremsstrahlung radiation. The beamlines are not just empty vacuum pipes, they may also contain optical components and other devices. A monochrometer crystal and its copper mounting block, for example, may be enough of an obstruction to bremsstrahlung radiation to cause shower build-up, with a significant amount of radiation emitted at large angles to the beam direction. This is a particular problem at insertion device beamlines where an appreciable amount of steady state gas bremsstrahlung may be produced in the long straight section of the storage ring. Lead shielding may be required around beamline devices on insertion device lines.

Bremsstrahlung radiation is most effectively shielded by materials with high atomic number and density such as lead and tungsten. A twenty to thirty centimeter thickness of lead is normally sufficient to stop a direct bremsstrahlung shower, and about 5 centimeters of lead on the sides will contain the large angle scattering. Tungsten is quite expensive, but has several useful properties. Because of its high density, the required thickness is only about two thirds that of an equivalent lead shield. Tungsten, which is used in the form of a sintered matrix with a small amount of copper and other metals, has good vacuum properties and can be used directly inside the vacuum chamber for beam stops and close-tolerance collimators. It is essentially impossible to melt, unlike lead. A high power density electron or photon beam can readily melt a hole through a lead shield, a very dangerous failure mode.

16.5.2 Neutron Radiation

As bremsstrahlung is absorbed by the shielding, neutrons are produced by two dominant processes. Giant resonance neutrons result from photonuclear excitation, are emitted almost isotropically and have an average energy of about 2 MeV. Electrons with energies above a few hundred MeV will provide a bremsstrahlung spectrum with photons above the pion production threshold and the resulting pions will initiate nuclear cascades resulting in the production of high energy neutrons with energies above 100 MeV. The high energy neutrons are not isotropic, but in many shielding configurations the transverse component is more important.[2] Electron machines are not "power" producers of neutrons, but this radiation is troublesome because neutrons are difficult to shield and quite hard to measure. Neutron shielding should contain a high concentration of hydrogen because the atomic weight is similar to the mass of the neutron. Lead is an ineffective neutron shield because the lead nucleus is so much heavier than the neutron that very little energy is exchanged. Water contains a lot of hydrogen and is cheap and abundant, but is difficult to lay up as shielding. Concrete contains water of hydrolyzation and is the most popular neutron shielding material. A storage ring tunnel will typically have concrete walls one half to one meter thick to provide shielding for neutrons and for scattered bremsstrahlung which misses the lead shielding.

16.5.3 Practical Considerations for Fixed Shielding

The design goals for the fixed shielding around a synchrotron radiation source represent a balance between potential radiation exposures on the one hand, and economic and operational factors on the other. One extreme would be to provide enough shielding so that the continuous loss of the electron beam at any part of the machine, from the injector to the storage ring, would not cause radiation levels of concern at any place outside the enclosure. Except for the smallest machines, this approach is not practical, not only due to the cost, but because the shielding would interfere with access to the technical components of the machine. A reasonable approach is to design the shielding (with a substantial safety factor) so that routine operation keeps radiation levels within continuously acceptable levels, and credible incidents of beam loss do not exceed radiation levels that are acceptable for short periods of time. A common practice is to survey carefully around the machines when they are first brought into operation, and identify points of electron loss and radiation hot spots. Local shielding is then installed as close to the loss point as possible to minimize the volume required. Lead is used for bremsstrahlung and polyethylene is suitable for neutron shielding. Because of the high density of hydrogen atoms, polyethylene is nearly twice as effective as concrete on a volume basis for neutron shielding, and its much lower weight means it can be installed without massive support structures. It is still possible for unexpected hot spots to occur where there is no extra shielding, so tight administrative control of machine operation and a comprehensive radiation monitoring program are necessary.

Much of the concrete shielding will be installed as part of the conventional construction and there should be appropriate supervision and quality control of this work. Contractors are accustomed to arranging imbedded conduits, ductwork, and other voids based on structural requirements, but the effect of this on shielding performance may be significant. Where space is at a premium, high density concrete using iron ore as aggregate can result in saving a third in required shield thickness.

16.5.4 Synchrotron Radiation

The synchrotron x-rays are much less penetrating than the high energy radiation discussed above, and a few millimeters of steel will often provide enough shielding. Much of this shielding will be provided by the beam pipes and the experimental apparatus, perhaps with a wrap of thin lead sheet depending on the x-ray energies involved. A computer program, PHOTON, developed at the NSLS by L.D. Chapman does calculations of source, scattering, and shielding parameters for x-ray beamlines and is useful for beamline shielding design.[3,4] The synchrotron beam contains x-rays at energies much higher than those used in experiments, and this part of the spectrum is what leaks through the shielding. The experience at the NSLS is that even at ten times the critical energy there are enough photons in the beam to warrant concern. A related phenomenon is that the amount of synchrotron radiation that penetrates the shielding will increase dramatically with small increases in storage ring energy. The high energy skirt of the intensity vs photon energy curve is very steep, and if the curve is shifted to the right by a small increase in storage ring energy, the number of high energy photons increases greatly. When the x-ray ring at the NSLS was increased in energy by about 8% in a recent test, the scattered synchrotron radiation coming through small openings in the storage ring shield increased by a factor of six.

The x-ray intensity from synchrotrons is extraordinarily high, typically 100 billion rads per hour in the beam, and the radiation from insertion devices in high energy storage rings can be orders of magnitude higher. Because the user requires repeated safe access to the beam location, and because the experimental equipment is part of the radiation shielding, interlocks and configuration control are particularly important at the beamlines.

16.5.5 Radiation Units

A fundamental unit of ionizing radiation dose is the rad which is a measure of the amount of energy deposited in matter by the production of ions. Different forms of radiation vary in the amount of biological damage caused by a given amount of energy deposited. Neutrons, for example, tend to deposit their energy in a small volume, comparable to the dimensions of a cell, rather than distributed through the tissue as do x-rays and gammas. The organism is less able to repair this concentrated damage, so the effects are more serious. This difference in biological damage is called the quality factor. Gammas and x-rays have a quality factor of one, while neutrons have a quality factor of two to ten depending on the energy. For the neutron spectrum found around electron accelerators the quality factor is assumed to be ten. The radiation unit which takes the quality factor into account is the dose equivalent, measured in rem. Common measuring instruments, for example the ionization chamber, measure rads. When measuring x-rays and gammas, rads on the meter equal rems, while a measurement of one rad of neutrons equals ten rems. For the radiation levels found in occupied areas the millirad and millirem are more convenient units. These units are in common use in the USA, but there will eventually be a conversion to standard international units. The SI unit for dose is the Gray with one mGy = 100 mrad. Dose equivalent is measured in Sieverts, with one mSv = 100 mrem.

16.5.6 Radiation Levels Significant to Personnel

The natural background radiation exposure varies from about 100 mrem to 300 mrem per year, with a few locations being much higher. This comes from cosmic radiation, geological sources, and natural radioactivity in the body. The largest single contribution, on average, comes from radon which varies widely from place to place. In the USA, the maximum allowed exposure for radiation workers is 5000 mrem, although administrative limits require exposure reduction actions before this level is reached. The maximum exposure that a radiation producing facility may cause to a member of the general public is 25 mrem per year, and the limit for non radiation workers in the facility is 100 mrem per year. The 5000 mrem limit for radiation workers is based on keeping the risk from radiation comparable to or below other generally accepted risks in the work place, while the 25 and 100 mrem levels are determined by public health considerations and are far below the level of significant risk for an individual. A general principle in radiation protection is to keep exposures As Low As Reasonably Achievable (ALARA). At most synchrotron radiation sources there should be no reason for elevated exposures other than leaky shielding or inefficient machine operation, and a reasonable design goal is to keep exposures below 100 mrem/year for all workers.

16.5.7 Radiation Monitoring

There are four general approaches to radiation monitoring: area, environmental, survey, and personnel. Each of these contributes to the knowledge and control of radiation at a synchrotron radiation source.

Area monitors are individual stations located around the facility which display radiation levels and alarms and may also provide readout and data recording at a central location. These devices, usually ion chambers, provide real-time information which can be correlated to machine operation and used to make operational decisions. An area monitor typically costs several thousand dollars, a limiting factor on the number deployed.

Environmental monitoring is done with small, inexpensive devices which are placed around the facility and collected and read periodically. Thermoluminescent dosimeters (TLDs) or film badges are commonly used, and these have a minimum readable dose of 5 to 10 mR. This means that they must be left in place at least a month to record radiation levels comparable to natural background. Because of their low cost (about $100 per location) these can used to monitor all parts of the facility, even those where radiation problems are not expected. These devices can also provide neutron information, and are one of the few ways to obtain comprehensive neutron coverage.

Radiation surveys with portable meters provide the only way of getting information which is spatially complete. A leak through the shielding may be between area or environmental monitors and not be registered by them, but can be picked up in a survey. On the other hand, a survey may be made at a time when radiation is not being produced at that location, so careful coordination of surveys with machine operation must be done. Routine surveys are a vital part of a comprehensive radiation protection program.

Personnel monitoring provides bottom line safety information: are people being exposed to radiation? Film badges and TLDs are used and are generally read each month. At a synchrotron radiation source, these readings should be consistent with natural background, and this is about the sensitivity level of badges or TLDs which are read monthly. Recently there has been discussion about the need for visiting users of a synchrotron radiation source to have personnel monitoring. The traditional practice has been to badge anyone who works with a source of radiation, without considering the actual risk of exposure, or the amount of useful information which can be derived from individual monitoring. This tradition may change with experience at user facilities and an increasing level of confidence in operating practices and protection systems.

16.5.8 Radiation Measuring Instruments

The basic instrument for measuring ionizing radiation is the ion chamber. These can provide accurate measurements of dose (rads) over a wide energy range. A hand-held survey meter can easily measure as low as 1/2 mrad per hour, while an area monitor which integrates data over a longer time can see changes in radiation of as little as .01

mrad in an hour. This high sensitivity is important, since in some circumstances radiation levels must be controlled to less than 100 mrad per year which is about .01 mrad per hour. Another advantage of the ion chamber is that, within limits, it is not sensitive to duty factor -- if the radiation comes in short pulses it is still measured accurately. Much of the radiation at a synchrotron radiation source occurs when beam is being injected into the storage ring, and this radiation is generated in very short pulses.

Geiger counters are small, inexpensive, sensitive monitors which are good for detecting radiation but less suitable for measuring it because the sensitivity depends on the energy of the particles making up the radiation. A more serious drawback is that Geiger counters are not tolerant of pulsed radiation, they may indicate levels many times below the actual radiation coming from a pulsed source like a linear accelerator or storage ring injector. A related problem is that at radiation levels well above the range of the counter, the instrument will not only saturate, it may "fold back" and indicate low or zero radiation. In summary, Geiger counters are very useful for detecting radiation leaks of synchrotron x-rays and steady state bremsstrahlung radiation around the storage ring and beamlines, but one must always be aware of their shortcomings.

Neutrons present a special problem because their ability to cause biological damage is large compared to the amount of ionization that they produce, with a quality factor usually taken as ten. This means that in a field of pure neutron radiation, there are 10 millirems for each millirad measured with an ion chamber. The problem is that around an electron accelerator, the radiation is generally an unknown mixture of bremsstrahlung and neutrons and the quality factor is uncertain. In addition, the calibration of TLDs and other neutron measuring devices varies, sometimes strongly, with the neutron spectrum. This spectrum depends not only on the source terms but also on the shielding at a particular location. Measurements of the neutron spectrum should be made, a fairly involved procedure, and with this information correction factors can be derived for neutron sensitive TLDs and other devices. These can then be used for continued monitoring of the neutron radiation. It is also helpful to arrange shielding so that radiation is far enough below critical levels so that precise measurements are not needed.

16.5.9 Configuration Control of Shielding

Much of the shielding at a synchrotron radiation source is movable: the lead bricks and concrete blocks around the storage ring and injector, and the beam pipes, optics enclosures and hutch walls at the beamlines. All of this must be properly in place for radiation protection to be effective. Purely administrative controls are usually sufficient for the shielding inside the enclosures for the electron machines since only a limited number of well trained staff have occasion to work there. On the experimental floor the situation is more difficult for several reasons. The number of components and people involved is much greater, the beamline components are opened or reconfigured frequently, and much of the work is done by those who are not part of the technical cadre at the facility. A formal and rigorously enforced program of configuration control must be in place. Components of this program include interlocks, locks, start-up inspections, checklists, photographs, drawings, and safety system work permits. Oversight of this program by the research operations staff must be essentially continuous, with operator control and verification of any beamline changes which could affect radiation safety. The NSLS beamline configuration control program is outlined in section 16.4.8.

16.6 Personnel Protection Interlocks

Interlock systems form an important part of radiation protection at light sources and other accelerators, and the design of an interlock is an interesting technical problem with the same kind of appeal as solving a puzzle or writing a computer program. It is important that the interlock not be considered only as a technical problem, however, since interlock operation is always embedded in some kind of activity by people: the conduct of an experiment at a beamline, or the operation of a storage ring, for example. The protection afforded by the system depends on these people doing the right things as well as the performance of devices and circuits. The design effort should look at the human interface and at the activity, not just the interlock mechanism.

16.6.1 Interlock Reliability Requirement

The design of an interlock system should start with a clear definition of the conditions and objectives. An evaluation of hazard severity can determine the reliability of protection required to provide an acceptable level of safety. A probability-consequence matrix is often used to evaluate acceptable risks, and a common criterion is that an event due to a failure of protection which could cause death or a life-threatening radiation exposure should be unlikely to occur during the life cycle of the facility. The radiation levels at most synchrotron radiation sources are capable of causing such an accident, and the interlock systems should meet that standard of reliability.

16.6.2 Technical Design Features

The following technical design features are important in the reliability of interlock systems:

1. The system should be fail safe
2. The system should be redundant
3. The redundant branches should be different
4. The system should be testable
5. The system should be simple

16.6.3 Fail Safe Design

Fail safe design means that the most likely failure events leave the system in a safe state. Such failures include loss of power, open circuits, and shorts to ground. In some cases it is not clear which failure mode is the more likely. For example, electromechanical relays usually fail to the de-energized state, and it is much less likely that the contacts will stick closed. A disadvantage of solid state components is that it is difficult to predict whether they are more likely to fail open or shorted.

16.6.4 Redundant Design

The dictionary definition of redundant is "exceeding what is necessary or normal." This is often taken to mean that active components are duplicated in an interlock system. A better functional definition of a redundant system is one "where no single failure will cause a loss of protection." To meet this definition, the redundant branches of an interlock circuit must have no common elements. The use of redundant systems can provide a very large decrease in the incidence of unsafe failures in an interlock, and the design of this feature deserves careful attention.

16.6.5 Contribution of Redundancy to Reliability

To illustrate this factor, we can use experience at the NSLS. There are now 65 independent interlock systems at the facility, covering the experimental hutches, storage rings, and injector. Each of these has one or more interlocked entry doors, one or two beam stops, interlock keys, search sequence logic, and emergency stops. Each uses 25 or more relays for the logic and has two independent chains of protection. The systems are interconnected in that each provides beam permit signals to the source upstream, but they are independent in the sense that a fault or combination of faults in one system cannot cause a failure of protection in an area covered by another system. When NSLS operations started in 1983, there were about ten interlock systems installed, and over the last 11 years this has increased to 65. On average, about 50 interlock systems have been in service during this time . During this period there have been 11 failures in operational systems which resulted in a reduction of protection, but no one was endangered because the redundant nature of the interlocks prevented a complete loss of protection. Eleven failures in 11 years among 50 independent interlock systems gives a rate of one failure per 50 years per system.

During this time there were roughly ten times as many "safe" failures which did not reduce protection but which shut down the radiation source, thus illustrating the failsafe nature of the design. Of the 11 failures, only two were component failures, with component failure contributing to one other incident. The rest were due to subtle wiring errors, design errors, poor work practice, and failure to follow QA procedures. There was even one "wire from the sky" incident where a metal chip got into a connector and several years later caused two conductors to short. These failures are difficult to predict, by Mean-Time-Between-Failure calculations or other methods, and collective experience is important. At the NSLS, the failure rate has decreased substantially as failure causes have been corrected.

When a failure occurs, that diminished protection may remain in place until the next periodic test of the system. At the NSLS we test each system every six months, so we can expect to find a reduced-protection failure once every 100 tests, and on average there should be two failures in a system once in every 10,000 tests. These figures mean that each system could potentially have a complete loss of protection due to double failures once each 5,000 years. This number is in line with the reliability criteria for radiation protection: with 65 interlocks in service the expected rate of loss-of-protection failures is about one per 70 years, well beyond the expected life of the facility.

16.6.6 Common Cause Failures and Diversity

There are two considerations which can modify this estimate. First, two random failures in a system may not combine to cause complete loss of protection, so in that regard the estimate is pessimistic. Second, and more important, the calculation assumes that failures are random and unrelated, while in fact there may be multiple failures due to a common cause. This has the potential of eliminating or greatly reducing the effect of redundancy, and seriously compromising system reliability. The best defense against common-cause failures is diversity and isolation between the redundant chains of protection. It is easy to think of the types of common failures which can be avoided by separating the two chains: shorts in both circuits caused by a pinched cable, or faults in both chains caused by a high voltage transient in a common power supply. By making the two protective chains different, we can protect against faults "we haven't thought of." Diversity also protects against a problem which is easy to predict: the subtle wiring error which is faithfully duplicated in both circuits.

One straight forward way to provide diversity is to make one chain much simpler than the other. The fundamental functions of a radiation protection interlock include perimeter security (door switches and locks), beam stop control, position sensing of beam stops, emergency crash, search sequence, warning signals, and status indicators. The first four functions are critical but the last three are not: the overall safety of an interlock system is not affected if a search sequence station fails once in 50 years rather than once in 500. Much of the complexity of the interlock logic comes from these last three functions, and if they are not included in one of the chains then the differences in circuit topology and number of logical elements will be substantial.

16.6.7 An Example Interlock System

The interlock logic diagram in figure 4 was for an early conceptual design for the Advanced Photon Source at Argonne National Laboratory. There are three separate interlocked areas: the linac/positron accumulator ring (PAR), the synchrotron, and the storage ring. These are separated by "partitions" which consist of a shield wall, beam pipe, and beam stops or switch magnet. When the partition is closed, an area may be safely occupied while there is beam in the adjacent area. The interlock systems are independent, with only the well defined permit signals connected from one area to the radiation source upstream. If these signals are disconnected, (and the beam stops locked closed) then maintenance or construction work can be done on an interlock system without fear of compromising the interlocks of adjacent areas.

The protective action of each of these systems is to turn off the source of radiation: the gun and klystrons in the linac, and the rf systems and dipole magnets in the synchrotron and the storage ring. The boxes marked "search-access management" for each area contain much of the system logic: the sequencing of check stations and door closings during the area search, warning devices and time-outs, status indications, and controlled access functions. The critical functions are included in this logic, perimeter security, beam stop sensing, and emergency stop. These critical functions are also duplicated in chain B on the lower part of the diagram, but here they are implemented in a simpler way: the door switches, crash buttons, and beam stop sensors are simply interconnected without auxiliary logic. The two chains are also completely separate from end to end. This not only makes it easier to test, but also minimizes the possibility that a fault in one chain can cause problems in the other. This also reduces the chance of failures in both chains resulting from a common event.

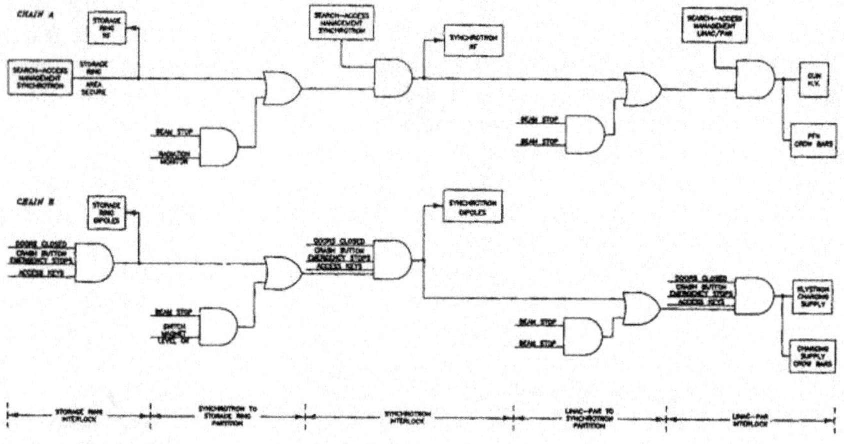

Fig. 16.4 Example interlock system logic diagram.

16.6.8 Interlock Testing

An installed interlock system is defined by a functional test. There is no other way to verify that wiring errors, component failures, or other faults do not compromise the system. From the discussion above, it can be seen that the reliability of a redundant system against double failures is directly proportional to the frequency of testing. A successful testing program depends on a system design which accommodates testing, a well designed series of tests, and the commitment of machine time and resources to accomplish the tests. In addition to periodic testing, the system must be tested before being placed in service, and after any maintenance or modification is performed. The test should exercise each system input and verify each protective response. Redundant chains should be exercised independently, and at least once during the test the system should be tested from end to end. For example, verify that opening an entry door causes a pulsing linac modulator to turn off, not just that a relay drops out or a ready light goes off.

16.6.9 Virtues of Simplicity

A simple interlock is likely to be more reliable, it will be easier to test, and it certainly will be more understandable. These are all desirable properties, and the temptation to add cleverly engineered features which deal with every contingency should be tempered with this realization.

16.6.10 Programmable Logic Controllers vs Relays[5]

A few years ago there would have been little choice on the type of logic to use for interlocks. Relay based systems have been the traditional approach to accelerator personnel protection, and there is a large body of experience available. Control systems based on computers, or more properly, programmable logic controllers (PLCs) are now widely used in industrial settings, and have found application in accelerator interlocks. The functional requirements discussed above apply to both systems, but there are qualitative differences which affect concerns and judgments in the two approaches.

PLC-based systems are inherently more complex and the failure modes more difficult to analyze. Consequently it will be more difficult to demonstrate a satisfactory level of reliability. PLCs may be suitable where the complexity of the logic would result in an unreasonably large number of relays, or if the facility components are widely separated so that serial or multiplexed signal transmission is more practical than conventional wiring. The system chosen should have a substantial and satisfactory history of operation in industrial control and related applications, and should show satisfactory response to power failures, electrical transients, environmental insults, and the like. Note that this may be difficult to establish for a non-commercial or custom designed system. The software system to be used should have a similar history of satisfactory performance.

Failure modes are difficult to predict in PLC based systems because of their complexity, and redundancy is particularly important. Common cause failures are also difficult to predict because linkages between failures can be subtle. Software bugs are a possible link, and independent software development is needed for the redundant chains. There may be communication links to the PLC for downloading software from a development unit or for delivering machine status information. It is possible for data or program corruption to occur through such links and they must be under tight configuration control. Configuration control of the software is even more important than for the physical components since software changes are often hard to detect.

Computer skill is common, but experience with sensor-based industrial control systems is not. Staff resources must be adequate for both hardware and software aspects during design, construction, operation, and maintenance phases.

16.7 The Integrated Safety Program

The best performance in almost any enterprise occurs when all the people involved are in a position to contribute their knowledge and insight and to accept responsibility for their part in the activity. This is certainly true of safety performance at a research facility, and the safety program should be arranged to take advantage of this. Safety should be part of everyone's goals rather than a set of restrictions imposed on activities. This integration of safety with other goals such as efficiency, productivity, and excellence is the culture mentioned at the beginning of this chapter. Although visiting users have a different relationship with the facility than the employees, they also must be engaged in the safety effort, and the success of the program depends in large part on their cooperation. This cooperation will more likely be forthcoming if the users perceive that the requirements are both reasonable and thoughtful, and their responsibilities are clearly spelled out.

A good way to reduce the effort required for enforcement is to make it easy for people to do the right thing. Designing forms to be easy to fill out and procedures so that safety formalities fit in as much as possible with the flow of the main activity are two examples. Putting information where it will be used is another way of avoiding frustration. A related concept is to provide pathways for the resolution of problems, and avoid bureaucratic brick walls. This is not to guarantee that all problems can be solved, but that the effort will be made.

References

1. U.S. Department of Energy, *Order DOE 5480.19 Conduct of Operations Requirements for DOE Facilities*, 1990.
2. H. J. Moe, *Advanced Photon Source: Radiological Design Considerations.* Argonne National Laboratory report APS-LS-141 revised, 1991. p. 5.
3. D. Chapman, *PHOTON: A User's Manual*, Brookhaven National Laboratory Informal Report #40822, 1988
4. E. Bräuer, and W. Thomlinson, *Experimental Verification of PHOTON: A Program for Use in X-Ray Shielding Calculations. Nucl. Inst. & Meth.* **A266** (1988) 195
5. U.S. Department of Energy, *Order DOE 5480.25 Safety of Accelerator Facilities, Guidance,* 1993. Part I F.2.a.

GLOSSARY

Note: In some cases, definitions relevant to synchrotron radiation sources are given, rather than more general definitions.

Absolute alignment: The adjustment of components to their design position as defined by the global reference grid.

Absolute position tolerance: Defines a maximum global shape distortion by specifying the tolerance on the alignment of a component relative to its design position.

Accelerating gap: A region across which a longitudinal RF electric field is established to transfer energy to the beam.

Airlock: An enclosed area attached to a room or building through which one must pass to enter or exit. Normal operation results in only one door of the area open at a time. Such a structure facilitates maintenance of cleanliness and control of temperature/humidity.

Aliasing: The phenomenon of translating signal information from frequencies lying above the Nyquist frequency of a sampled system to frequencies lying below the Nyquist frequency. Aliasing is avoided by filtering out frequencies higher than the Nyquist limit prior to sampling, or it can be employed intentionally to shift high frequencies to a lower frequency processing band. In the latter case, ambiguous frequency information is avoided if the range of frequency components of the undersampled, high-frequency signal is less than the Nyquist bandwidth, if frequencies outside of this range are filtered out prior to sampling, and if the sampling frequency is selected so that two frequencies within this range are not aliased to a single frequency.

AM/PM conversion technique: A technique for processing signals from pickup assemblies in which the amplitude difference is converted into a phase angle, which is proportional to the beam position, independent of beam current.

Ampere-turns: The total effective current flowing in a coil made up of several turns powered in series. The product of the current in each turn by the number of turns making up the coil.

Analog-to-digital conversion (ADC): The conversion of an analog signal into a discrete sequence of numbers by periodic sampling.

Application programs: Software packages designed to perform specific accelerator control, monitoring, or data manipulation tasks.

Bake-out: Conditioning the vacuum system to achieve lower base pressure in a shorter time by heating vacuum components to accelerate thermal desorption, particularly of water vapor, from the surface.

Bandwidth: A range of frequencies (or wavelengths, or energies). In the context of electronics, it is the range of frequencies to which a process or instrument will respond with amplitude deviations less then a specified fraction (typically 3 dB) of the maximum response amplitude.

Bayard–Alpert gauge: A hot wire filament ionization gauge that enables pressure measurement below one nanotorr by using a fine wire collector instead of a large cylindrical collector.

Beam loss monitor: A monitor for the detection of particles lost from the stored beam. It consists of distributed detectors which sense electromagnetic showers created by lost electrons.

Beam position monitor (BPM)—electron beam: A system for the measurement of charged particle beam position by sensing the fields created by the beam. The monitor usually consists of a set of four pickup electrodes, some signal processing electronics, and a computer controlling the processing and display of the data.

Beam position monitor (BPM)—photon beam: A system for the measurement of the position of a photon beam. Typically part of the photon beam interacts with the monitor producing a signal related to the photon beam position.

Beamline: The system for conditioning and transporting synchrotron radiation from the storage ring to an experimental station.

Beryllium window: A thin (few to few hundred microns thick) foil of beryllium metal which transmits X-rays but which absorbs longer wavelength radiation. It can be used as a high-pass filter or as a vacuum boundary.

Beta function: A periodic function of the distance s along the accelerator. This function, together with the emittance, determines the maximum transverse beam envelope.

Beta particles or beta radiation: Electrons

Betatron oscillations: The transverse (horizontal and vertical) oscillations of particles around the closed, or reference, orbit.

Betatron resonances: The values of the betatron tunes that satisfy given numerological relationships which may cause growth of betatron oscillations due to reinforcing of magnetic perturbations at each ring revolution. See **Working point.**

Betatron tunes: The number of horizontal and vertical betatron oscillations per ring revolution. Particular non-integer values of the tunes are chosen to avoid resonances. See **Working point**.

Blue line survey: The layout of anchor bolt positions in the empty tunnel. Traditionally, blue snap lines were used in the process.

Bragg crystal: A single crystal which scatters X-rays (by Bragg diffraction) in a very specific pattern according to X-ray wavelength and angle of incidence. Used as a dispersing element in crystal monochromators.

Bragg–Fresnel lens: An optical component based on a system of microstructures (Fresnel zones) machined into the surface of a crystal.

***Bremsstrahlung*:** *Bremsstrahlung*, or 'braking radiation,' is the dominant source of unwanted high energy radiation produced by electron accelerators. *Bremsstrahlung* is produced when high energy electrons encounter matter; the rapid deceleration results in gamma-ray production, a shower of electron–positron pairs, and other particles.

Brightness: The number of photons emitted per second, per square millimeter of source size, per square milliradian of opening angle, within a given spectral bandwidth usually chosen as 0.1%. Brightness is a measure of the concentration of the radiation and increases as the electron beam emittance decreases, until the diffraction limited emittance is reached. Undulators produce the highest brightness beams of synchrotron radiation. Brightness is often referred to as *brilliance.*

Broadband impedance model: A model used to characterize the entire storage ring impedance.

Broadband resonator: A resonator impedance with low quality factor.

Bucket: See **RF bucket**.

Bump: An orbit perturbation localized to a specific region in the storage ring.

Bunch clock: The electronics to provide a trigger pulse which is synchronized with a selected bunch circulating in a storage ring.

Bunch purity: A measure of the number of electrons in unwanted bunches. The purity of a single bunch is important for time resolved experiments requiring a dark time window between the light pulses.

CAMAC: Computer Aided Measurement And Control. An IEEE standard for mechanical and electrical characteristics of crate and modules. Allows for modular design and mixture of components from different vendors.

Capacitive impedance: An impedance whose reactive component is positive.

Causal function: Any function of time that vanishes for times earlier than a specified time.

Cavity: An electromagnetic resonator. In an accelerator the cavity contains a gap across which a longitudinal electric field accelerates particles.

CCD camera: A TV camera using a charge coupled device (CCD) instead of a vidicon tube as the light sensing element.

Chromatic behavior: The dependence of parameters such as tunes, beta functions, etc. on the energy of the particles.

Chromaticities: The horizontal and vertical chromaticities express the variation of tune with electron energy.

Clean room: An enclosed area which is maintained at ultra-clean conditions by applying rules of utilization (dress, type of machinery, etc.) and by continuous mechanical processes (air recirculation and air-lock entries, filtering/precipitation, and material transfer, etc.). The class of a clean room is a number which specifies the number of particles per unit volume and implies the largest particle size, according to industry standards.

Closed orbit: The electron trajectory that exactly repeats itself on every revolution. The electrons in a storage ring oscillate around the closed orbit.

Coherence: A measure of the correlation in phase of emitted photons, either transversely or longitudinally. The transverse coherence of a synchrotron radiation beam is proportional to the brightness and hence is highest for undulator beams. Highly coherent beams are useful in making interference patterns and holographic images. Synchrotron radiation is partly coherent. Lasers, including free electron lasers, have very high coherence.

Coil (for magnet excitation): The assembly of windings, placed around a magnet yoke, to conduct the currents that will generate the required magneto-motive force in the magnet.

Coil (for magnetic measurement): A set of windings used for magnetic measurements. The different types include bucking, flipping, harmonic, radial, rotating, and tangential coils.

Cold cathode gauge: A vacuum gauge that ionizes the gas by field ionization rather than electron emission. The gauge is robust and does not cause thermal decomposition due to hot filaments.

Collective phenomena: Phenomena arising from the beam acting back on itself as a result of its interaction with the vacuum chamber environment.

Combined function magnet: Any magnet in which different field types are combined. Usually applied specifically to a magnet that combines dipole and quadrupole field within the same gap.

Complex coherent frequency shift: The frequency shift of the collective motion of the particles in a single bunch, or of all the bunches in the beam, due to the impedance. The real part is the actual frequency shift; the imaginary part is the growth rate. If the imaginary part is negative, the motion is damped; if positive, it is 'antidamped,' and it is unstable if the growth rate overcomes the damping mechanisms.

Compton scattering: The scattering of photons by charged particles. For example, the head-on collision of relativistic electrons with visible light produces high energy photons.

Commutation: The change of the conduction current from one phase to another. Natural commutation refers to the automatic changeover in diode arrays.

Conductance (gas): The ratio of quantity of gas flow, or throughput, between two surfaces, divided by the pressure difference between the surfaces. Units are liters/second.

Configuration control: Administrative systems designed to insure the integrity of accelerator systems. These may include procedures, checklists, key control, and work permits.

Control system: The collective set of hardware and software components needed for accelerator control and operation. Includes front-end interfaces, processors, networks, server and workstation hardware.

Conventional facility: The buildings and mechanical and electrical systems supporting and facilitating all activities and equipment associated with the laboratory.

Correction magnets: The small magnets located in the lattice of a particle accelerator to allow corrections and adjustments to the beam direction, focusing, etc. Correction magnets can be dipoles, quadrupoles, multipoles, etc.

Coulomb scattering: The deflection of one charged particle (e.g., an electron in a stored beam) due to its interaction with the charge of another particle (e.g., the nucleus of a residual gas atom).

Coupled-bunch mode: An oscillation pattern of a system of many bunches that interact via wakefields and cavity resonances in the vacuum chamber. Complex multibunch oscillations can be decomposed into a system of decoupled, orthogonal modes, each having a particular frequency.

Coupled-bunch instability: A beam instability resulting from an exponential growth of one or more coupled-bunch mode amplitudes.

Coupling coefficient: A complex quantity describing the effects of elements producing linear coupling.

Coupling resonance: A linear difference resonance coupling horizontal and vertical betatron motions of the particles.

Cultural noise: The motion of the ground due to man-made sources.

Current transformer: An instrument for the measurement of beam current based on the detection of the magnetic field created by beam. The transformer is a toroidal sensor mounted around an insulated piece of the beam tube. The signal is induced in a secondary winding.

Cutoff frequency: The lowest frequency of mode propagation in a waveguide; also the frequency above which there are no more resonant cavity modes. It is usually defined by $\omega = c/b$, where c is the velocity of light and b is the vacuum pipe radius.

Cycloconverter: A power supply that has AC input and output.

Damper/damping kicker: A transverse or longitudinal beam kicking device used in feedback systems to counteract the growth of oscillations.

Damping time constant: The time needed for the exponential decay of a damped beam oscillation to reach $1/e = 37\%$ of its original amplitude.

Database: The central part of a control system where parameters are stored in structures adapted to the specific needs of the accelerator. The database can reside in memory or on disk.

DC current transformer (DCCT): An instrument for the high precision measurement of the stored beam current based on the precise compensation of the magnetic field created by the beam current in a transformer core.

Decimation: A method that obtains a reduced, but still greater than Nyquist, data rate in an oversampled system by periodically ignoring samples.

Deflection of the vertical (geodetic): The angle of divergence between the gravity vector and the ellipsoid normal.

De-Qing a mode: Reducing the quality factor of an oscillatory mode by mechanical or electromagnetic means.

Desorption (thermal): The departure of molecules from a surface into the gas phase due to thermal excitation of molecular vibrational energy levels. Vacuum system heating, or bakeout, hastens this process and is an important part of conditioning the vacuum system.

Desorption (electron stimulated): The departure of atoms, molecules, and ions from a surface upon electron bombardment of that surface.

Desorption (photon stimulated): The departure of atoms, molecules, and ions from a surface upon irradiation of the surface with light. The gas emitted from the surface of the storage ring vacuum chamber when irradiated by synchrotron light is a major source of gas in a synchrotron light source.

Difference-over-sum processing: A technique for processing signals from four-electrode beam position monitors. The current independent beam positions are obtained from the ratio of difference signals to sum signals.

Differential leveling: A method of determining the elevation difference between two points using a spirit or automatic level.

Diffraction: The change in direction or phase of a wave due to the interaction with an object or aperture. Diffraction of synchrotron radiation by regular spaced lines in a diffraction grating or periodic arrays of atoms in crystals is used to select particular wavelengths from a broader spectrum.

Diffraction limit: The lower limit of the product of the transverse rms size and the transverse rms angular divergence of a beam of electromagnetic radiation of a given wavelength. This product (also called the emittance) cannot be less than about one tenth of the wavelength. This sets the ultimate limit on the brightness that can be achieved and on the resolution of images.

Digital filter: A software or hardware implementation to process a digital input signal to generate a modified digital output signal.

Digital signal processor (DSP): A device which processes a digitized signal via digital computation.

Digital-to-analog conversion (DAC): Reconstruction of the analog signal from a discrete sequence of numbers.

Dipole beam breakup instability: A transverse instability in linacs in which the transverse displacement of the tail of the bunch grows with the distance traveled by the bunch.

Dipole coupled-bunch oscillations: Transverse or longitudinal oscillations of a multibunch beam characterized by the motion of the bunches about their nominal centers as if they were rigid 'macroparticles.' These oscillations are typically the dominant concern, from the perspective of coupled-bunch stability, in the design of any multibunch circular machine.

Dipole magnet: A magnet in which two poles produce a magnetic field across the path of the beam. In most cases the dipole magnet produces a uniform field. In some cases a gradient is introduced to provide some focusing. The main dipoles in a particle accelerator will produce a vertical field to bend the electron beam radially, whilst correction dipoles can be configured for horizontal or vertical fields.

Dispersion function: A function characterizing the trajectory of off-momentum particles.

Dispersion relation: (1) An integral relation that connects the real and the imaginary parts of an impedance (or, more generally, the real and the imaginary parts of the Fourier transform of any causal function). (2) A consistency condition that relates the complex coherent frequency shift and the impedance that causes it via an integral over the frequency spectrum of the bunch.

Distributed database (versus Central database): A database distributed over several computers (servers, workstations or minicomputers). It has global synchronization mechanisms for overall consistency.

Distributed processing (versus Central processing): Algorithms and programs running in parallel on different computers connected via a local area network.

Downsampling: See **Decimation.**

Dynamic alignment system: A system for automatically maintaining alignment by monitoring motion and applying corrections.

Dynamic aperture: Central region of a magnetic guide field where particles can circulate on stable trajectories. Strong focusing lattices with high chromaticity require strong sextupoles that reduce the dynamic aperture.

Dynamic range: The range of input signal amplitudes that can be usefully processed by a system. Usually dynamic range is defined at the low end by measurement noise, and at the high end by saturation or non-linearity of processing components.

E_{010} mode: An electromagnetic field configuration which has a longitudinal component to change the energy of a charged particle passing through a cavity in which such a field is present.

Eddy currents: The electric currents that are caused to flow by the application of a time-varying magnetic field to an electrically conducting material. In some cases, such as for a cycling synchrotron injector, eddy currents can cause significant power loss and heating in bus bars, coils, magnet cores, and vacuum chambers, as well as a weakening and distortion of the magnetic field that would be present in the absence of eddy currents. They can also be a problem in storage ring corrector magnets if rapid changes are made in these correctors.

Effective impedance: The convolution of an impedance with the bunch spectrum. The effective impedance pertains to the bunch as a whole, while the impedance pertains to individual particles.

Effective length: The length of a magnet used in an accelerator model. It differs from the physical length of the magnet because it takes into account the effect of fringe fields at the ends of the magnet.

Eigenvector/value: A set (or vector) of variables in a linear system of equations that yields a solution that is that same vector times a constant (eigenvalue). Any system vector can be decomposed into a linear combination of decoupled (orthogonal) eigenvectors; system analysis using eigenvectors is simplified because the equations are decoupled.

Electrical break: A non-conducting spool piece in the vacuum system, typically made out of ceramic, to provide electrical isolation.

Electromagnetic spectrum: The broad range of wavelengths extending from long wavelength radio waves to progressively shorter microwaves, to infrared (IR), visible, and ultraviolet light, to vacuum ultraviolet light (VUV), to soft and hard X-rays and then to gamma rays. The energy of each photon of electromagnetic radiation increases as the wavelength decreases. Synchrotron radiation is the most intense source of electromagnetic radiation in the VUV, soft X-ray and hard X-ray parts of the spectrum, roughly from photon energies of 10 eV to several hundred keV. It is also more intense than other sources in some regions of the IR.

Electron gun: The part of an electron accelerator where the electrons are liberated from a cathode, pre-accelerated, and focused into a spatially well defined stream of particles suitable for injection into another accelerator which then brings the beam to a higher energy.

Ellipsoid: In general, a geometrical body whose intersection with an arbitrary plane is an ellipse. In a geodetic context, the regular figure that most closely approximates the shape of the earth.

Emittance: The area of phase space (position–angle in transverse coordinates x or y) of a stored electron beam divided by π. Together with the beta-function, the emittance characterizes the maximum amplitude of the beam envelope. At a symmetry point in a storage ring, the emittance is the product of the transverse rms size and the transverse rms angular divergence of the beam. Photon beam brightness increases with decreasing electron beam emittance until the diffraction-limited emittance is reached.

Equipotential surface: A surface on which the potential is constant. The force is always normal to this surface.

Experimental (end) station: The terminus of a beamline where the synchrotron radiation interacts with the experimental sample, detectors, etc.

Faraday cup: A cup-like structure made from high-Z material for the collection and measurement of the total beam charge.

Fast head–tail instability: See **Transverse mode-coupling instability.**

Feedback system: A system that regulates the action of a process in response to a detected deviation of that process from a desired action. Feedback systems function in a closed loop, with information from a detector being fed through appropriate filters and amplifiers to stably control a process.

Ferromagnetic material: Material (steel, iron, ferrite etc.) which has a very high relative permeability. Within such material, low magnetic fields will generate high magnetic inductions.

Fiducial: A reference point, marker, or value used for the control or monitoring of a component or system. Also a standard of reference for measurement or calculation.

Firing angle: The angular phase between a thyristor or SCR trigger and the line voltage.

Fluorescence: The emission of photons from a material at an energy corresponding to that given up by electrons as they drop from one energy state to a lower one.

Fluorescent screens: Phosphor-coated screens which can be moved into the path of a beam of electrons or photons so that the emitted visible light produces an image of the distribution of the incident radiation. Other names are target, paddle, flag.

Flux: The number of photons emitted per second, per horizontal milliradian, integrated over all vertical angles, within a certain spectral bandwidth, usually chosen as 0.1%. The emitted flux depends on the stored current and energy of the electron beam, but not on the emittance.

Forced centering: A mechanical interface which allows the removal and the subsequent replacement of the same or of a different instrument to exactly the same location.

Fourier transform: A mathematical technique used to determine the steady-state frequency content of a time-domain function. Time domain functions can be expressed as a summation of sinusoidal functions of different frequencies having amplitudes given by the Fourier transform.

Free electron laser (FEL): A source of intense, coherent radiation based on the interaction of a relativistic electron beam with a superimposed beam of radiation in the presence of a spatially periodic magnetic field such as produced by a wiggler magnet.

Free space impedance: The impedance seen by a charged particle in uniform circular motion in free space, whose origin is the synchrotron radiation emitted by the particle. This impedance should not be confused with the vacuum impedance.

Freewheeling diode: A diode placed in parallel with a power supply rectifier bridge or a reactive load which provides an alternative path for circulating current, thereby preventing a large induced reverse voltage on, and the backflow of energy to, the power supply.

Frequency response: The response of a system as a function of the frequency of a sinusoidal input, commonly depicted as graphs of response magnitude and phase versus the logarithm of input frequency.

Fume hood: A small partially or totally enclosed area connected to the chemical-fume exhaust system. In this area, chemicals producing noxious and toxic fumes can be safely handled.

Fundamental mode: The lowest-frequency resonant mode of a cavity.

Gamma radiation: High energy electromagnetic radiation produced by nuclear transition and *bremsstrahlung* processes. Since gamma radiation is a major component of *bremsstrahlung* radiation, adequate shielding measures must be incorporated into the design of an electron accelerator.

Gap (magnet): The space at the center of a magnet in which the vacuum vessel and the circulating electron beam are located. The magnet generates a field within the gap to influence the electron trajectories.

Generations of synchrotron radiation sources: A classification of synchrotron radiation sources. First generation sources are those that were designed for high energy physics purposes. Second generation sources are the first sources to be designed as dedicated sources of synchrotron radiation. Third generation sources are dedicated sources with lower electron beam emittance and generally more straight sections for insertion devices than first and second generation sources. Fourth generation sources would be sources which produce radiation with brightness, coherence, peak power, etc. which are significantly higher than third generation sources. Free electron lasers are one example of a fourth generation source.

Geocentric space: A coordinate system defined by the rotation axes of the ellipsoid.

Geodetic: Pertaining to *geodesy*, the science of the size and shape of the earth and its gravitational field.

Geoid: The gravitational equipotential surface at mean sea level.

Getter: A material which has a large trapping probability for gases in the vacuum system. The gas is adsorbed on the surface of the material.

Girder: A strongback or platform onto which a group of components of a storage ring can be mounted and prealigned.

Global orbit feedback system: A system consisting of steering magnets and beam position monitors that is designed to reduce the deviation of the stored electron beam (and hence also the deviations of synchrotron radiation beams) from their optimum values everywhere in a storage ring.

Global Positioning System (GPS): A system of satellites and portable receivers to determine position.

Green-field site: The site for a facility built away from the campus of a university, private or government laboratory, industrial complex or other installation which can provide utilities and other operational support. In control system parlance, 'green field' refers to a design opportunity that does not have to adhere to existing configuration standards at the facility.

Head–tail damping (antidamping): The damping (antidamping) of the transverse motion of the head and tail of a bunch due to the resistive part of the impedance for storage rings operating with nonzero chromaticity.

HEPA filter: Air filters and associated air-recirculating systems used to maintain a specified degree of air cleanliness with respect to suspended particulate matter. HEPA is the acronym for 'high-efficiency particulate air.'

Heterodyne receiver: A device for detection of amplitude- or frequency-modulated sinusoidal radio signals. It makes use of a mixer to convert the input signal to a lower-frequency signal that is more easily filtered. Heterodyne receivers can selectively detect one of several signals having nearly the same frequency. For this reason heterodyne receivers can be used for beam position monitoring or to detect synchrotron oscillations in the presence of a strong stable-beam signal from a beam position monitor.

Higher order modes: A resonant mode of a cavity with a frequency higher than that of the fundamental, e.g., the electromagnetic modes in cavities of higher resonant frequency than the E_{010} mode.

Hutch: A secured, usually interlocked, enclosure surrounding an experimental station or beamline component which prevents exposure to radiation inside the enclosure.

Hysteresis: A general term for the dependence of the state of a system on its previous history. In the case of the magnetic field of an iron core electromagnet, hysteresis refers to the fact that the magnetic field is not a unique function of the current in the coil, but also depends on the precise path in time which was used to achieve the current. In a rapidly cycling magnet, such as in a booster synchrotron, hysteresis effects can result in significant energy loss and heating.

Ideal component coordinate: One of a set of three values describing the design location of a component with respect to the reference system.

Impedance: Similar to electron flow through a common circuit, charged particle motion in an accelerator is influenced by the impedance of the surrounding vacuum chamber (via electromagnetic fields). Both the longitudinal and the transverse impedance have resistive and reactive components which, when evaluated as a function of frequency, are related to the spatial structure of the wake potential via the Fourier transform. The transverse impedance has units of Ω/m, while the longitudinal impedance has units of Ω.

Impedance budget: The maximum allowable ring impedance. The sum of impedances should not exceed the budget.

Inductive impedance: An impedance whose reactive component is negative.

Injection efficiency: The ratio of the rate of electron capture in the ring to the rate of electrons delivered to the ring by the injector.

Inscribed radius (of a magnet): The radius of the circle, centered in the magnet gap, which is tangential to the poles at their mid points; it is therefore the radius of the largest circular vessel that can be located in the magnet.

Insertion device: A periodic array of magnets with alternating magnetic field that is inserted in the space between the main lattice magnets of a storage ring to generate enhanced synchrotron radiation. Insertion devices can be categorized as either wigglers (generally with high magnetic fields and longer period lengths) and undulators (generally lower field and shorter period lengths). Wiggler insertion devices produce a broad spectrum similar to that of a bending magnet. Undulators produce a more concentrated, brighter, quasi-monochromatic spectrum.

Interlock: A personnel protection interlock acts to protect personnel from a hazard such as radiation. It usually includes access control and shut-off features. An equipment interlock senses a condition (e.g. high temperature, excessive current, poor vacuum) that could result in damage to equipment, and causes appropriate action (turn off of a power source, closing valves) to prevent or mitigate the damage.

Intrabeam scattering: Multiple small-angle Coulomb scattering of the particles within the bunch.

Ion pump or sputter-ion pump: A pump that evacuates gas by both ionizing the gas molecules and getter trapping.

Ion trapping: Trapping of positively charged ions in the electrostatic potential of an electron beam. Ion trapping has detrimental effects on emittance and lifetime. It can be avoided by using positrons as the stored particle. High current, low emittance electron beams cause an overfocusing of ions which prevents them from being trapped. A gap left in the fill pattern of the ring also mitigates ion trapping effects since it causes the ions to drift out of the potential well.

Ionization: The process by which an atom loses (or gains) an electron, thereby changing its charge state.

Ionizing radiation: Particles (including photons) with sufficient energy to ionize matter. Alpha, beta, gamma, and X-rays are examples. As counter examples, infrared and microwave radiation can have physical effects, but do not produce ionization.

Isolator/circulator: A device which, when inserted in a transmission line (waveguide or coaxial), allows power to pass in one direction with little attenuation, but absorbs strongly (isolator) or deflects into a load (circulator) any power traveling in the opposite direction (e.g. reflected from a mismatched load). Commonly used to protect klystron power sources from reflected power due to a load mismatch in the ring.

Jig transit: A precision optical instrument used to establish a vertical reference planes. When an optical mirror is attached to the horizontal axle, it can also be used to establish perpendicular planes to a line of sight.

Keil–Schnell criterion: A relation that establishes the approximate threshold for the negative mass instability or the longitudinal microwave instability.

Kicker: A component of a storage ring used to deflect or 'kick' the particle beam. Kickers are used to inject/extract particles into/from a circular accelerator. A kicker is distinguished from a corrector magnet by the rapidity with which its strength can be changed. Kickers can change strength in times shorter than that necessary for the beam to travel around the storage ring a few times. Wide-band kickers for feedback control of instabilities can change strength in a time short compared to one rotation period of the ring. Transverse kickers may be designed to deflect the beam vertically or horizontally. Longitudinal kickers can increase or decrease the energy of the beam.

Kinematic mount: Mechanical mount in which the number of constraints balances the number of degrees of freedom.

Knobs: Used in accelerator control systems (along with graphical 'sliders') to adjust hardware set points.

Landau damping: A damping mechanism that effectively transforms the coherent motion of the beam into incoherent motion of the particles via phase mixing induced by the oscillation frequency spread.

Longitudinal microwave instability: A single-bunch instability that arises above a certain current threshold that leads to an increase in the bunch length and the energy spread.

Laplace transform: A mathematical technique for transforming time-domain functions into the complex s-frequency domain ($s = \sigma + j\omega$). It is similar to, but more general than, the Fourier transform in that it permits analysis of the transient (as well as steady-state) response and stability of systems that are suddenly activated with a set of initial conditions.

Lasertracker: A servo-controlled tracking laser interferometer measuring tool. A laser interferometer measures the linear distance to the target, and two angular encoders measure the azimuth and elevation angles of the laser beam. The tracker follows a retroreflective target, providing real-time coordinate information of the target center location.

Lattice coordinate system: Usually a three dimensional Cartesian coordinate system with an arbitrary origin used to define the design position of every machine component.

Lattice of a storage ring: Sequence of magnetic elements designed to keep the electron beam focused and on the design trajectory. The lattice is the major determinant of properties of the stored electron beam such as emittance, and hence also the major determinant of the performance of the ring as a light source. Several types of lattices are used in synchrotron radiation sources.

Lifetime: The time it takes for a stored beam to decay in intensity to $1/e = 37\%$ of its initial current (*e*-folding lifetime).

Linac (linear accelerator): A particle accelerator which accelerates the particles along a straight (linear) path.

Linear regulator: A power transistor working like an adjustable resistor.

Local orbit feedback: The periodic correction of the angle and/or displacement of an X-ray beam using a local bump to change the electron beam path in the vicinity of the source point for an individual beam line. The bump can be controlled by signals from photon or electron beam position monitors.

Log-ratio technique: A signal processing method for signals from pickup electrodes based on logarithmic amplifiers. The normalized beam positions are obtained from the difference of the logarithms of the individual signals.

Longitudinal motion: The variation in the time, or RF phase, at which an individual particle crosses the accelerating gap, relative to the nominal crossing time or phase with respect to the RF field. This time, or phase, variation causes the particle to move around the bunch in the horizontal plane, varying its position both longitudinally and radially, and also varying its energy. This motion is due to the interaction of the particle with the RF system and is quite separate from the betatron motion which is purely transverse and is due to the transverse focusing effects of the magnetic fields which form the lattice. It is also called phase motion or synchrotron motion.

Loss impedance: The real part of the effective impedance.

Magnetic field *(H)*: The field generated either by electric currents in the coil of an electromagnet, or, in specialized insertion magnets, by permanently magnetized ferromagnetic material; magnetic field has units of Amperes per meter.

Magnetic induction *(B)*: The magnetic flux density generated in a material or in free space by a magnetic field. Magnetic induction has units of Tesla.

Mask: An aperture in a photon beamline through which the X-ray beam passes and which has sufficient cooling to withstand exposure to the photon beam. The aperture size and placement of a mask is chosen to prevent damage to more fragile beamline components in case the beam is incorrectly steered.

Microseismic noise: The motion of the ground due, for example, to the coupling of ocean waves to the continents.

Microtron: An RF particle accelerator where a DC magnetic field is used to recycle the particles several times through the same accelerating section on paths of increasing bending radius.

Mixer: An electronic component with two inputs and one output. The output, called the 'intermediate frequency' or 'IF' is proportional to the product of the voltages applied to the two inputs. One input, called the 'radiofrequency' or 'RF' input, accepts the signal to be measured, e.g., from a beam position monitor. The other input, called the 'local oscillator' or 'LO' input, is used to cause an attenuated, modulated, or frequency- translated version of the RF input signal to appear at the IF output.

Model: A mathematical set of equations used to simulate the response of a physical system (e.g. storage ring optics, photon beamline optics).

Modulation: The time-domain variation of a signal's amplitude (AM), frequency (FM), phase (PM), or other characteristic. A single-frequency carrier signal can be modulated to transmit information.

Modulator: A power supply which delivers voltage or current in short pulses as needed by pulsed klystrons or other devices.

Momentum compaction factor: The fractional change in the closed orbit length divided by the fractional change in energy which caused it.

Monochromator: A component of a beamline that can be adjusted to select a particular photon wavelength to be used for an experiment from the broader synchrotron radiation spectrum. At X-ray wavelengths, monochromator elements are usually crystals such as silicon. At longer wavelengths, ruled gratings are used. Synthetic layered materials, or multilayers, are also used to monochromatize a photon beam.

Monolithic slab: A concrete slab which behaves as a structure created from a single continuous pouring of concrete. It behaves as a 'single rock.'

Multibunch feedback: A feedback system that acts to suppress coupled-bunch instabilities, either by acting on specific coupled-bunch modes in the frequency domain, or by acting on each bunch individually in the time-domain.

Multibunch instabilities: See coupled-bunch instabilities.

Multibunch operation: An operational mode in which many bunches of particles are stored in a storage ring. This mode provides the highest average current.

Multilayer (ML): A sandwich of many layers with different refractive indices. Through constructive interference the ML acts as a wavelength selective mirror and hence a monochromator element.

Multiplexed receiver: A device commonly used to analyze the signals from the beam position monitor pickup assemblies. The signals are switched one after the other (RF multiplexed) to the receiver.

Multipole magnets: A term loosely applied to magnets with a large number of poles, such as sextupoles, octupoles etc., or, more specifically, to a magnet that is capable of generating a magnetic field with a number of such higher order harmonics.

Narrow-band resonator: A resonator impedance with high quality factor, Q, characteristic of resonant cavities. Typical Q-values are in the range of one hundred to ten thousand for room-temperature cavities, and one million to one billion for superconducting cavities.

Negative mass instability: An instability found in coasting (unbunched) proton beams below transition energy, if the impedance is capacitive, that tends to bunch the beam.

Neutron: An uncharged sub-atomic particle. Neutrons can be generated when high energy electrons or gamma rays strike matter. Accelerators must be shielded to protect personnel against exposure to neutrons.

Non-oriented steel: Steel intended for magnetic purposes in which there is high isotropy of the magnetic domains. The magnetic and loss parameters of such steel are nearly independent of the direction of flux within the steel. Such steel is preferred for accelerator magnets.

Nyquist frequency: Half of the sampling frequency. The frequency characteristic of a digital filter is symmetric about the Nyquist frequency.

One-turn map: Describes the evolution of phase space coordinates after traversing the accelerator for one full revolution.

Open-loop gain: A frequency-dependent function that defines the relationship between the input signal and output signal of a circuit or system. The characteristics of the open-loop gain of a feedback loop can be modified as part of the design of a closed-loop feedback system.

Optical tooling: A system of measurement and alignment based on the use of telescopic sights to define precise reference lines and planes from which accurate measurements are made with optical micrometers and an optical tooling scale.

Oriented steel: Steel intended for magnetic purposes in which there is very low isotropy of the magnetic domains. The magnetic and loss characteristics are strongly dependent on the direction of the flux relative to the direction of rolling of the material at the mill. Along the rolling direction superlative properties are displayed, whilst normal to this direction, the properties are very poor.

Outer-ground insulation: An additional layer of insulation wound around the outside of a coil, to provide a mechanically and electrically robust outer finish.

Overconstrained: The number of constraints exceeds the number of degrees of freedom.

Overvoltage factor: The ratio of the peak RF voltage applied to the accelerating gaps (integrated round the circumference) to the minimum voltage needed to exactly maintain the particles at the correct energy if they were to traverse the gaps at the peak of the RF field.

Parasitic loss: Power dissipated by the beam into the vacuum chamber components due to the impedance.

Phase space: The two- or more-dimensional space having coordinate axes corresponding to different canonical parameters of a system. For example the horizontal phase space of an ensemble of electrons in a bunch at a specific point in the storage ring has transverse orbit position for one axis and orbit angle for the other. A more complete phase space defining an electron bunch has six dimensions: four for horizontal and vertical position and angle, one for longitudinal phase (time-of-arrival), and one for electron energy.

Phase stability: The phenomenon by which particles which traverse the accelerating gaps at the incorrect phase (i.e., time), oscillate around the correct phase in a stable manner (and due to radiation damping move towards it) rather than being lost. This depends on there being an overvoltage, and applies to particles which are within a finite acceptance angle or 'bucket.'

Photoemission: The emission of electrons from a surface caused by energy transfer from incident photons.

Photon absorber: A component of the vacuum system that intercepts the synchrotron radiation light to shield sensitive parts of the accelerator vacuum system, such as the chamber, flanges, bellows, etc.

Pickup electrode: A conducting surface exposed to the electromagnetic field of the beam and on which an electrical signal is capacitively and/or inductively induced. This signal is related to the position and intensity of the beam.

PID (proportional-integral-derivative) filter: A filter that adds a pole at zero frequency and two adjustable zeros, as well as overall gain, to the open-loop transfer function of a feedback system and that is used to optimize stable operation of the closed-loop system.

Pinhole imaging system: An optical system that uses a small aperture to produce an image of a light source. Synchrotron radiation pinholes can be used to observe the particle beam cross-section. Pinholes are particularly useful in the short wavelength region where other imaging systems are more complicated. X-ray pinholes are typically tens of microns in diameter.

Polarimeter: A device which measures the polarization of the spins of the stored electron beam or the polarization state of a photon beam.

Polarization: The direction of oscillation of the electric field vector of synchrotron radiation, or any electromagnetic radiation, is called the polarization. In the median plane, synchrotron radiation from bending magnets, and most insertion devices, is highly linearly polarized with the electric vector oscillating in the plane of the orbit. Out of the plane a vertical component of polarization enters and the radiation becomes elliptically and circularly polarized; i.e., the electric vector rotates in an elliptical or circular fashion along the direction of propagation.

Poles (magnetic): The extremities of the yoke, the profiles of which determine the amplitude and distribution of the magnetic flux within the gap.

Potential well distortion: A distortion of the RF bucket away from its harmonic shape due to the longitudinal impedance. Potential well distortion causes bunch lengthening or shortening.

Power factor: The ratio between active and reactive power. Active power is the part of the power delivered by the source that is dissipated in the load. Reactive power is the component of power that circulates between the source and load.

Pressure (total): The force/unit area determined by all molecular species crossing a conceptual surface. Common units are torr, bar, and pascal.

Pressure (partial): The pressure of a particular molecular component of a gas mixture.

Principle curve analysis: A method to fit a one-dimensional nonparametric curve through a field of three dimensional points.

Pumping speed: The volumetric displacement rate of gas. Units are liters/second.

Quadrupole magnet: A magnet in which four poles generate a field with a constant gradient in both the horizontal and vertical planes, the magnitude of the field being zero at the center of the magnet.

Quality factor *(Q)*: For an oscillator, 2π times the energy stored divided by the energy loss per cycle.

Quantum lifetime: The time constant for beam loss due to the emission of discrete quanta or photons. Sometimes a radiated quantum has such high energy that the energy lost by the circulating electron is sufficient to place it outside the stable energy/phase acceptance of the ring and the particle is lost. Quantum lifetime depends critically on the RF overvoltage.

Radiation damping: The damping of beam oscillations in a storage ring caused by the energy lost to synchrotron radiation.

Ramping: The process of increasing the beam energy in a storage ring from the injection level to the operating level. With full energy injection, ramping is not required.

Reactive component: The imaginary part of an impedance.

Reflective memory: A daisy-chained network of replicated shared memory. Data written to a part of the memory is automatically duplicated in all other units in the network.

Relative alignment: The adjustment of components relative to their neighboring elements.

Relative positioning tolerance: Defines the alignment quality of a component relative to adjacent components.

Resistive component: The real part of an impedance.

Resistive wall impedance: The impedance whose origin is the finite resistivity of the vacuum chamber. It is typically important only at low frequencies.

Resonant frequency: The frequency at which the real part of the resonator impedance is a maximum (and at which the imaginary part vanishes).

Resonant spin depolarization: A method for the precise determination of the energy of an electron/positron storage ring based on the resonant depolarization of the electron (or positron) spin vectors by an external magnetic field.

Resonator impedance: A simple form for an impedance that is used to model, in practice, a component (or a set of components) of the ring. The simplest realization of a resonator impedance is an RLC circuit in which all three elements are in parallel.

RF: Radio Frequency. For accelerators, the frequency of the electromagnetic field applied to the accelerating gaps. The term is often applied to the total system; i.e., the RF system is the accelerating gaps (the cavities) plus the generators of the RF field and associated hardware.

RF bucket: The potential well, or region of longitudinal oscillation phase space around the nominal value, within which a stored electron executes stable synchrotron oscillations.

Right (upright) magnets: A magnet, with any number of poles, which generates a field distribution in which the magnetic flux cutting the x-axis is normal to that axis (mid-plane symmetry). The standard dipole, quadrupole, etc. in a particle accelerator have such distributions. The alternative distribution is produced by a 'skew' or 'rotated' magnet.

Robinson damping (antidamping): The damping (antidamping) of the longitudinal dipole coupled-bunch oscillations caused by a slight detuning of the fundamental mode of the RF cavity.

Rotation harmonic: An integer multiple of the revolution frequency of a storage ring.

Safety envelope: The formally defined limits on parameters important to the safe operation of a facility. These may include beam energy and current, duty factor, operating hours, discharges to the environment, and staff requirements. These limits are established in the safety analysis done during facility design.

Safety orientation: Information which focuses on awareness of hazards, safety procedures, rules, and individual responsibility. Access to information and resources is emphasized rather than detailed skills training.

Sagitta: The distance between the straight line chord connecting the endpoints of an arc and a line which is parallel to the chord and tangential to the arc. Accelerator dipole magnets are often curved in the beam-bending plane so that the pole width need not accommodate the sagitta of the beam arc.

Sampling: The discrete acquisition of an analog or digital signal.

Sampling frequency: The number of samples per unit time.

Saturation: Nonlinear (and usually maximal) response of an otherwise linear processing component (amplifier, ADC, DAC, etc.) to high input signal levels.

Scraper: A variable mechanical aperture used as a stored beam diagnostic.

Secondary electron yield: Electron flux emitted from a surface due to inelastic scattering of a high energy electron or photon with the bulk material.

Septum magnet: The first or last bending magnet of a transfer line to or from a circular accelerator. It is positioned very close to the aperture of the circular accelerator in order to minimize kicker magnet strengths. It deflects the extracted or injected beam with little or no effect on the circulating beam.

Sextupole magnet: A magnet in which six poles generate a field which varies in amplitude as the square of the distance from the magnet's center, the field magnitude being zero on the center line. Sextupoles are used to control the chromaticities.

Shim: A small permeable protuberance added to the pole face, usually close to an edge, to fine tune the shape of the field, for example to compensate for the reduction in field that is due to the finite width of the pole.

Shuffling: A randomizing technique commonly used when assembling laminated magnets. Steel laminations from all batches produced at the rolling mill are distributed throughout the individual magnets that make up a particular family. The objective is to produce magnets which are very similar in their magnetic properties by averaging the differences in magnetic properties among different batches of steel.

Shunt impedance (of a cavity): The resistive component of the cavity, regarding it as a combination of inductance, capacitance and resistance all in parallel. Its value is given by the square of the rms voltage appearing across the accelerating gap, divided by the power dissipated in the cavity.

Silicon steel: Steel produced specifically for transformer and magnet applications, in which controlled quantities of silicon (usually a few percent) considerably reduce the AC losses (eddy and hysteresis) in the steel whilst, at the same time, causing some reduction in the magnetic properties.

Single-bunch instabilities: Instabilities arising from the interaction of the particles within a given bunch via short-range wakefields.

Single-bunch operation: An operational mode in which one single bunch of particles is stored in the ring. It provides the highest peak currents (up to several hundred amperes) and is used for experiments needing time resolution.

Singular Value Decomposition (SVD): A numerically robust technique that solves a linear or nonlinear system of equations with a vector of minimal length in the least squares sense. If the linear system matrix is sufficiently well-behaved (of full rank), SVD finds the unique solution obtained by inverting the system matrix. Otherwise, SVD separates the system vectors into eigenvectors that contribute to the equation and those that don't, and creates a 'pseudoinverse' of the contributing eigenvectors.

Skew magnets: The alternative distribution to a 'right magnet.' Whatever the order of the magnet, the flux lines will be parallel to the x-axis. This would normally imply that there will be a magnet pole located on the x-axis.

Skin depth: The characteristic depth (at a given frequency) in a conducting material beyond which the electromagnetic field does not penetrate appreciably.

Smoothing: A technique to remove residual systematic effects from component alignment or experimental data.

SPEAR scaling: A power-law dependence of the bunch length as a function of the bunch current when the latter exceeds the threshold of the longitudinal microwave instability. This translates into a power law dependence of the impedance upon frequency that is valid only within a limited range of frequencies.

Spectrum analyzer: Device for the analysis of electrical signals in the frequency domain (spectrum).

Spheroid: See **Ellipsoid**.

Spirit level: An instrument to establish a horizontal line of sight, e.g., to determine the elevation difference between points.

Steering magnet: A dipole magnet which imparts an angular kick to the charged particle beam. Also called a corrector magnet.

Steinmetz circuit: A circuit for compensation of active and reactive power in a three phase system.

Stripline electrode: An extended linear electrode in the longitudinal direction. It can be used as a kicker for the transverse and longitudinal planes.

Strut: An adjustable length member with a spherical joint at each end.

Superconducting magnets: Magnets which include coils fabricated from special material (usually niobium–titanium or niobium–tin) and cooled to very low temperatures, usually by liquid helium. At these temperatures the conductor material has no electrical resistivity and can conduct very high currents with no consequential power loss. The resulting magnetic flux densities can be three to four times higher than the saturation flux density in steel, and, in some cases, even higher.

Surface network: The system of surface survey stations and their geometry. The stations are usually survey pillars or towers mounted above ground markers.

Survey–monument grid: A grid of permanently installed survey targets which establishes the coordinate system for the building.

Survey reference frame: A mathematical model of the space in which the surveyor takes measurements and performs data analysis.

Symplectic integration: An integration technique for tracking particle motion within a Hamiltonian mechanics formalism.

Synchrotron integrals: Derived from optical properties of the accelerator (Twiss parameters). Synchrotron integrals are used to calculate the statistical properties of the beam (emittance, damping times, etc.).

Synchrotron: An RF particle accelerator in which the particles increase in energy as they pass through the same accelerating section(s) many thousands of times on the same recycling path. As the bending and focusing magnetic fields increase in value, the particle will follow the same path with increasing energy, due to the principle of phase stability, as long as there is enough RF voltage to provide the required extra energy to maintain a constant, or nearly constant, orbit in the increasing magnetic field.

Synchrotron oscillations: The longitudinal oscillations (sinusoidal for small displacements) in phase/energy coordinates which particles execute about the nominal values as a result of initial displacements from the nominal values. The principle of phase stability assures that such oscillations are stable over a certain range of amplitudes. See **Longitudinal motion** and **Phase stability**.

Tachymetry: The simultaneous measurements of direction and distance by the same instrument.

TEM mode: The propagating electromagnetic wave in which both the electric and magnetic fields are transverse to the propagation direction. In regions where TEM waves can exist, propagation velocity is independent of frequency.

Theodolite: A precise surveying instrument used to measure horizontal and vertical directions.

Throughput: The quantity of gas flowing through a conceptual surface per unit time. Units are torr–liters/second.

Topping up: A mode of operating a storage ring and injector such that, by repeated injections at time intervals which are short compared to the beam lifetime, the stored current is kept nearly constant.

Total station: An instrument which measures directions and distances electronically.

Touschek effect: The process by which two electrons within the same bunch scatter off each other via their mutual Coulomb force and thereby exchange some of their transverse momentum into the longitudinal direction. If the resulting longitudinal momentum is larger than the momentum acceptance of the ring, the particles fall out of the RF bucket and are lost. The contribution of this loss mechanism to the overall lifetime is called the Touschek lifetime. It can be a limitation on the lifetime of rings with high current, low emittance and/or low energy.

Tracking error: The difference between the requested current and the real output current during a ramp.

Tracking generator: As part of a spectrum analyzer the tracking generator produces a sinusoidal signal with a frequency locked to the frequency of the analyzer.

Transfer function: The complex ratio of the output and input signals for a circuit element analyzed in the frequency domain.

Transit time factor: The ratio of the energy actually gained by a particle as it crosses the accelerating gap to the energy it would have gained if the field had been constant (instead of varying sinusoidally with time).

Transport matrix: Matrix formalism for propagation of charged particle beam (or photon beam) trajectories.

TRANSPORT: A computer program which performs particle trajectory simulations.

Transuranics: Elements beyond uranium in the periodic table, generated as by-products in nuclear reactors. They are not found in nature, but may be produced in nuclear reactions, usually as fission products. Plutonium, neptunium, and americium are examples. They are dangerous because of their radioactivity and chemical toxicity.

Transverse mode-coupling instability: A transverse single-bunch instability occurring in storage rings when the current exceeds a certain threshold. Typically, it leads to fast beam loss.

Transverse turbulent instability: See **Transverse mode-coupling instability.**

Triangulation: A method of surveying in which coordinates are calculated from angle measurements. Scale is provided either by the calculated distance between two known points or by at least one measured distance.

Trigonometric leveling: A method to determine the elevation difference between two points by measuring the slope distance and the vertical angle.

Trilateration: A method of surveying in which coordinates are calculated from distance measurements.

Tunnel network: The system of tunnel survey stations and their geometry. The stations are usually floor markers or wall brackets. The tunnel network is usually tied to a surface network by observations through vertical survey shafts.

Turbulent bunch lengthening instability: See longitudinal microwave instability.

Turbulent bunch lengthening: Growth of the bunch length and energy spread when the current exceeds the threshold of the longitudinal microwave instability. This growth stops when a new equilibrium situation is reached, and typically does not lead to beam loss.

Twiss parameters: The three functions (alpha, beta, gamma) used to describe the transverse motion of an individual electron in the linear optics formalism and hence also the beam envelope and other properties of the beam.

Vacuum impedance: A measure of the relationship between the electric and magnetic field components of an electromagnetic wave propagating in vacuum and whose numerical value is 377 Ω (SI units). This impedance should not be confused with the free space impedance.

Vacuum impregnation: The process of flooding a coil immediately after winding with a liquid resin under vacuum. In the absence of air, the resin is able to permeate through the coil, giving a solid monolithic construction after the resin is cured.

Vibration isolation: The reduction of ground motion by passive dampening or active suppression.

VME: Standard (IEEE-1014) that defines mechanical and electrical characteristics for crate and modules. Allows for modular design and mixture of components. Features high speed communication between modules over a backplane bus, interrupt capabilities, bus arbitration, and memory mapping.

Wake fields: Electromagnetic fields produced by the interaction of a beam of relativistic charged particles with nearby conducting surfaces, such as the walls of the vacuum chamber. These fields remain in the 'wake' of the leading particles and provide driving forces that act on trailing particles, potentially causing longitudinal and transverse beam instabilities.

Wake potential: The wake potential can be thought of as the net voltage impressed on a test particle traveling behind a leading particle (evaluated as a function of distance behind the leading particle) as the pair travels through a section of vacuum chamber. The Fourier transform of the wake potential yields the complex impedance of the vacuum structure.

Wall current monitor: A non-destructive beam current monitor based on the measurement of the image current flowing in the beam pipe.

White circuit: A resonant circuit used for the excitation of the magnets in a cycling synchrotron.

Wire calibration technique: The simulation of the stored beam by a current flowing through a wire threading through a device in order to measure certain properties of the device before installation. For example it can be used to measure the impedance, electrical centre and sensitivity of pickup assemblies.

Working point: A point on a two-dimensional horizontal and vertical diagram of betatron tune space corresponding to the operational or design values of the tunes. The working point must be chosen to avoid the many resonances that occur in the tune diagram. Damaging resonances occur when certain linear combinations (with integer coefficients) of the tunes equal an integer. Longitudinal resonances can be described by plotting the lines representing linear combinations of the two betatron tunes and the synchrotron tune.

Workstation: A medium to high power processor with high resolution graphic screen, keyboard and mouse or track ball. Workstations are usually connected via local area networks with servers and other workstations.

X-ray lithography: The process of using X-ray beams to transfer a pattern from a mask to a photosensitive surface.

X-ray microscopy: Microscopy utilizing X-rays instead of visible light or electrons to image structures.

X-Windows: A widely adopted standard for computer display graphics, originating from MIT. Contains subroutine library, tool kit and presentation standards (MOTIF).

Yoke (magnetic): The ferromagnetic material in a magnet. The yoke conducts the magnetic flux through the poles and into the gap.

Z-transform: A complex transformation of a discrete sequence of numbers; analogous to the Laplace transform for analog signals.

INDEX

Absolute positioning 295
Absolute zero reading 267
Absorber 203
Accelerating gap 87
Access control 411
Accuracy 250, 255
  absolute 161
Acidic detergents 211
ADC 219
Adjustment screws 279
Adjustment systems 277
AE firm 415, 429
Aesthetics 410, 411
Air 426
  conditioning 417, 424
  ducts 421, 422, 429
  flow 421
Air-core EM devices, pulsed 392
Air-core magents 353
Air-core microundulators, pulsed 398
Airlocks 418
Alarm 222
Aliasing 358
Alignment 275, 288, 418, 419
Alkaline detergents 211
Aluminum 200, 202, 210, 211
Aluminum conductor 146
Aluminum oxide 204
Al–Zr 206
AM to PM conversion 253
Ampere-turns 142, 143
Amplifier 380
Amplitude modulation (AM) 372
Analog-to-digital conversion (ADC) 357
Annular wiggle devices 400
Anocast stands 277
Antidamping 48, 322, 324, 330, 355, 365
Aperture 212, 344, 345, 360, 378, 422
Application programs 221
Ar 205
Architects and engineers 410, 411, 413
Artificial intelligence 218
Assembly areas 417
Automated procedures 223
Azimuthal modes 325

B-factory 203
B-function 35
Backstreaming 207
Bakeout 211, 426
Bandwidth 338, 353, 357, 361, 365, 370, 372, 374, 382
Bayard–Alpert ionization gauge 207
Be windows 427
Beam based alignment 236
Beam
  breakup 329
  dynamics 306, 403
  envelope 232
  focusing 403
  lifetime 88, 199, 202, 212, 403
  loss monitor (BLM) 262
  monitoring and control 353, 361, 362, 371, 372
  motion categories 346–348
  motion sources 348, 349
  position measurement 250
  position monitor (BPM) 225, 250, 349–353, 361, 362, 370
  position monitor, photon 225, 349–352, 361, 362
  power 98
  scraping 400
  size 44
  stability 88
  steering 386, 391, 402, 405
Beam-induced photons 199
Beamline 344, 345, 409–427
  diagnostic 259

optical components 345, 348, 362
optics 345
sourcepoint 353, 360
Bellows 204, 209, 210, 421
Bend magnet 3, 9, 33, 379, 425
Bend magnet radiation 199, 346, 351, 352, 393
Beta functions 35, 198, 215, 228, 265, 399
Betatron oscillations 31, 34, 347, 366–368, 370, 371, 374, 403
Binary information 216
Biological materials 427
Bitbus 220
Blade electrodes, photoemission monitor 351, 352
Blue line survey 295
Booster synchrotron 417, 418
Booster synchrotron magnets 123
Borosilicate glass 204
Bragg-crystal 422
Brazes, Ag–Au 204
Brazing 210
Breakers 426, 428
Bremsstrahlung 198, 435, 447, 448
Brightness 12, 309, 382, 383, 401, 410
Broadband average 312
Building codes 410, 412, 427, 429, 474
Building contractor 429
Bump 238, 360–363, 365
Bump interaction 362
Bunch
clock 82
frequency 324, 325
length 88, 95, 96, 259, 309
lengthening 270, 308, 318, 326
Bunching 383
Buttons, pickup 250

Cables 429
Calibrated model 235
Calibration 159, 207, 215, 216, 226
CAMAC 219
Camshaft, eccentric 287
Capital cost of magnet systems 124
Capital investment 412
Capture pumps 204
Cartesian coordinate system 290
Causal functions 312
Cavity construction 109
Cavity design, computer codes for 106
Cavity dissipation 98
Cavity, low frequency 102
Cavity tuning 108
CCD camera 256
CEBAF cartridge 283
Cell 51, 225
Centroid, beam 350
Ceramics 210
CERN adjuster system 282
CERN LEP dipole support 284
Chamber, vertical aperture 198
Charge-coupled devices (CCDs) 350
Chasman–Green lattice 53
Chicane 396, 397
Chillers 421
Chopper, DC 157
Chromatic aberrations 46
Chromaticity 42, 265
Chromaticity correction 45
Circumference 410, 416
Clay 413, 422, 424
Clean room 411, 415, 417, 421, 424, 430
Cleaning 210, 211
Climate 410, 416
Closed orbit 31, 33, 215, 255, 307, 346, 347, 353, 365, 371
Coherence 12, 380
Coherence length 383, 399
Coherent oscillations 325, 326, 330
Coherent radiation 380
Coil
bucking 187
flipping 193
harmonic 181
radial 181
rotating 183
tangential 189
Coil manufacture 147

Coil testing 148
Cold cathode gauges 207, 208
Collective effects 323
Collective phenomena 306
Combined function dipoles 130
COMFORT 230
Commissioning 218, 411
Communications 411, 427
Commutation 150–152
Complex coupling coefficient 270
Complex frequency shift 336, 338
Compressors 421
Compton scattering 261
Computer architecture 217
Computer interface 216
Computer network 216, 427
Computer rooms 411
Computer simulations 420
Concrete girders 276
Concrete shielding 449
Conductor material 146
Conference rooms 409–417
Configuration control 445, 452
Configuration, machine 222
Construction and operational cost 412
Construction oversight 429
Construction schedule 411
Contamination 210, 211
Control room 416, 417, 427
Control signals 216
Control system 215, 216
Convector, resonant 157
Conventional facility 409
Converter, static 149
Cooling 412
Copper 146, 201–203, 210, 211
Correction scheme 158
Corrector, dipole 122
Correlator filter 371
Corrosion 204, 205, 209
Coulomb scattering 198, 210, 212
Coupled-bunch mode 336, 347, 373
Coupling, horizontal–vertical 39
Coupling resonance 39, 265
Courant–Snyder coordinates 347
Crane 411, 424, 425, 428, 430
Crash buttons 427
Critical energy 381
Critical frequency 317
Cryogenic 416, 425
Crystals 418
Cultural noise 301
Current density 143
Current transformer 248
Current-related parameters 249
Cutoff frequency 312, 316
Cycloconverter 165
Cyclotron 418

DAC 219, 357, 374
Dampers 316
Damping, kicker 365–370
Damping time 47, 49, 233, 309, 320
Darkroom 411
Darwin width 422
Data acquisition 220
Data communication 219
Database 218, 220, 221
  distributed 220
  object-oriented 217
  relational 221
DC interlink 165
DCCT 159
De-Qing 316
Debunching 404
Decimation (downsampling) 374
Dedicated storage rings 410
Deflection of the vertical 289
Deflection parameter 381
Depolarisation 261
Depth of focus 258
Desorption 200
DESY PETRA single component support system 286
Detectors 427
Diagnostic beamlines 259
Dichroism 393, 394, 396
Difference-over-sum processing 251
Differential leveling 291
Diffraction 257
Diffraction-limited emittance 15
Diffusion pumps 207

Digital filter 358, 374
Digital signal processing (DSP) 357, 364, 374
Digital-to-analog convertor (DAC) 219, 357, 374
Diode pump 205
Dipole 3, 128
Dipole correction fields 122, 138
Dipole coupled-bunch oscillations 331
Dipole field 120
Dipole magnets 119, 418
Dipole yoke 128
Dispersion 42, 43, 231
  horizontal 266
  vertical 267
Dispersion relation 312, 322
Displacement 414, 423
Distortion 420
Distributed database 220
Distributed ion pumps 205
Distributed NEG pumps 206
Distribution functions, measurement 257
Distribution, longitudinal 259
Divergence 44, 345, 346, 360
Doppler shift 382
Dose equivalent 450
Double-bend achromat (DBA) 53
Dowels 419
Downsampling (decimation) 374
Dual frequency systems 97
Duct banks 422, 426
Dynamic alignment system 302
Dynamic aperture 45–47, 226, 403
Dynamic range feedback 353, 361, 362, 374

Earthquake 418, 428
Easy-axis orientation error 387, 405
Eddy current 145, 353, 352, 362
Eddy current compensation 356, 357, 359, 360
Eigenfrequencies 421
Eigenvector 239, 365
Elastic coulomb scattering 268
Electric field 380, 383, 394
Electric power 426, 430
Electrical breaks 204
Electrical installations 430
Electrocution 427
Electrodes
  button 250
  clearing 340
  positioning of 250
  signals 251
  stripline 264, 369, 374
Electromagnetic field perturbations 348
Electromagnetic noise 219
Electromagnetic spectrum 1
Electromagnets 119
Electron equation of motion 381
Electron-stimulated desorption (ESD) 200
Ellipsoid 289
Emergency 429, 430
Emission patterns 9
Emittance 12, 35, 47, 49, 229, 232, 346, 380, 422
Emittance measurements 258, 260
End-stations 416
Energy acceptance 268
Energy consumption 411
Energy determination 261
Energy deviations 367, 369, 370, 372
Energy gain per turn 93
Energy oscillations 347
Energy spread 44, 233, 308, 309, 330
Energy variation 265
Energy widening 270
Engineers 410
Environment 411, 430
Environmental impacts 411, 435
Equipotential surface 289
Errors 401
Ethernet 221
Ethylene glycol 426
Ethylene-propylene 204
Evaporable getters 206
Exhaust systems 427–430

Expansion 412
Experiment hall and areas 413–429
Expert systems 218
Extraction 69

Facility services 412
Failsafe 453
Fan-coil units 424
Fans 424, 430
Faraday cup 248
Feedback 224, 309, 316, 349–375, 419
  bunch-by-bunch 372, 373
  global 240, 362–365
  local steering 357, 360–365
  longitudinal 374, 375
  mode 373
  multibunch 365–375
Feeder losses 100
Feedthroughs 204, 209
FEL See *Free electron laser*
Few bunch mode 81
Fiber optic 221
Fiducial 293
Field bus 220
Field distributions 125
Field errors
  antisymmetric 404
Field index 130
Field strength 395
Filter 358, 424, 426, 427
  active/passive 155
  Finite impulse response (FIR) 358
  Infinite impluse response (IIR) 358
Fire 427–429
Fire extinguishers 429
Fire resistant insulation 429
Fire security systems 412
Firing angle 151
First-generation storage rings 6
Flange seals 210
Floor 416, 419, 424
Floor marks 292
Floor stability 410, 424
Flux, magnetic 384, 389, 391
Foci 422
Focusing
  vertical 386
FODO 53
Following error 161
Forklifts 412, 424, 428, 430
Fourier harmonics 364
Fourier transform 355, 358, 363, 364
Fractional tune 265
Free electron laser (FEL) 13, 14, 380, 383, 390, 392, 398, 402
Freewheeling diode 151
Frequency 95
Frequency control 116
Frequency modulation (FM) 372
Frequency response
  correctors, kickers 353, 370
  feedback systems 355, 356
Frequency spectrum 316, 321
Frequency spread 309
Front-end, control system 215, 217, 219
Fume hoods 428
Fundamental mode 315

Gamma radiation 411
Gamma radiation shielding 413
Gap, magnetic 378, 385, 386, 395, 400
Gas
  cylinders 428, 429
  poisonous 428
  reactive 426
Gas desorption rate 201
Gas flow 202
Gas load 202
Gated multibunch mode 81
Geoid 289
Geological factors 411–413, 419
Geometric view factor 200
Getter pumps 204, 205
Getters
  non-evaporable 206
Girders 275
Glass
  borosilicate 204

machinable 204
Glid-cop 203
Global harmonic feedback 363, 364
Global orbit correction algorithms 362–364
Global orbit feedback 240, 362–365
Global positioning system 291
Glow discharge 207, 208
Good field aperture 124
Government 412
GPIB 220
Graphic displays 219
Gratings 418
Gravity vector 289
Green field 410, 412
Grids 207
Ground motion 299, 348
Grounding 428
Growth time 320

Hall plate 173
Handedness 394–397
Hangers, flexible 421
Harmonic
field 385, 403
revolution 332
Harmonic generation 398
Harmonic number 309
Harmonic RF system 323
Hazardous materials 411, 428, 429
Hazards 441
Hazards, natural 434
Head–tail damping 268, 330
Head–tail mode 330, 338
Health 430
Heat-exchanger 426
Heat-transfer limitation 391
Heaters 424
Helical insertion devices 394, 395
Helium 205, 427
HEPA filters 424
Heterodyne receiver 372–374
Hierarchical structure 218
High frequency cavities 103
High voltage 428, 429
High-speed networks 217
Higher harmonics 154, 162
Higher order fields 140
Higher order mode (HOM) 53, 101, 306, 307, 309, 315, 316, 324, 325, 338
Higher order mode damping 109
Hutches 424
Hysteresis 145, 353, 362

Imaging systems, pinhole 259
Imaging with synchrotron radiation 257
Impedance 307, 309–312, 314, 318, 323, 330, 336, 338
beyond cutoff 317
broadband 319, 324, 326
effective 311, 324
free space 317
high-frequency 317
loss 324
narrowband loss 324
resistive wall 312, 319, 338
resistive wall loss 325
shunt 94, 315, 318, 319
simple resonator 314
vacuum 317
Impedance budget 319
Impregnation of coil 147
Inconel 202, 204
Induction, magnetic 172
Inductive output tubes 113
Industrial production 410
Industrial users 412
Inelastic rest gas scattering 269
Inflammable materials 429
Infrastructure 412, 413
Injection 69
Injection bump 70
Injection efficiency 249
Injector 413
Input coupling 107
Inscribed radius 121
Insertion device 40, 206, 212, 377, 403, 424
asymmetric 397
end effects 404
error fields 404
fields 401

helical 394, 395
hybrid 386
millimeter-period 401
power 405
power density 405
pure permanent magnet 384, 365
submillimeter-period 399
warm electromagnetic 389
Inspection 425, 430
Instability 96, 306, 311, 337
coherent beam 307
coupled-bunch 309, 315, 316, 331, 338, 339, 347
fast head–tail 323, 329
longitudinal 338
longitudinal coupled-bunch 334
longitudinal microwave 309, 323, 324
microwave 327
multibunch 307, 345, 365
negative mass 327
single-bunch 307, 309
thresholds 318
transverse coupled-bunch 338
transverse mode-coupling 308, 323, 329
Integrated circuits 414
Integrator 173
Inter-lamination insulation 145
Interfacing of BPMs 253
Interlock 435, 453
failures 453
switches 428
testing 456
International schools 412
Intrabeam scattering 199, 307, 340
Inverse free electron lasers 401
Inverter
bridge 157
resonant 167
Ion chamber position monitor 351, 354
Ion clearing 340
Ion ladder 340
Ion trapping 263, 307, 309
Ion-clearing gap 340
Ionization gauge, Bayard–Alpert 207
Ionizing radiation 411, 435
Iron, low-carbon 388

Keil–Schnell criterion 327
Key and channel structures 419
Kicker, damping 365, 366, 367, 369, 370
Kicker magnet 69, 73, 368
Klystrons 111, 416
Knobs 216
Krypton 205

Laboratories 411, 415–417
Laboratory–office buildings 417
Laminations 123, 143–146
Landau damping 309, 320, 321, 327, 328, 339
Laplace transform 355, 358
Lasertracker 296
Lateral motions 422
Lattice 3, 33, 225
Laws and codes 410, 412, 429
Layout description reference frame 292
Leak detectors 209
Library 411, 412, 417
Lifetime 60, 249, 268, 415
Linac 65, 416, 417, 426
Line management 432, 436
Line variation 162
Line-of-sight electrons 207
Linearity 161
Linewidth 402
Lithography, X-ray 8, 415, 416, 421, 424
Local area networks 217
Local pressure 202
Local steering feedback systems 357, 360–365
Lock-in 268
Log-ratio 253
Logging 222
Long-term stability 419, 424
Longitudinal feedback systems 367, 370, 374, 375

Longitudinal motion 90, 346, 365
Lorentz force 310, 378
Lorentz transformation 382
Loss characteristics of electrical steel 144
Loss factor 314, 315, 324
Loss parameter 314
Low emittance 410
Low-level control 216
Lubricants, dry 211

Machine control 215
Machine shop 417
MAFIA 318
Magnet 419, 421, 422, 426
 AC and DC 122
 air core 351
 bend 3, 9, 33, 379, 425
 booster synchrotron 123
 C-core 119, 128
 combined function 55
 corrector 138, 353
 dipole 119, 418
 encapsulated 212
 H-core 128
 multipole 122, 140
 octupole 323, 339
 permanent, reverse biasing 391
 quadrupole 3, 33, 121, 133, 418
 septum 69, 75
 sextupole 3, 45, 122, 136
 skew 126
 steel 143
 undulator 9
 wiggler 9
Magnet ends 140
Magnet materials and fabrication 143
Magnet supports 274
MAGNET 128
Magnetic access cards 429
Magnetic axis 185
Magnetic centre 121, 185
Magnetic charges 404
Magnetic field measurement 171, 294
Magnetic field strength 385, 387–390
Magnetic flux 384, 389, 391
Magnetic gap 378, 385, 386, 395, 400
Magnetic induction 172
Magnetic length 178
Magnetic saturation 390, 391, 405
Magnetic scalar potential 388, 390
Magnetic sector filters 208
Magnetic sector RGAs 208
Magnetization strength 405
Maintenance 415, 425
Matching 52
Mechanical equipment 419, 421, 424, 427
Mechanical room 417
Mechanical stability 415
Mechanical stress 422
Medical applications 8
Medical suites 411
Meteorological changes 419, 420
MICADO 238
Micro-channel plate (MCP) 350
Microcircuit 413, 422
Micromechanics 8
Microprocessor 215, 219
Microscopy, X-ray 422
Microtrons 66
Microundulator 392, 393
Microwave acceleration 400
Microwave instability 327
Microwave power 201
MIL-STD-1553B 220
Minicomputers 219
Mirrors 345, 418, 427
Mixer 352, 372
Model 224
Model calibration 233
Model-based control 237
Modes of operation 81
Molecular biology 411
Momentum acceptance 198, *341*
Momentum compaction 44, 91, 240, 262, 265, 327, 367
Momentum transfer pumps 207
Monitoring 222

Monochromator 344, 345, 397, 399, 422, 427
Monolayers 201
Monolithic floor slab 419
Motorized adjustment systems 286
Multi-turn injection 71
Multi-undulators 401
Multibunch feedback systems 365–375
Multibunch instabilities 307, 345, 365
Multibunch mode 81, 199, 309, 312, 332, 373
Multibus 220
Multipole expansions 225
Multipole magnet 122, 140

National laboratories 412
Natural ground motion 300
Neon 205
Network 217, 218, 221
Neutron 411, 413, 435
Neutron radiation 448
Neutron shielding 448, 449
NIJI III 53
Nitrogen 426
NMR 175
Noble gases 205
Noise, beam stability 349, 352
Noise spectrum 345
Non-destructive photon position monitors 351–352
Nonlinearity 307, 323
Nyquist frequency 358

Occupancy 415, 430
Octupole magnets 323, 339
Off-momentum 231
Offices 409, 411, 416, 417
Ohmic losses 325
One-turn map 226
Online calculations 217
Online model 226
Open-loop gain 354, 355
Open-loop gain transfer function 355
Opening angle, half 257
Operation 411, 415, 430
  modes of 81
  parasitic 419
Operational budget 412
Operational readiness review (ORR) 430, 436
Operator input 219
Operator interface 219, 220, 222
Optical components, beamline 345, 348, 350, 362
Optics control 240
Optics, linear 225
ORACLE 221
Orbit
  closed 31, 33, 215, 255, 307, 346, 347, 353, 365, 371
  control 238
  measurement 236, 352
  perturbations 231
Organic contamination 210
Oscillation
  betatron 31, 34, 347, 366–368, 370, 371, 374, 403
  coherent 325, 326, 330
  dipole-coupled bunch 331
  energy or synchrotron 39, 91, 95, 96, 325, 334, 346, 347, 367, 370, 371, 374, 402
  transverse 346, 365
  transverse multibunch 309
Oscillator 380
Outgassing 210
Overshoot 161
Overvoltage factor 93, 95
Oxygen-free, electrolytic (OFE) copper 201

Padlocks 428
Parametric current transformers 249
Parasitic loss 324
Parasitic modes 315
Parasitic operation 410, 419
Partial pressures 208
Particulation 210
Path length 233
Peak cavity voltage 93
Permanent magnet overhanging and

packing 388
Permanent magnet, reverse biasing 391
Permeability 384, 385, 387, 388, 390, 405
Personal computers 222
Phase acceptance 91
Phase advance 229
Phase advance, measurement 268
Phase control 115
Phase error, synchronous 371, 374
Phase locked loop (PLL) 264
Phase modulation (PM) 372, 374
Phase shifter 397, 398
Phase space 345
Phase space ellipse 229
Phase stability 90
Phase-slip factor 327
Photodiodes 248
Photoelectron emission 207
Photon beam position monitor 225, 349–352, 361, 362
Photon beam stops 209
Photon emission 232
Photon intensity 199
Photon-stimulated desorption (PSD) 199, 200, 210
Photons, beam-induced 199
Physical aperture 268
Physical plant 412
Pipes 421, 422
Planar Hall effect 175
Plant room 416
Pneumatic devices 426
POISSON 128
Polarimeter 262
Polarization 12, 261
circular 393–397
Polarization time 261
Pole contouring in sextupole 137
Pole geometry for dipole 129
Pole tapering 391
Poles, floating 131
Poles, transfer function 355, 356
Police 412
Polyethylene 204
Polymer 204
Position tolerance 292
Positron 2, 77
Potential, magnetic scalar 388, 390
Potential well distortion 325
Potentiometers 216
Power 413, 425
reactive 152
real 152
Power factor 152
Power input limitations 109
Power requirement 97
Power source 97, 165
Power supply 117, 150–168, 416, 417, 425–429
Prealignment 294
Preparation areas 415
Pressure 207
Principal curve analysis 298
Process area 416
Processing equipment 415
Programmable logic controllers 457
Project center 417
Proportional-integral-derivative (PID) control 358–359
Public access 411
Pulse width modulation (PWM) 156, 167
Pulsed-beam method 318
Pumping speed 204
effective 202
Pumps 427
momentum transfer 207
Pure permanent magnet insertion devices 384, 385

Quadrant operation 151
Quadrupole electrostatic filter RGA 208
Quadrupole energy filters 208
Quadrupole families 121, 135
Quadrupole magnet 3, 33, 121, 133, 418
Quality assurance 436
Quality control survey 297
Quality factor (Q) 314
Quantum lifetime 94, 198
Quartz 204

Radial moments 325
Radiated energy and power 10
Radiation 413, 427
Radiation compatible resin 147
Radiation cone 381
Radiation damping 367
Radiation dose 450
Radiation exposure 450
Radiation harmonics 401
Radiation monitoring 451
Radiation protection 434, 447
Radiation safety 427
Radiation safety hutches 428
Radiation shielding 411, 421
Radiation spectra 383
Radio transmitter 425
Radioactive 411, 428
Random field errors 401
Random walk 402
Reactive component 315, 336
Readback values 216
Real-time control 217
Rectifier bridge 150
Reflected power 98
Refrigeration 426
Regulator, linear 155
Relational database 221
Relative positioning (smoothing) 297
Relativistic electron 380, 381
Repetition rate 163, 399
Reproducibility 256
Requirements, dynamic 149, 162
Requirements, regulation 161
Research facilities 412
Research operations 433, 439
Research operations staff 439, 440
Residential area 413
Residual gas 198, 207, 208
Residual gas analyzers 208
Resistive component 315, 330, 336
Resolution 255, 257
Resolution, current 162
Resonance 37, 390, 400
  nonlinear 38
Resonant circuit 164
Resonant depolarisation 261
Resonant frequency 314, 382, 384
Resonant mode 334, 335
Resonator
  broadband 315, 319, 328
  narrowband 315
Resonator bandwidth 324
Resonator impedance 315
Response matrix 228, 234, 235, 267, 364
Revolution harmonics 332
RF acceleration cavity 201
RF bucket 91, 198
RF cavity 87, 96, 100, 306, 307, 309, 312, 316, 334, 338, 339
RF equipment room 416
RF gun 65
RF multiplexing 252
RF photocathode guns 398
RF protection 117
RF quadrupole 339
RF system 87, 88, 425, 426
RGA 208
  magnetic sector 208
  quadrupole electrostatic filter 208
Rigidity, beam 347
Ripple, power supply 349
Roads 419
Robinson damping 334
Robinson instability 309
Robinson wiggler 402
Rock geology 424
Roof 419, 422
Rooms 412
Rotary translators 209
Rotated quadrupoles 40
Rotation harmonics 372, 373
RS232 220
Rubber mats 421

Safety 411, 427–436
Safety facilities 437
Safety orientation and training 445
Sampling 357, 365
Sampling frequency 357, 358
Sapphire 204

Saturation 404
Saturation field 388
Scalar potential 126, 404
Scalar potential, magnetic 386, 389
Scattering monitor, fluorescence 350, 351
Scraper 268
Screen
 fluorescent 256, 349, 350
Screw jacks 279
Scrubbers 428
Second-generation storage rings 7, 410
Secondary electron yield 200
Security 411, 427, 429
Self-amplified spontaneous emission (SASE) 380, 398
Semiconductor 416
Semiconductor processing 428
Septum magnets 69, 75
Set point values 216
Settling 424
Sextupole magnet 3, 45, 122, 136
Shielding 416, 419, 421, 422, 425, 427, 447
Shielding
 gamma radiation 413
 neutron 448, 449
Shim stacks 279
Shim used in a quadrupole 136
Shops 409, 412, 415
Shuffling 124
Shunt impedance 94, 315, 318, 319
Shutters 427
Sideband 332, 335, 336
Sigma matrix 229
Signal processing 251
Signals from pickup electrodes 251
Single bunch 65
Single bunch mode 81
Single bunch purity 245
Single turn mode 255
Singular value decomposition (SVD) 239, 363, 365
Site 411, 412, 424
Six-strut system 280
Size, beam 345, 346, 360
Skin depth 319, 325, 338
Slab 421, 422
SLAC 3-d stage 284
SLAC damping ring girder support 283
SLAC FFTB magnet positioners 286
SLAC final focus girder support 284
Slits 418, 422
Slow beam bump 69
Smoke detectors 429
Smoothing algorithm 298, 299
Software 217, 219, 221
Soil 413, 422, 424
Soldering 210
Source 413
Source depth 379
Sourcepoint, beamline 353, 360
SPEAR scaling 328
Spectral broadening 401, 402
Spectral distribution 11
Spectrum analyser 264
Spheroid 289
Spin tune 261
Split in magnetic yoke 133
Spontaneous emission 377
Spring supports 422
Springs 421
Sprinkler systems 429
Sputter-ion pump 204
Stability 418
 angular 422
 beam 88
 current 162
 floor 410, 424
 long-term 419, 424
 mechanical 415
 phase 90
Stability boundary 322
Stability criteria
 beam 345, 346
 criteria feedback system 355, 356, 371
Stabilizing capability, feedback system 361, 362, 364

Stable beam signals, suppression 371, 372, 374
Stable phase 372
Staff 410, 415
Stainless steel 201–203, 210, 211
Standard deviation 422
Stands 277
Steel
  grain-oriented 145
  silicon 144
  stainless 201–203, 210, 211
Steinmetz circuit 165
Steppers 415, 422
Stock room 417
Stop-band width 38
Storage cabinets 428
Stores 412
Streak camera 259
Stress, mechanical 422
Striplines 264, 369, 374
Structural-steel box girder 276
Sub-station 425
Substations, electrical 417
Subsystems 220
Sump pump 430
Superconducting 425
  Cavities 101
  dipoles 120, 132
  insertion devices 392
  storage ring 415, 416
Superferric undulators 398
Supply, ramping 163
Support floor 419
Support space 416
Supports 275
Surface network 291
Survey 275
Survey and alignment team 288
Survey and alignment toolbox 299
Survey network 291, 419
Survey reference frames 288
Switch mode 155
Symmetric field errors 404
Symmetry constraints 127
Symplectic integration 226
Synchro-betatron resonances 311
Synchronous energy 34
Synchrotron 68
Synchrotron frequency 325
Synchrotron integrals 215
Synchrotron oscillation 39, 91, 92, 95, 96, 325, 334, 346, 347, 367, 370, 371, 374, 402
Synchrotron period 329
Synchrotron radiation 199, 377, 378
  properties of 8
Synchrotron radiation pioneers 410
Synchrotron radiation resolution 350
System infrastructure 217
Systematic field errors 402

Technical infrastructure 412
Teflon 204
Tefzel 204
Television, closed-circuit 427
Television transmitter 416
TEM mode 369
Temperature 420
Temperature control 410, 415
Terrain surface 289
Tetrodes 110
Theodolites 419
Thermal 210
Thermal desorption 202
Thermal effects on component stability 348, 353
Thermal fluctuations 422
Thermal-stimulated desorption 211
Third generation 7, 410, 416, 418, 422
Threshold currents 270
Thyristor 150
Time structure 12
Time-related signals 217
Titanium 206
Token-ring 221
Tolerance, position 292
Tolerances 401, 405
Tool core 416
Topping up 82
Total photon dose 201

Total pressure gauges 207, 208
Touschek effect 340
Touschek lifetime 61, 94, 269
Touschek scattering 199, 262, 307
Toxic exhaust systems 437
Tracking generator 264
TRACY-II 226
Traffic control 411
Transfer function 356
  poles and zeros 355, 356
Transformation matrix 293
Transformer 426, 427
Transit time factor 103
Transmission 114
Transmission line 369, 370, 372
Transport lines 79
Transport matrices 215, 227
TRANSPORT 226
Transverse coupled-bunch instabilities 338
Transverse damper 368, 369, 371
Transverse feedback systems 366, 370
Transverse mode-coupling instability 308, 323, 329
Transverse multibunch oscillations 309
Transverse oscillations 346, 365
Transverse turbulent instability 329
Trapping of dust particles 263
Triangulation 291
Trigger signals 217
Trilateration 291
Triode pump 205
Triple bend achromat (TBA) 55
Truck airlock 417
Truss 421, 422
Tunability enhancement 391
Tune measurements 264
Tune meter 264
Tune related parameters 265
Tune spread 322, 323
Tuned receiver 252
Tunes 36, 233
Tunnel 413, 416, 429
Tunnel net 292
Turbomolecular pumps 207
Turbulent bunch lengthening 323, 327, 328
TV transmitter 416
Twiss parameters 229
Two-dimensional magnetostatic equations 126

UHV gauge, nude 207
Ultra-high vacuum 204
Ultrasonic baths 211
Undulator 346, 351, 352, 361, 379, 422
  superferric 398
Undulator magnet 9
Undulator radiation 260, 382
Users, offices and facilities 410, 415, 439
Utilities 410–430

Vacuum 411
  chamber 197
  failures 201
  impedance 317
  state equation 202
  systems 426
  ultra-high 204
Valves 204, 209, 427
Vanadium permendur 388
Vertical dipole field in a quadrupole 138
Vibrating equipment 421
Vibration 302, 348, 411–424, 430
Vibration criteria 422
Vibronic molecular levels 200
Viton 204
VME 219
Voltage control 115
Voltage ripple 154

Wake field 306, 307, 310, 323, 325–328, 330, 331, 336, 338, 347, 365, 367
Wake function 310, 311, 318, 325
Wall current monitor 248
Warning tags 428
Water 412, 421, 426, 427

Water pumps 421
Wavelength shifter 379, 389, 392
Wavenumber 384
Wedge jack adjusters 279
Wedged poles 388
Welding 210
White circuit 164
Wiggler 346, 361, 379
Wiggler magnet 9
Wiggler radiation 381
Window-frame magnet 128
Windows 204
Windows, synchrotron radiation 351
Wire technique 250
Working diagram 39
Workstations 219

X-ray lithography 8, 415, 424
X-Windows 222

Z-transform 357, 358
Zeros, transfer function 355, 356
Zirconium 206

www.ingramcontent.com/pod-product-compliance
Lightning Source LLC
Chambersburg PA
CBHW061142040426
42445CB00013B/1509
* 9 7 8 9 8 1 0 2 2 4 2 4 0 *